Grains

Grains

Engineering Fundamentals of Drying and Storage

Fuji Jian and Digvir S. Jayas

CRC Press is an imprint of the
Taylor & Francis Group, an **informa** business

First edition published 2022
by CRC Press
6000 Broken Sound Parkway NW, Suite 300, Boca Raton, FL 33487-2742
and by CRC Press
2 Park Square, Milton Park, Abingdon, Oxon, OX14 4RN

Library of Congress Cataloging-in-Publication Data

Names: Jian, Fuji, author. | Jayas, Digvir S., 1958- author.
Title: Grains : engineering fundamentals of drying and storage / Fuji Jian, Digvir S. Jayas.
Description: First edition. | Boca Raton : CRC Press, 2021. | Includes bibliographical references.
Identifiers: LCCN 2021032664 (print) | LCCN 2021032665 (ebook) | ISBN 9781032013985 (hardback) | ISBN 9781032029993 (paperback) | ISBN 9781003186199 (ebook)
Subjects: LCSH: Grain--Drying. | Grain--Storage.
Classification: LCC SB189.77 .J53 2021 (print) | LCC SB189.77 (ebook) | DDC 633.1/046--dc23
LC record available at https://lccn.loc.gov/2021032664
LC ebook record available at https://lccn.loc.gov/2021032665

ISBN: 978-1-032-01398-5 (hbk)
ISBN: 978-1-032-02999-3 (pbk)
ISBN: 978-1-003-18619-9 (ebk)

DOI: 10.1201/9781003186199

Typeset in Century Old Style Std
by KnowledgeWorks Global Ltd

Contents

PREFACE XVII
ACKNOWLEDGEMENTS XIX
ABOUT THE AUTHORS XXI

1 Physical Properties of Bulk Grain 1
Introduction 1
1.1 Density and Porosity 1
1.1.1 Density 1
1.1.1.1 True and Bulk Density 1
1.1.1.2 Bulk Density Measurement 2
1.1.1.3 Physical Properties Influencing Bulk Density 6
1.1.1.4 True (Kernel) Density Measurement 9
1.1.1.5 Physical Factors Influencing True Density 11
1.1.2 Porosity 11
1.1.2.1 Porosity (ϵ) and Measurement 11
1.1.2.2 Physical Properties Influencing Porosity 12
1.2 Geometrical Properties of Grain Kernels 13
1.2.1 Dimensions (Size) and Measurement 13
1.2.2 Surface Area and Volume 15
1.2.3 Shape and Shape Factors 16
1.2.3.1 Elongation, Lengthiness, and Flakiness 16
1.2.3.2 Shape Factor 17
1.2.3.3 Cross Section 17
1.2.3.4 Sphericity 17
1.2.3.5 Roundness and Aspect Ratio 19

1.3 Hardness and Measurement 21
1.3.1 Hardness 21
1.3.2 Measurement of Hardness 23
1.3.2.1 Particle Size Index (PSI) Method 24
1.3.2.2 Near-Infrared (NIR) Spectroscopy Method 24
1.3.2.3 Single Kernel Characterization System 25
1.3.2.4 Methods Measuring the Force to Crush, Shear, or Peal Kernels 25
1.3.2.5 Flour Yield, Particle Size, and Surface Area 26
1.3.2.6 Starch Properties 26
1.4 Shrinkage and Swelling 26
1.4.1 Shrinkage and Swelling 26
1.4.2 Shrinkage and Swelling of Grain Kernels 28
1.5 Frictional Properties of Grain Bulks 28
1.5.1 Surface Difference of Grain Kernels among Crops and Cultivars 28
1.5.2 Coefficient of (Sliding) Friction and Measurement 31
1.5.2.1 Coefficient of (Sliding) Friction 31
1.5.2.2 Measurement of Coefficient of Friction 32
1.5.3 Repose Angle 33
1.5.3.1 Repose Angle and Measurement 33
1.5.3.2 Relationship between Repose Angle and Other Physical Properties 37
1.5.3.3 Relationship between Repose Angle and Internal Friction 39
1.6 Drag Force and Terminal Velocity 40
1.6.1 Drag Force and Influencing Factors 40
1.6.1.1 Drag Force 40
1.6.1.2 Influencing Factors 41
1.6.2 Terminal Velocity and Measurement 42
1.6.2.1 Terminal Velocity 42
1.6.2.2 Measurement 42
1.6.2.3 Empirical Equations 42
1.6.2.4 Factors Influencing Terminal Velocity 43

1.7 Behaviour of Grain Bulks in Silos 45
1.7.1 Flow Pattern of Unloading Grain Bulks 45
1.7.2 Stress in Silos with and without Grain Discharge 46
1.7.2.1 Stress in Silos without Grain Discharge 46
1.7.2.2 Lateral Stress Ratio 48
1.7.2.3 Stress in Bins During Unloading 49
1.8 Airflow through Grain Bulks 50
1.8.1 Airflow Velocity through Porous Medium 50
1.8.2 Mathematical Models of Airflow 51
1.8.2.1 Darcy Law 51
1.8.2.2 Darcy–Forchheimer Law 53
1.8.2.3 Ergun Equation 55
1.8.2.4 Brinkman Form of Darcy's Law 55
1.8.2.5 Effect of Acceleration, Wall Friction, and Compression 56
1.8.3 Prediction of Airflow Resistance in Bulk Grains 57
1.8.3.1 Most Used Equations 57
1.8.3.2 Measurement of Pressure Drop 59
1.8.3.3 Grain Conditions Influencing Pressure Drop 60
1.9 Dielectric Properties and Measurement 68
1.9.1 Dielectric Properties 68
1.9.2 Parameters of Dielectric Properties 70
1.9.3 Factors Influencing Dielectric Properties of Grain 73
1.9.3.1 Frequency 73
1.9.3.2 Temperature 74
1.9.3.3 Moisture Content 74
1.9.3.4 Bulk Density 75
1.9.3.5 Chemical Composition 76
1.9.4 Measurement of Dielectric Properties 76
1.9.5 Dielectric Properties of Grain Bulks and Applications 77
1.10 Remarks on the Application of the Physical Properties 78
Nomenclature 79
References 83

2 Thermal Properties and Temperatures in Grain Bulks **101**
Introduction 101
2.1 Thermal Properties 102
2.1.1 Heat Content and Heat Capacity 102
2.1.2 Specific Heat and Measurement 103
2.1.2.1 Specific Heat 103
2.1.2.2 Measurement Methods of Specific Heat 104
2.1.2.3 Prediction of Specific Heat 107
2.1.2.4 Relationship between Specific Heat and Other Physical Properties 107
2.1.3 Thermal Conductivity and Measurement 109
2.1.3.1 Thermal Conductivity 109
2.1.3.2 Measurement Method of Thermal Conductivity 112
2.1.3.3 Prediction of Thermal Conductivity 114
2.1.3.4 Relationship between Thermal Conductivity and Other Physical Properties 114
2.1.4 Thermal Diffusivity and Measurement 116
2.1.4.1 Thermal Diffusivity 116
2.1.4.2 Measurement Method of Thermal Diffusivity 117
2.1.4.3 Relationship between Thermal Diffusivity and Other Physical Properties 118
2.1.5 Latent Heat of Vaporization and Isosteric Heat 118
2.1.5.1 Definition 118
2.1.5.2 Factors Influencing Net Isosteric Heat of Sorption 120
2.1.5.3 Determination of Net Isosteric Heat of Sorption 120
2.1.5.4 Net Isosteric Heat and Latent Heat of Vaporization of Cereal Grain 126
2.1.6 Convective Heat Transfer 127

2.1.6.1 Determination of Convective Heat Transfer Coefficient 128
2.1.6.2 Selection of Determination Methods 134
2.1.6.3 Convective Heat Transfer Coefficient of Crop Grain 136
2.2 Temperature of Grain Bulks 137
2.2.1 Temperature Patterns 137
2.2.1.1 Temperature Lag 137
2.2.1.2 Temperature Gradient Patterns without Aeration 138
2.2.2 Factors Affecting Grain Temperatures in Bins 143
2.2.2.1 Headspace, Bin Diameter, and Grain Depth 144
2.2.2.2 Bin Shape and Wall Material 146
2.2.2.3 Geographical Location 149
2.3 Remark on Application of Grain Thermal Properties 150
Nomenclature 153
References 155

3 Water in Biomaterials and Relationship with Its Environment 163
Introduction 163
3.1 Moisture Content and Water Activity 164
3.1.1 Moisture Content and Measurement 164
3.1.1.1 Moisture Content 164
3.1.1.2 Measurement of Moisture Content 165
3.1.2 Water Distribution, Status, and Mobility 167
3.1.2.1 Distribution 167
3.1.2.2 Status 167
3.1.2.3 Water Mobility 170
3.1.3 Vapour Pressure, Relative Humidity, and Water Activity 171
3.1.3.1 Vapour Pressure 171
3.1.3.2 Relative Humidity (RH) 173
3.1.3.3 Measurement of RH 174
3.1.3.4 Water Activity 175

3.1.3.5 Relationship between Moisture Content and Water Activity 176
3.1.3.6 Water Activity and Microorganism Multiplication 178
3.1.3.7 Surface Properties and Water Activity 178
3.1.3.8 Water Activity Measurement 180
3.2 Sorption (Adsorption and Desorption) Isotherm 180
3.2.1 Sorption and Desorption 180
3.2.1.1 Adsorption and Absorption 180
3.2.1.2 Sorption and Desorption 181
3.2.2 Sorption Isotherm 183
3.2.2.1 Sorption and Desorption Isotherm 183
3.2.2.2 Isotherm Shape and Hygroscopic Property 185
3.2.3 Factors Influencing Isotherm 187
3.2.3.1 Temperature and RH Effect 187
3.2.3.2 Pressure Effect 190
3.2.3.3 Food Composition Effect 191
3.3 Hysteresis of Food Materials 192
3.3.1 Hysteresis 192
3.3.1.1 Mechanism of Hysteresis 192
3.3.1.2 Types of Hysteresis 196
3.3.1.3 Hysteresis in Grain Mixture and Successive Sorption – Desorption Cycles 196
3.3.2 Laboratory Determination of Sorption and Desorption Isotherms 197
3.3.2.1 General Approach 197
3.3.2.2 Salt Solution Static Method 198
3.3.2.3 Dynamic Vapour Sorption Method 199
3.3.2.4 Thin Layer Dynamic Method 200
3.3.2.5 Dynamic Dew-Point Isotherm Method 200
3.3.2.6 Thermal Cell Method 201
3.3.2.7 Difference Between Isotherm Methods 201

3.4 Soaking of Crop Seeds 202
3.4.1 Soaking 202
3.4.2 Hard to Cook Phenomenon 204
3.5 Practical Considerations on the Application of Isotherm Theories 205
Nomenclature 206
References 207

4 Fundamental Principles of Aeration, Drying, and Rewetting 217
Introduction 217
4.1 Heat Supply and Air Properties 218
4.1.1 Methods of Heat Supply and Drying 218
4.1.2 Energy Efficiency of Convection 219
4.1.3 Energy Consumption During Grain Drying 219
4.1.4 Air Properties Considered During Ventilation 220
4.1.5 Adding Heater 222
4.2 Principles of Aeration, Drying, and Rewetting 223
4.2.1 Drying Mechanism 223
4.2.1.1 Capillary Theory 223
4.2.1.2 Liquid Movement Due to Gravitational Effects 224
4.2.1.3 Liquid Diffusion 225
4.2.1.4 Vapour Diffusion 225
4.2.1.5 Liquid and/or Vapour Migration 225
4.2.1.6 Effusion (Knudsen) Flow 226
4.2.2 Combination of the Mechanisms 226
4.3 Thin Layer Drying and Parameters 227
4.3.1 Thin Layer Drying 227
4.3.2 Drying Rate 228
4.3.3 Drying Phase 228
4.3.4 Water Activity During Drying 231
4.3.5 Material Temperature During Drying 231
4.3.6 Convection Drying Energy Source 232
4.4 Factors Influencing Grain Drying and Cooling Rate 233
4.4.1 Physical Property of Drying Materials 233

4.4.2 Properties of Supplied Air and Initial Grain Condition 234
4.4.3 Airflow Rate (Air Velocity) 234
4.4.4 Tempering and Intermittent Drying 235
4.5 Drying Resistance 235
4.5.1 Resistance During Constant Rate of Drying Phase Period 236
4.5.2 Resistance in Falling Rate Period 236
4.5.3 Resistance Due to Low Efficiency of Mass and Heat Transfer 237
4.5.4 Parameters Used to Evaluate Drying Resistant 237
4.5.4.1 Convective Mass Transfer Coefficient 238
4.5.4.2 Biot Number 238
4.5.4.3 Lewis Number 239
4.5.4.4 Water (Moisture) Diffusivity 241
4.6 Glass Transition 250
4.6.1 Principle of Glass Transition 250
4.6.2 Measurement of Glass Transition 251
4.6.2.1 DSC Method 252
4.6.3 Factors Influencing Glass Transition 252
4.6.4 Glass Transition in Grain Drying 252
4.7 Collapse, Breakage, Volume Change, and Pore Formation 254
4.7.1 Forces Influencing Volume Change and Pore Formation 255
4.7.1.1 Surface Tension 255
4.7.1.2 Glass Transition 255
4.7.1.3 Unbalanced Internal Forces 256
4.7.2 Determination of Stress-Cracked Kernels 259
4.8 Effect of High Temperature and Safe Drying Temperature 259
4.8.1 Effect of High Temperature 259
4.8.2 Safe Drying Temperature 261
4.9 Deep Bed Drying, Zones and Fronts 263
4.9.1 Deep Bed Drying 263
4.9.2 Zones and Fronts 264

4.9.3 Possible Psychrometric Processes During Grain Drying, Wetting and Aeration 267
4.9.4 Example of Aeration and Drying 268
4.9.4.1 Aeration 268
4.9.4.2 Natural Air Drying 270
4.9.4.3 Dryeration and Combination Drying 272
4.10 Remark on Application of Drying Theories 274
Nomenclature 275
References 276

5 Mathematical Modelling of Isotherm, Drying, and Wetting 287
Introduction 287
5.1 Sorption (Equilibrium) Isotherm Equations (Models) 288
5.1.1 Theoretical and Semi-Theoretical Models 288
5.1.1.1 Brunauer-Emmett-Teller (BET) Equation 288
5.1.1.2 Bradley Equation 292
5.1.1.3 Clausius–Clapeyron Equation 292
5.1.1.4 D'Arcy and Watt Equation 295
5.1.1.5 Freundlich Adsorption Equation 296
5.1.1.6 GAB (Guggenheim, Anderson, de Boer) Equation 297
5.1.1.7 Harkins–Jura Equation 300
5.1.1.8 Langmuir Isotherm Equation 301
5.1.1.9 Lewicki Model 303
5.1.1.10 Park Equation 303
5.1.1.11 Peleg Equation 303
5.1.1.12 Raoult's Law and Henry's Law 305
5.1.1.13 Smith Equation 306
5.1.1.14 Young and Nelson Equation 306
5.1.2 Empirical Models 307
5.1.2.1 Chen Equation 307
5.1.2.2 Halsey Equation 309
5.1.2.3 Henderson Equation 310
5.1.2.4 Modified Chung–Pfost Equation 311
5.1.2.5 Oswin Equation 311

5.1.3 Model Selection and Most Used Models in Grain Storage 312
5.1.3.1 Temperature Effect on Model Selection 312
5.1.3.2 Selection of Composite or Single Component Models 313
5.1.3.3 Selection of Models With More or Less Parameters 314
5.1.3.4 The Most Used Equations 315
5.2 Drying and Wetting Models 316
5.2.1 Theories of Water Movement in Solids 316
5.2.1.1 Fick's Law of Diffusion and Effective Diffusivity 316
5.2.1.2 Capillary Theory 321
5.2.1.3 Receding Front Theory and Modelling 322
5.2.2 Types of Mathematical Models and Assumptions 326
5.2.2.1 Types of Mathematical Models 326
5.2.2.2 Model Assumptions 327
5.2.3 Mathematical Modelling of Drying and Wetting 328
5.2.3.1 Theoretical Model 328
5.2.3.2 Semi-Theoretical and Empirical Model 335
5.2.4 Models of Soaking (Wetting) 341
5.3 Models of Deep Bed Drying and Aeration 342
5.3.1 Relationship of Mathematical Modelling between Thin Layer and Deep Bed Drying 342
5.3.2 Types of Deep Bed Drying Models 344
5.3.2.1 Non-Equilibrium Models 344
5.3.2.2 Equilibrium Models 347
5.3.2.3 Logarithmic Models 350
5.3.3 Numerical Modelling and CFD Analysis 351
5.3.3.1 Numerical Modelling and CFD 351
5.3.3.2 Governing Equations of CFD 353
5.3.3.3 Procedure of CFD Model Development 355

5.3.3.4 Advantages and Disadvantages of CFD 355
5.3.3.5 Application of CFD in Grain Drying 358
5.4 Remarks 359
Nomenclature 361
References 367

6 Moisture Migration and Safe Moisture Content for Storing Grain in Bins 383
Introduction 383
6.1 Hygroscopicity of Grains 384
6.1.1 Hygroscopicity of Different Crop Kernels 384
6.1.2 Hygroscopicity of Different Classes and Cultivars of the Same Crop Type 386
6.1.3 Hygroscopicity of Different Parts of a Kernel 387
6.1.4 Factors Influencing Hygroscopicity 387
6.2 Initial Grain Moisture Content and Modes of Water Entering into Bins 388
6.2.1 Initial Moisture Content 388
6.2.1.1 Moisture Content at Harvest 388
6.2.1.2 Moisture Content after Filling 389
6.2.1.3 Moisture Content after Drying 389
6.2.1.4 Moisture Content after Aeration 389
6.2.2 Entrance of Moisture into Grain Bulks 390
6.2.2.1 Snow Water 390
6.2.2.2 Grain Wetting Due to Humid Air 390
6.3 Grain Moisture Migration 391
6.3.1 Diffusion under Moisture Gradients at a Uniform Temperature 391
6.3.1.1 Mechanism 391
6.3.1.2 Process 392
6.3.1.3 Application 393
6.3.2 Diffusion under Temperature Gradients at a Uniform Moisture Content 393
6.3.2.1 Mechanism 393
6.3.2.2 Process 394
6.3.2.3 Application 396

6.3.3 Moisture Migration Due to Free Convection Currents 396
6.3.3.1 Mechanism 396
6.3.3.2 Experimental Proof 399
6.3.3.3 Application 400
6.3.4 Condensation of Water on Grain Surface 400
6.3.4.1 Condensation During Turning and Loading 400
6.3.4.2 Condensation inside Bins 401
6.3.4.3 Water Sorption During Air Exchange 401
6.3.4.4 Freezing Due to Condensation 402
6.3.5 Factors Affecting Moisture Migration 402
6.3.5.1 Temperature and Moisture Distribution 402
6.3.5.2 Resistance to Airflow 403
6.3.5.3 Infection by Fungi and Infestation by Insects 403
6.4 Safe Storage Moisture Content 404
6.4.1 Moisture Content Required by Pests Infesting Grain 404
6.4.2 Moisture Content Required by Milling 405
6.4.3 Moisture Content Required by Microorganism 405
6.4.4 Moisture Content Required by Grain for Respiration and Germination 408
6.4.5 Moisture Content Required by Chemical Reactions 409
6.4.6 Safe Storage Time and Moisture Content 411
6.5 Remark on Ecosystem Consideration 413
Nomenclature 414
References 415

APPENDIX 423
INDEX 469

Preface

Drying and storage are two significant unit operations in the food industry and are applied to both raw and processed products including cereal grains, oilseeds, legumes, flour, noodle, coffee, and cornstarch. The common characteristic of these materials is that all of them are hygroscopic and contain water. The hygroscopic properties are influenced by their physical properties, which are influenced by their storage environments such as bins, warehouses, bunkers, and temporary storage structures. This book focuses on the storage and drying of bulk products in these storage structures.

On many occasions in our work with the grain storage and drying personnel especially our graduate students and industry contacts, we found a book explaining the fundamental principles of grain storage and drying is needed. Therefore, the primary objective of this book is to help readers understand the fundamental principles of grain storage and drying and develop a well-informed approach to solve grain storage and drying problems. Technologies for grain storage and drying are advanced through research; therefore, literature review and background on each topic has also been included. The book is generally intended for grain storage and drying students, engineers, and scientists. As reflected in the contents which are presented at several levels of depth, this book will serve well readers with different backgrounds and interests. An effort has been made to allow for independent reading of different sections, and to make a large part of this work accessible to a non-mathematical audience.

The authors have combined their experience of teaching grain storage and drying to undergraduate and graduate students in the faculties of Agricultural and Food Sciences and Engineering. Material in the book is organized into broad topic areas: physical properties (Chapters 1 and 2), grain temperature and moisture (Chapters 2 and 6), water in biomaterials and relationship with its environment (Chapter 3), fundamental principles of aeration, drying, and rewetting (Chapter 4), and mathematical modelling of isotherm, drying, and re-wetting (Chapter 5).

We hope our readers will benefit from the contents of the book for many decades.

Acknowledgements

Humans have stored grains since they started cultivating grains, thus the field of grain storage and drying is mature but research and development in different aspects of storing and drying grains continues to advance with time. There are a huge number of publications in literature. To bring this established field into a single book, we relied heavily on the published articles (listed at the end of each chapter), and gratefully acknowledge the contributions of each researcher and their teams around the globe whose work we have cited. Great care has been given in paraphrasing their works and hope we have succeeded in doing so. We ask for their forgiveness if we have not done it correctly. Even though the published works have been selectively cited based on the relationship to the presented topics, all the published works are critical to the growth of this established but ever-expanding field. We hope our presentation will continue to inspire readers to explore field further. We thank readers for reading whole or parts of the book and we hope with the enhanced understanding of the subject matter you will be designing better systems to manage stored grains safely around the globe and helping feed the growing world population.

Writing of a book requires dedicated time in our daily busy schedules and when focusing on writing, the family gets less attention. We sincerely thank our spouses (Mrs. Suping Yang Jian and Mrs. Manju Jayas) and our children (Annie Jian, Erick Jian, Drs. Rajat Jayas, Ravi Jayas, and Rahul Jayas) for their understanding and support. Dr. Jayas also thanks his grandchildren (Priya Jayas, Isabella Jayas, Rohan Jayas, Gabriel Jayas, and Leon Jayas) for not spending as much time he would like to spend with them. During busy times of writing, we may have not unconsciously completed some of our assigned duties at work in a timely fashion, and for this we thank our employer, the University of Manitoba, our immediate supervisors, and close working colleagues for their understanding and encouragement.

We acknowledge gratefully the assistance we received from Sonal Motla in copyediting the book, Siobhán Greaney and Ashraf Reza, for spearheading the project through the production process, and Taylor & Francis Group for giving us an opportunity to write this book.

Last but not the least, we thank the readers of the book in whole or in part and hope they will find it useful and will let us know how we could have done it better.

Fuji Jian
Digvir S. Jayas

About the Authors

Dr. Fuji Jian is a professional engineer and associate professor in the Department of Biosystems Engineering at the University of Manitoba. Fuji was educated at the HuaZhong Agricultural University (China), Henan University of Technology (China), University of Greenwich (UK), and University of Manitoba (Canada). Fuji received his Ph.D. at the University of Manitoba in 2003, and excellence of his Ph.D. studies was recognized with the "Governor General's Gold Medal for outstanding graduate studies academic achievement in Canada" and the "CSAE/SCGR 2004 Ph.D. Thesis Award." After working for a grain storage company (OPI System, Calgary, Canada) for 5 years, he returned to academia as an instructor in 2010 and assistant professor in 2015. He is the recipient of the prestigious 2019 John Clark Award of CSBE/SCGAB and 2020 Merit awards in the category of research, scholarly work and creative activities, University of Manitoba. Fuji's research interests are in the area of post-harvest grain quality, stored-product protection, drying and aeration of grains, and sensor development with the sole purpose to enhance food safety and security. He is the leading authors in 85 of more than 100 technical articles in scientific journals, conference proceedings and books. These publications covered: (1) mathematical modelling of grain storage ecosystems; (2) insect biology and ecology inside stored grain bulks; (3) sampling inside grain bins; (4) physical property of stored grain bulks; (5) safe storage of grain and monitoring; (6) insect control and pesticide resistance; (7) grain aeration and drying; (8) sensor development; (9) biomass processing; and (10) particle segregation. Fuji published more than 30 mathematical models and his expertise in mathematical modelling and grain storage ecosystem has been instrumental in revitalization of grain storage monitoring industry. Drs. Jayas and Jian received the John Ogilvie Research Innovation Award (CSBE/SCGB) in 2021 for their work on "Mathematical models of stored-grain ecosystems for management of stored grain." Fuji has taught Grain Storage/Crop Preservation, Plant and Animal Physiology for Engineers, Unit Operations, Transfer (mass and heat) Phenomena, and Modelling and Simulation of Biological Systems. He is currently supervising/co-supervising more than 10 M.Sc./Ph.D. students.

Distinguished Professor Dr. Digvir S. Jayas was educated at the G.B. Pant University of Agriculture and Technology in Pantnagar, India; the University of Manitoba; and the University of Saskatchewan. Since 2011, he is serving as Vice-President (Research and International) at the University of Manitoba. Before assuming the position of Vice-President (Research and International), he was Vice-President (Research) for two years, and Associate Vice-President (Research) for eight years. Prior to this, he was Associate Dean (Research) in the Faculty of Agricultural and Food Sciences, Head of the Department of Biosystems Engineering, and Interim Director of the Richardson Centre for Functional Foods and Nutraceuticals. For a year, he has served as Interim President of Natural Sciences and Engineering Council of Canada (NSERC). He is a Registered Professional Engineer and a Registered Professional Agrologist. Dr. Jayas is a former Tier I (Senior) Canada Research Chair in Stored-Grain Ecosystems. He conducts research related to drying, handling and storing of grains and oilseeds and digital image processing for grading and processing operations in the Agri-Food industry. He has collaborated with researchers in several countries and has had significant impact on the development of efficient grain storage, handling and drying systems in Canada, China, India, Ukraine and the United States. He has authored or co-authored over 950 technical articles in scientific journals, conference proceedings and books dealing with issues of storing, drying, handling and quality monitoring of grains. Dr. Jayas has received awards in recognition of his research and professional contributions from the Agriculture Institute of Canada, Applied Zoologists Research Association (India), American Society of Agricultural and Biological Engineers (ASABE), Canadian Institute of Food Science and Technology, Canadian Academy of Engineering, Canadian Society for Bioengineering, Engineers Canada (formerly Canadian Council of Professional Engineers), Engineers and Geoscientists Manitoba (formerly Association of Professional Engineers and Geoscientists of Manitoba), Engineering Institute of Canada, Indian Society of Agricultural Engineers, Manitoba Institute of Agrologists, National Academy of Agricultural Sciences (India), National Academy of Sciences (India), and Sigma Xi. He was the recipient of the 2017 Sukup Global Food Security Award from

ASABE, and the 2008 Brockhouse Canada Prize from NSERC. In 2009, he was inducted as a Fellow of the Royal Society of Canada and in 2018, he was appointed as an Officer of the Order of Canada. Dr. Jayas serves on the boards or committees of many organizations including: ArcticNet, Cancer Care Manitoba Projects Grants and Awards Committee, Churchill Marine Observatory (CMO), Centre for Innovative Sensing of Structures (SIMTReC), Genome Prairie, GlycoNet, Manitoba Centre for Health Policy, National Coordinating Centre for Infectious Diseases (NCCID), North Forge Technology Exchange, NSERC Council, Oceans Research in Canada Alliance Council, Research Manitoba, Research Institute of Oncology and Hematology, RITHIM Steering Committee, and TRIUMF. He has served as the President of the Agriculture Institute of Canada, the Canadian Institute of Food Science and Technology, the Canadian Society for Bioengineering, Engineers Canada, Engineers and Geoscientists Manitoba, and the Manitoba Institute of Agrologists. He currently chairs NSERC Council and the board of directors of TRIUMF and RESOLVE, a prairie research network on family violence. He also chairs the Smartpark (University's Research and Technology park) Advisory Committee.

Chapter 1

Physical Properties of Bulk Grain

Introduction

Granular materials or "bulk solids" are any material composed of many individual solid particles, irrespective of particle size. Crop grain kernels are bulk solids and have to be transported, conveyed, handled, stored, and processed. Thus, characterization of bulk solids regarding their physical properties at static and dynamic conditions plays an important role for product development, optimization, and operation in cleaning, grading, drying, handling, and processing of the crop grains. This chapter mainly discuss the physical properties of bulk of grain kernels.

1.1 Density and Porosity

1.1.1 Density

1.1.1.1 True and Bulk Density

Density of porous grain bulks could be expressed either as true density or bulk density of the grain bulk. The true density of the grain bulk is the weight mass of the grain kernels over the volume occupied by the grain kernels and not including the volume of pore spaces among the kernels in the grain bulk. The true density is also termed as kernel density, which is

DOI: 10.1201/9781003186199-1

similar to the particle density used in the mechanical science. Kernel density is also termed as absolute density or real density.

$$\rho_k = \frac{m}{V_k} \tag{1.1}$$

where ρ_k is the kernel density (kg/m^3), m is the mass (kg), and V_k is the volume occupied by the kernels (m^3). The measured volume includes the volume of the closed pores inside each kernel, but not the externally connected pores among kernels. True densities of cereal grains and oilseeds are in the range of 1110–1450 kg/m^3, while true density of some seeds with large portion of fibre hull (such as sunflower seeds) is less than 1000 kg/m^3 (Table 1.1). Therefore, cereal grain seeds or kernels will sink into pure water because its true density is higher than water (1000 kg/m^3).

Bulk density of grain bulk is mass of grain divided by volume occupied by grain including intergranular space typically filled with air. Bulk density of the porous grain bulk can be changed by packing with force or through vibrations but true density is not affected by packing. Bulk density of the porous grain bulk can be calculated as:

$$\rho_b = \frac{m}{V_b} \tag{1.2}$$

where ρ_b is the bulk density (kg/m^3), and V_b is the volume occupied by the grain bulk (m^3). The volume (V_b) includes the volume of the intergranular air among the grain kernels and the volume occupied by the kernels in the bulk grain. This means that the volume in the bulk density calculation includes both externally connected pores among the kernels and closed pores inside each kernel. Bulk densities of crop grains are in the range of 430–900 kg/m^3 and seeds with high oil content such as canola, hemp, and sunflower usually have a lower bulk density (Table 1.1). When the porosity of grain bulks is reduced, the bulk density of some seeds can be higher than 1000 kg/m^3. Carmen (1996) reported the bulk density of lentil was 935–1190 kg/m^3 at porosity of 27–32%, while Irvine et al. (1992) reported 783–825 kg/m^3 at porosity of 40–43% (Table 1.1).

1.1.1.2 Bulk Density Measurement

Even though different researchers use different methods to measure the bulk density, the main measurement principle is the same, which is to measure the mass of the grain kernels in a specified container with a determined volume. The main differences among different methods are the filling method of the tested grain bulk, removing or retaining dockage and foreign materials, and the temperature of the grain. These different

Table 1.1 Bulk and True Densities and Porosity of Seeds of Cereal Grains, Pulses (Legumes), Oilseeds and Speciality Crops

Crop	*MC (%)*[a]	ρ_b *(kg/m³)*[b]	ρ_k *(kg/m³)*[c]	ε *(%)*[d]	*Reference*
Barley	12.7–16.4	649–664	1346–1372	44–47	Muir and Sinha (1988)
Caper	5.7–14.1	438–399	678–806	41–46	Dursun and Dursun (2005)
Canola[e]	8.1	664–687	1093–1129	33–34	Muir and Sinha (1988)
Corn	4.7–22.0	649–710	1250–1325	43–51	Seifi and Alimardani (2010)
Corn (sweet)	8.4–14.6	698–765	1274–1315	42–43	Karababa and Coskuner (2007)
Dry bean[f]	12.0–18.0	789–826	1385–1434	42–43	Senthilkumar et al. (2018)
Dry bean[g]	12.0–18.0	814–827	1416–1443	42–43	Senthilkumar et al. (2018)
Durum	12.7–16.4	709–744	1372–1377	41–42	Muir and Sinha (1988)
Durum[h]	9.3–41.5	675–711	1241–1333	46–47	Al-Mahasneh and Rababah (2007)
Faba bean	12.6–21.9	761–815	1373–1393	42–45	Irvine et al. (1992)
Flaxseed	7.0–15.1	574–624	1143–1148	45–50	Irvine et al. (1992)
Hemp	9.0–15.0	466–470	1110–1135	55–59	Jian et al. (2018)
Lentil	11.4–18.0	783–825	1392–1409	40–43	Irvine et al. (1992)
Lentil	6.1–24.6	935–1190	1288–1750	27–32	Carmen (1996)
Linseed	7.6–18.2	545–691	1010–1020	32–47	Selvi et al. (2006)
Oats	12.7–16.4	537–555	1315–1329	52–54	Muir and Sinha (1988)
Okra	7.5–46.7	558–592	986–1107	43–46	Sahoo and Srivastava (2002)
Pigeon pea	5.9–22.0	745–806	1251–1305	38–40	Shepherd and Bhardwaj (1986)
Pumpkin kernel	4.0–27.5	481–554	1080–1143	52–56	Joshi et al. (1993)
Red kidney bean	9.8–12.8	582–661	1234–1393	46–58	Isik and Unal (2007)

(Continued)

Table 1.1 Bulk and True Densities and Porosity of Seeds of Cereal Grains, Pulses (Legumes), Oilseeds and Speciality Crops (*Continued*)

Crop	*MC (%)*[a]	ρ_b *(kg/m³)*[b]	ρ_k *(kg/m³)*[c]	ε *(%)*[d]	*Reference*
Rough rice[i]	10.0	471–554	1126–1300	53–63	Varnamkhasti et al. (2008)
Soybean	8.0–24.0	630–720	1250–1149	37–45	Kashaninejad et al. (2008)
Soybean	8.0–20.0	708–735	1124–1216	37–40	Deshpande et al. (1993)
Sorghum	12.0–18.0	757–868	1136–1160	33–36	Mwithiga and Sifuna (2006)
Sunflower[j]	3.8–16.7	434–462	706–765	34–43	Gupta and Das (1997)
Sunflower[k]	3.8–16.7	574–628	1050–1250	45–50	Gupta and Das (1997)
Vetch	9.6–17.1	826–861	1286–1370	33–40	Yelcin and Ozarslan (2004)
Wheat[l]	12.7–16.4	725–763	1370–1384	39–40	Muir and Sinha (1988)
Wheat[m]	9.2–11.1	721–749	1343–1491	46–48	Markowski et al. (2013)

[a] Moisture content (wb, %).
[b] Bulk density or test weight measured using Canadian Grain Commission method, or a similar method.
[c] True (kernel or particle) density.
[d] Porosity.
[e] Canola or rapeseed.
[f] Black bean (*Phaseolus vulgaris* L.).
[g] White bean (*Phaseolus vulgaris* L.).
[h] Un-matured durum (green wheat).
[i] Paddy.
[j] Sunflower seed.
[k] Sunflower kernel (without hull).
[l] Three cultivars of hard red spring wheat.
[m] Four cultivars of winter wheat.

measurement methods result in different bulk densities of the same grain variety (cultivar). To avoid this inconsistency, measurement of bulk density in different countries is standardized and termed with different names. Canada and the USA standardize it as Test Weight, but use different test kits. Also, the temperature of the grain should be recorded when the bulk

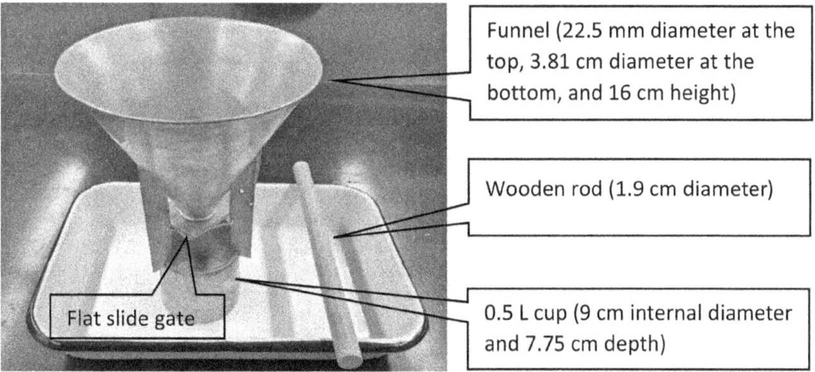

Figure 1.1 Test weight kit specified by the Canadian Grain Commission. The white tray to capture spilled or excess grain from levelling is not included in the kit.

density is measured because temperature slightly influences the bulk density (Jian et al. 2013).

The test weight, specified by the Canadian Grain Commission, is measured by using a specified kit after dockage inside the grain bulk is removed. There is a specified procedure to remove the dockage before the Test Weight is determined (Canadian Grain Commission 2018). The kit includes a 0.5 L cup, wooden rod, and funnel with a flat slide gate at the bottom of the funnel (Fig. 1.1). After the funnel is placed on the top of the 0.5 L cup, the flat slide gate is located 4.41 cm above the cup. The grain to be tested is loaded into the funnel. Then the flat slide gate is opened, the grain drops into the cup, fills it and flows over the sides. The grain at the top of the cup is stroked off with the wooden rod in three zig-zag motions at an angle of 45° and the remaining mass inside the cup is weight. The measured test weight is expressed as kilograms per hectolitre (kg/hL), kilograms per 100 L of volume, or grams per half-litre (g/0.5L), which may commonly be referred to as "bushel weight" and expressed as pounds per bushel (lb/bu), or called "Avery" lb/bu. The "Avery" bushel weight is based on the British "Imperial" bushel volume and one bushel is 36.369 L. One Avery bushel weight in Canadian system is 12.472 kg/m^3. The Canadian Grain Commission recommends the following equation to convert the measured mass in the 0.5 L cup to mass per hectolitre:

$$Y = \frac{1.93736m + 46.4962}{10} \text{ for wheat, oats, barley, rye, and flax} \quad (1.3a)$$

$$Y = \frac{1.91816m + 35.35844}{10} \text{ for rapeseed, caola, and mustard} \quad (1.3b)$$

where Y is the test weight (kg/hL), and m is the mass of grain in the 0.5 L cup (g). The test weight in the USA grading system is measured on the

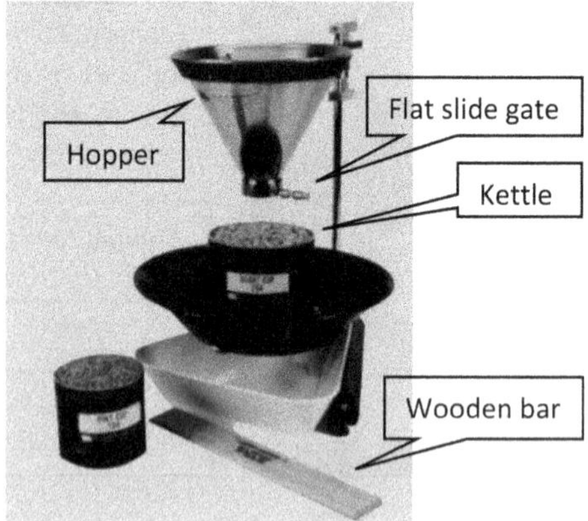

Figure 1.2 Test weight kit approved by the United States Department of Agriculture (USDA).

sample as a whole or after the removal of dockage. A test kit (Fig. 1.2) is approved by the Federal Grain Inspection Service (FGIS), United States Department of Agriculture (USDA). The kit and measurement procedure is similar to the Canadian grading system except the size of the funnel (hopper) and the cup (a quart kettle). The distance between the hopper's base and the top of the kettle is 2 in. The USA grading system uses a wooden striker bar in three zig-zag motions to remove the extra grain from the top of the kettle. Weight of grain in the kettle is measured in pounds and this value is multiplied by 32 (the number of quarts in a volume or Winchester bushel) to obtain the grain test weight in units of pounds per bushel, which is also called pounds per bushel "Winchester" (lb/bu W). The Winchester bushel weight is based on the US bushel volume and one bushel is 35.239 L. One bushel weight in the USA system is 12.872 kg/m^3. Therefore, a bushel of wheat in Canadian system is not equal to a bushel of wheat in the US system. This bushel is not the same as that used to define the weight bushel of crop in the US system. A "weight" bushel of corn is exactly 56 pounds, a soybean bushel is 60 pounds and a wheat bushel is 60 pounds, regardless of the test weight.

1.1.1.3 Physical Properties Influencing Bulk Density

Bulk density is influenced by many factors including grain type, cultivar, dockage and foreign materials in the bulk grain, moisture content, temperature, method of filling (height of fall, flow rate, and vibration), and compaction effect inside the grain bulk because these factors influence the shape and sizes of pores among kernels.

For clean grain, the grain moisture content is the most important factor influencing the bulk density. Bulk density of grain linearly or quadratically decreases with the increase of moisture content in certain range of the grain moisture content and also decreases when grain is over dried (Sun et al. 2014; Jian et al. 2018; Senthilkumar et al. 2018). Compared with wheat, the bulk density of canola is affected less by moisture content, increasing from 690 kg/m^3 at 14.5% moisture content to 700 kg/m^3 at 6.5% moisture content (Jayas et al. 1989). Whereas the bulk density of hard red winter wheat increased from 650 kg/m^3 at 24% moisture content to 770 kg/m^3 as moisture content decreased to 11% (Nelson 1980).

Compared with the moisture content, temperature has a small influence on the bulk density. However, this influence will increase if the grain has a higher than the recommended safe storage moisture content and temperature is lower than 0°C (Jian et al. 2013). Bulk density of canola had a parabolic relationship with temperature and moisture content of the canola seeds.

Bulk densities increase with the applied pressure. Soybean bulk densities varied with pressure in a manner similar to creep behaviour in other agricultural materials (Milani et al. 2000). Thompson and Ross (1983) reported the change in bulk density at 7 and 35 kPa was caused by the rearrangement and the packing characteristics of grain in mass. The change in bulk density at 35–172 kPa was mainly a function of the elasticity and deformational capabilities of the grain kernels. As pressure increases, the voids diminish and the change of bulk density per unit pressure increase decreases. Therefore, the bulk densities approximately linearly increase with pressure above 14 kPa. The bulk density changes of wheat caused by elastic behaviour of individual kernels are higher at high pressures (34–172 kPa) and moisture contents (16–24%). At higher pressures, the increase of bulk density is due to individual kernels deformation (Milani et al. 2000) because grain kernels are more elastic at higher moisture content of soybeans (10.5–20%) at 0–55.2 kPa.

Increase of dockage percentages usually decreases the bulk density (Bian and Subramanyam 2015) because the dockage may include broken kernels, weed seeds, stems (chaff), and leaves. These dockage materials usually have a low true density and irregular shapes, which usually is not the same as the grain kernels, and these different shapes and sizes increase the porosity of the grain bulk. Highly elongated particles with random orientations have 5% more voids than circular particles, while very elongated particles have in 7% lower of bulk density than circular particles (Cleary and Sawley 2002). Vibration can decrease the bulk volume and hence the increase of the bulk density. Cleary and Sawley (2002) found non-circular particles reach to higher bulk densities after vibration. Dockage with smaller dimensions than the grain kernels may increase the bulk density because the dockage will fill in the pore spaces among grain kernels. The bulk density of hemp seeds with moisture content from 9 to 15% significantly increased when 5–15% dockage (more than 45% of the dockage was smaller than the hemp seed kernels) is mixed with the seeds (Jian et al. 2018). Bulk density of corn mixed with fines increases (McNeill et al. 2004). However, hand shelled corn having little mechanical damage can have a higher bulk density than the corn threshed in a combine due to a higher broken ratio. Therefore, chaff and fines usually have more influence on the bulk density than moisture content and temperature, and this influence depends on the components of the chaff and fines.

Bulk density is a good indicator of grain quality and higher bulk density normally means higher quality because bulk density decrease as grain deteriorates (Jian et al. 2012). The bulk density of canola seeds with 5.5–15.5% moisture contents decreased about 3% after 60 days storage at 30°C (Jian et al. 2012). Compaction will increase the bulk density, and newly harvested canola with moisture content from 5.5% to 15.5% has a compaction ratio of 3.6–7.7%, while the same seeds after 60 days storage period at 30°C have a compaction ratio of 3.5–10% (Jian et al. 2013). Compressibility index (C_I) and Hausner ratio (H_R) can be calculated as:

$$C_I = \frac{\rho_c - \rho_b}{\rho_c} \tag{1.4a}$$

$$H_R = \frac{\rho_c}{\rho_b} \tag{1.4b}$$

where C_I is the compressibility index, H_R is the Hausner ratio, ρ_b is the bulk density (kg/m^3), and ρ_C is the compacted density (kg/m^3). Bulk density in stored grain bin can be estimated when grain mass loaded in the bin is known and volume occupied by grain can be calculated. Bulk densities in a grain bin can be different from the lab measured bulk density. The density in a bin also varies with height of fall and method of filling, and the location of the grain inside the bin. The bulk density of wheat filled with a spout is 3–5% higher than the test weight (Stephens and Foster 1976). Chang et al. (1981) observed an increase in the bulk density from 5 to 9% for wheat, 6 to 10% for corn, and 11 to 12.5% for sorghum when spreaders were used. Bulk density of grain filled by spout has a wider variation than that filled by spreader (Jayas et al. 1989). During grain drying, there will be a reduction in volume due to shrinkage, which will result in the increase of bulk density. Bowden et al. (1983) reported a 10% reduction in grain volume due to drying of barley from 22 to 14% moisture content. Variation of bulk density in a bin results in uneven distribution of porosity. Mathematical models are developed to estimate the bulk density inside bins (Milani et al. 2000; Bhadra et al. 2015; Bhadra et al. 2018). The developed model usually considers the effect of moisture content, grain type, test weight, bin geometry, and bin wall material.

1.1.1.4 True (Kernel) Density Measurement

There are two general methods used to measure the true density: liquid displacement or gas displacement. The most used liquids are toluene, alcohol, tetrachloroethylene, oil, and mercury. In liquid displacement method, the weighed kernels are directly dropped in the liquid, and the volume of the displaced liquid is assumed as the volume of the tested kernels. Selection of the liquid depends on the requirement of measurement accuracy and the dissolving property of the kernels in the liquid. It is required the liquid will not penetrate into the grain kernels and the chemicals inside the kernels will also not dissolve into the liquid. This requires that the used liquids should have a low surface tension and be absorbed very slowly by the grain seeds. Coating of kernels with films or paints may be required to prevent liquid absorption. The surface tension of most liquids, like water or mercury, poses a problem for the determination of void space volumes for most grain and seed kernels (Chang 1988). Even though toluene displacement method has a low accuracy, it is used by researchers in the literature because it is simple to operate. If some chemical compounds inside grain kernels can be dissolved into toluene, the accuracy of toluene method also decreases. For a quick estimation of the true density, water can be used instead of toluene. The volume of the weighed sample inside the water should be measured immediately after the sample is dropped in the water (Jian et al. 2018).

The gas pycnometer is specially designed for the measurement of true density by employing Archimedes' principle of fluid (gas) displacement and the technique of ideal gas expansion. The gas pycnometer method usually has a higher accuracy than the liquid displacement method because helium used in the pycnometer can penetrate faster in every surface and flaw down to about 1 Å, thereby enabling the measurement of volumes with a high accuracy (can go up to 0.02% of the sample volume). The measurement of true density by helium displacement often can reveal the presence of impurities and occluded pores which cannot be determined by other methods because it (1) can rapidly penetrate into small pores due to its small molecules; (2) is not adsorbed due to its inert property; (3) can be considered as an ideal gas for pressures and temperatures usually employed in the test; and (4) provides a useful means for determining porosity of low permeability grain due to its higher diffusivity than other gases (CO_2, N_2) or liquid (such as water). High concentration nitrogen can also be used to replace helium with a slight lower accuracy. If there are pores in which the gas cannot penetrate, the measured pore volume will be less than the true pore volume.

The gas pycnometer is developed based on the principle of porosity tank method (Figura and Teixeira 2007). The gas can be helium, nitrogen, or air (accuracy is lower when air is used). The porosity tank method includes two tanks with the same volume, three valves, and a manometer for the pressure measurement (Fig. 1.3). After the sample is loaded inside

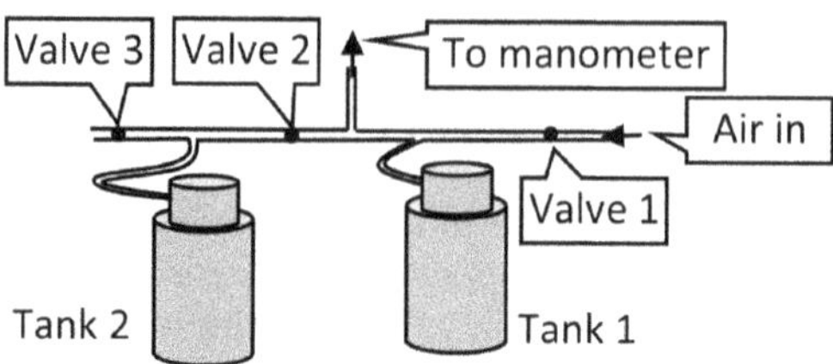

Figure 1.3 Schematic of porosity measurement set up (porosity tank method). (Adapted from Figura and Teixeira, 2007).

tank 2 and the initial pressure (P_1) in the system is measured, the valve 2 and valve 3 are closed. Then compressed gas is introduced in tank 1. After valve 1 is closed and the pressure inside the tank 1 is measured (P_2), valve 2 is opened and the pressure inside the system is measured again (P_3) by using the manometer. Based on the idea gas law, the gas mass inside tank 1 and 2 before valve 2 is opened is:

$$m_a = \frac{P_1(V_2 - V_s)}{RT_1} + \frac{P_2V_1}{RT_2} \tag{1.5a}$$

After the valve 2 is opened, the same amount of gas mass is distributed in tank 1 and 2, and can be calculated as:

$$m_a = \frac{P_3(V_2 - V_s)}{RT_3} + \frac{P_3V_1}{RT_3} \tag{1.5b}$$

where m_a is the mass of the gas (kg), R is the specific gas constant (J·kg^{-1}·k^{-1}), T_1, T_2, T_3 are the temperatures measured before and after the compressed gas is introduced, and after tank 1 and tank 2 reach equilibrium, respectively (K); P_1, P_2, P_3 are the pressures measured in tank 1 at the very beginning, in tank 2 after the compressed air is introduced, and in both tank 1 and tank 2 after valve 2 is opened, respectively (Pa); and V_1, V_2, V_s are the volumes of tank 1, tank 2, and the sample, respectively (m^3).

The volume of the sample is:

$$V_s = V_2 - \frac{\frac{P_2}{T_2} - \frac{P_3}{T_3}}{\frac{P_3}{T_3} - \frac{P_1}{T_1}} V_1 \tag{1.5c}$$

If $V_1 = V_2$, $T_1 = T_2 = T_3$, then

$$V_s = V_2\left(1 - \frac{P_2 - P_3}{P_3 - P_1}\right) \tag{1.5d}$$

If the tank 2 is filled entirely with the sample, then the porosity of the sample (ε) is:

$$\varepsilon = \frac{P_2 - P_3}{P_3 - P_1} \tag{1.5e}$$

The true density (kernel density) can be calculated as:

$$\rho_k = \frac{m}{V_s} = \frac{m}{V_2\left(1 - \frac{P_2 - P_3}{P_3 - P_1}\right)} \tag{1.5f}$$

where m is the mass of the sample.

1.1.1.5 Physical Factors Influencing True Density

Kernel density is mostly affected by crop types, classes within a type and cultivars within a class, growing conditions, harvesting methods, and the moisture content of the grain kernels (Table 1.1). The growing conditions determine the plumpness of the kernels and harvest method determines the mechanical damage to kernels. It is usually assumed kernels are incompressible inside stored bins. Therefore, the kernel density will not be influenced by methods of filling and compaction effect inside the grain bins. True density of cereal grain kernels is linearly decreased with increase of moisture content (Isik and Unal 2007; Jian et al. 2012; Sun et al. 2014; Senthilkumar et al. 2018) because density of water is usually less than that of the dry matter of the grain kernels. Since oils are lighter than water, seeds with high oil content also have low kernel density. However, for the grain kernels like hemp seeds, which has gaps between the seed coat and cotyledon inside the kernel, the true density linearly or quadratically increases with the moisture content increase. The reason is that seed coat may not enlarge much when seed moisture content is increased, and the water is absorbed by the cotyledon and the expanded cotyledon replaces the air space. The gaps between the cotyledon and coats become smaller or disappear after the cotyledon expansion (Jian et al. 2018). The seed coat of the hemp seeds is harder than pulses and cereal grains.

1.1.2 Porosity

1.1.2.1 Porosity (ϵ) and Measurement

Porosity of a grain bulk is the ratio of the volume of intergranular pores to the total volume occupied by the grain bulk. Porosity can be reported as percentage or decimal from 0 to 1.

There are many methods to determine the porosity, pore size, and pore structure of any porous material such as rock, soil, and drug in different fields. The most used methods are gas or water vapour sorption, pore saturation (gas or liquid replacement), pycnometer (gas expansion), mercury intrusion, microscopy, and imaging technique (Anovitz and Cole 2015). Image processing method such as QEMSCAN imaging, nuclear magnetic resonance, CT scan, and scattering methods can calculate the porosity by determining the microscopic view of a random section of the porous medium in two or three dimensions. Helium porosimetry using pycnometer is an easy and established technique that can measure only open pores. Due to the limitation of some methods, the most used methods to determine the porosity of grain bulks are pore saturation (gas or liquid replacement) and pycnometer (gas expansion) methods. Only a few publications using imaging technique (Neethirajan et al. 2008) are available. These methods are also used to measure the true density of grain bulks.

The concept of effective porosity has been used in the study of airflow resistance, which is defined as the ratio of the interconnected (effective) pores to the total volume of the porous material. The effective or open pore is the pore that allows fluids or volatiles to flow through it, while the ineffective or closed pore is the non-conducting pore space, includes the occluded pores and the dead-end pores. Effective porosity is most commonly considered in grain industry because it is related to airflow resistance and distribution. Even though pycnometer method cannot be used to directly measure the effective porosity, the effective porosity should be similar to the porosity determined by helium – pycnometer method because it only measures the open pores and there are negligible disconnected pores in grain bulk (Neethirajan et al. 2006).

Porosity ($\mathcal{E}$) is usually calculated from bulk and true density in literature as:

$$\varepsilon = \left(1 - \frac{\rho_b}{\rho_k}\right) \times 100\% \tag{1.6}$$

It should be noted that the porosity does not give any information concerning pore sizes, their distribution, and their degree of connectivity. Thus, the porosity of porous materials consisting of mono-sized spheres range from 0.26 to 0.48 (Woodcock and Mason 1987). The crop grains have about 40% porosity and it ranges between 35 and 55% even though different grain kernels have different kernel sizes and shapes (Table 1.1).

1.1.2.2 Physical Properties Influencing Porosity

Any factor influencing bulk density of a grain bulk affects the porosity. These factors include moisture content, temperature, percentages of dockage, and method of filling. For most crops (such as wheat, corn, canola), porosity linearly decreases with the increase of bulk density (Chung and Converse 1971). Porosity of crop grain usually increases with the increase of moisture content (Muir and Sinha 1988) because increased size of the grain kernels with the increase of moisture content will enlarge the pore size among the grain kernels. Porosity of sweet corn seeds increase from 57.5% to 61.3% with the increase of moisture content from 11.5 to 19.7% (Coskun et al. 2006). Porosity is higher at higher moisture content of white wheat, red wheat, and corn but lower at higher moisture content of soybeans (Molenda et al. 2005). Porosity of grain seeds with hard seed coat may not be affected by moisture content because the size of the grain kernels may not change much as the grain moisture content changes. Therefore, the change of porosity with change of moisture content depends on grain types. Different sizes of dockage have different effect on the porosity. If the dockage can fill in the pore spaces among the grain kernels, porosity will be decreased; otherwise, it will increase (Jian et al. 2018). Grain inside a storage bin is usually loaded from certain heights. This results in compact forces on the grain bulks during loading and storage. Bulk density inside storage bins is not uniformly distributed. This variation results in the variation of porosity inside grain bulk which would cause non-uninform distribution of airflow resistance (Jian et al. 2019).

1.2 Geometrical Properties of Grain Kernels

1.2.1 Dimensions (Size) and Measurement

Dimensions (size) of grain kernels are usually measured along three mutually perpendicular axes because measurements in more than three dimensions are usually difficult and not necessary (Jian et al. 2018). The three mutually perpendicular axes account for some 93% of variation in volume. Of this total percentage, the intermediate axis contributes only 4% to volume prediction (Mohsenin 1986). The three mutually perpendicular axes are termed as major (principal), medium, and minor axes (diameter), which corresponds to length, width, and thickness (depth) of the kernel, respectively.

The length, width, and thickness can be directly measured by using a calliper. After the kernel freely rests on a levelled flat surface, the longest dimension in the horizontal plan is the length, the shortest is the width, and the longest dimension in the vertical plane is the thickness (Fig. 1.4). Dimension (size) of grain kernels can also be determined using the projected area method produced by a light source located in different

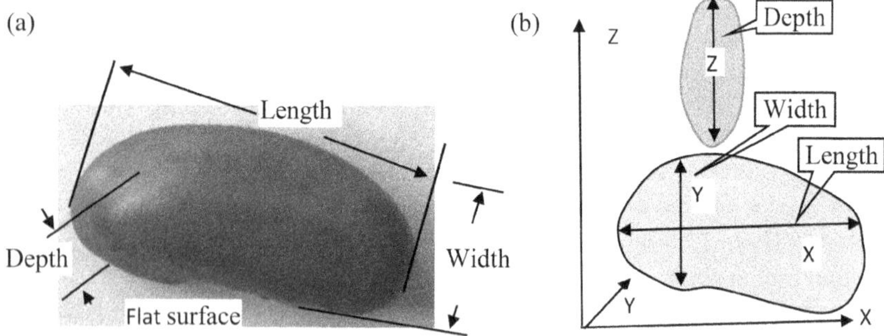

Figure 1.4 (a) Representation of length, width, and depth of a typical red kidney bean rested on a flat surface (Jian and Jayas 2018) and (b) two maximum project areas generated by light sources on the X-Y and Y-Z planes.

angles or a shadowgraph tracing the area on a light sensitive or ordinary graph paper. Two maximum project areas can be generated by using a light source at the top (generating the projected area in the X-Y plan, Fig. 1.4) and side (generating the projected area in the X- or Y-Z plan, Fig. 1.4) of the rest kernel, respectively. The major diameter (length) is the longest dimension of the maximum projected area on the X-Y plane (Fig. 1.4). Intermediate diameter (width) is the longest dimension (diameter) on the X-Y plane and in the direction perpendicular to the major length line. Minor diameter (thickness) is the longest dimension on the X- or Y-Z plane (Fig. 1.4).

Indirect methods, such as digital image processing (Senthilkumar et al. 2018) and particle size distribution analysis, are also used to determine the dimension of grain kernels. Particle size distribution analysis determines both size of particles and their status of distribution. Particle size distribution analysis method is most used for powder and dust particles and this method cannot exactly measure the dimension of grain kernels.

After the dimension of grain kernels are measured, the geometric mean, equivalent, and arithmetic diameters can be estimated as (Mohsenin 1986):

$$D_g = (LWŦ)^{1/3} \tag{1.7a}$$

$$D_e = \left[L \frac{(W + Ŧ)^2}{4} \right]^{1/3} \tag{1.7b}$$

$$D_a = \frac{L + W + Ŧ}{3} \tag{1.7c}$$

where D_g, D_e, and D_a are geometric mean diameter (m), equivalent diameter (m), and arithmetic diameter (m), respectively; and L, W, Ŧ are the length (m), width (m), and thickness (m), respectively.

1.2.2 Surface Area and Volume

After dimension of grain kernels are measured, surface area can be predicted by using empirical equations (Mohsenin 1970; Deshpande et al. 1993; Avena-Bustillos et al. 1994; Jain and Bal 1997).

$$S = 2\pi L^2 + \pi \frac{W^2}{e} ln \frac{1+e}{1-e} \tag{1.8a}$$

$$S = \pi \text{Ŧ}^3/2 + \frac{L\text{Ŧ}}{2e} \sin^{-1} e \tag{1.8b}$$

$$S = 2\pi W^2 + 2\pi \frac{LW}{e} \sin^{-1} e \tag{1.8c}$$

$$e = [1-(\text{Ŧ}/L)^2]^{1/2} \tag{1.8d}$$

$$S = \frac{\pi\sqrt{W\text{Ŧ}}\,L^2}{2L-\sqrt{W\text{Ŧ}}} \tag{1.8e}$$

$$S = \pi D_g^2 \tag{1.8f}$$

$$S = \frac{\pi L L_m}{2}\left(\frac{L_m}{L} + \frac{1}{u}\arcsin U\right),\quad L_m = \frac{W+\text{Ŧ}}{2},\quad U = \frac{\sqrt{L^2 - L_m^2}}{L} \tag{1.8g}$$

where S is the surface area (m^2). Surface area can also be directly measured by using light project method or by using an optical planimeter. An external coat of silicon rubber solution or other similar material can be applied to the surface of grain kernels. The surface area can be determined by peeling off the external coat (Jindal et al. 1974). Coating method is mostly used in the literature published before 1980s.

Different researchers suggest different empirical equations to estimate volume (Deshpande et al. 1993; Jain and Bal 1997; Markowski et al. 2013). Most of these empirical equations are developed by using the relationship among volume and dimension of regular shapes such as sphere and ovoid. The most used empirical equations are:

$$V = \frac{4}{3}\pi L W^2 \tag{1.9a}$$

$$V = 0.25[(\pi/6)L(W+\text{Ŧ})^2 \tag{1.9b}$$

$$V = \frac{\pi W\text{Ŧ}L^2}{6\left(2L-\sqrt{W\text{Ŧ}}\right)} \tag{1.9c}$$

$$V = \frac{\pi LW^2}{6} = \frac{\pi D_g^3}{6} \text{ Prolate spheroid (e.g., lemon)} \tag{1.9d}$$

$$V = \frac{\pi L^2 W}{6} = \frac{\pi D_g^3}{6} \text{ Oblate spheroid (e.g., grapefruit)} \tag{1.9e}$$

$$V = \pi h\left(r_1^2 + r_1 r_2 + r_2^2\right)/3 \text{ Right circular cone (e.g., carrot)} \tag{1.9f}$$

$$V_k = \frac{\pi D_e^3}{6} \tag{1.9g}$$

$$V = \frac{\pi}{6} LW\text{Ŧ} \tag{1.9h}$$

where r_1, r_2, and h are the radii of the base, top, and height of the cone. Volume can also be directly measured by using the liquid displacement method mentioned in the measurement of true density and porosity. These methods cannot provide enough accuracy for a single kernel but can measure the average volume of kernels after a known number of kernels are used. If high accuracy of the volume measurement is required or suitable liquid is not available, gas displacement method or pycnometer is mostly used.

1.2.3 Shape and Shape Factors

Different terms have been used to define the shape of particles. These terms include elongation, lengthiness, flakiness, shape factor, sphericity, roundness, and aspect ratio. Shape of particles can be judged by the measured length, width, and thickness.

1.2.3.1 Elongation, Lengthiness, and Flakiness

$$E_l = \frac{L}{W} \tag{1.10a}$$

$$f = \frac{W}{\text{Ŧ}} \tag{1.10b}$$

where E_l is the elongation or lengthiness, and f is the flakiness. Plate-like materials such as leaves and kennels of lentil have large values of f, while needle-like materials such as chaff and kernels of rye have large values of E_l.

1.2.3.2 Shape Factor

Shape factor can be calculated as (Jain and Bal 1997):

$$S_v = V/W^3 \tag{1.11a}$$

$$S_s = \frac{S}{6W^2} \tag{1.11b}$$

$$\lambda = S_v/S_s \tag{1.11c}$$

where λ is the shape factor, S_v is the volume-based shape factor, and S_s is the surface-area-based shape factor. Gorial and O'callaghan (1990) found the volume shape factors for different types of grain lie between 0.27 and 0.49. The term of shape factor has been misused and has different definitions under different applications. When the terminal velocity is measured, the shape factor is termed as the ratio of the geometric mean diameter to the diameter of an equivalent sphere. This definition is the same as one of the definitions of sphericity.

1.2.3.3 Cross Section

The cross section of a particle is usually the projected cross section when viewed in the direction of the thickness. The diameter of the projected cross section is the diameter of the circle having the same area as this projected cross section.

$$d_p = \left(\frac{4}{\pi}WL\right)^{1/2} \tag{1.12}$$

where d_p is the diameter of the projected cross section (m).

1.2.3.4 Sphericity

Sphericity is defined in three ways: (1) the ratio of the surface area of a sphere, which has the same volume as the kernel, to the actual surface area of the kernel; (2) the ratio of the diameter of a sphere of the same volume as the object to the diameter of the smallest circumscribing sphere; and (3) the ratio of the diameter of largest inscribed sphere to the diameter of the smallest circumscribed sphere (Fig. 1.5). Based on these definitions, the

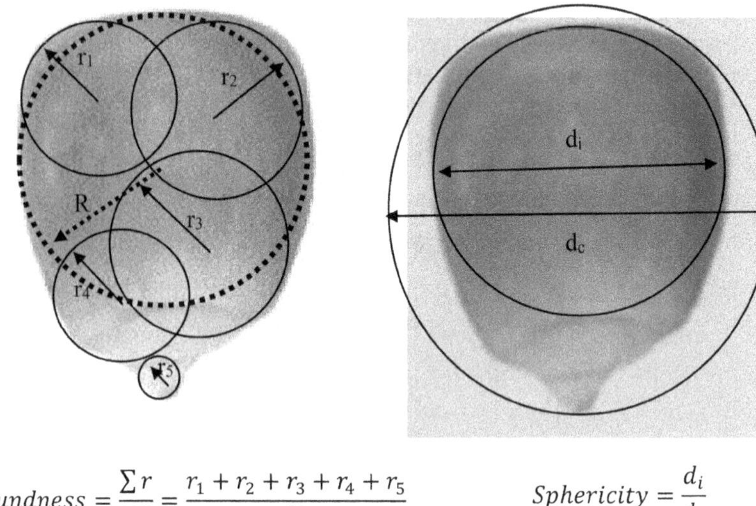

$$Roundness = \frac{\sum r}{NR} = \frac{r_1 + r_2 + r_3 + r_4 + r_5}{5R} \qquad Sphericity = \frac{d_i}{d_c}$$

Figure 1.5 Conceptual diagrams of roundness described by Wadell (1932) and sphericity described by Curray (1951). The thickness of the corn kernel is larger than $2R$ and less than d_c. N = number of spheres, r = radius of a sphere, d_i = diameter of the minimum inscribed sphere, and d_c = diameter of the maximum inscribed sphere.

sphericity of a perfect sphere is 1. It is usually difficult to exactly measure the surface area of a kernel. Therefore, sphericity is usually estimated by using the measured dimensions along three mutually perpendicular axes.

$$Ø = \frac{D_g}{L} \tag{1.13a}$$

$$Ø = \frac{D_p}{D} \tag{1.13b}$$

$$Ø = \frac{D_e}{L} \tag{1.13c}$$

$$Ø = \left[\frac{\sqrt{WŦ}\left(2L - \sqrt{WŦ}\right)}{L^2}\right]^{1/3} \tag{1.13d}$$

$$Ø = \frac{6\,V_k}{D_e S}, \quad \text{and} \quad S = \pi D_e^2 \tag{1.13e}$$

where D is the diameter of the circle circumscribing the project area (m). Different equations may not necessarily yield the same results. Sphericity of grain kernels is from 0.40 to 0.88 (Table 1.2).

1.2.3.5 Roundness and Aspect Ratio

Roundness (circularity or roundness ratio) is used to measure the smoothness of the particle. Roundness is usually not measured in the grain industry due to the difficulty of its measurement. The most accepted definition is proposed by Wadell (1932) and the definition is the ratio of the average curvature radius of the particle edges and corners to the radius of the maximum inscribed sphere (Fig. 1.5). However, it is difficult to measure the roundness by using this definition because so many spheres inside the kernel should be identified and measured (Fig. 1.5). Different definitions and equations are provided in literature.

$$R_c = \frac{2r_c}{D_L} \text{ (Mitchell and Soga 2005)} \tag{1.14a}$$

$$R_c = \frac{r_c}{r_{av}} \text{ (Mitchell and Soga 2005)} \tag{1.14b}$$

$$R_c = \frac{4\pi A_P}{P_o^2} \text{ (Li et al. 2002)} \tag{1.14c}$$

$$R_c = \frac{A_l}{A_c} \text{ (Curray 1951)} \tag{1.14d}$$

$$R_c = \frac{r_{cs}}{r_m} \tag{1.14e}$$

$$R_c = \frac{A_p}{A_c} \times 100 \text{ (Mohsenin 1986)} \tag{1.14f}$$

where R_c is the roundness or circularity (decimal or percentage), r_c is the curvature radius of the maximum convex part of the particle (m), D_L is the longest diameter through the most convex part of the particle (m), r_{av} is the mean radius (m), A_p is the particle projected area (m^2), P_o is the overall projection perimeter (m), A_l is the largest projected area of object in natural rest position (m^2), A_c is the area of smallest circumscribing circle (m^2), r_{cs} is the radius of curvature of the sharpest corner, and r_m is the mean radius of the object.

The aspect ratio (A_R) is calculated as:

$$A_R = \frac{W}{L} \tag{1.15}$$

Table 1.2 Mass of Thousand Kernel, Surface Area, Volume, and Sphericity of Seeds of Cereal Grains, Pulses (Legumes), Oilseeds and Speciality Crops

Crop	*MC (%)*[a]	*MTK (g)*[b]	*Surface (mm^2)*	*Volume (mm^3)*	*Sphericity*	*Reference*
Caper	5.7–14.1	6.6–7.8	15.8–23.1		0.72–0.74	Dursun and Dursun (2005)
Corn	4.7–22.0	271–321	133–160	202–265	0.59–0.62	Seifi and Alimardani (2010)
Corn (sweet)	8.4–14.6	220–268	120–183	94–194	0.63–0.69	Karababa and Coskuner (2007)
Durum[c]	9.3–41.5	32–52	47.1–61.0	29–42	0.69–0.72	Al-Mahasneh and Rababah (2007)
Flaxseed	9.1–15.2	7.2–7.4			0.47–0.48	Bhise et al. (2013)
Hemp	9.0–15.0	12.4–14.1	68.6	13.3 ± 0.6	0.80	Jian et al. (2018)
Kidney bean[d]	8.9–16.4	522–560	278–329		0.73–0.76	Isik and Unal (2007)
Linseed	7.6–18.2	6.0–6.7	15.8–18.6		0.49–0.50	Selvi et al. (2006)
Okra	7.5–46.7	66–130		67–124	0.70–0.77	Sahoo and Srivastava (2002)
Pigeon pea	5.9–22.0	76 ± 1	67.4 ± 2.9	50–65	0.82 ± 0.53	Shepherd and Bhardwaj (1986)
Rough rice	10.0	21.6 ± 0.7	31.8 ± 2.3	20.3 ± 2.3	0.40 ± 1.5	Varnamkhasti et al. (2008)
Soybean	8.0–24.0	145–250	115 – 155		0.81–0.88	Kashaninejad et al. (2008)
Sunflower[e]	5.8	34 ± 1			0.43–0.68	Gupta and Das (1997)
Sunflower[f]	5.8	49 ± 12			0.46–0.67	Gupta and Das (1997)
Vetch	9.6–17.1	55.5–59.0			0.84–0.86	Yelcin and Ozarslan (2004)
Wheat[g]	8.6–9.8	39.7 ± 9.8			56.5 ± 2.8	Gürsoy and Güzel (2010)
Wheat[h]	9.2–11.1	46–52	55.7–67.0	33–42	0.54–0.61	Markowski et al. (2013)

[a] Moisture content (wb, %).
[b] Mass of thousand kernels.
[c] Un-matured durum (green wheat).
[d] Red kidney bean.
[e] Sunflower kernel.
[f] Sunflower seed.
[g] One cultivar (Nurkent) of bread wheat.
[h] Four cultivars of winter wheat.

The roundness and aspect ratio are highly dependent upon the sharpness of the particles and are usually related to their angularity. Krumbein and Sloss (1963) developed the roundness and sphericity chart for visually determining the roundness. A visual measurement of the roundness of grain seeds is not applicable. The following methods that are used in other research areas have also been used by some researchers: fractal analysis, Fourier methods, and image analysis and processing.

The decrease in the roundness or increase in the angularity of the particles will increase the particle interlocking, thereby increasing the angle of repose and decreasing the rolling and flow property of the solid.

1.3 Hardness and Measurement

1.3.1 Hardness

Grain hardness or toughness is most used in the grain processing industry and there is no universally accepted definition. The most used definition of kernel hardness is the resistance of the kernel to penetration of foreign matter or resistance to destruction and breakdown to particles. Hardness is also known as rupture force, is mainly related to plastic deformation of material by indentation. The peak force occurred during the first compression is termed as hardness. Grain hardness has both qualitative and quantitative meaning because hardness is not only related to the kernel firmness, but also related to milling and baking qualities; and for pulses to dehulling and cooking quality. During milling, hardness influences the sieving, energy consumption, fineness of the finished product, and the milling extraction. Milling energy is linearly correlated with kernel hardness (Szabo et al. 2016). Optimal dry milling requires hard kernels, whereas optimal wet milling requires soft kernels.

Hardness is widely used by the wheat industry at the very beginning because different classes of wheat have different hardness (in Canada, there are eight wheat classes and multiple cultivars [varieties] within each class). Studies have been conducted on the effect of hardness on quality in barley (Psota et al. 2007; Turuspekov et al. 2008), triticale (Ramiärez et al. 2003), durum (Anjum and Walker 1991), sorghum (Anglani 1998), millet (Abdelrahman and Hoseney 1984), and maize (Dombrink-Kurtzman and Bietz 1993). Wheat is still the main crop for which its new cultivars are classified on the basis of hardness. The level of hardness often determines the end use of the wheat. Soft, intermediate, and high hardness wheat's are usually used for cookies, bread, and pasta, respectively (Ross 1997; Morris 2002). Soft wheats are more friable, require less energy to mill and produce flours and meals with reduced particle-size distribution, including many free starch granules. Hard kernels are more difficult to reduce to flour-sized particles, so hard grain flour is coarser, has more broken and damaged starch granules, a larger mean particle size than soft grain flour, but flow more easily. Hard wheat flour has a higher damaged starch produced during the milling process, and damaged starch has a higher water absorption capacity and is more readily hydrolyzed by amylase enzyme.

Hardness is related to the kernel size, kernel density, and vitreousness. Lower density indicates soft kernels. The texture is significantly affected by quantity, quality and the ratio of protein and starch, size of cells and their mutual connections within individual tissues. The interaction between carbohydrates and proteins strongly influences the hardness of endosperm. Texture, i.e., the organization of individual grain components (water-soluble pentosans, lipid content, endosperm, and hulls) determines whether the grain will be hard or soft. Some seeds having high protein content will be hard, while total protein content in some crop seeds do not show significant correlation with kernel hardness (Pasha et al. 2010). Increase in protein can be linked to an increase in vitreousness and hardness (Morris 2002; Morris and Massa 2003). Grain hardness is related to the degree of adhesion between the starch granules and the protein matrix, whereas the endosperm density is affected by the fracture force. Therefore, hardness is a genetic and physicochemical property related to classes regardless its moisture content, even though a grain kernel usually becomes soft with the increase of its moisture content.

The relationship of hardness with the texture, genetic, and physicochemical properties is not completely explicit because many factors influence the hardness. For example, hardness is significantly influenced by classes and farming practice such as different breeding programs and growing locations (Psota et al. 2007). Hardness of wheat and barley is linked to a specific gene family with allelic variation related to variability in hardness and quality. For maize, a number of quantitative trait loci have been linked to a range of hardness measures, including vitreousness. The puroindoline proteins a and b form the molecular basis of wheat grain hardness (Morris 2002). When both a and b are in their "functional" wild state, grain texture is soft, otherwise it is hard if either one of the puroindolines is absent or altered by mutation (Takata et al. 2010). The locus hardness is located on the short arm of the 5D chromosome (Greffeuille et al. 2006).

Two proposed theories explain hardness of wheat: continuity of the protein matrix and physical contact between starch and protein (Moss et al. 1980), and the bond or adhesion between starch and protein (Simmonds 1972; Simmonds et al. 1973). Abdelrahman and Hoseney (1984) found the substances responsible for hardness in those grains are extractable and sensitive to heat. Glenn and Saunders (1990) reported the existence of intracellular space around the starch granules of soft, but not of hard wheat.

This physical discontinuity provides a natural path for shearing forces when the kernel is under pressure, leading to softer material that is more easily reduced in particle size. The degree of adhesion between starch granules and endosperm protein matrix affects hardness (Barlow et al. 1973). Softer wheat has a higher content of amylose bound with lipids and has a lower content of total starch than the harder wheat (Gaines et al. 2000). Classes with softer endosperm texture have larger ratio of 25 μm particles after milling than the ones with harder endosperm (Devaux 1998). These studies further concluded that water-soluble material acts as cement between starch granules and storage protein, across the amyloplast membrane interface. Hardness is also related to lipid and pentosane concentration (Bettge and Morris 2000).

1.3.2 Measurement of Hardness

It is generally recommended that, prior to hardness measurements, samples should be equilibrated to a desired moisture content or a correction factor should be applied (Armstrong et al. 2007). More than 20 methods have been reported and most of these methods are empirical. These methods include measuring compressive strength, impact and shear resistance, resistance to grinding, yield of grits, and starch gelatinization properties (Chandra et al. 2001). The mechanical properties of yield stress, yield strain, and Young's modulus are usually measured to understand the milling properties (Kang et al. 1995). Taylor and Duodu (2009) provide the testing methods for maize, sorghum and other non-wheat cereals. There is no standardized method to measure hardness because each method has advantages and disadvantages and none can provide a complete characterization. Fox and Manley (2009) reviewed these methods. The three most commonly used methods are the particle size index (PSI), near-infrared spectroscopy (NIR), and the Single Kernel Characterization System (SKCS). Some methods such as the SKCS use single kernels and some such as the PSI use bulk grain sample. All those tests respond in different ways to change in grain hardness. A number of factors affect breakage including temperature, moisture, genetics, growing environment, and firmness. Therefore, the measured condition of hardness should be specified. Measurement of grain hardness requires reference materials for instrument standardization. The American Association of Cereal Chemists (AACC) Approved Methods employ reference materials prepared by the US Department of Agriculture Federal Grain Inspection Service (USDA-FGIS). The material is comprised of genetically pure commercial grain lots of five soft and five hard wheat cultivars and is made available through the National Institute of Standards and Technology.

1.3.2.1 Particle Size Index (PSI) Method

This method provides only an indication of the total hardness of a sample and does not provide any information on possible variation within kernels. The grain sample is milled and then fractionated through sieves. The PSI is calculated as:

$$PSI = \frac{m}{\text{Sample mass}} \times 100 \tag{1.16}$$

where m is the mass of the particles passing through a specified sieve (g); and sample mass is the mass of the sample used to conduct the sieving test (g). Researches used different equipment and different number of sieves of different mesh sizes. To test the hardness of the barley, Psota et al. (2007) used a laboratory mill (LM 3303, Perten Instruments, GmbH-Hamburg, Germany) with head number two and 0.075 mm sieve. Ten grams of the milled sample was taken and placed into the Sifter Swing 200 (Mezos, Czech Republic) and sieved for 10 min. Morris and Massa (2003) used Tecator Cemotec 1090 burr mill (Foss-Tecator, Eden Prairie, MN) and AACC 55-30 method to measure the wheat hardness. The ground meal was sifted (Ro-Tap, W.S. Tyler Co., Mentor, OH) using 210 μm opening, 20.3 cm diameter sieves, for 5 min/sample. The weight of the meal recovered from the bottom pan was used to calculate the PSI.

1.3.2.2 Near-Infrared (NIR) Spectroscopy Method

The principle using the level of vitreousness (the proportion of the endosperm that is translucent) as an indirect measure of hardness is that vitreous endosperm has a glassy appearance and is denser than opaque floury endosperm, which contains more airspaces. Endosperm that is translucent against the non-translucent areas shows differences in protein content and composition. Light at particular wavelengths in the near infrared region is absorbed by some bonds such as C-H, O-H, and N-H, which vibrate in proportion to their concentration in the grain. Correlations between vitreousness levels and end-use quality have also been investigated, both in reflectance and transmission modes, and extensively used for indirect measurement of maize kernel hardness. The more vitreous samples are harder than the less vitreous samples. Near-Infrared (NIR) spectroscopy has been used to estimate maize hardness by relating the measured wavelength to the measured values from the Tangential Abrasive Dehulling Device (model 4E-220, Venables Machine Works, Saskatoon, SK) (Wehling et al. 1993), PSI value (De Alencar Figueiredo et al. 2006), Glenn Mill (Glenn Mills Inc, Clifton, NJ) (Armstrong et al.

2007), floaters method (Eyherabide et al. 1996), kernel density (Siska and Hurburgh 1994), and coarse/fine ratio (Robutti 1995) as reference methods. The accuracy and precision of the reference method are critical to the development of any NIR calibration (Hoffman et al. 2010). The use of a single wavelength (860 or 1680 nm) as well as the maximum absorbance/reflectance between 620 and 680 nm have been used to relate to hardness (Robutti 1995). Light Transflectance Meter (LTm) developed by Brewing Research International (Chandra et al. 2001) is used to measure the quantity of light transmitted through a kernel, and low and high LTm values indicate mealy and steely grain texture, respectively. The GrainCheck™ 310 instrument (Hillerød, Denmark) is used to measure the total light reflectance of kernels. The NIR reflectance spectroscopy can be used to measure the baseline shift (BLS) because different grinding-induced particle size will produce different BLS.

1.3.2.3 Single Kernel Characterization System

The SKCS developed by Martin et al. (1993) is most used to measure the single kernel hardness of wheat. The SKCS works by crushing 300 individual grains in a sample between a serrated rotor and a crescent. Both the mean and the standard deviations for the 300 kernels in the sample are reported. The force required to crush individual grains of a sample between two surfaces are measured by taking into account the weight, diameter, and moisture content of the grain. Bean et al. (2006) used a SKCS to evaluate the hardness and moisture content of sorghum and found the SKCS should be suitable for measuring sorghum grain attributes with a lower moisture prediction than the oven method (described in Chapter 3).

1.3.2.4 Methods Measuring the Force to Crush, Shear, or Peal Kernels

Methods measuring resistance to grinding include Brabender hardness test, pressure rod, Tangential Abrasion Dehulling Device (TADD), resistance to pearling, and single-kernel compression. Two steps of Brabender hardness tester measure average energy input and particle size by recording the torque value during grinding (Brabender units). The TADD measures the amount of material removed from kernels when the kernels are abraded for a defined period of time, with higher values indicating softer kernels. Resistance to pearling (Chung et al. 1977) is another method to evaluate the hardness and the milling properties can be studied at various dehulling degrees (Reichert and Youngs 1976). Different researchers used different equipment and procedures to measure the resistance of kernels when a rod is pressed into the kernels, or compressed under a universal testing machine. Speed reduction and vibration of a mill during grinding can also be used to evaluate the required force.

1.3.2.5 Flour Yield, Particle Size, and Surface Area

These methods include Stenvert method, Roff milling index or milling index, and particle size distribution analysis. Stenvert (1974) method records the time required to fill a 17 mL tube attached to a grinder. A longer time indicate harder kernels. Stenvert method has been modified by some researchers (Mestres et al. 1995) by adding different sieves and computer-controlled data capture to report parameters such as energy usage and time of grinding. The Roff milling index or milling index is calculated from the meal and bran fractions obtained from milling a sample through a roller mill system. A maize sample (preconditioned to 14% moisture) is milled through the roller mill system of three rollers with gaps of 0.3, 0.38, and 0.08 mm. The method can distinguish between maize cultivars with different hardness and identify cultivar and environmental effects on maize hardness. Hard and soft wheats gave different size distribution of the flour and Wu et al. (1990) found mean flour particle size might measure wheat hardness more reliably than PSI.

1.3.2.6 Starch Properties

Starch property methods include the starch gelatinization and diastatic activities. Methods of starch gelatinization properties include ViscoAnalyser (RVA). The RVA is the method that relates biochemical components to hardness of maize. After the sample is ground, the ground material is mixed with water and constantly stirred while the sample is heated to 100°C. The method provides information on starch properties, including paste viscosity, gelatinization temperature, and time (Almeida-Dominguez et al. 1997). The starch damage and diastatic activities are correlated with hardness (Stenvert 1972).

1.4 Shrinkage and Swelling

1.4.1 Shrinkage and Swelling

Shrinkage and swelling are the decrease or increase in volume of grain kernels after grain kernel loses or gains water, respectively (Fig. 1.6). Uniform shrinkage or swelling in all dimensions of kernels are called isotropic

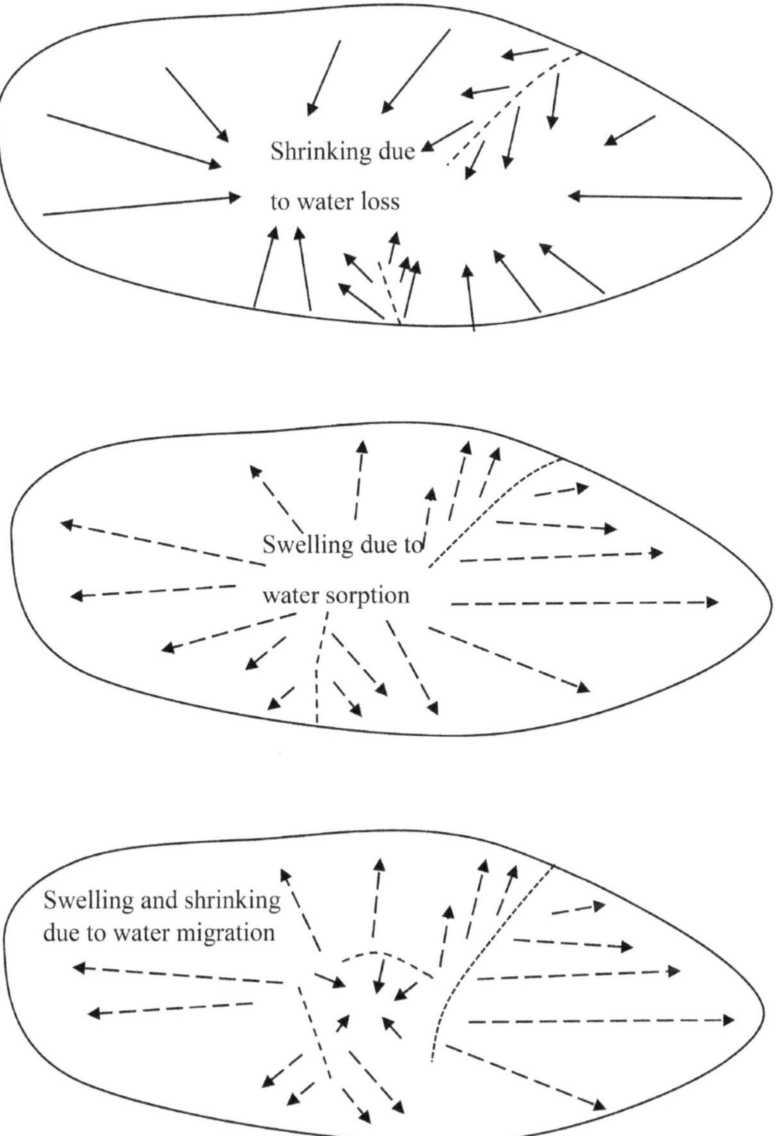

Figure 1.6 Schematic of seed shrinking (top), swelling (middle), and shrinking and swelling (bottom). Solid arrows show the shrinkage, and dashed arrows show the swelling. Length of the arrows shows the uneven shrinkage or expansion. Dashed lines show the crack or fissure locations.

shrinkage or swelling, respectively, otherwise termed as anisotropic shrinkage or swelling. Apparent and percentage shrinkage or swelling are usually used to describe this volume change.

$$Sh_{app} \quad \text{or} \quad Sw_{app} = \frac{V_{app}}{V_i} \tag{1.17a}$$

$$Sh_{\%} = \frac{V_i - V_{app}}{V_i} \times 100\% \tag{1.17b}$$

$$Sw_{\%} = \frac{V_{app} - V_i}{V_i} \times 100\% \tag{1.17c}$$

where Sh_{app} is the apparent shrinkage (decimal), $Sh_{\%}$ is the percentage of shrinkage (%), Sw_{app} is the apparent swelling (decimal), $Sw_{\%}$ is the percentage of swelling (%), V_{app} is the volume of the kernel after shrinkage or swelling (m^3), V_i is the volume of the kernel before shrinkage or swelling (m^3).

1.4.2 Shrinkage and Swelling of Grain Kernels

Grain seeds usually have an uneven shrinkage or swelling due to different physical and chemical properties of different parts of seeds (Fig. 1.6). Red kidney bean and hemp seeds have anisotropic shrinkages or swelling during thin layer drying or water sorption, respectively (Jian et al. 2017a,b, 2018). The shrinkage or swelling percentage of grain kernels are usually less than 15% (Misra and Young 1980, Jian et al. 2017a,b, 2018). Red kidney beans decreased approximately 7% of their saturated kernel volumes after being dried (Jian et al. 2017a). The volume of grain seeds or kernels is linearly or nonlinearly expanded or reduced with the increase or decrease of its moisture contents (Shepherd and Bhardwaj 1986; Karababa and Coskuner 2007; Varnamkhasti et al. 2008; Markowski et al. 2013; Jian et al. 2018).

1.5 Frictional Properties of Grain Bulks

1.5.1 Surface Difference of Grain Kernels among Crops and Cultivars

As the seed ripens, the pericarp (seed coats) firmly attaches to the wall of the seed. The pericarp, nucellus, and aleurone layer form bran. The floral envelopes (modified leaves known as lemma and palea), or chaffy parts, within which the seed (caryopsis) develops, persist to maturity in the grass family such as cereal grain. If the chaffy structures (hull) envelopes the caryopsis so closely that they remain attached to it when the kernel is threshed (as in rice and most varieties of oats and barley), the outer layer of the harvested seed is the hull. If the caryopsis readily separates from the hull on threshing, the outer layer of the harvested kernel is bran as the kernel of wheat, rye, hull-less barley, and the common classes of corn. The endosperm and germ (embryo) are covered by the bran. In a well-filled wheat kernel, the germ comprises about 2–3% of the kernel, bran 13–17%, and starchy endosperm the remainder. In a typical dent corn, the pericarp comprises 6%, germ 11%, and endosperm 83% of the kernel. In light thin oats, hulls may comprise as much as 45% of the grain; in very heavy or plump oats, these may represent only 20%.

A legume is a plant from the family Fabaceae (Leguminosae) which produces seeds (termed as bean or pea) in a pod. A pulse is another term used for legume. The external structures of the bean or pea are the testa (seed coat), hilum, micropyle, and raphe. The testa is the outer most part of the seed and covers almost all of the seed surface. The hilum is an oval scar on the seed coat where the seed was attached to the stalk. The micropyle is a small opening in the seed coat next to the hilum. The raphe is a ridge on the side of the hilum opposite the micropyle. Cotyledon and endosperm are covered by the seed coat. The endosperm in some crops like soybean is a membrane-like, semi-transparent, and fully active tissue located between the seed coat and cotyledon.

Different plant families or types have distinctive features. For example, the endosperm in some crops like canola and rapeseed from crucifers family disappears inside matured seeds and only aleurone layer left. Kent and Evers (1994) provided structural details of grain types. Evers and Millart (2002) reviewed cereal grain structure and development. The chemical and physical structures of different types of crops are different. The structure and chemical composition of the outer layer (hull or bran) are important in protecting the germinating grain and determining the fracture properties of the kernel, hence the physical properties of grain bulks such as water sorption and friction (Evers and Millart 2002). Pericarps (hull or bran) vary considerably among species but they are generally dry empty cells, some maintaining their shape, others being shrunken and distorted, and leaving large spaces. Compared with bran, hull has higher concentration of silica which provides hard and coarse surface structure and can protect seeds from mechanical damaging during harvesting and subsequent handling, slow insect attack, and increase friction with other surface.

Chemical compositions of different crops and their varieties vary widely because it is influenced by genetic, soil, cultural, and climatic factors. Legumes are characterized by relatively high protein and high oil contents (Samson et al. 2005; Morris and Beecher 2012). The wheat bran (pericarp) which is about 50 g/kg of the kernel consists of non-starch polysaccharide, cellulose, protein, ash, and fat of 715, 200, 60, 20, and 5 g/kg of bran, respectively (Delcour and Hoseney 2010). A total of 1 kg of maize bran has approximately 200, 100–130, 90–230, 40, 20–30, and 20 g of cellulose, protein, starch, phenolic acids, lipid, and ash, respectively (Carvajal-Millan et al. 2007). Konopka et al. (2015) analyzed chemical and physical differences of four winter wheat cultivars cultivated in Poland between kernels with a vitreous and mealy appearance. They found the ratio of vitreous kernels, which were darker and slightly heavier and harder than mealy kernels in the cultivars, ranged from 39.18% to 76.28%. Additionally, these kernels were 2.13% more abundant in proteins, 4.02% richer in phenolic compounds, and 4.53% less abundant in carotenoids.

The various components are also not uniformly distributed in the kernel. The hulls and bran of cereal grain have high concentration of cellulose, pentosane and ash; the germ is high in lipid and rich in proteins, and sugars. The endosperm contains the starch and has a lower protein content and low fat and ash. Furthermore, the various proteins, carbohydrate, and lipid are also not distributed uniformly in the kernel and outer layer (seed coats, hull, and bran).

Studies have shown that agricultural practices can influence the chemical composition of the seeds. It is proved that soil nitrogen, solar radiation, degree of plant maturation, application of fertilizer influence chemical components (Juliano and Bechtel 1985; Graham et al. 1999), hence the hull and bran physical properties. Senadhira et al. (1998) reported that iron and zinc content were influenced by nitrogen application and soil quality.

These chemical and structural differences result in different physical properties of the grain such as the roughness of kernels. Even though chemical and morphological structure of bran and hull are extensively studied (Rosa-Sibakova et al. 2015), the physical properties of the bran and hull are rarely determined. For example, roughness of kernel surface is usually not measured because the small surface area of grain kernels, and the most evaluated physical properties of kernel outer layers are their friction on different material surfaces. Kaliniewicz et al. (2018) measured external friction angle of wheat, rye, barley, oats, and triticale on steel friction plates with different roughness. The angle of external friction ranged from 17.56° for rye kernels to 34.01° for oat kernels. The external friction angle between different crops is different and the greatest similarities were between wheat and triticale kernels, whereas the greatest differences were between barley and oat kernels and between barley and triticale kernels. These results indicate that different roughness and friction forces of bran among different crops. Therefore, physical properties such as coefficient of friction should be measured individually for seeds of each crop type.

1.5.2 Coefficient of (Sliding) Friction and Measurement

1.5.2.1 Coefficient of (Sliding) Friction

The internal friction is the friction between the normal and resultant forces that occur at failure due to shear stresses within a substance. Interface friction is the force resisting the relative motion of solid surfaces sliding against each other. Therefore, interface friction is used to determine the skin frictional resistance between different substances. The friction between solids is also termed as dry friction that is subdivided into static friction and kinetic friction. With the exception of atomic or molecular friction, dry friction generally arises from the interaction with surface features such as surface roughness. Static friction is the friction between two or more solid objects (surfaces) that are not moving relative to each other. Kinetic friction, also known as dynamic friction or sliding friction, occurs when two objects are moving relative to each other and rubbing together. For example, static friction can prevent an object from sliding down a sloped surface. The coefficient of static friction is usually higher than the coefficient of kinetic friction. This explains the fact that it is easier to further the motion of a moving body than to move a body at rest. The maximum possible friction force between two surfaces is the friction before the moment that the materials just start to slide. After the instant sliding occurs, static friction is no longer applicable because the friction between the two surfaces then transfers to kinetic friction. Coefficient of friction is the ratio of the force of friction between two bodies and the force pressing them together. The static coefficient of friction or coefficient of static friction is the ratio of the maximum static force of friction and the pressing force, while the dynamic coefficient of friction is the ratio of the dynamic force and the pressing force. Mohsenin (1986) provides a detailed explanation of basic concepts of solid friction, factors influencing friction, and friction in agricultural materials. In literature, coefficient of sliding friction is usually misused with the coefficient of friction and both refer to the maximum coefficient of static friction because the coefficient of (sliding) friction is the measured maximum friction before the moment that the materials just start to slide. In this book, the coefficient of friction is used to refer to the maximum coefficient of static friction.

Internal coefficient of friction of grain is higher than the coefficient of friction because the bulk grain is not homogeneous material and grain kernels at the surface are more loosely packed than those inside the bulk (Stewart 1968). Internal friction of a granular material is caused by a combination of grain interlocking, rolling resistance, and sliding resistance. There might be difference in the resistance among grain seeds from the resistance between grain bulk and a surface of building material. The internal coefficient of friction of the beans, maize, peanuts, rough rice, soybean, and wheat in the moisture contents of 10–25% is higher than the coefficients of friction on the following surfaces: galvanized steel, concrete, plywood, and hardboard (Zaalouk and Zabady 2009). The peak angle of internal friction is 0.2–1.0 fold higher than the angle of friction against surfaces.

1.5.2.2 Measurement of Coefficient of Friction

There are several published methods to measure the coefficient of friction of agricultural materials (Mohsenin 1986). Coefficient of friction against structural surfaces is usually measured by forcing the bulk grain to slide on an inclined surface (inclined plane method). During test, the surface is fastened to a tilting table and the tested bulk grain is filled inside a wooden frame, which is located on the surface (Fig. 1.7). During test, the frame does not contact the surface. The table is tilted slowly by an automatically driven screw device until movement of the whole grain bulk and frame is detected. The coefficient of friction is the tangent of the slope angle of the table, which can be measured by using protractor.

$$\vartheta = \tan(\theta) = \frac{P_s}{P_n} \tag{1.18}$$

where ϑ is the coefficient of (sliding) friction, θ is the angle of sliding friction (°), P_s is the force causing slip on any plane (N), and P_n is the normal force acting perpendicular to the plane of slip (N). Rolling friction is defined as the force that resists movement of a particle rolling on a surface. The rolling friction coefficient is much smaller than the sliding friction. Even though different models and equations are presented in the literature for specific conditions and applications, rolling friction is difficult to measure and characterize. Rolling resistance is directly proportional to the weight of the rolling object and the coefficient of rolling resistance, which depends on

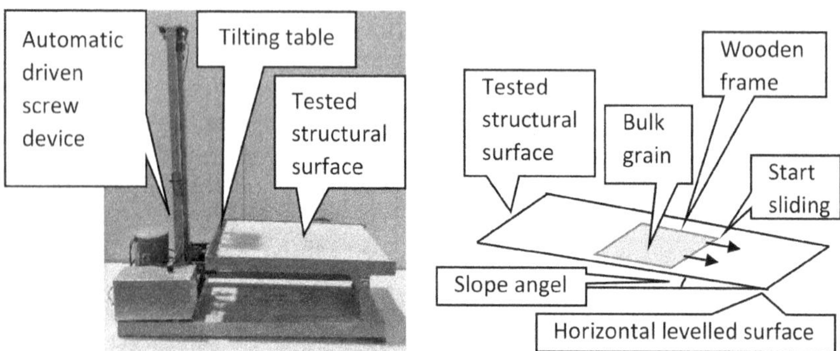

Figure 1.7 Schematic (right) and device (left) used for the measurement of coefficient of (sliding) friction against structural surfaces.

the rigidity of the supporting surface, and is indirectly proportional to the effective radius of the rolling object. The weight of grain kernel is small and the supporting surface in grain industry is usually firm. Therefore, rolling friction is much smaller than the surface friction and is not used in grain industry.

For a crop type, or a class within the crop type, or even a cultivar within a class, coefficient of friction against a structural surface is mostly influenced by the structural surface, moisture content, dockage, foreign material inside the grain bulk, and the physical properties of the grain bulk (such as seed size, shape, and weight). Jian et al. (2018) found the coefficient of friction against structural surfaces was not in the same order as the roughness of the tested surfaces. Generally speaking, coefficient of friction against structural surfaces linearly or quadratically increases with the increase of the moisture content (Jian et al. 2018). The reason for this increase is that the roughness of the kernel surfaces in a certain moisture content range increases with the increase of moisture content. However, it might be opposite to this conclusion in some situations. Sun et al. (2014) found the coefficient of high oil content canola either increased or decreased with the increase of moisture content. Coefficient of friction usually increased with the increase of dockages material (Jayas et al. 1992, Jian et al. 2018) because increase of dockage usually results in the increase of roughness of the grain mass.

Researchers use different surface materials to measure the coefficient of friction. The materials selected are usually building and handling materials related to grain storage and processing. The most tested surfaces are galvanized steel, galvanized iron, concrete, rubber, hardwood, plywood, and aluminium sheet. Published coefficients of friction of various agricultural products in literature have shown rather wide ranges. These measured data are not comparable because the roughness of the materials is usually not reported. The sliding direction related to the distribution of roughness is also not reported. For example, the coefficient of friction of sliding grain in the parallel direction to the wood grain is different than that in other directions (Dursun and Dursun 2005). The coefficient of friction of grain bulks on galvanized steel or iron sheets is in the range of 0.16–0.75 (Table 1.3).

1.5.3 Repose Angle

1.5.3.1 Repose Angle and Measurement

Different types (including static and dynamic angle of repose) of repose angles are defined and one of the most used definitions is the steepest slope of the unconfined material, measured from the horizontal plane on which the material can be heaped without collapsing. If the heap is formed

Table 1.3 Coefficient of Friction and Repose Angle of Seeds of Cereal Grains, Pulses (Legumes), Oilseeds and Speciality Crops on Galvanized Steel or Iron Sheet

			Repose angle (°)		
Crop	*MC (%)*[a]	*Coefficient*	*Filling*	*Emptying*	*Reference*
Barley	12.7–16.4		24–26	26–32	Muir and Sinha (1988)
Buckwheat	13–17.9	0.16–0.29	22–28	22–27	Parde et al. (2003)
Caper	5.7–14.1	0.40–0.47	21–32		Dursun and Dursun (2005)
Canola[b]	8.1		24	26	Muir and Sinha (1988)
Canola[c]	8.0–14.0	0.25–0.35	18–27	24–32	Sun et al. (2014)
Corn	4.7–22	0.38–0.65		49–58	Seifi and Alimardani (2010)
Corn (sweet)	8.4–14.6	0.61–0.74		30–35	Karababa and Coskuner (2007)
Dry bean[d]	12.0–18.0		26–28	31–33	Senthilkumar et al. (2018)
Dry bean[e]	12.0–18.0		26–28	32–33	Senthilkumar et al. (2018)
Durum	12.7–16.4		23–29	21–30	Muir and Sinha (1988)
Faba bean	12.6–21.9		29–30	28–29	Irvine et al. (1992)
Flaxseed	7.0–15.1	0.27–0.66	25–35	30–38	Irvine et al. (1992)
Hemp	9.0–15.0	0.15–0.30	24–28	28–36	Jian et al. (2018)
Lentil	11.4–18.0	0.27–0.66	26–29	25–29	Irvine et al. (1992)
Oats	12.7–16.4	0.27–0.39	27–36	25–32	Muir and Sinha (1988)
Okra	7.5–46.7	0.37–0.49	28–40		Sahoo and Srivastava (2002)
Pigeon pea	5.9–22.0	0.26–0.37		22–25	Shepherd and Bhardwaj (1986)
Pumpkin kernel	4.0–27.5	0.3–0.65		34–42	Joshi et al. (1993)
Rough rice	10.0	0.32 ± 0.01		38 ± 1	Varnamkhasti et al. (2008)
soybean	8–24	0.34–0.53		32–36	Kashaninejad et al. (2008)

(*Continued*)

Table 1.3 Coefficient of Friction and Repose Angle of Seeds of Cereal Grains, Pulses (Legumes), Oilseeds and Speciality Crops on Galvanized Steel or Iron Sheet (*Continued*)

Crop	*MC (%)*[a]	*Coefficient*	*Repose angle (°)* *Filling*	*Emptying*	*Reference*
Sorghum	12.0–18.0			20–26	Mwithiga and Sifuna (2006)
Sunflower kernel	3.8–16.7	0.41–0.75		27–38	Gupta and Das (1997)
Sunflower seed	3.8–16.7	0.39–0.52		34–41	Gupta and Das (1997)
Wheat[f]	12.7–16.4		26–31	27–34	Muir and Sinha (1988)
Wheat[g]	9.2–11.1	0.43–0.51	18–21		Markowski et al. (2013)

[a] Moisture content (w.b., %).
[b] Rapeseed or canola with low oil content.
[c] Canola with high oil content.
[d] Black bean (*Phaseolus vulgaris* L.).
[e] White bean (*Phaseolus vulgaris* L.).
[f] Three cultivars of hard red spring wheat.
[g] Four cultivars of winter wheat.

by filling the material, the angle is the filling (dynamic) repose angle of the material. If the angle is formed by emptying the material, the angle is defined as the emptying (static) angle of repose of the material.

Mixtures with different portions of different size particles have different repose angles (Jian et al. 2019). This makes the prediction of repose angle difficult. Therefore, repose angle of grain bulks is measured experimentally. The following five methods can be used to measure repose angles: tilting box, fixed funnel, revolving cylinder/drum, hollow cylinder method, and tilting cylinder (Al-Hashemi and Al-Amoudi 2018). These methods are neither standardized nor consistent in the literature; therefore, comparison among results from these methods is difficult because each measurement method is targeted for a specific application (Rackl et al. 2017). These methods are extensively reviewed by Al-Hashemi and Al-Amoudi (2018).

In the study of grain bulks, repose angle is usually measured by mimicking the filling and emptying of grain bulks to measure static and dynamic repose angles (Muir and Sinha 1988). In the fixed funnel method (Fig. 1.8), the granular materials are poured from a funnel at a certain height onto a wood or metal board. The funnel is fixed while the conical shape of the

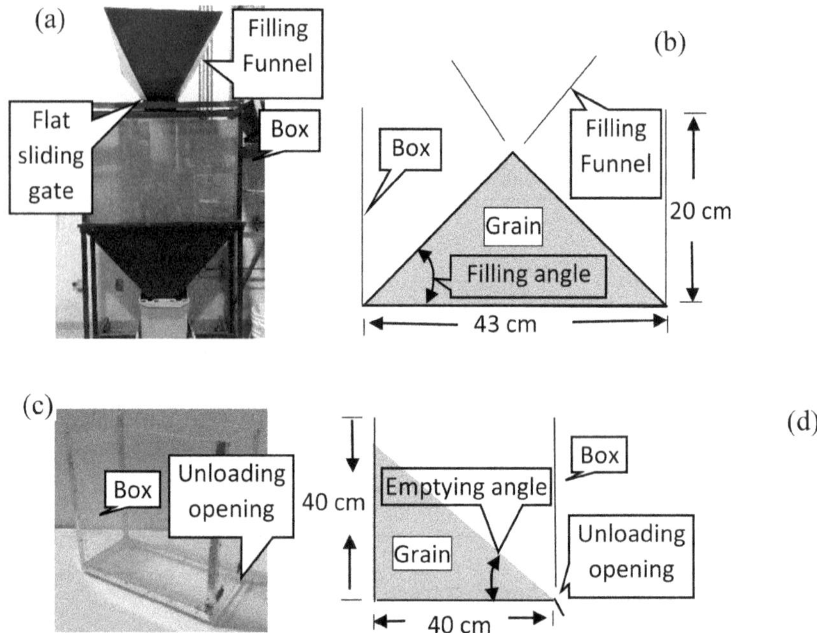

Figure 1.8 Devices and schematics for measurement of filling (top) and emptying (bottom) angles of repose of grains. The boxes used for the filling and emptying angle tests are transparent. Filling angle of repose is measured by loading the grain from the filling funnel to the box. There is a lid at the bottom of the box used for the emptying angle test, and the emptying angle is formed after the lid is opened.

material heap is forming to minimize the effect of the falling particles. The pouring of the material is stopped when the heap reaches a desired height or width. Then, the angle of repose (or termed as filling angle of repose, or external angle of repose) is determined by measuring the average radius of the formed conical base and the maximum height of the grain heap (Fig. 1.8a and b). The emptying angle of repose (or termed as internal angle of repose) is usually measured inside a box with a lid at the bottom of one side of a box (Fig. 1.8c). When the lid is opened, the grain inside the box can freely flow out of the box and the emptying angle forms inside the box after the grain in the box stops flowing out (Fig. 1.8c and d). These measured angles of reposes tend to decrease with an increase in the amount of the material, the size of the conical heap, and tends to increase with slower funnel loading speeds, and increase of the base roughness. These phenomena were observed on sulfathiazole materials for a pharmacology application

(Nelson 1955). Chukwu and Akande (2007) used the following base types with different roughness and various granular materials to measure the filling angle of repose of 20 agricultural materials including seed bulks: plywood, wood, glass, galvanized metal sheet and iron rod, and flat bar. However, they did not report the difference among the different base types. They reported differences between their values and the values in the literature. When the fixed funnel method is used, it is assumed the steepest slope of the unconfined material is in a straight line. Fraczek et al. (2007) measured the angle of repose for four corn cultivars within the moisture range 7–35% using the fixed funnel method. They proved that most of the tested materials exhibited deviations from the assumed conical shape with either truncation of the upper part of the pile, convexity of the slope, or concavity of the slope. The truncation of the upper part of the pile is common in grain industry especially when grain has a medium to low moisture contents.

To overcome the influence of different amounts of materials, base materials, and filling or emptying methods, different methods to normalize these influencing factors have been developed. Fraczek et al. (2007) summarized the following fours methods that are similar to the fixed funnel method but these influencing factors are normalized: emptying, submerging, piling, and pouring methods (Fig. 1.9). In the emptying method, material is unloaded through the outlet in the middle of container bottom (Fig. 1.9a). The method of "submerging" is to form a cone on a circular plate located inside the container by slow emptying of the container (Fig. 1.9b). Static slope angle is determined on the basis of the height and diameter of the cone. The "piling" method (Fig. 1.9c) is similar to the fixed funnel method. The "pouring" method (Fig. 1.9d) is to load the test materials into a cylinder and then lift the cylinder vertically up under a constant velocity controlled by the crank system.

1.5.3.2 Relationship between Repose Angle and Other Physical Properties

Angle of repose can be influenced by many factors and different researches have different conclusions on the parameters influencing repose angles. Al-Hashemi and Al-Amoudi (2018) conducted an extensive review on these factors influencing the repose angle. Lee and Herrmann (1993) found that the repose angle strongly depends on the friction coefficient but is insensitive to other parameters of the granular particles such as particle size, shape, and surface roughness of particle material. However, these parameters influence the friction coefficient. Lumay et al. (2012) found that when the particle sizes of rice, flours, and silicon carbide abrasives were smaller than 50 μm, the cohesion between the particles tends to affect the angle of

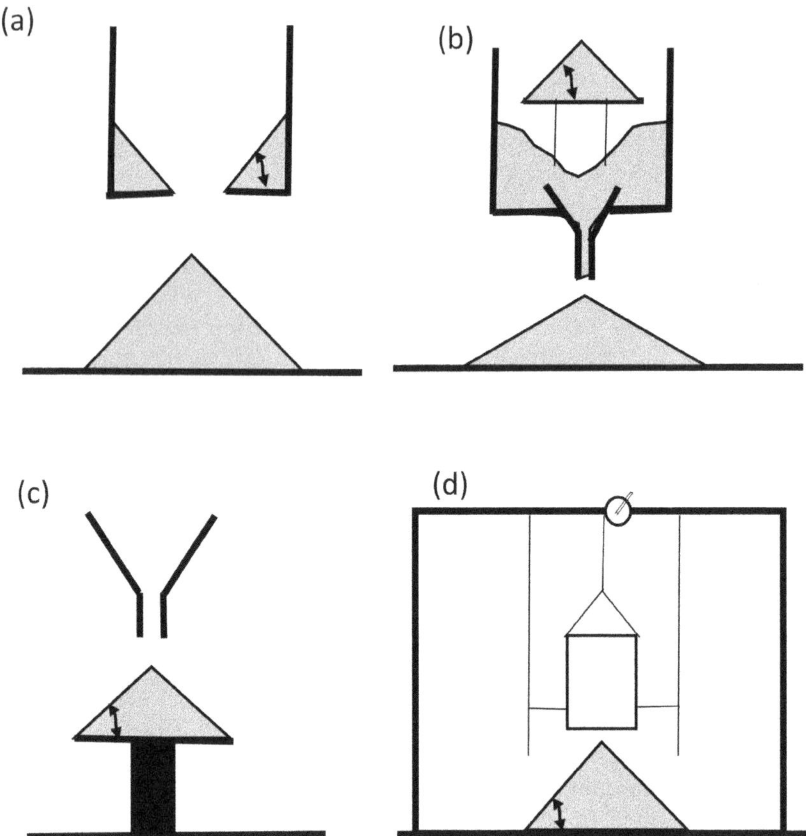

Figure 1.9 Methods of angle of repose determination: (a) emptying, (b) submerging, (c) piling, and (d) pouring. Arrow shows the repose angle. (Reprinted with permission from Fraczek et al. (2007)).

repose, and the angle of repose would increase with the increase of cohesion. However, for particle sizes larger than 50 μm, cohesion effect was not significant. For non-uniform sediments, the angle of repose increases slightly when the mean sediment size increases (Yang et al. 2009). The difference between the dynamic (emptying) and static (filling) angles of repose increases with decrease of mean size of particles (Santos et al. 2017). Makse et al. (1998) found angle of repose does not depend on the size of the grains, and is a function of the shape of the grains: The rougher the shape of grains the larger the angle of repose because for more faceted grains the packing of grains is less dense than for more rounded grains. The angle of repose decreases when the circularity and/or aspect ratio increase, and when the sphericity increases (Matuttis et al. 2000). Therefore, most researchers agree surface roughness plays an important role on repose angle. Miller and Byrne (1966) and Guo et al. (2014) developed empirical equations for predicting angle of repose of particles inside water. The empirical equation shows angle of repose increases with decrease in size, departure from sphericity and increased angularity. Guo et al. (2014) found the angle of repose of blends has a better linear relationship with the secondary particle mass fraction. This empirical equation is not used by researchers because it has low prediction accuracy and it is also difficult to determine the parameters in the empirical equation.

For a crop grain, repose angle is mostly influenced by its moisture content and dockage inside the grain bulk (Jayas et al. 1992). If grain is too dry or too wet, the repose angle might increase or decrease. The emptying angle of grain bulks is usually larger than its filling angle (Table 1.3) because of the impact and compact forces during grain filling. The filling repose angle of crop grain is in the range 22–45° with some exceptions (Table 1.3). The filling angle of repose in stored corn farm bins is from 15.7° to 30.2° (Bhadra et al. 2017). The reported value of repose angle for the same grain type has a wide variation because different material sources are used by different researchers (Table 1.3). The differences in materials are caused by different cultivars, different production countries, and different production years.

1.5.3.3 Relationship between Repose Angle and Internal Friction

In general, the angle of repose increases with an increase in the friction coefficient (Al-Hashemi and Al-Amoudi 2018). Stewart (1968) found that using angles of repose for sorghum to replace the angle of internal friction produced errors because the friction of particles under low confining pressure is considerably different than that under zero confining pressure (Deng et al. 2012). Pressure inside grain bulk is higher than that at the surface of the grain bulk. Moreover, for dry fine and granular material with a small angle of repose and a particle size of less than 5 μm, no defined relationship exists between the repose and internal friction angles (Lanzerstorfer 2017). The heap of materials usually does not form a perfect conical shape (Fraczek et al. 2007), which results in different angles of repose. These two angles (angle of repose and angle of internal friction) are equal when the pure materials inside the heap have the uniform density and moisture content and the particles of the material have a uniform size, which rarely occur. Therefore, the angle of repose is not equal to the angle of internal friction of the heaped material but is close to the internal friction of the same material at the surface of the grain bulk because the materials

Table 1.4 Classification of Flowability of Grain Bulk or Powder Based on Emptying Repose Angle

Description	*Repose Angle (°)*
Very free flowing	<30
Free flowing	30–38
Fair to passable flow	38–45
Cohesive	45–55
Very cohesive (non-flowing)	>55

Source: Riley and Hausner (1970).

on the surface of the heap, which mainly form the angle of repose, are under the loosest condition (Metcalf 1966).

Repose angle is strongly associated with flowability of grain kernels (Table 1.4) and grain bulks will flow at angles greater than the angle of repose. The flowability is mainly influenced by internal friction of the grain bulk. Angle of repose is greater than the internal friction angle for cohesive materials but smaller for non-cohesive materials (Teunou et al. 1995). The difference between them increases with growing cohesion forces. It is difficult to determine a strict relationship between these angles because either cohesion or the friction force cannot be easily determined without influencing other factors.

1.6 Drag Force and Terminal Velocity

1.6.1 Drag Force and Influencing Factors

1.6.1.1 Drag Force

The terminal velocity and drag force are the most important aerodynamic properties for the design of processing tools for pneumatic conveying, separation, cleaning, harvesting and drying. The study of these aerodynamic properties will help to understand particle segregation and distribution of grain kernels during grain handling. The fundamental forces experienced by moving particles in air stream are the weight of the particle and the aerodynamic drag force. Aerodynamic drag force includes the pressure drag and skin friction and is related to relative velocity of the particle with fluid (most case it is air), the density of air and the particle, sphericity and frontal area of the particle because shape, orientation to the air stream and Reynolds number influence the drag force.

$$F_D = \frac{1}{2} C_D \rho_a A_f v_r^2 \tag{1.19}$$

where F_D is the drag force (N), C_D is the drag coefficient, ρ_a is the air density (kg/m^3); A_f is the frontal area (m^2), and v_r is the relative velocity (m/s).

1.6.1.2 Influencing Factors

At *Re* greater than 4000, the drag will be mainly the pressure drag. The effect of skin friction increases with the decrease of Reynold number. It is difficult to directly measure drag force of grain kernels due to the small size of grain kernels and irregular shape that result in the difficulty of the estimation of C_D and A_f. Therefore, drag force is usually calculated or measured after C_D and A_f are determined. The C_D can be estimated after terminal velocity is defined (Zewdu and Solomon 2007).

$$C_D = \frac{2mg}{\rho_a A_f v_t^2} \tag{1.20}$$

where g is the gravitational acceleration (m/s^2), and v_t is the terminal velocity (m/s). Drag force is usually estimated by relating the drag coefficient to the sphericity of kernels (West 1972). Due to the irregular shape of most agricultural materials, the frontal area and sphericity are the most important factors, which should be considered in the drag coefficient and drag force calculation (West 1972). Different empirical equations have been reported to estimate the drag coefficient (Zewdu and Solomon 2007).

$$C_D = 5.31 - 4.88\phi \quad 2000 < Re < 200{,}000 (\text{West } 1972) \tag{1.21a}$$

$$C_D = 2.53 - 0.283Re^{2\phi} \quad 2000 < Re < 200{,}000 (\text{Tabak and Wolf } 1998) \tag{1.21b}$$

where *Re* is the Reynolds number, and ϕ is sphericity (decimal). The drag coefficients of grains with nearly spherical and cylindrical shapes are close to 0.44 and 1.0, respectively. Based on the average drag coefficient, Gorial and O'callaghan (1990) classify grains into four categories:

- Group 1 (average drag coefficient is 0.47) includes soybean, adzuki bean, mung bean, sorghum, millet, and oilseed rape.
- Group 2 (average drag coefficient is 0.6) includes lentil, sesame, pinto, marrowfat, and buckwheat.
- Group 3 (average drag coefficient is 0.8) includes wheat, chickpeas, beans, rye, maize, and rice.
- Group 4 (average drag coefficient is 1.0) include barley and green lentil.

In each group, drag coefficient does not change with Reynolds number, and different sizes of the grain kernels do not have different drag coefficients

(Gorial and O'callaghan 1990). Therefore, the drag coefficients of different grains depend mainly on shape.

1.6.2 Terminal Velocity and Measurement

1.6.2.1 Terminal Velocity

If an object is dropped from a sufficient height, it will accelerate until the drag force exerted by the air, balances the gravitational force. When the drag force becomes equal to the gravity force, the object has zero acceleration and it will fall at a constant velocity which is termed as the terminal velocity. Therefore, terminal velocity is the maximum velocity attainable by an object as it falls through a fluid.

1.6.2.2 Measurement

Based on the definition of the terminal velocity, it can be measured by drop and suspension methods (Gorial and O'callaghan 1990). By dropping the particle from certain height, the terminal velocity can be determined from the dropping distance versus time curves where it begins to become linear. Suspension method is to measure the terminal velocity by dropping the object in a vertical air stream, and measure the airstream velocity when the object is suspended in the airstream. The air stream can be created by using a vertical column and fan which provide different air velocities in the upward direction in the column. This measurement system is called elutriator. An elutriator usually includes a tunnel, a fan, velocity regulator, and velocity meter. The air velocity that kept the grain float in the air stream is reported as the terminal velocity.

Compared with suspension method, dropping method only needs a long vertical tunnel and do not need a control system. However, dropping method is not suitable for measuring terminal velocity of longer materials such as straws and heavy materials such as larger grain kernels because straws can slip (rotate) with a sidewise velocity (Gorial and O'callaghan 1990) and heavy materials require a longer tunnel. Therefore, terminal velocity of grain kernels is mostly measured by using suspension method.

1.6.2.3 Empirical Equations

When Reynolds numbers is less than 50, irregular particles can be assumed as sphere and the diameter of an equivalent sphere can be used. When Reynolds numbers are greater than 50, the drag coefficient curves level off and the assumption of sphere will result in considerable errors. A correction empirical factor ($\pi/6Z$) is introduced to account for deviations from the spherical shape (Heywood 1948).

$$v_t = \frac{4gD_e\rho_k\left(\frac{6Z}{\pi}\right)}{3\rho_a 0.44} \quad \text{and} \quad Z = \frac{\pi}{6}\phi\left(\frac{D_e}{D_g}\right)^3 \tag{1.22}$$

The value of Z for a range of granular particles was 0.4, and was $\pi/6$ for a sphere (Heywood 1948). Based on the measured terminal velocity of 20 crop seeds, Gorial and O'callaghan (1990) developed the following empirical equation:

$$v_t = 7\sqrt{Z\rho_k D_s} \quad 500 < R_e < 200{,}000 \tag{1.23}$$

where D_s is the diameter of silo (m). Song and Litchfield (1991) suggested the theoretical terminal velocities of grain kernels can be calculated by using Lapple's equation as:

$$v_t = 1.74\sqrt{gD_g\left(\rho_b - \rho_a\right)/\rho_a} \tag{1.24}$$

1.6.2.4 Factors Influencing Terminal Velocity

Increasing moisture content from 7 to 20%, the terminal velocity of wheat kernels increases linearly from 6.81 to 8.63 m/s due to the mass and size increase (Khoshtaghaza and Mehdizadeh 2006). Other researchers (Gupta and Das 1997; Tabak and Wolf 1998; Zewdu and Solomon 2007) also arrived at the similar conclusion. The terminal velocity of cottonseeds decreased, when the turbulence intensity increased from 3.8 to 14.4% (Tabak and Wolf 1998).

The terminal velocity of grain kernels is in the range of 3–12 m/s (Table 1.5). Grain kernels rotate inside wind tunnel and this rotation affects values of the measured terminal velocity (Bilanski and Lal 1965; Song and Litchfield 1991). Straws are neither symmetrical nor uniform (due to tapering) in density (due to nodes) and this lack of symmetry causes aerodynamic instability (Zewdu and Solomon 2007) including rotation. Rotation of particles in an air stream caused a higher drag coefficient and lower terminal velocity. Therefore, straws have a wider range of suspension velocities than grains and an overlap in terminal velocities between grain and straw materials occurs, which results in incomplete pneumatic separation

Table 1.5 Terminal Velocity of Seeds of Cereal Grains, Pulses (Legumes), Oilseeds and Speciality Crops

Crop	*MC (%)*[a]	*Velocity*[b] *(m/s)*	*Reference*
Barley	16.7–17.1	7.3–8.6	Song and Litchfield (1991)
Caper	5.7–14.1	6.1–7.7	Dursun and Dursun (2005)
Corn	9.3–9.7	9.8–11.3	Song and Litchfield (1991)
Cotton seed	5.7–20.6	9.1–10.2	Tabak and Wolf (1998)
Lentil	6.1–24.6	11.1–12.1	Carmen (1996)
Linseed	7.6–18.2	2.5–3.8	Selvi et al. (2006)
Oat	13.4–14.1	7.0–8.3	Song and Litchfield (1991)
Pumpkin kernel	4.0–27.5	4.3–5.0	Joshi et al. (1993)
Red kidney bean	8.9–16.4	8.1–8.8	Isik and Unal (2007)
Sunflower kernel	3.8–16.7	3.5–5.8	Gupta and Das (1997)
Sunflower seed	3.8–16.7	5.8–7.6	Gupta and Das (1997)
Soybean	10.2–11.2	11.2–11.9	Song and Litchfield (1991)
Tef grain	6.5–30.1	3.1–4.0	Zewdu and Solomon (2007)
Vetch	9.6–17.1	9.9–10.3	Yelcin and Ozarslan (2004)
Wheat[c]	14.4–14.7	7.7–9.6	Song and Litchfield (1991)
Wheat[d]	9.2–11.1	9.3–10.1	Markowski et al. (2013)

[a] Moisture content (wb, %).
[b] Terminal velocity.
[c] Three cultivars of wheat.
[d] Four cultivars of winter wheat.

of straw from the grain. Terminal velocity of wheat straw increased by increasing length from 1 to 2 cm and decreased by increasing length from 2 to 10 cm and terminal velocity of wheat straw depended on node position and the end node position had the highest terminal velocity (Khoshtaghaza and Mehdizadeh 2006).

The amount of material removed by aerodynamic means in factories is much less than that removed under laboratory conditions because there is a greater difference between the terminal velocities of different shapes of materials. Partly threshed heads of wheat kernels are the most variable components (Bilanski and Lal 1965). Higher air velocities were required to separate chaff from grain than to suspend chaff on its own (Gorial and O'callaghan 1990). Complete pneumatic separation of corn cobs from stalks was impractical because of the overlapping of the terminal velocities of some stalks with those of the cobs (Smith and Stroshine 1985).

1.7 Behaviour of Grain Bulks in Silos

1.7.1 Flow Pattern of Unloading Grain Bulks

Two primary flow patterns can occur during grain unloading from a bin or a silo: mass (plug) flow and funnel (core) flow (Fig. 1.10). When solid materials are discharged from a bin with a mass flow, the entire grain bulk is in motion and the order of the material discharge is "first-in, first-out" (Fig. 1.10). This operation eliminates the formation of stagnant regions at the walls of the silo. The uniform velocity profile also helps to reduce the segregation of the grain mixture. The velocity is generally uniform in the cylinder section of a mass flow bin. However, the velocity in the centre could be faster than that along the walls inside the hopper (Prescott and Hossfeld 1994), which depends on the configuration of hoppers (Prescott and Hossfeld 1994). Inside bins with a funnel (core) flow, an active flow channel forms above the outlet of the hopper (called ratholes), and the stagnant material remains at the walls of the bin. The order of the discharged material is "first-in, last-out" except the materials inside the formed funnel at the very beginning of the unloading. Funnel flow can cause erratic flow, risks of formation of arches (bridge) and exacerbate segregation, and particle degradation (leading to caking and spoilage) in the stagnant regions (Arteaga and Tuzun 1990). The funnel flow can also induce high loads on the structure due to collapsing ratholes and the formation of eccentric flow channels. The advantages of the funnel flow are the larger storage capacity in the hopper part and the reduced abrasion of hopper walls since grain kernels at walls do not move during the most time of the grain loading.

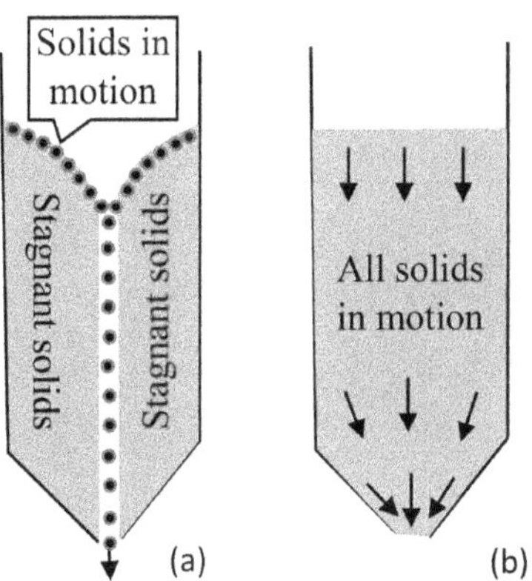

Figure 1.10 Funnel (core) flow (a) and mass (plug) flow (b) in hopper bins.

The flow pattern of unloaded grain bulks depends on the angle of the hopper, wall fiction, and grain physical properties. It is designed that bulk grain at a safe storage condition has a funnel flow in hopper bottom grain silos. Flow inside flat bottom bins usually is the funnel flow if only small openings are opened at the bottom of the flat bin.

1.7.2 Stress in Silos with and without Grain Discharge

1.7.2.1 Stress in Silos without Grain Discharge

Inside solid bulks, positive normal stress is exerted on the solid bulk due to the weight of solid grain kernels. Liquid behaves like a Newtonian fluid, the stresses in the horizontal and vertical directions (in any direction) inside the liquid will equal. The pressure increases linearly with the increase of depth in a liquid container. The behaviour of a bulk solid is quite different from that of a Newtonian fluid (liquid), but bulk solids like crop grain bulks at rest can also transmit shear stresses (contrary to Newtonian fluids). The bulk solid is compressed vertically by the weight of the material above it, but strain in the horizontal direction is prevented due to the existence of silo walls. Since the silo walls are not frictionless, the vertical stress is the principal stress along the silo axis, but not on silo walls. The wall normal stress (σ_w) and the mean vertical stress (σ_v) will not linearly increase in the downward direction (Fig. 1.11), and the major principal stress orients vertically along the silo axis and deviates more and more from vertical towards the silo walls (Fig. 1.11). Therefore, stress inside a solid silo depends on the wall friction, and the friction on the bin walls carries part of the bulk solids weight (Fig. 1.11). The stress in silos in the vertical direction under rest condition (no surcharge pressure) can be estimated as (Janssen 1895):

$$\sigma_v = \frac{\rho_b A}{k \tan\theta_w U}\left[1 - e^{\frac{-k\tan\theta_w U z}{A}}\right], \text{ not cylinder silo} \tag{1.25a}$$

$$\sigma_v = \frac{\rho_b D}{4k \tan\theta_w}\left[1 - e^{\frac{-4k\tan\theta_w z}{D_s}}\right], \text{ cylinder silo} \tag{1.25b}$$

$$\sigma_{v\infty} = \frac{\rho_b A}{k \tan\theta_w U}, \text{ when } z \to \infty \tag{1.25c}$$

$$\theta_w = \frac{\tau_w}{\sigma_h} \tag{1.25d}$$

$$k = \frac{\sigma_h}{\sigma_v} \tag{1.25e}$$

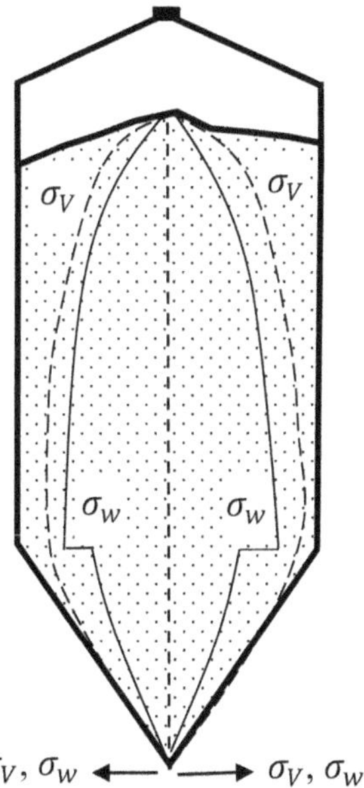

Figure 1.11 Qualitative distributions of wall normal stress, σ_w, and mean vertical stress, σ_v. (Modified from Schulze (1994)).

where σ_v is the stress in vertical direction (Pa), $\sigma_{v\infty}$ is the stress in vertical direction in an infinite depth of silo (Pa), θ_w is the wall friction angle (°), k is the lateral stress ratio, A is the surface area of the wall covered with bulk solid (m^2), U is the perimeter of the storage structure (m), z is the depth of the silo (m), D is the diameter of the bin, D_s is the solid particle diameter (m), τ_w is the shear stress on the wall (Pa), and σ_h is the stress in horizontal direction (Pa).

Equation 1.25 indicates the vertical stress varies proportionally with the bulk density of the stored material and bin diameter, and can be expressed in dimensionless form $(\sigma_v/(\rho_b D))$. This equation does not consider the surcharge stress, which can be a significant contribution to stresses during loading and unloading. The maximum possible stress in both vertical and horizontal directions inside a cylindrical section is proportional to the silo diameter and height of the silo. The normal stress on wall is taken as the horizontal stress (σ_h) at that height and can be calculated as:

$$\sigma_h = k\sigma_v \tag{1.26}$$

Singh et al. (2010) showed that vertical stress at the same radius is almost same when the angle of friction of the material varies from 25° to 45°. The above equations have good agreements with laboratory test data for a low wall roughness (most grain storage structures have a low wall roughness). Therefore, it is possible to estimate vertical stress in silos containing different crop grain bulks using Janssen's equation. When the wall surface becomes rough, the analytical equation (Janssen equation) tends to underestimate the vertical stresses and incorporating additional parameters such as angle of dilation in this analytical formulation may improve the predictions.

The stress in hoppers is more complex than that in the vertical section of storage structures. At the transition to the hopper the wall normal stress suddenly increases due to the change of wall inclination. Further downwards in the hopper, the stresses in the hopper can decrease continuously from the transition to the apex as in Fig. 1.11.

1.7.2.2 Lateral Stress Ratio

The prediction accuracy of the Janssen equation depends on the estimated k value. To estimate the k, the stresses at different locations and different conditions were used by different researchers. Janssen (1895) used the horizontal stress at the wall and vertical stress at the centre, but Jaky (1948) used both stresses at the centre of silos. Pipatpongsa and Heng (2010) confirmed that Jaky's assumption can be applied at rest conditions. Koenen (1896) proposed the following equation to calculate k:

$$k = \frac{1 - \sin\theta_e}{1 + \sin\theta_e} \tag{1.27}$$

where θ_e is the effective angle of internal friction (°). The assumption of Koenen's equation is that the vertical and horizontal stresses in the vertical section are principal stresses, and the bulk solid is in an active plastic state of stress that is equivalent to steady state flow. The steady-state flow could only be achieved if the bulk solid in the vertical section has the possibility to sufficiently dilate in the horizontal direction. This is normally not possible due to the stiffness of the silo walls (Schulze and Schwedes 1994). Kézdi (1962) proposed the following equation:

$$k = 1 - \sin\theta_i \tag{1.28}$$

where θ_i is the angle of internal friction (°). Kézdi equation is valid for a particle bulk having infinite dimension in the horizontal direction. Silo codes such as German code (DIN 1055 part 6, 1987, Loads in silos) is based on a slightly modified form of Kezdi equation.

$$k = 1.2\left(1 - \sin\theta_i\right) \tag{1.29}$$

This equation leads to larger wall normal stresses and shear stresses in the upper silo part, which leads to safer load assumptions compared to Kezdi equation. The k value of bulk solids stored at rest (such as stored grain bulk in bins) is mostly in the range 0.3–0.6. For rough estimates, $k = 0.4$ is often recommended (Schulze 2008).

1.7.2.3 Stress in Bins During Unloading

When solid materials are discharged and after the materials move downward, the bulk solid is compressed horizontally due to the convergent flow zone in the hopper, while it dilates in the vertical direction due to the downwards flow (Fig. 1.10). This results in the larger stresses in the horizontal direction. This stress field is called "arched" because the major principal stress in the hopper is toward to the walls. Bulk discharge inside the hopper can be assumed as a steady-state flow at constant stresses and constant bulk density. The ratio of the minor and major principal stresses at steady-state flow cannot be described with the lateral stress ratio, k, which is valid only for uniaxial compression.

For free-flowing grain kernels, these arches are not strong enough to stop the grain flow, but may slow down the grain kernel moving and produce shearing force within the moving grain kernels. Shearing causes the granular material to dilate (Smith and Lohnes 1980), which results in the increase of the wall pressure when dilation in the lateral direction is restricted by the bin walls. The deceleration also results in an inertia force. The arch also carries some of the material weight above it. Both the inertia force and the additional material weight are transferred through the arch to the bin walls, causing the increase of pressure on the bin walls.

Arching can occur in two forms: cohesive arching and mechanical blockage. Mechanical blockage occurs when the hopper outlet is blocked by large blocks of materials. Cohesive arching occurs when cohesive force between particles is large. The cohesion and adhesion of wet and/or spoiled grain kernels is higher than that of dry and sound kernels. Materials with high cohesion and adhesion have high unconfined strength which can result in the enlarged and stable arch. The stable arch stops the flow of materials above it, while the material below it continues to flow. As a consequence, some of the material above the arch stops moving and the additional forces (such as the weight of the material above the arch) will be transferred to the bin walls which causes an increase in lateral pressure on the bin walls. The stable arch can cause collapse of the bin and human life loss if it is not handled correctly. Rotter (2008) reported that increases in pressure (from the static value) during discharge ranged from 10% to 30% of the stable values, but short-term pressure increases two or three times the static value. At the same time, the pressure underneath the arch becomes lower than the static pressure because some material weight above the arch is transferred to the bin wall by the arch.

To date, no theories are universally accepted for predicting the dynamic pressures of discharge in bin design (Zhang and Cheng 2014). The most widely accepted theory is the pressure switch (the active stress state changes to the passive stress state) suggested by Jenike et al. (1973). Smith and Lohnes (1980) hypothesized that dilation of the stored material was the main cause of the overpressure during discharge.

1.8 Airflow through Grain Bulks

1.8.1 Airflow Velocity through Porous Medium

When air is forced to flow through a porous medium (such as grain bulks), air has to squeeze through the pores and has tortuous paths (Fig. 1.12). The interstitial velocity (true velocity) of the air within the porous medium is greater than the superficial (mean) velocity (Fig. 1.12). This results in interfacial frictions between the fluid and the particles (including kernels, dockage, and foreign materials) and among fluid molecules within the body of the porous medium, which results in airflow resistance and drag as a result of energy losses caused by the friction and turbulence. This phenomenon is termed as resistance to airflow, commonly defined as the pressure drop per unit bed depth. These two terms (airflow resistance and pressure drop) are usually interchanged in the published literature and so are in this chapter.

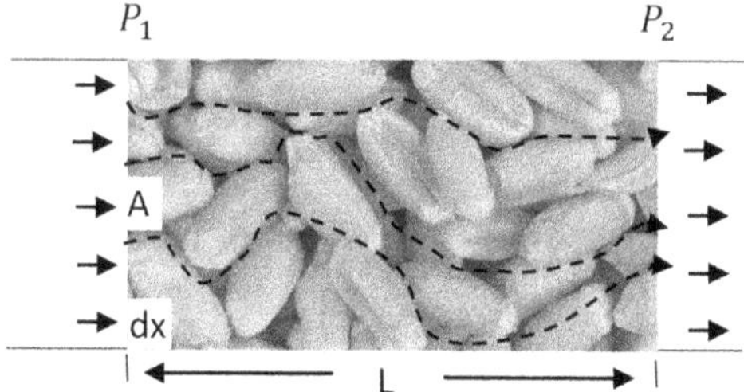

Figure 1.12 Schematic diagram of airflow through a wheat bulk (porous medium). Dashed lines show possible air stream lines.

1.8.2 Mathematical Models of Airflow

1.8.2.1 Darcy Law

Figure 1.13 schematically shows the physical model of airflow through a wheat bulk (porous medium). The flow is assumed to be steady, fully developed, and only in one direction (the arrow direction). The distribution of pores in the wheat bulk is not uniform and the shape and size of the pores are irregular. The flow velocity and pressure are also uneven. To simplify mathematical models, the quantities of the flow are usually described using space-averaged quantities for representative volumes. The pressure and velocity are considered to be uniformly distributed along a tiny length (*dx*) in the representative volume. The mean flow velocity in each *dx* can be calculated as:

$$q = \frac{G}{\rho_a A} \tag{1.30}$$

where q is the mean flow velocity (superficial velocity, or filtration velocity) (m/s), and G is the mass flow rate (kg/s). In the absence of gravitational forces and in a homogeneously permeable porous medium with permeability K, the proportional relationship between the instantaneous flow rate through the medium and the pressure drop over a given distance can be described by Darcy's law.

$$Q = -\frac{KA\Delta P}{\mu L} \tag{1.31}$$

where Q is the total discharge (or volume flow rate, m^3/s or L/s), K is the (intrinsic) permeability (m^2), ΔP is the pressure drop between two locations (Pa), μ is the dynamic viscosity of the flow medium (Pa.s). Darcy's law can be simplified as:

$$q = -\frac{K}{\mu}\nabla P \tag{1.32}$$

where $q = Q/A$ = Darcy flux, Darcy velocity, mean velocity, or superficial velocity of the fluid (m/s); and $\nabla P = \frac{\Delta P}{L}$ = pressure gradient (Pa/m). The q is not the true velocity of the fluid traveling through the pores (Fig. 1.12). The average fluid velocity inside the medium is:

$$v = q/\varepsilon \tag{1.33}$$

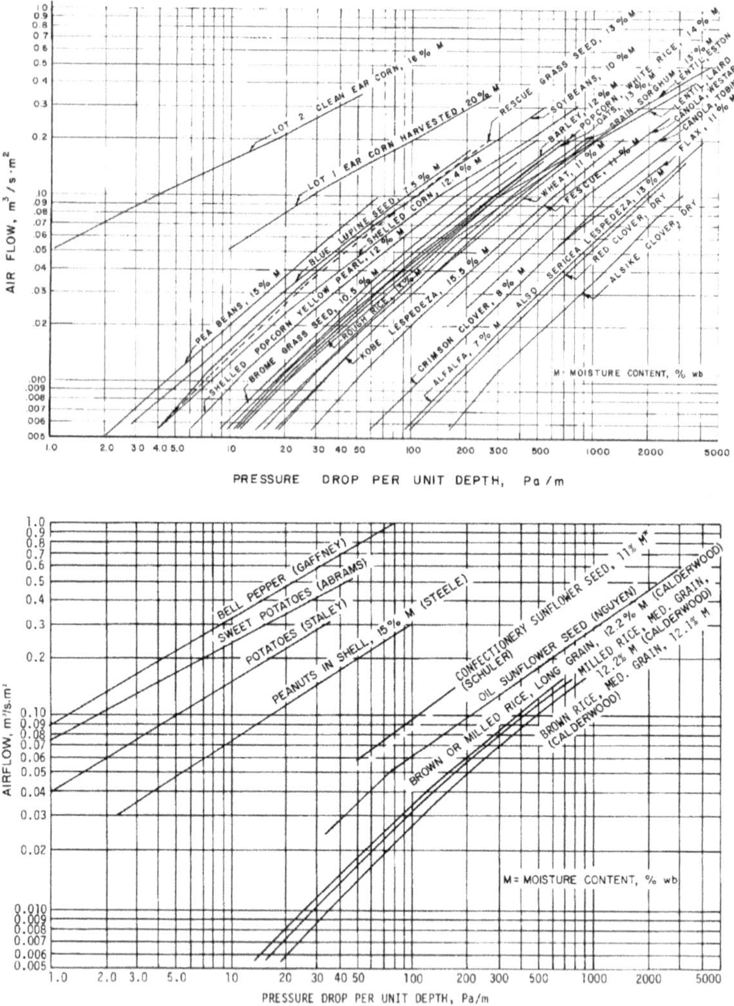

Figure 1.13 Resistance to airflow of loose fill (not packed), clean, and dry grain seeds (ASABE 2016) (top), and other agricultural products (bottom). For a loose filled clean and wet grain (≥85% equilibrium relative humidity), use only 80% of the indicated pressure drop for a given rate of airflow. Packing of the grain in a bin may cause 50% higher resistance to air flow than the values shown. The pressure drop in the vertical direction may be higher or equal to that in the horizontal direction, please consult the literature (Kumar and Muir 1986; Jayas 1987; Kay et al. 1989; Sokhansanj et al. 1990; Jayas et al. 1991b; Smith et al. 1992; ASABE 2016) for more information. (Note: White rice is a variety of popcorn). (Reprinted with permission from ASABE (2016)).

where ε is the porosity (decimal). In these equations, the q and P are also referred to as the macroscopic velocity and pressure, respectively. The term macroscopic refers to a spatial average on a scale that is large as compared to the pore geometry. This macroscopic scale can be evaluated by the Reynolds number.

$$R_e = \frac{\rho_a v D_s}{\mu} \tag{1.34}$$

where v is the propagation velocity or fluid velocity inside the porous medium (m/s), and D_s is the diameter of pore (m). Darcy's law is only valid for slow and viscous flow, which has a Reynolds number of 0.1–70 (Govindarajan 2019), and indicates that $\rho v D_s < \mu$. Therefore, Darcy law can predict flow behavior properly when the flow is dominated by viscous effect. During grain aeration, the mean air velocity can be lower than 0.0001 m/s. The mean air velocity for grain drying can be higher than 0.5 m/s. For a 10 mm diameter pore, the Reynolds number is 0.06 at 0.0001 m/s, and 314.7 at 0.5 m/s. Therefore, Darcy law cannot be used to predict flow behaviour properly under any grain aeration or drying condition. To quantify airflows in grain bulks, Darcy's law is modified by different researchers.

1.8.2.2 Darcy–Forchheimer Law

For flows in porous media with Reynolds numbers between 1 and 10, inertial effects can be significant and an inertial term (Forchheimer term) is usually added to the Darcy's equation. Darcy–Forchheimer law is named after the Forchheimer term is added to the Darcy's equation.

$$\nabla P = -\frac{\mu}{K} q - \frac{\rho F}{\sqrt{K}} q^2 \tag{1.35}$$

where F is the Forchheimer constant (non-Darcy coefficient), a parameter to be determined from experiments and to represent the inertial effect. The Forchheimer-Darcy equation employs the square root of permeability as equivalent length characteristics in order to agree with the linear Darcy law at low velocities. For a bed of uniform diameter rigid spheres, the following correlation has been reported in the literature (Nield and Bejan 2013):

$$F = \frac{1.8}{\left(180\varepsilon^5\right)^{0.5}} \varepsilon \tag{1.36}$$

Experimental tests have shown that water flow with Reynolds numbers up to ten may not need the Forchheimer term; but the gas flow is usually high

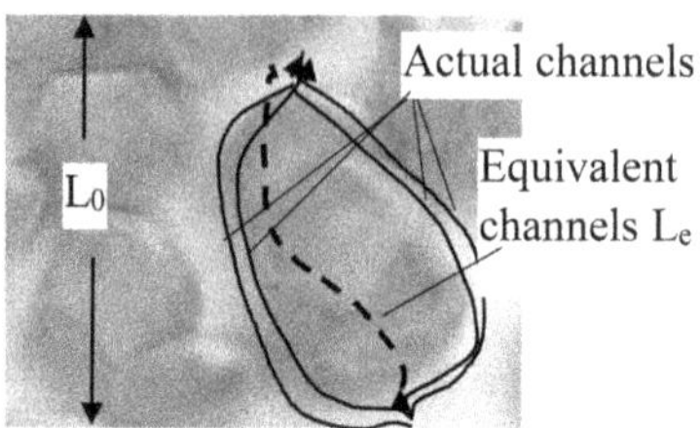

Figure 1.14 Geometrical interpretation of the equivalent path.

enough and the Forchheimer term should not be ignored because gas viscosity is low and the high gas velocities can form drag on solid particles.

The permeability of a porous medium is a property that depends on the pore size, shape and structure and vary from one geometry to another, which can be predicted by the empirical Karman-Cozeny equation:

$$K = \frac{D_p^2 \varepsilon^3}{36C(1-\varepsilon)^2} \tag{1.37}$$

where C is the constant of the packed porous bed. When a fluid flows from one location to another in a porous medium, there are more than one possible channels which results in more than one actual lengths of flow paths (Fig. 1.12). A single "imaginary" (equivalent) channel that has the same "conducting capacity" as the sum of all microscopic channels is defined by Sobieski et al. (2012) (Fig. 1.14). The K value is also influenced by tortuosity of the fluid path, which is defined as the ratio of the length of equivalent channel of actual flow paths to the physical depth of a porous bed (Sobieski et al. 2012).

$$\tau = \frac{L_e}{L_0} \tag{1.38}$$

where τ is the tortuosity (decimal). Another definition of the tortuosity is (Sobieski et al. 2012):

$$\tau = \left(\frac{L_e}{L_0}\right)^2 \tag{1.39}$$

where L_0 is the depth of bed (m), and L_e is the length of equivalent channel of flow paths (m).

This determined tortuosity was defined by Sobieski et al. (2012) as the overall (equivalent) hydraulic tortuosity, which is difficult to be calculated mathematically or measured directly. Therefore, no relationship between K and τ has been reported.

Inside grain bulks, the arithmetic diameter of grain kernels (D_s) and ε are not uniform and pore size, shape, and structure are also not the same from one place to another, this makes the applicability of Carman–Kozeny equation and Forchheimer–Darcy equation quite restrictive. One of the main advantages of the Forchheimer–Darcy law is that it makes the momentum equation linear (one order less than the Navier–Stokes equation) and thus removes a great amount of difficulty in solving the governing equations. However, the no-slip hydrodynamic boundary condition cannot be applied and therefore the velocity at the impermeable surface is predicted to be the maximum. In a multidimensional flow field, the local pressure gradient is also an unknown. Therefore, Forchheimer–Darcy equation should be used with the mass balance equation to obtain the flow and pressure distribution in the porous medium.

1.8.2.3 Ergun Equation

Ergun (1952) proposed an empirical relation for describing the pressure drop through porous media based on the porosity and a geometrical length scale as:

$$\nabla P = -\alpha \frac{(1-\varepsilon)^2 \mu}{\varepsilon^3 \Phi^2 D_s^2} q - \beta \frac{(1-\varepsilon)\rho}{\varepsilon^3 \Phi D_s} q^2 \tag{1.40}$$

where α is the factor for the viscous drag portion of the pressure drop, β is the constant for the form drag portion, ρ is the density of the fluid (kg/m^3). In the case of fluid flow through a porous bed consisting of spherical particles, $\alpha = 150$, $\beta = 1.75$ (Niven 2002). Ergun (1952) equation is the same as Darcy–Forchheimer law if

$$k = \frac{D_s^2 \varepsilon^3}{\alpha(1-\varepsilon)^2} \quad \text{and} \quad F = \frac{\beta}{\sqrt{\alpha\varepsilon}} \tag{1.41}$$

1.8.2.4 Brinkman Form of Darcy's Law

An additional factor that should be considered is the variation of permeability of the porous medium near the solid wall. The wall boundary layers are thickened and are under no-slip condition. At this condition, the flow may involve large shear. To correct this error, Brinkman's approximation introduces a viscous shear stress term in Darcy's law. The viscous shear stress is $\frac{\tilde{\mu}}{\varepsilon}\nabla^2 q$, which is based on the assumption that the force on a particle situated in a swarm of particles could be calculated as if these solid particles were imbedded in a porous mass; where $\tilde{\mu}$ is the effective viscosity (Pa.s). The Brinkman and Forschheimer-corrected Darcy's law is written as:

$$\nabla P = -\frac{\mu}{K}q - \frac{\rho F}{\sqrt{K}}q|q| + \frac{\tilde{\mu}}{\varepsilon}\nabla^2 q \tag{1.42}$$

where $\tilde{\mu}$ is the effective viscosity (Pa.s). This equation requires the sum of all external forces to be zero, a condition applicable only for non-accelerating flow fields. The uncertainty associated with the effective viscosity hampers the application of the Brinkman form of Darcy flow model. Almost all authors using Brinkman's equation have assumed that the effective viscosity is equal to the pure fluid viscosity (Kolodziej 1988). The value of effective viscosity was determined experimentally or mathematically and $\tilde{\mu}$ is higher or lower than μ under different conditions (Kolodziej 1988; Givler and Altobelli 1994). There is no report of $\tilde{\mu}$ in stored grain bulks.

1.8.2.5 Effect of Acceleration, Wall Friction, and Compression

The expression for fluid acceleration in a Lagrangian frame of reference can be written as:

$$a = \frac{1}{\varepsilon}\frac{dq}{dt} = \frac{1}{\varepsilon}\left(\frac{\partial q}{\partial t} + \frac{1}{\varepsilon}q \cdot \nabla q\right) \tag{1.43}$$

Therefore, the non-Darcy model of fluid flow in a porous medium is:

$$\nabla P = -\frac{\mu}{K}q - \frac{\rho F}{\sqrt{K}}q|q| + \frac{\tilde{\mu}}{\varepsilon}\nabla^2 q - \frac{\rho}{\varepsilon}\left(\frac{\partial q}{\partial t} + \frac{1}{\varepsilon}q \cdot \nabla q\right) \tag{1.44}$$

where t is the time (s). This equation does not consider the compressed fluid and wall friction effects.

Any stored grain is bounded by confining walls of storage structure, so that an influence of the walls on the pressure drop is to be expected due to an additional resistance caused by the wall friction. At the same time, the walls force the particles to order in such a way that a region of increased void fraction is formed, extending approximately half a particle diameter from the walls into the bed. The wall friction effect exists when the column to particle diameter ratio is less than 50 and it is especially significant when the ratio is less than ten (Eisfeld and Schnitzlein 2001; Zhong et al. 2016). The column diameter is the tube (bin) diameter in which the air passes through. In grain aeration and drying, this wall friction effect is negligible because the ratio is much larger than 50.

For the compressible case, integrating the pressure with respect to length, with the boundary conditions (pressures at $x = 0$ and $x = L_o$, Fig. 1.14), the following modified Ergun equation can be obtained:

$$\beta \frac{(1-\varepsilon)RT}{\varepsilon^3 D_s A^2} G^2 + \alpha \frac{(1-\varepsilon)^2 \mu RT}{\varepsilon^3 A D_s^2} G + \frac{P_2^2 - P_1^2}{2L_o} = 0 \tag{1.45}$$

where T is the temperature (°C), and p_1 and p_2 are the pressure at location 1 and 2, respectively (Pa). Compression effect is not usually considered due to the low air velocity during grain aeration and drying.

1.8.3 Prediction of Airflow Resistance in Bulk Grains

1.8.3.1 Most Used Equations

Ergun equation is widely used to predict airflow resistances of grain bulks. This equation requires the bulk porosity and particle equivalent diameter. Bulk porosity can be influenced by loading method, percentage of dockage, and grain moisture content. Particle equivalent diameter can be influenced by grain moisture content and percentage of dockage. All these variations result in various predictions of airflow resistance. The resistance to airflow in dense filled flax seeds is 1.3–1.5 times of loose filled flax seeds (Pagano et al. 1998). Therefore, different modified Ergun equations can be found in literature for different grains such as corn (Bern and Charity 1975; Kay et al. 1989), cherry pits (Bakker-arkema et al. 1969), rye (Matthies and Petersen 1974), sorghum (Yang and Williams 1990), wheat with different percentage of fines, alfalfa, and lentil (Li and Sokhansanj 1994). The most accepted modified Ergun equation is proposed by Li and Sokhansanj (1994) and Yang and Williams (1990) as:

$$\nabla P = k_1 \frac{(1-\varepsilon)^2 \mu}{\varepsilon^3 d_p^2} q + k_2 \frac{(1-\varepsilon)\rho}{\varepsilon^3 d_p} q^2 \tag{1.46}$$

where k_1 and k_2 are the product-dependent constants obtained from experiments. Equation 1.46 is fitted to the data of wheat, alfalfa seed, laird lentil, asi-like clover, red clover, flax, oats, barley and shelled corn over an airflow range from 0.754×10^{-4} to about 0.9 $m^3/(m^2{\cdot}s)$, and it is a good general model to predict the airflow resistance of the tested materials (Górnicki and Kaleta 2015a).

The main parameters affecting resistances of bulk materials to airflow are air properties (velocity, viscosity, and density), pore properties (porosity, tortuosity, pore size, shape, and distribution), and particle properties (particle size, shape, surface roughness, orientation of the particles). Most parameters such as pore properties are extremely difficult to measure. Therefore, it is difficult to predict k_1 and k_2. For this reason, parameters in Eq.1.45, except airflow superficial velocity, are lumped in one or two parameters for each material, and these developed models are referred to as empirical models. The empirical models include the following formats:

$$\Delta p/L_o = aq + bq^2 \text{(Ergun equation)} \tag{1.47a}$$

$$\Delta p/L_o = aq + bq^2 - \text{cMq (Haque et al. equation)} \tag{1.47b}$$

$$\Delta p/L_o = aq^b \text{(Shedd equation)} \tag{1.47c}$$

$$\Delta p/L_o = \frac{aq^2}{\ln(1+bq)} \text{(Hukill and Ives equation)} \tag{1.47d}$$

where M is the grain moisture content (%, wb). The parameters of a, b, c are product dependent and only can be determined from experiments. Equation of Haque et al. (1982) (Eq. 1.47b) is developed by using clean grains (corn, sorghum, and wheat) at 12.4–25.3% moisture content (wb) for 0.01–0.22 m/s airflow. Shedd (1953) equation (Eq. 1.47c) is the most accepted model. Hukill and Ives (1955) model (Eq. 1.47d) can handle wide range of air velocity (0.01–2.0 $m^3 \cdot m^{-2} \cdot s^{-1}$) and is used in the standard of the American Society of Agricultural and Biological Engineers (ASABE). Hukill and Ives equation has been modified by different researchers to predict airflow resistance of oilseed such as canola or grain mixed with fines and chaff (Jayas and Sokhansanj 1989), flax seed (Jayas et al. 1991a), and lentils (Sokhansanj et al. 1990). Some non-linear regression models are also developed and these empirical models are reviewed by Górnicki and Kaleta (2015a). Jayas and Mann (1994) regressed airflow resistance of 22 seeds using Shedd's equation and found the a value in the Shedd equation (Eq. 1.47c) did not change for different crop seeds when the airflow was divided into two sub-ranges as low: 0.004–0.05 $m^3 \cdot s^{-1} \cdot m^{-2}$; and high: 0.05–0.35 $m^3 \cdot s^{-1} \cdot m^{-2}$. They summarized the airflow resistance data into one equation (Eq. 1.48) with different parameters for each seed type (Table 1.6).

$$\frac{\Delta p}{L_o} = 5178 C_1 q^{1.11} \quad 0.004\frac{m}{s} < q \le 0.05 \text{ m/s} \tag{1.48a}$$

$$\frac{\Delta p}{L_o} = 6368 C_2 q^{1.67} \quad 0.05\frac{m}{s} < q \le 0.35 \text{ m/s for four specified seed types} \tag{1.48b}$$

$$\frac{\Delta p}{L_o} = 8404 C_3 q^{1.28} \quad 0.05\frac{m}{s} < q \le 0.35 \text{ m/s for other seed types} \tag{1.48c}$$

The values of C_1, C_2, and C_3 are presented in Table 1.6.

Table 1.6 Values of Parameters in Eq. 1.48 (Jayas and Mann 1994)

Crop[a]	C_1	C_2	C_3
Alfalfa	4.73		4.07
Barley	0.56		0.63
Brome grass	0.49		0.51
Canola (Tobin)	2.29		2.21
Canola (Westar)	1.6		1.61
Clover (Alsike)	7.73		7.23
Clover (Crimson)	3.12		2.88
Clover (Red)	5.17		4.67
Ear corn (Lot 1)		0.25	
Ear corn (Lot 2)		0.05	
Shelled corn	0.29	1	
Fescue	1.46		1.41
Flax	3.25		3.36
Lentil (Eston)	0.75	2.14	
Lentil (Laird)	0.68	2.3	
Lespedeza (Kobe)	0.96		0.91
Lespedeza (Sericea)	4.73		4.07
Lupine (Blue)	0.2		0.23
Oats	0.61		0.69
Popcorn (White rice)	0.63		0.69
Shelled popcorn (Yellow)	0.37		0.45
Rescue	0.23		0.26
Rough rice (Paddy)	0.67		0.72
Sorghum	0.83		0.83
Soybeans	0.22	0.66	
Wheat	1		1

[a] Text in the parentheses is the cultivar or another name of the crop.

1.8.3.2 Measurement of Pressure Drop

Pressure drops (airflow resistance) of different grain types are usually measured under the in-bin drying condition, so different factors influencing airflow resistance can be determined. The in-bin drying condition is usually mimicked at the lab condition. The measurement system usually consists of three parts: a variable-flow blower and its control, airflow measurement equipment, and the bin (column) filled with grain (Fig. 1.15). Manometer, pressure tap, or Pitot tube is usually used to measure the pressure. Piezometer orifices, Rotameter, nozzle, or orifice plate is usually used to measure the airflow. Different models of the equipment are used to achieve high measurement accuracy in different ranges of airflow (Jayas et al. 1991b). To decrease the turbulent effect of the introduced air, airflow

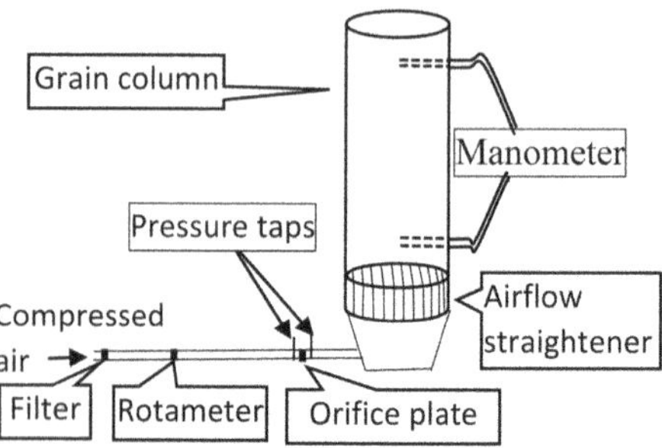

Figure 1.15 Schematic of experimental apparatus used for measurement of airflow resistance. (Adapted from Jayas et al. (1991a) with permission).

straightener is installed at the bottom of the grain bin (column) (Jayas et al. 1991b). To measure the effect of airflow direction, the grain bin (column) is usually specially designed so the same grain bulk can be ventilated in the horizontal or vertical direction (Kumar and Muir 1986). A spreader or spout can be installed at the top of the grain bin (column), so the effect of different filling method can be measured (Jayas et al. 1991a).

1.8.3.3 Grain Conditions Influencing Pressure Drop

Different grain types have different airflow resistances and kernel with a smaller kernel size usually has a higher airflow resistance at the same airflow (Table 1.7). Of the major grains and oilseeds, corn has one of the lowest resistances to airflow and flax the highest (Fig. 1.13). The ASABE Standard, ASAE D272.3 gives the airflow resistance of clean, dry, loosely-filled (not packed) seeds of 22 crops dried or aerated at low or medium airflows (Fig. 1.13). The binned grain is usually not at conditions of clean, dry, and loosely-fill. Therefore, the pressure drop estimated from this standard should be modified based on the stored grain condition and airflow. Any factor influencing air properties, grain physical properties, and pore properties will influence airflow resistance. Alagusundaram and Jayas (1990) reviewed the earlier works on this research topic. Górnicki and Kaleta (2015b) reviewed the effect of these factors on airflow resistance. This book only outlines the effects on airflow resistance from following factors: airflow and direction, fine and chaff concentration, grain depth, grain moisture content, crop type, and filling method and fall depth (drop height).

1.8.3.3.1 Airflow and Direction Equation 1.47 describes that the relationship between pressure drop and superficial velocity is the second-degree polynomial. Actually, most published charts show the logarithm

Table 1.7 Comparison of Airflow Resistance (Pa/m) of Different Crop Types at Conditions of Dry, Clean, and Loose Fill

	Crop[a]						
Airflow ($m^3 \cdot s^{-1} \cdot m^{-2}$)	*Barley*	*Canola*[b]	*Shelled Corn*	*Lentil*[c]	*Rough Rice*[d]	*Soybean*	*Wheat*
0.005	25	37	4	8	10	3	16
0.01	52	74	8	17	21	7	32
0.02	109	154	17	39	44	15	67
0.03	173	238	29	66	69	23	104
0.04	242	327	42	96	97	33	144
0.05	316	421	56	130	127	43	186
0.06	395	519	72	168	159	55	230
0.08	568	729	107	253	228	79	325
0.1	760	955	148	352	305	107	429
0.2	1981	2326	423	1023	796	284	1066
0.3	**3597**	4060	805	1964	**1444**	522	**1884**
0.4	**5573**	**6128**	**1285**	3154	**2238**	**815**	**2869**
0.5	**7889**	**8509**	**1858**	4578	**3168**	**1161**	**4009**
0.6	**10529**	**11190**	**2520**	**6228**	**4228**	**1555**	**5298**
0.7	**13481**	**14159**	**3268**	**8095**	**5413**	**1998**	**6731**
0.8	**16735**	**17407**	**4100**	**10175**	**6720**	**2487**	**8302**
0.9	**20284**	**20925**	**5013**	**12461**	**8145**	**3022**	**10008**
1	**24121**	**24708**	**6006**	**14950**	**9686**	**3600**	**11846**

Source: Data are produced from the Hukill and Ives equation and the parameter values recommended by ASABE (2016).

[a] Bolded values are the extrapolated values calculated using the parameters recommended by ASABE (2016). However, these extrapolations agree well with available data of shelled corn and wheat at airflows up to 1.0 $m^3 \cdot s^{-1} \cdot m^{-2}$ (Stark and James 1982).

[b] Tobin.

[c] Laird.

[d] Paddy.

relationship (Figs. 1.13 and 1.16 are plotted on log-log scale) because airflow used for aeration and natural air drying is low and b value in the Shedd equation (Eq. 1.47c) is less than two (Eq. 1.48). However, the quadratic airflow resistance will be obvious when the airflow is high (Reynolds numbers between one and ten). For the case of turbulent flow, the pressure drop is proportional to the velocity squared. Therefore,

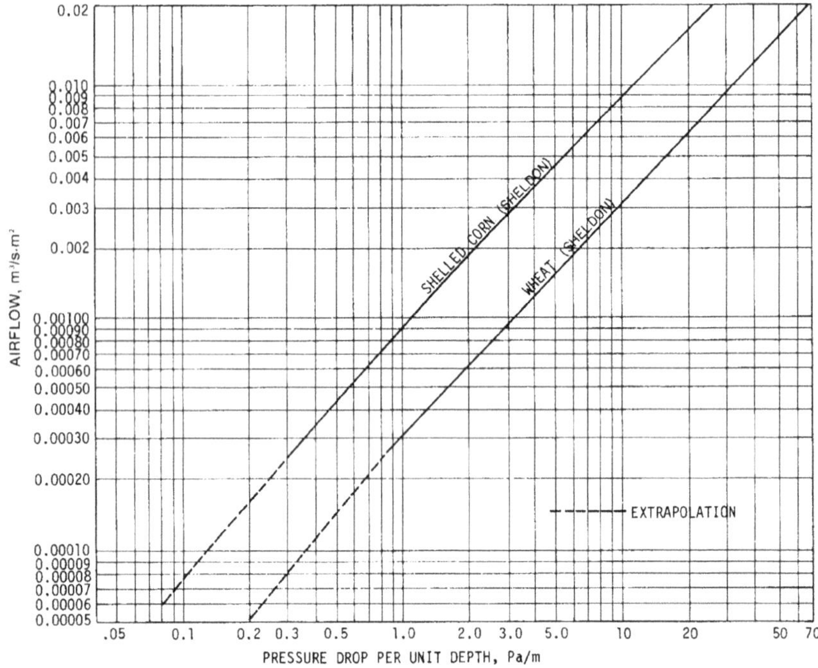

Figure 1.16 Airflow resistance of shelled corn and wheat at low airflows. (Reprinted with permission from ASABE (2016). In the graph, Sheldon shows the data source from Sheldon et al. (1960)).

these empirical equations (Eqs. 1.46–1.48) and charts (Figs. 1.13 and 1.16) should be used in the recommended ranges.

Lower resistance to airflow in the horizontal direction than that in the vertical direction has been reported for many grains (Table 1.8). The main reason for this difference is the different orientation distributions of the pores and grain kernels in the horizontal direction than that in the vertical direction. Wheat, kidney beans, and soybeans in grain mixtures have their major dimensions along the radius of the pile at any loading height (Boopathy 2018). Air path areas along the horizontal direction are 100% higher than those along the vertical direction for non-spherical grain bulks (wheat, barley, and flax seed) and 30% for spherical kernels (pea and mustard) (Neethirajan et al. 2006). The measured airflow resistance in the horizontal direction is about 0.4 to 0.7 times that in the vertical direction (Table 1.8). Hood and Thorpe (1992) measured the airflow resistance of ten crop types and found that the airflow resistance in vertical direction was approximately two times higher than in horizontal direction at the airflow

Table 1.8 Ratio of Resistance to Airflow in the Horizontal Direction to Vertical Direction in Grain Bulks

Crop	*Condition of the Bulks*[a]	*Airflow*[b]	*Ratio*	*Source*
Barley	Clean and unclean, 11.8% MC, test weight 610.6 kg/m³, porosity 55%, angle of repose 27°	0.077	0.47–0.62	Kumar and Muir (1986)
Canola	7.0% MC, 0–100% chaff or fines and filled by spout or sprinkle	0.0158–0.1709	0.5–0.7	Jayas et al. (1987b)
Shelled corn	Test weight 752.7 kg/m³	0.0127–0.4767	0.45–0.58	Kay et al. (1989)
Alfalfa pellets	8% and 12% MC, porosity 0.45–0.53, bulk density 472–668 kg/m³	0.003–1.0		Sokhansanj et al. (1993)
Flax	6–12% MC mixed with up to 25% fines, 0–15% chaff or fines, bulk density 559.4–677.4 kg/m³, porosity 0.43–0.49	0.036–0.25	0.38–0.65	Jayas et al. (1991b)
Lentil	12.1–12.8% MC, bulk density 759 kg/m³, porosity 0.47, 5–25% fines, dense or loose fill	0.003–0.6	0.52	Sokhansanj et al. (1990)
Oat	10.1–31.7% with 3–25% dockage and foreign materials	0.0097–0.459	0.04–0.17	Pagano et al. (1998)
Paddy rice	11% MC with broken grain from parboiled rice	0.01–0.1	0.19–0.31	Nalladurai et al. (2002)
Wheat	Clean and unclean, 8.8–9.7% MC, test weight 772–787 kg/m³, porosity 44–46%, angle of repose 44–55°	0.077	0.63	Kumar and Muir (1986)

[a] Different physical properties of the grain bulks in different published studies.
[b] Airflow velocity (m/s) or flow rate ($m^3 \cdot s^{-1} \cdot m^{-2}$).

velocity of 0.2 m/s. ASABE (2016) standard recommends the airflow resistance in the horizontal direction is 60–70% of that in the vertical direction for non-spherical shaped seeds (wheat, barley, and flaxseeds). Different empirical models have been developed for different crop types such as flax (Jayas et al. 1991a) and shelled corn (Kay et al. 1989).

1.8.3.3.2 Fine and Chaff Concentration The concentration of fines (particles smaller than seeds) and chaff (particles larger than seeds) in grain bulks significantly influences pressure drops. There is no specific correction recommended because resistance to airflow usually increases if the fines increase in grain, and resistance to airflow decreases if the chaff increases in the grain. The increase in airflow resistance due to the addition of fine material become greater as the size of the fines is decreased. Jayas et al. (1991c) observed that the resistance to airflow through fines was 2.3–3.1 times the resistance of clean canola, while through chaff was 0.12–0.15 times the resistance of clean canola, and this resistance depended on the airflow rate. For loose filled canola, resistance to airflow increased from 5–24% with increase in fines from 5 to 25% for airflow from 0.0243 to 0.2633 $m^3 \cdot m^{-2} \cdot s^{-1}$ (Jayas and Sokhansanj 1989). The fan power requirement of corn mixed with 3% fines is about 110% of those for clean corn for in-bin dryer, but requires up to 147% for a typical column dryer, to 299% for a typical counterflow bin dryer, and to 364% for a natural air in-bin dryer (Grama et al. 1984). Similar results are reported for oat (Pagano et al. 1998), flax (Jayas et al. 1991b), paddy and its by-products (Nalladurai et al. 2002), shelled maize (Haque et al. 1978), long grain rough rice (Siebenmorgen and Jindal 1987), canola (Jayas et al. 1991a), and lentils (Sokhansanj et al. 1990). Therefore, the estimation of pressure drop due to fine and chaff concentration is mostly based on grain storage practical experience such as the dockage distribution in bins and average concentration of fine and chaff associated with each harvest method. Since fines and chaff can have large particle distributions and characteristics, it is not possible to develop a correction factor. Therefore, empirical models are developed by different researchers for different crop types such as bulk canola (Jayas and Sokhansanj 1989), flax seed (Jayas et al. 1991a), rough rice (Siebenmorgen and Jindal 1987), shelled corn (Haque et al. 1978; Grama et al. 1984), oat (Pagano et al. 2000), and lentils (Sokhansanj et al. 1990).

1.8.3.3.3 Grain Depth Increasing grain depth causes a rapid increase in the total airflow resistance. For example, consider a cylinder with 1 m^2 cross section area (A) and 1 m deep containing 1 m^3 wheat (clean, dry and loose fill). Assuming the airflow is 10 $L \cdot s^{-1} \cdot m^3$ and this airflow is fixed for any condition (because this is usually the requirement for grain in-bin drying), then the airflow resistance under following three conditions can be estimated as:

1. a cylinder with 1 m^2 perforated floor area and 1 m deep.

$$Q = 1\ \text{m}^3 \times 10\ \frac{\text{L}}{\text{s}\cdot\text{m}^3} = \frac{10\ \text{L}}{\text{s}} = 0.1\ \text{m}^3/\text{s}$$

$$q = \frac{Q}{A} = \frac{0.1\ \text{m}^3/\text{s}}{1\ \text{m}^2} = 0.1\ \text{m/s}$$

$$\nabla P = 429\ \text{Pa/m}\ (\text{from Table 1.7}).$$

$$\Delta P = 1\ \text{m} \times 429\ \frac{\text{Pa}}{\text{m}} = 429\ \text{Pa}$$

2. a cylinder with 2 m^2 perforated floor area and 1 m deep.

$$Q = 2\ \text{m}^3 \times 10 \frac{\text{L}}{\text{s}\cdot\text{m}^3} = \frac{20\ \text{L}}{\text{s}} = 0.2\ \text{m}^3/\text{s}$$

$$q = \frac{Q}{A} = \frac{0.2\ \text{m}^3/\text{s}}{2\ \text{m}^2} = 0.1\ \text{m/s}$$

$$\nabla P = 429\ \text{Pa/m}\ (\text{from Table 1.7}).$$

$$\Delta P = 1\ \text{m} \times 429\ \frac{\text{Pa}}{\text{m}} = 429\ \text{Pa}$$

3. a cylinder with 1 m^2 perforated floor area and 2 m deep.

$$Q = 2\ \text{m}^3 \times 10 \frac{\text{L}}{\text{s}\cdot\text{m}^3} = \frac{20\ \text{L}}{\text{s}} = 0.2\ \text{m}^3/\text{s}$$

$$q = \frac{Q}{A} = \frac{0.2\ \text{m}^3/\text{s}}{1\ \text{m}^2} = 0.2\ \text{m/s}$$

$$\nabla P = 1066\ \text{Pa / m}\ (\text{from Table 1.7}).$$

$$\Delta P = 2\ \text{m} \times 1066 \frac{\text{Pa}}{\text{m}} = 2132\ \text{Pa}$$

Therefore, doubling the quantity of grain by doubling the grain depth, doubles the total airflow from 0.1 to 0.2 m^3/s, but the airflow resistance increases by more than four times ($2132/429 \approx 5$). However, doubling the quantity of grain by doubling the base area, doubles the total airflow from 0.1 to 0.2 m^3/s, but the airflow resistance remains the same. The reason is all the 0.2 m^3/s of air must pass through the grain located at the base. The passage of the 0.2 m^3/s of air doubles the air velocity through the grain, and doubling the air velocity approximately doubles the airflow resistance.

A fan that provides 0.2 m^3/s against a pressure of 429 Pa is usually much cheaper than a fan that moves the same amount of air against a higher pressure. To produce higher pressure, an axial fan with a larger motor or an expensive centrifugal fan may be needed, which results in a higher operational cost. Therefore, it is the best to increase the floor area from the view of airflow resistance. However, increasing floor area must be balanced against the increased capital cost of the steel walls, perforated flooring, foundation of a bin with increased floor area and cost of longer augers and other grain handling equipment.

The grain depth at the top of a peaked or coned pile of grain is higher than the lower portions of the cone of peak. Air will take the path of least resistance or shortest distance. Therefore, the airflow through the top coned pile is much less than through the lower portions of the cone. To pass a drying or cooling front completely through the top of a grain cone may take >80% more time than if the grain is levelled (Friesen and Huminicki 1987). Friesen (1989) developed an empirical equation to predict this extra time (t_P) required to dry or aerate grain piles with a cone top:

$$t_P = \left[1 + \frac{1.7h_P^{1.5}}{h_e}\right] t_L \tag{1.49}$$

where t_p is the time for air to pass through a grain pile with a peaked surface (h), h_P is the height of the grain cone (not the total height of the grain) (m), h_e is the equivalent (total) depth of grain if the cone is levelled off (m), and t_L is the time for air to pass through a levelled grain pile of the same volume and diameter (h).

If the design of a natural-air drying system is to be made based on a completely perforated floor and a levelled grain pile, the fan run time for the coned or peaked pile must be $1 + \frac{1.7h_P^{1.5}}{h_e}$ times the airflow time for the equivalent levelled pile.

1.8.3.3.4 Grain Moisture Content and Crop Types The resistance to airflow generally decreases with an increase in grain moisture content (Haque et al. 1982) because of an increase in porosity and change of surface characteristics with an increase in moisture content. The effect of moisture on void fraction is more prominent than surface characteristics on the resistance to airflow. The change in resistance attributed to moisture content is minor compared to the more than doubled resistance due to packing (Siebenmorgen and Jindal 1987). At moisture content of 14.5%, the pressure drop is about 85% of the pressure drop across dry canola (6.5% moisture content) (Jayas and Sokhansanj 1989). Different crop types at different moisture contents have been studied and these studies reach the same conclusions (Górnicki and kaleta 2015b). Haque et al. (1982) developed a nonlinear regression model that described the relationship among static pressure drop, moisture content and airflow rate for corn, sorghum, and wheat. Empirical equations related pressure drop and moisture content are developed for different crop types such as rough rice (Siebenmorgen and Jindal 1987), yellow dent corn, sorghum, and hard red winter wheat (Haque et al. 1982), and oat (Pagano et al. 2000).

Different crop types have different airflow resistance (Figs. 1.13 and 1.16) at the same test condition due to the different physical seed characteristics, such as shape, size, surface roughness, and porosity. This physical difference also explains the difference of resistance among different classes of the same crop type and sometimes different cultivars in the same class. Jayas and Mann (1994) found ear corn lot 1 and lot 2 had different airflow resistance. Therefore, airflow resistance of grain inside grain bins might be different from these published values due to the complex combination of the physical factors influencing airflow resistance.

At the same depth of grain ventilated at the same superficial velocity, the airflow resistance of corn is about one-third of the resistance of wheat, and canola resistance is about double that of wheat (Table 1.7). Therefore, the same configuration of a grain storage system (the same bin, same fan, same floor) filled with different grain types to the same grain depth will have different drying effects.

1.8.3.3.5 Filling Method and Drop Height Method of filling affects the grain bulk porosity, density, pore properties (pore size, shape), grain orientation, and compactness, hence, the resistance against airflow. Use of spreader increases 6–10% of bulk density of corn, 11–12.5% of sorghum, and 5–9% of wheat. This increased bulk density increases the resistance to airflow more than doubling in corn, and about 55–67% in wheat at 0.08 $m^3 \cdot m^{-2} \cdot s^{-1}$ (Chang et al. 1981). Horizontal airflow resistance of layer filling (similar to spreader filling) is higher than end filling (similar to spout filling) by 25–35% at air velocity of 0.077 m/s (Kumar and Muir 1986). Sprinkle filling generates consistently higher bulk densities for canola by approximately 7% compared to spout filling, which results in a significant difference in airflow resistance (Jayas et al. 1987a). Researchers have the similar conclusion for sorghum (Yang and Williams 1990), green gram (Nimkar and Chattopadhyay 2002), poppy seeds (Sacilik 2004), and lathyrus grain (Kengh et al. 2013).

Different drop heights of the loaded grain produce different bulk densities, pore structures and orientation of grain kernels, hence affect airflow resistance. Lukaszuk et al. (2009) concluded that gravitational axial filling of the wheat column from three heights (0.0, 0.95, and 1.8 m) resulted in the pressure drops of 1.0, 1.3, and 1.5 kPa at the airflow velocity of 0.3 m/s. Therefore, the same grain at different depths inside bin might have different airflow resistance due to the different drop height when grain is loaded.

1.9 Dielectric Properties and Measurement

1.9.1 Dielectric Properties

Some materials do not possess free charges, and become polarized upon the application of an external electric field. These materials with negligible electrical conductivity and high polarizability are referred to as dielectric materials. Dielectric properties of these materials are defined as molecular properties that are capable of impeding electron movement resulting in polarization within the material on exposure to an external electric field. Many biomaterials conduct electric currents to some low levels and are classified as dielectric materials. The dielectric properties determine how rapidly a material will heat in radio-frequency or microwave dielectric heating applications. Their dielectric response in electric fields also provides a means for sensing certain of their physical properties such as moisture content.

Dielectrics are often confused with insulators. The insulator typically implies the electrical obstruction and none or low electrical conductivity. However, dielectric is typically referred to the material with the ability to store energy and superior polarizability, which are expressed numerically as relative permittivity. The common example is a capacitor that has two insulated plates (Fig. 1.17). When a capacitor is connected with a DC (direct current) supply voltage (under the electric field), the capacitor's surface charge is raised by polarising the dielectric plates (Fig. 1.17), and charges up to the value of the applied voltage at a rate determined by

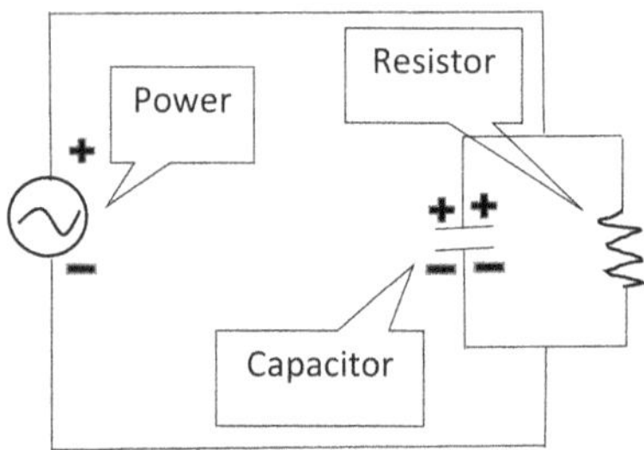

Figure 1.17 A simple pure capacitor and resistance circuit.

its time constant. During this charging process, a charging current (I_C) flows into the capacitor and gradually decreases to 0 A as the voltage on the plate gradually reaches to maximum. Therefore, the charging current (I_C) is opposed to the voltage change at a rate which is equal to the rate of change of the electrical charge on the plates, which can be mathematically expressed as:

$$I_C = C_C \frac{dV_V}{dt} \tag{1.50}$$

where I_C is the charging electrical current (A), C_C is the capacitance value of capacitor (F), and V_V is the power potential (V). When an alternating sinusoidal voltage is applied to plates of the capacitor, the capacitor is first charged in one direction and then in the opposite direction, which changes polarity at the same rate as the alternating current (AC) supply voltage. This instantaneous change in voltage across the capacitor is opposed by the fact that it takes a certain time to deposit this charge onto the plates, and the current in the circuit continuously flows and always leading the voltage by 1/4 of a cycle or $\pi/2 = 90°$ "out-of-phase" (Fig. 1.18). The total current can be calculated as:

$$I = I_C + I_R = j\omega C_c V_V + \frac{V_V}{R_R} \tag{1.51}$$

where I is the total current (A), I_R is the current through resistance (A), j is the operator employed for complex function ($j = \sqrt{-1}$), ω is the angular frequency, and R_R is the resistance (ohm). The angle (θ), separating the total current and the impressed voltage is called the phase angle (Fig. 1.18), and δ is the loss angle of the dielectric. The tangent of δ, also called the loss

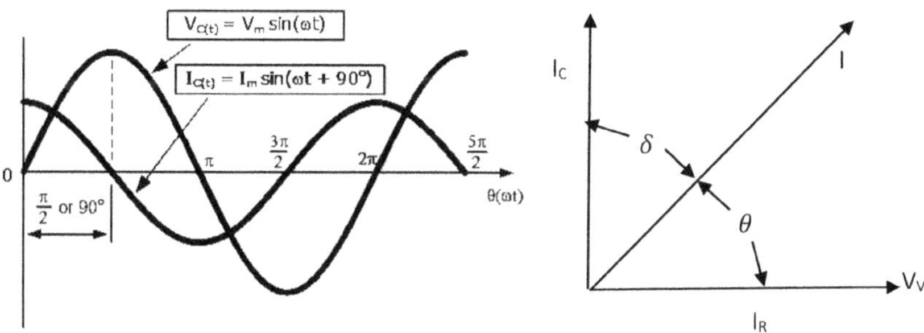

Figure 1.18 The current is always leading the voltage by 1/4 of a cycle or $\pi/2 = 90°$ "out-of-phase" for the circuit presented in Fig. 1.17 (left). Angles separating the total current (I), charging current (I_C), and resistor current (I_R).

tangent or dissipation factor, may be expressed in terms of the displacement and conduction currents:

$$\tan(\delta) = \frac{I_R}{I_C} = \frac{1}{\omega C_c R_R} \tag{1.52}$$

where δ is the loss angle of dielectric (°). When biomaterials such as a grain bulk are put between two metal capacitor plates, which are alternatively charged positively and negatively by a high frequency alternating electric field, polar molecules, such as water, try to align themselves with the polarity of the electric field (termed as "dipole rotation"). Inside the materials, positive ions try to move towards negative regions of the electromagnetic field and negative ions try to move towards positive regions of the field (termed as "ionic migration"). Since the polarity changes rapidly, the molecules try to continuously realign themselves with the electric field by flip-flop motion (including dipole rotation and ionic migration). This flip-flop motion results in frictional interaction between molecules. The resulting kinetic energy and friction generate heat within the material. This is the reason that materials, which have polar molecules, can be heated by using microwave (between 300 MHz and 300 GHz) and radio frequency waves (13.56 and 40.68 MHz). Since these waves are in the radar range and can interfere with communication systems, only the following selected frequencies are permitted for domestic, industrial, scientific, and medical applications: 13.56, 27.12, 40.68, 915, 2450 MHz, and 5.8 and 24.124 GHz (Piyasena et al. 2003). Biomaterials are composed of electrons, atoms, molecules and ions. Under a high frequency alternating electric field, electric displacements of these constituents occur. Therefore, biomaterials such as grain can be heated and dried by using microwaves and radio frequency waves.

1.9.2 Parameters of Dielectric Properties

Between the capacitor plates, the electric field intensity and direction vary with time and is transmitted by electromagnetic field, instead of by the action of charge carriers such as electrons or ions. Therefore, this produced current through the medium between the capacitor plates is called displacement current. A displacement current can even be produced in a vacuum. However, a suitable dielectric material does intensify the transfer of electromagnetic energy in the electric field. If it is a vacuum between two capacitor plates, the current increases in intensity and a minimal amount of electrical energy is lost between the plates. If a dielectric material fills the vacuum, the current decreases in intensity and a larger amount of electrical energy is lost to the material and absorbed by the material. The electrical permittivity through the medium between plates of the capacitor depends on the medium. The capability of an electric field to permeate a vacuum between two separated electric charges with spherical symmetry is given by Coulomb's law:

$$\varepsilon_0 = \frac{1}{4\pi F_c}\frac{q_1 q_2}{L^2} \tag{1.53}$$

where ε_0 is the vacuum permittivity (permittivity of free space, 8.854×10^{-12} F/m, or 8.854×10^{-12} $C^2 \cdot N^{-1} \cdot m^{-2}$), and F_c is the electrical force (N). The L is the distance between the two separated electric charges (m), the unit of the electrical charge (q_1, q_2) is Coulomb (C), and the unit of the vacuum permittivity (ε_0) is F/m, $C^2 \cdot N^{-1} \cdot m^{-2}$, $C \cdot Volt^{-1} \cdot m^{-1}$, or $C^2 \cdot s^2 \cdot kg^{-1} \cdot m^{-3}$. The vacuum permittivity (ε_0) can also be determined as:

$$c\mu_0\varepsilon_0 = 1 \tag{1.54}$$

where μ_0 is the magnetic constant (1.26×10^6 H/m), and c is the velocity of propagation in free space (2.997925×10^8 m/s). The key property of a dielectric material (the medium between the two separated electric charges) is the absolute dielectric permittivity (ε_{abs}), which is related to the vacuum permittivity.

$$\varepsilon_{abs} = (1+\Psi)\varepsilon_0 = \sigma\varepsilon_0 \tag{1.55}$$

where ε_{abs} is the absolute dielectric permittivity (F/m, F/m, or $C^2 \cdot N^{-1} \cdot m^{-2}$), Ψ is the electric susceptibility, and σ is the relative permittivity or dielectric constant (dimensionless).

The electric susceptibility (Ψ) is a measure of the degree of polarization of a material in response to an electric field, and $\Psi = 0$ for vacuum. Equation 1.55 indicates $\sigma \geq 1$. In solid, liquid, and gas, the permittivity has higher values and therefore the dielectric constant (σ) is usually expressed relative to the value in vacuum.

The relationship describing the dielectric behaviour of a material such as the permittivity is:

$$\vec{D} = \varepsilon_{abs}\vec{E} = \sigma\varepsilon_0\vec{E} \tag{1.56}$$

where $\vec{D}$ is the electric flux density (or called electric displacement, C/m^2), and $\vec{E}$ is the electric field strength (V/m). Therefore, the key parameter describing the dielectric behaviour of a material is the relative permittivity or termed as dielectric constant (σ), which measures how easily the medium is polarized.

The dielectric conductivity (K_C, volume conductivity, or specific conductivity) is defined as the conductance between two 1 cm^2 plates separated by a 1-cm cube of the dielectric if no current flows on the faces of the cube. It follows from Eq. 1.56 as:

$$K_C = \omega \varepsilon_0 \sigma \tan(\delta) \tag{1.57}$$

where K_C is the dielectric conductivity ($WV^{-2}m^{-1}$). The power dissipated per cm^3 in a dielectric is $\vec{E}^2 K_C = 55.63 f \vec{E}^2 \varepsilon'' \times 10^{-12}$ W/m^3. Therefore, the power dissipation in the dielectric is proportional to $\sigma \tan(\delta) = \varepsilon''$ (or $\tan(\delta) = \varepsilon''/\sigma$), which is called the dielectric loss factor. The dielectric loss factor measures the dissipation of electric energy into heat. A material with high values of dielectric loss factor will absorb energy at a faster rate than materials with lower loss factors because electrical power transferred to the unit volume of load as heat is directly proportional to the loss factor as well as the frequency and the square of electric field. Materials that have substantially high loss factors are usually called "lossy" because they absorb energy when is exposed to radio frequency (RF) or microwave fields.

The dielectric properties usually of interest are the dielectric constant and dielectric loss factor. The parameters permittivity and conductivity represent the properties of the media through which the waves propagate. The dielectric constant is associated with the ability of a material to store energy in the material in the electric field, while the loss factor is associated with the ability of the material to convert electric energy into heat energy. Dielectric loss factor is a measurement of the energy loss in a dielectric material through conduction, slow polarization currents, and other dissipative phenomena. The lost energy is absorbed in the medium as an electromagnetic wave passes through that medium. The dielectric loss factor, for example, is an index of a material's tendency to warm up in a microwave oven.

Relative complex permittivity can be expressed as:

$$\varepsilon_r = \sigma - j\varepsilon'' \tag{1.58}$$

where ε_r is the relative complex permittivity (dimensionless), and ε'' is the dielectric loss factor (dimensionless). Since ε_r is the complex permittivity of the material relative to free space or vacuum, relative complex permittivity influences the velocity of propagation of electromagnetic waves in a dielectric material and can be calculated as $v = c/\sqrt{\varepsilon_r}$.

Microwave power can only penetrate to certain depth of the material by polarizing the dielectric material. The penetration depth, d_p is usually defined as the depth into a sample where the microwave power has dropped to 36.8% of its transmitted value, or defined as the distance at which the microwave power has been attenuated to 50% of transmitted power. The d_p can be estimated as:

$$d_p = \frac{\lambda_0 \sqrt{\sigma}}{2\pi \varepsilon''} \tag{1.59}$$

where λ_0 is the free space microwave wavelength (for 2450 MHz, λ_0 = 122 mm). Most food materials have $\varepsilon'' < 25$ and σ is in the range of 2.2–4.5, which implies a d_p of 6–10 mm.

1.9.3 Factors Influencing Dielectric Properties of Grain

The main factors influencing dielectric properties of grain bulks are frequency, temperature, water content, chemical composition of the grain mixture, and factors which can influence bulk density. There are extensive review papers published in different times on this topic (Nelson 1981, 1982; Piyasena et al. 2003; Venkatesh and Raghavan 2005; Nelson 2010; Ramli et al. 2019; Zhou and Wang 2019). Therefore, this section only provides a general introduction.

1.9.3.1 Frequency

The same material will have different dielectric constants and dielectric loss factors at different frequencies. The relative permittivity can be mathematically estimated by using equation developed by Debye (1929) for a pure polar material:

$$\sigma = \varepsilon_{r\infty} + (\varepsilon_{rs} - \varepsilon_{r\infty}) / (1 + j\omega\tau_t) \tag{1.60}$$

where $\varepsilon_{r\infty}$ is the relative dielectric constant at high frequency, ε_{rs} is the static or direct current (DC) value of the relative dielectric constant (the value at zero frequency). The relaxation time (τ_t) is the period associated with the time for dipoles to revert to random orientation when the electric field is removed. The angular frequency is equivalent to $2\pi f$, and f is the frequency of the alternating field. If the frequency is too high, the voltage is applied in one direction for a very short time interval. If this time is shorter than τ, the displacement will be minimal and the power absorption will be negligible because the molecule has no time to turn. If the frequency is too low, the voltage is applied for long intervals relative to τ_t, and the displacement will reach its steady value for each pulse. Therefore, the power absorption will be negligible. If the frequency is of the same order as τ_t, the displacement never has time to reach its full value (because of the lag of δ angle) but is still appreciable. Hence there must be a maximum power factor in between the very high and very low frequency, and the loss tangent and the dielectric loss factor may either increase or decrease when frequency is increased, and this depends on the relaxation time of the chemical components in the material.

The dielectric constant of hard red winter wheat decreases with increasing frequency from 1 to 5.5 GHz, and being more dependent upon frequency at higher moisture contents. The dielectric loss factor exhibits little frequency dependence in this frequency range, and the variation of the dielectric loss factor is much less regular due to the influence of dielectric relaxation (Nelson 1973a). Nelson (1979a) found variability of the dielectric constant among 21 lots of corn was about the same at 20 MHz, 300 MHz, and 2.45 GHz. Nelson and his colleagues (Nelson 1965; Stetson and Nelson 1972; Trabelsi and Nelson 2012) measured dielectric properties of more than 12 crops in the range of 1–50 MHz and found dielectric constant either remains constant or decreases as frequency increases at any measured moisture contents while dielectric loss factor and loss tangent may either increase or decrease with moisture content or with frequency. They also reported the exponential increase of loss factor for hard red winter wheat in the measured frequency range with the increase of moisture content in the range of 6–21%.

1.9.3.2 Temperature

As temperature increases, the relaxation time decreases and the loss factor peak will shift to higher frequencies. In a region of polar dispersion, the dielectric constant will increase with increasing temperature while outside such a region it will decrease with temperature because the increase or decrease of the loss factor will depend on whether the operating frequency is higher or lower than the relaxation frequency. A non-linear increase in dielectric constant of the grain as temperatures increase is observed (Nelson 1991). Dielectric constant of corn at frequency of 20, 300, and 2450 MHz, increases approximately linearly with increasing temperature, and varies from linearity at higher moisture contents and lower frequencies (Lawrence et al. 1990). However, temperature dependence is not seen for dried solids but increases dramatically at higher moisture contents (Stuchly and Stuchly 1980).

1.9.3.3 Moisture Content

There is a strong positive correlation between dielectric constant and water content. Dielectric properties of liquid water are well described (Hasted 1973) and show a single dipolar relaxation at about 17.9 GHz at 22°C (Kaatze 1989). The dielectric loss factor generally increases with increasing moisture content of grain products, but the variation is not uniform with frequency. In the frequency range from 1 to 5.5 GHz, both dielectric constant and the dielectric loss factor of hard red winter wheat increase almost linearly with moisture content between 6 and 21% (Nelson 1973a). At the same frequency, dielectric constant and loss factor increased with the increase of moisture content, and the changes due to the variations in moisture content are greater at lower frequencies. At any frequency, $\tan(\delta)$ and loss factor will depend on the frequency and moisture content range. The relaxation frequencies also depend on moisture content. These dependences have been obtained over wide frequency range from 250 Hz to 12 GHz (Nelson 1981). These conclusions are found for wheat, corn, soybean over 1–200 MHz (Jones et al. 1978).

The status of water inside the kernels influences the dielectric properties. Water may be associated with other components forming layers of water and the water might have different status such as bound water or free water, and these statuses affect its dielectric properties. Chemically bound water exerts less influence on the dielectric properties of the material than free water. Ban (1974) reported that the dielectric constant and the loss tangent of rough rice varied with time after grain was dried. During grain drying, dielectric constant and loss factor increase with time and reached an equilibrium value after 2–3 h at 80°C (Sokhansanj and Nelson 1988) because grain kernels lose free water from porous structures. Immediately after water is added to kernels, the absorbed water fills the intercellular spaces in the outer layers of the kernel. A large portion of the absorbed water in this region remains unbound. This results in the change of dielectric properties. Large differences in the properties of frozen and unfrozen samples are observed (Ohlsson et al. 1974), so the status of water inside grain kernels can be determined. Energy absorption at 1.0 and 3.0 GHz increased at higher moisture contents and temperatures (Pace et al. 1968).

1.9.3.4 Bulk Density

Both the dielectric constant and the loss factor are greater for grain and seed samples with high bulk densities at high moisture contents. Dielectric constant of shelled corn at 300 MHz and 2.45 GHz increases linearly with bulk density at 10–40% moisture content (Nelson 1979a). Over wider ranges of bulk density (0–1600 kg/m^3), the square and cube roots of the dielectric constant are linear with bulk density (Nelson 1983, 1984). Researchers also found $(\sigma-1)/\sigma$ was a density-independent function (Meyer and Schilz 1980; Meyer and Schilz 1981; Kent and Meyer 1982; Nelson et al. 1998; Nelson et al. 2003). This finding provides the possibility for on-line and real-time monitoring for grain moisture content measurement without density calibration for several kinds of cereal grain (Nelson and Kraszewski 1990; Nelson et al. 2011). Density-independent moisture calibration function has been developed (Nelson et al. 2011) and meters based on this concept have been developed (Meyer and Schilz 1981).

1.9.3.5 Chemical Composition

Different chemical compounds have different dielectric properties. A wide variation in dielectric behaviour due to differences of chemical composition, physical state, and temperature is observed (Bengtsson and Risman 1971) because interactions among chemicals influence dielectric properties. For example, the dielectric loss factor remains nearly constant for barley and oats and decreases with increasing frequency for wheat and grain sorghum at 5–15 GHz, while corn shows a slight increase with increasing frequency (Trabelsi and Nelson 2012). Dielectric loss factors increase with ash content and decrease with fat content. Dielectric properties (relative complex permittivity) of solid mixture can be closely estimated from the density of the solid and the dielectric properties of the air-particle mixture, which has granular or powdered forms of the solid material. Water content and chemical composition of foods also reflect the temperature dependence of the dielectric properties. Increasing the moisture content of straw particles and reducing the frequency has significant effects on dielectric constant (Jafari et al. 2020). Kudra et al. (1992) found predictions of milk dielectric loss factor based on conductivities implied by ash contents should be corrected for binding and non-binding interactions of milk salts.

1.9.4 Measurement of Dielectric Properties

Models for estimating the dielectric constants of rough, brown and white rice and winter wheat of different bulk densities and moisture contents at 5 MHz to 12 GHz have been developed by Nelson (1992). Although dielectric properties of agricultural products can now be predicted over a wide range of frequencies for many foods and processing conditions by using developed models, dielectric properties of agricultural products are usually measured for different applications because a quantitative model for the coupling of electrical energy in foods by radiative transfer has not yet been found (Venkatesh and Raghavan 2005) and models usually do not reach the required accuracy.

Earlier reports on polar dielectrics and modelling studies date back to 1950s and the permittivity measurement is based on DC electrical resistance to determine grain moisture content (Field 1954). Venkatesh and Raghavan (2005) reviewed the history of the measurement of dielectric properties of agricultural materials and applications such as the development of dielectric heating and moisture meter development. Measurement methods of the dielectric properties of granular and powdered materials at microwave and RF frequencies and the factors affecting the dielectric properties of materials were reviewed by several research groups (Field 1954; Nelson 1991; Nelson 1994; Venkatesh and Raghavan 2005; Nelson 2010; Ramli et al. 2019).

If the frequency circuit parameters, such as impedance or admittance, can be measured appropriately, the dielectric properties of that material at that particular frequency can be determined from equations that properly relate the permittivity of the material to those circuit parameters. Several methods and different sensors can be used to measure these circuit parameters such as: lumped circuit, resonator, transmission line, and free-space methods (Nyfors and Vainikainen 1989). The most used instruments are: the parallel plate capacitor, coaxial probe, waveguide, resonant structure, inductance, capacitance-resistance meter (LCR meter), impedance analyzer, and scalar and vector network analyzer (Venkatesh and Raghavan 2005; Jafari et al. 2020). The challenge to measure permittivity or dielectric property accurately is the design of the material sample holder (Nelson 1979b). Current developments are aimed at eliminating the need for some expensive yet versatile accessories (Nelson 1991). Biological materials are transient in nature; therefore, it is difficult to standardize the tools for dielectric measurements. Applying proper calibration and mathematical routines can minimize errors and generate useful information on the measured materials. The techniques for permittivity measurements in the low, medium, and high frequency ranges, including the use of several bridges and resonant circuits was reviewed by Field (1954). Venkatesh and Raghavan (2005) provided detailed review on principles and methods of dielectric property measurement. They also compared the microwave dielectric measurement systems based on data in literature and the authors' own experiences. Nelson (2010) illustrated the fundamental principles and methods of dielectric properties measurements and agricultural applications. Principles of these systems and techniques are beyond scope of this book.

The choices of equipment and sample holder design depend upon the dielectric materials to be measured, the extent of the research, available equipment, and resources for the studies. For example, grain samples can be measured with a precision bridge at 250 Hz to 20 kHz by using a coaxial sample holder (Corcoran et al. 1970). Grain and seed tested using a Q-meter based on resonant circuit have been reported in the 1–50 MHz range (Nelson 1991). Microwave dielectric properties of wheat and corn have been measured at several frequencies by using free-space measurements with a vector network analyser (Trabelsi et al. 1997). The open-ended coaxial probe method has been used for many agricultural products over different frequencies, temperatures, and moisture contents.

1.9.5 Dielectric Properties of Grain Bulks and Applications

The dielectric constant of hard red winter wheat bulk at 11.5% of moisture content is from 2.45 to 4.35 at 9.4 GHz. The dielectric constant of the ground wheat at 10.9% moisture content and at 11.7 GHz is from 2.07 to 3.70, and for ground white rice with 12.2% moisture content at 11.0 GHz is 2.38 to 4.02. The loss factor of the above mentioned materials is 0.22 to 0.69. Nelson (1973b) reviewed the dielectric properties of agricultural products and materials. The dielectric properties of many materials have been tabulated (Nelson 1973a, Tinga and Nelson 1973, Kent 1987).

The main application areas of dielectric properties are grain moisture content sensor development, microwave heating and drying, seed treatment and separation in product processing, and insect control. The significant relationship between dielectric properties and moisture content has been used to develop the method of moisture content measurement and different models of moisture meter have been developed (Ban 1974; Nelson et al. 2003; Nelson 2010; Nelson et al. 2011; Ramli et al. 2019). Most of these grain moisture meters in the early development utilized frequencies between 1 and 20 MHz and sensed changes of capacitance when grain samples were introduced between electrodes (Nelson 1977b). To measure moisture content by the meters require corrections for the influence of temperature, bulk density, and test weight. There are many modifications in later development (Ramli et al. 2019). Microwave and RF drying has been widely used in different industries even though non-uniform heating and sometimes runaway heating are still major problem with the application of RF and microwave drying (Piyasena et al. 2003; Zhou and Wang 2019). Different dielectric properties of grain kernels with different moisture content and different materials in a grain mixture have been used to separate these kernels and materials (Jafari et al. 2020). The loss factor of insects is more than 5 times greater than that of the wheat in the frequency range of 10–100 MHz, and those differences provide selective heating of insects. These heating difference has been used to control insects (Nelson 1977a; Vadivambal et al. 2010; Ling et al. 2014; Sosa-Morales et al. 2015). Detail design and techniques of these applications are out of scope of this book.

1.10 Remarks on the Application of the Physical Properties

The fundamental physical properties related to grain storage and processing are explained in this chapter. These fundamental physical properties include the single kernel and bulk grain properties. The single kernel properties include geometrical properties, hardness, and cracks due to shrinkage and swelling. The bulk grain properties include density and porosity, frictional properties, dielectric properties, and airflow resistance. These properties are intertwined and interdependent. In grain storage practice and processing, application of the physical properties should consider the interdependent properties of the grain bulk as a whole.

The main goal of a grain storage manager or the owner of grain is to safely store the grain with a minimum investment of both time and money. Therefore, using the simplest and cheapest method to monitor grain physical properties is always the best choice in the practice of grain storage, transportation, and processing. One of the disadvantages of the simplest and cheapest method might be that the physical properties of the grain as a whole or as an ecosystem might not be monitored. For example, grain storage bins are usually only equipped with the temperature cables or no equipment at all. The physical properties such as the change of moisture content, bulk density, true density, porosity, and airflow resistance are usually not monitored due to the high cost of these monitoring. Without the information of these physical properties, only monitoring temperature might provide an incomplete information. Using the theories related to the change of these physical properties might help to complete the missing information. For example, the change of these physical properties will influence the temperature change because physical properties such as the change of bulk density and moisture content will influence the thermal conductivity. The measured grain temperature is the temperature of the grain bulk as a whole which is affected by the change in these physical properties. Understanding the theories behind the change of these physical properties is the key to understanding the temperature change. Therefore, understanding theories related to grain physical properties is the fundamental requirement to safely store grain with a minimum investment.

Nomenclature

α: factor for the viscous drag portion of the pressure drop, or acceleration (m/s^2);

β: constant for the form drag portion;

ρ: density (kg/m^3);

ρ_a: air density (kg/m^3);

ρ_C: compacted density (kg/m^3);

ρ_b: bulk density (kg/m^3);

ρ_k: kernel density (kg/m^3);

ε: porosity (%, or decimal);

ε_{abs}: absolute dielectric permittivity (Farads/m, F/m, or $C^2 \cdot N^{-1} \cdot m^{-2}$);

ε_r: relative complex permittivity (dimensionless);

ε_{rs}: static or direct current (DC) value of the relative dielectric constant (the value at zero frequency);

$\varepsilon_{r\infty}$: relative dielectric constant at high frequency;

ε_0: vacuum permittivity (permittivity of free space, 8.854×10^{-12} Farads/m, or 8.854×10^{-12} $C^2 \cdot N^{-1} \cdot m^{-2}$);

ε'': dielectric loss factor (dimensionless);

λ:	shape factor;
T:	thickness (m);
Ψ:	electric susceptibility;
ϕ:	sphericity (%, or decimal);
δ:	loss angle of dielectric (°);
ϑ:	coefficient of (sliding) friction;
θ:	angle of sliding friction or angle of phase angle (°);
θ_e:	effective angle of internal friction (°);
θ_i:	angle of internal friction (°);
θ_w:	wall friction angle (°);
σ:	relative permittivity or dielectric constant (dimensionless);
σ_h:	stress in horizontal direction (Pa);
σ_v:	stress in vertical direction (Pa);
$\sigma_{v\infty}$:	stress in vertical direction at an infinite depth of silo (Pa);
σ_w:	wall normal stress (Pa);
τ:	tortuosity (decimal);
τ_w:	shear stress on the wall (Pa);
μ:	dynamic viscosity (Pa·s);
$\tilde{\mu}$:	effective viscosity (Pa·s);
μ_0:	magnetic constant (1.26×10^6 H/m);
ω:	angular frequency;
λ_0:	free space microwave wavelength (for 2450 MHz, $\lambda_0 = 122$ mm);
τ_r:	relaxation time (s);
Ω:	angular velocity (rad/s);
f:	frequency of the alternating field (Hz);
A:	surface area, cross section area of the silo, cross section of flow channel (m^2);
A_c:	area of smallest circumscribing circle (m^2);
A_f:	frontal area (m^2);
A_1:	largest projected area of object in natural rest position (m^2);
A_p:	particle projected area (m^2);
A_R:	aspect ratio;
C:	constant of the packed porous bed;
C_c:	capacitance value of capacitor (farads, F);
C_D:	drag coefficient;
C_I:	compressibility index;
C_1, C_2, and C_3:	constants in the airflow resistant equations;
c:	velocity of propagation in free space (2.997925×10^8 m/s);
D:	diameter of the circle circumscribing the projected area or diameter of bin (m);
$\vec{D}$:	electric flux density (or called electric displacement, C/m^2);
D_a:	arithmetic diameter (m);
D_e:	equivalent diameter (m);

D_g: geometric mean diameter (m);
D_L: longest diameter through the most convex part of the particle (m);
D_p: projected diameter (m);
D_s: diameter of silo (m), diameter of pore, or solid particle diameter (m);
d_p: diameter of the projected cross section (m);
E_l: elongation or lengthiness;
$\vec{E}$: electric field strength (V/m);
e: eccentricity;
F: constant or Forchheimer constant (non-Darcy coefficient);
F_c: electrical force (N);
F_D: drag force (N);
f: flakiness;
G: mass flow rate (kg/s);
g: gravitational acceleration (m/s^2);
H_R: Hausner ratio;
h: height of the cone (m);
h_e: equivalent (total) depth of grain if the cone is levelled off (m);
h_p: height of the grain cone (not the total height of the grain) (m);
I: total current (A);
I_R: current through resistance (A);
I_C: charging electrical current (A);
j: operator employed for complex function ($j = \sqrt{-1}$);
K: (intrinsic) permeability (m^2);
K_C: dielectric conductivity (WV^{-2}m^{-1});
k: lateral stress ratio;
k_1, k_2: product-dependent constants obtained from experiments;
L: length or distance (m);
L_e: length of equivalent channel of flow paths (m);
L_o: depth of bed (m);
M: grain moisture content (%, w.b.);
MC: grain moisture content (%, w.b.);
m: mass (g, or kg);
m_a: mass of the gas (kg);
P: pressure (Pa);
PSI: particle size index;
P_1, P_2, P_3: pressure measured in tank 1 at the very beginning, in tank 2 after the compressed air is introduced, and in both tank 1 and tank 2 after valve 2 is opened, respectively (Pa);
P_o: overall projection perimeter (m);
P_s: force causing slip on any plane (N);
P_n: normal force acting perpendicular to the plane of slip (N);
ΔP: pressure drop between two locations (Pa);
∇P: pressure gradient (Pa/m);
Q: total discharge (or volume flow rate, m^3/s or L/s);

q: mean flow velocity (superficial velocity, or filtration velocity) (m/s);
q_1, q_2: the electrical charges (Coulomb, C);
R: specific gas constant ($J \cdot Kg^{-1} \cdot k^{-1}$);
R_c: roundness or circularity (decimal or percentage);
Re: Reynolds number;
R_R: resistance (ohm);
r: radius (m);
r_{av}: mean radius (m);
r_c: curvature radius of the maximum convex part of the particle (m);
r_{cs}: radius of curvature of the sharpest corner (m);
r_m: mean radius of the object (m);
r_1, r_2: the radii of the base, and top of the cone (m), respectively;
S: surface area (m^2);
Sh_{app}: apparent shrinkage (decimal);
$Sh_{\%}$: percentage of shrinkage (%);
S_s: surface area based shape factor;
S_v: volume based shape factor;
Sw_{app}: apparent swelling (decimal);
$Sw_{\%}$: percentage of swelling (%);
T: temperature (°C);
T_1, T_2, T_3: temperatures measured before and after the compressed gas is introduced, and after tank 1 and tank 2 reach equilibrium, respectively (K);
t: time (s);
t_L: time for air to pass through a levelled grain pile of the same volume and diameter (h);
t_p: time for air to pass through a grain pile with a peaked surface (h);
U: perimeter of the storage structure (m);
V: volume (m^3);
V_{app}: volume of the kernel after shrinkage or swelling (m^3);
V_b: volume occupied by the grain bulk (m^3);
V_i: volume of the kernel before shrinkage or swelling (m^3);
V_k: volume occupied by the kernels (m^3);
V_V: power potential (voltage, V);
v: propagation velocity or fluid velocity inside the porous medium (m/s);
v_r: relative velocity (m/s);
V_s: volume of sample (m^3);
V_1, V_2: volumes of tank 1, 2, respectively;
v_t: terminal velocity (m/s);
W: width (m);
Y: test weight (kg/hL);
z: depth of the silo (m).

References

Abdelrahman, A. A., and R. C. Hoseney. 1984. Basis for hardness in pearl millet, grain sorghum, and corn. Cereal Chemistry 61: 232–235.

Alagusundaram, K., and D. S. Jayas. 1990. Airflow resistance of grains and oil seeds. Postharvest News and Information 1: 279–283.

Al-Hashemi, H. M. B., and O. S. B. Al-Amoudi. 2018. A review on the angle of repose of granular materials. Powder Technology 330: 397–417.

Al-Mahasneh, M. A., and T. M. Rababah. 2007. Effect of moisture content on some physical properties of green wheat. Journal of Food Engineering 79: 1467–1473.

Almeida-Dominguez, H. D., E. L. Suhendro, and L. W. Rooney. 1997. Factors affecting rapid visco analyser curves for the determination of maize kernel hardness. Journal of Cereal Science 25: 93–102.

Anglani, C. 1998. Sorghum endosperm texture – a review. Plant Foods for Human Nutrition 52: 67–76.

Anjum, F. M., and C. E. Walker. 1991. Review on the significance of starch and protein to wheat kernel hardness. Journal of the Science of Food and Agriculture 56: 1–13.

Anovitz, L. M., and D. R. Cole. 2015. Characterization and analysis of porosity and pore structures. Reviews in Mineralogy & Geochemistry 80: 61–164.

Armstrong, P. R., J. E. Lingenfelser, and L. McKinney. 2007. The effect of moisture content on determining corn hardness from grinding time, grinding energy, and near-infrared spectroscopy. Applied Engineering in Agriculture 23: 793–799.

Arteaga, P., and U. Tuzun. 1990. Flow of binary-mixtures of equal-density granules in hoppers size segregation, flowing density and discharge rates. Chemical Engineering Science 45: 205–223.

ASABE. 2016. Resistance to airflow of grains, seeds, other agricultural products, and perforated metal sheets, pp. 7. In ASABE Standard. American Society of Agricultural and Biological Engineers, St. Joseph, MI.

Avena-Bustillos, R. D. J., J. M. Krochta, M. E. Saltveit, R. D. J. Rojas-Villegas, and J. A. Sauceda-Perez. 1994. Opyimization of edible coating formulations on zuchini to reduce water loss. Journal of Food Engineering 21: 197–214.

Bakker-arkema, F. W., R. J. Patterson, and W. G. Bickert. 1969. Static pressure– airflow relationship in packed beds of granular biological materials such as cherry pits. Transactions of the ASAE 12: 134–140.

Ban, T. 1974. Grain moisture meter and moisture detector for flowing grain. Japan Agricultural Quarterly 8: 177–182.

Barlow, K. K., M. S. Buttrose, D. H. Simmonds, and M. Vest. 1973. The nature of the starch-protein interface in wheat endosperm. Cereal Chemistry 50: 443–454.

Bean, S. R., O. K. Chung, M. R. Tuinstra, J. F. Pedersen, and J. Erpelding. 2006. Evaluation of the single kernel characterization system (SKCS) for measurement of sorghum grain attributes. Cereal Chemistry 83: 107–113.

Bengtsson, N. E., and P. O. Risman. 1971. Dielectric properties of food at 3 GHz as determined by a cavity perturbation technique. II. Measurements on food materials. Journal of Microwave Power 6: 107–123.

Bern, C. J., and l. F. Charity. 1975. Airflow resistance characteristics of corn as influenced by bulk density. In Annual of the ASAE. ASAE, St. Joseph, MI.

Bettge, A. D., and C. F. Morris. 2000. Relationships among grain hardness, pentosan fractions, and end-use quality of wheat. Cereal Chemistry 77: 241–247.

Bhadra, R., M. E. Casada, S. A. Thompson, J. M. Boac, R. G. Maghirang, M. D. Montross, A. P. Turner, and S. G. McNeill. 2017. Field-observed angles of repose for stored grain in the United States. Applied Engineering in Agriculture 33: 131–137.

Bhadra, R., M. E. Casada, A. P. Turner, M. D. Montross, S. A. Thompson, S. G. McNeill, R. G. Maghirang, and J. M. Boac. 2018. Stored grain pack factor measurements for soybeans, grain sorghum, oats, barley, and wheat. Transaction of the ASABE 61: 747–757.

Bhadra, R., A. P. Turner, M. E. Casada, M. D. Montross, S. A. Thompson, J. M. Boac, S. G. McNeill, and R. G. Maghirang. 2015. Pack factor measurements for corn in grain storage bins. Transacrions of the ASABE 58: 879–890.

Bhise, S., A. Kaur, and M. R. Manikantan. 2013. Engineering properties of flaxseed (Lc 2063) at different moisture. Journal of Postharvest Technology 1: 52–59.

Bian, Q., and B. Subramanyam. 2015. Effect of chaff on bulk flow properties of wheat. Journal of Stored Products Research 64: 21–26.

Bilanski, W. K., and R. Lal. 1965. Behaviour of threshed materials in a vertical wind tunnel. Transactions of the ASAE 8: 411.

Boopathy, N. R. B. 2018. Segregation of Canola, Kidney Bean and Soybean in Wheat During Bin Loading. MSc, University of Manitoba Winnipeg, Canada.

Bowden, P. J., W. J. Lamond, and E. A. Smith. 1983. Simulation of near-ambient grain drying: i. Comparison of simulations with experimental results. Journal of Agricultural Engineering Research 28: 279–300.

Canadian Grain Commission. 2018. Official grain grading guide. Canadian Grain Commission. https://www.grainscanada.gc.ca/oggg-gocg/27/oggg-gocg-27-eng.htm.

Carmen, K. 1996. Some physical properties of lentil seeds. Journal of Agricultural Engineering Research 63: 87–92.

Carvajal-Millan, E., A. Rascón-Chu, J. A. Márquez-Escalante, V. Micard, N. P. León, and A. Gardea. 2007. Maize bran gum: extraction, characterization and functional properties. Carbohydrate Polymers 69: 280–285.

Chandra, S., L. Wheaton, K. Schumacher, and R. Muller. 2001. Assessment of barley quality by light transmission – the rapid LTm meter. Journal of the Institute of Brewing 107: 39–47.
Chang, C. S. 1988. Measuring density and porosity of grain kernels using a gas pycnometer. Cereal Chemstry 65: 13–15.
Chang, C. S., L. E. Shackelford, F. S. Lai, C. R. Martin, and B. S. Miller. 1981. Bulk properties of corn as affected by multiple-point grain spreaders. Transactions of the ASAE 24: 1632–1636.
Chukwu, O., and F. B. Akande. 2007. Development of an apparatus for measuring angle of repose of granular materials. AU Journal of Technology 11: 62–66.
Chung, C. J., S. J. Clark, M. C. Lindholm, R. J. McGinty, and C. A. Watson. 1977. The pearlograph technique for measuring wheat hardness. Transactions of the ASAE 18: 185.
Chung, D. S., and H. H. Converse. 1971. Effect of moisture content on some physical properties of grains. Transactions of ASAE 14(4): 612–614.
Cleary, P. W., and M. L. Sawley. 2002. DEM modeling of industrial granular flows: 3D case studies and the effect of particle shape on hopper discharge. Applied Mathematical Modeling Journal 26: 89–111.
Corcoran, P. T., S. O. Nelson, L. E. Stetson, and C. W. Schlaphoff. 1970. Determining dielectric properties of grain and seed in the audio frequency range. Transactions of the ASAE 13: 348–351.
Coskun, M. B., I. Yalçın, and C. Özarslan. 2006. Physical properties of sweet corn seed. Journal of Food Engineering 74: 523–528.
Curray, J. K. 1951. Analysis of Sphericity and Roundness of Quartz Grain. MSc, The Pennsylvania State University University Park, Pa.
De Alencar Figueiredo, L. F., F. Davrieux, G. Fliedel, J. F. Rami, J. Chantereau, M. Deu, B. Courtois, and C. Mestres. 2006. Development of NIRs equations for food grain quality traits through exploitation of a core collection of cultivated sorghum. Journal of Agricultural and Food Chemistry 54: 8501–8509.
Debye, P. 1929. Polar Molecules, The Chemical Catalogue Co., New York.
Delcour, J. A., and R. C. Hoseney. 2010. Structure of cereals, pp. 1–22, Principles of Cereal Science and Technology. AACC International Inc., St. Paul, MN.
Deng, Z., A. Smolyanitsky, Q. Li, X. Q. Feng, and R. J. Cannara. 2012. Adhesion-dependent negative friction coefficient on chemically modified graphite at the nanoscale. Nature Materials 11: 1032–1037.
Deshpande, S. D., S. Bal, and T. P. Ojha. 1993. Physical properties of soybean. Journal of Agricultural Engineering Research 56: 89–98.
Devaux, M. F. 1998. Particle size distribution of break, sizing and middling wheat flours by laser diffraction. Journal of the Science of Food and Agriculture 78: 237–244.
Dombrink-Kurtzman, M. A., and J. A. Bietz. 1993. Zein composition in hard and soft endosperm of maize. Cereal Chemistry 70: 105–108.

Dursun, E., and I. Dursun. 2005. Some physical properties of caper seed. Biosystems Engineering 92: 237–245.

Eisfeld, B., and K. Schnitzlein. 2001. The influence of confining walls on the pressure drop in packed beds. Chemical Engineering Science 56: 4321–4329.

Ergun, S. 1952. Fluid flow through packed columns. Chemical Engineering Progress 48: 89–94.

Evers, T., and S. Millart. 2002. Cereal grain structure and development: some implications for quality. Journal of Cereal Science 36: 261–284.

Eyherabide, G. H., L. Robutti, and F. S. Borras. 1996. Effect of nearinfrared transmission-based selection on maize hardness and the composition of zeins. Cereal Chemistry 73: 775–778.

Field, R. F. 1954. Lumped circuits and dielectric measuring techniques, pp. 12–22. In A. von Hippel (ed.), Dielectric Materials and Applications. John Wiley and Sons, New York, NY.

Figura, L. O., and A. A. Teixeira. 2007. Food Physics, Physical Properties – Measurement and Applications, Springer, Berlin Heidelberg, Germany.

Fox, G., and M. Manley. 2009. Hardness methods for testing maize kernels. Journal of Agricultural and Food Chemistry 57: 5647–5657.

Fraczek, J., A. Złobecki, and J. Zemanek. 2007. Assessment of angle of repose of granular plant material using computer image analysis. Journal of Food Engineering 83: 17–22.

Friesen, O. H. 1989. Airflow patterns in grain bins with partially perforated floors. American Society of Agricultural Engineers 6040: 1–26.

Friesen, O. H., and D. N. Huminicki. 1987. Grain aeration and unheated air drying. Manitoba Agriculture 732–1: 1–31.

Gaines, C. S., M. O. Raeker, M. Tilley, P. L. Finney, J. D. Wilson, D. B. Bechtel, R. J. Martin, P. A. Seib, G. L. Lookhart, and T. Donelson. 2000. Associations of starch gel hardness, granule size, waxy allelic expression, thermal pasting, milling quality, and kernel texture of 12 soft wheat cultivars. Cereal Chemistry 77: 163–168.

Givler, R. C., and S. A. Altobelli. 1994. A determination of the effective viscosity for the Brinkman- Forchheimer flow model. Journal of Fluid Mechanics 258: 355–370.

Glenn, G. M., and R. M. Saunders. 1990. Physical and structural properties of wheat endosperm associated with grain texture. Cereal Chemistry 67: 176–182.

Gorial, B. Y., and J. R. O'callaghan. 1990. Aerodynamic properties of grain/straw materials. Journal of Agriculture Engineering Research 46: 275–290.

Górnicki, K., and A. Kaleta. 2015a. Resistance of bulk grain to airflow – a review. Part I: equations for airflow resistance. Annals of Warsaw University of Life Sciences 65: 31–41.

Górnicki, K., and A. Kaleta. 2015b. Resistance of bulk grain to airfow – a review. Part II: effect of process parameters. Annals of Warsaw University of Life Sciences 65: 43–51.

Govindarajan, S. K. 2019. An overview on extension and limitations of macroscopic Darcy's law for a single and multi-phase fluid flow through a porous medium. International Journal of Mining Science 5: 1–21.

Graham, R., D. Senadhira, S. Beebe, C. Iglesias, and I. Monasterio. 1999. Breeding for micronutrient density in edible portions of staple food crops: conventional approaches. Field Crops Research 60: 57–80.

Grama, S. N., C. J. Bern, and J. C. R. Hurburgh. 1984. Airflow resistance of mixtures of shelled corn and fines. Transactions of the ASAE 27: 268–272.

Greffeuille, V., J. Abecassis, M. Rousset, F. X. Oury, A. Faye, and V. Lullien-Pellerin. 2006. Grain characterization and milling behaviour of near-isogenic lines differing by hardness. Theoretical and Applied Genetics 1141: 1–12.

Guo, Z., X. Chen, H. Liu, Q. Guo, X. Guo, and H. Lu. 2014. Theoretical and experimental investigation on angle of repose of biomass–coal blends. Fuel 116: 131–139.

Gupta, R. K., and S. K. Das. 1997. Physical properties of sunflower seeds. Journal of Agricultural Engineering Research 66: 1–8.

Gürsoy, S., and E. Güzel. 2010. Determination of physical properties of some agricultural grains. Research Journal of Applied Sciences, Engineering and Technology 2: 492–498.

Haque, E., Y. Ahmed, and C. W. Deyoe. 1982. Static pressure drop in a fixed bed of grain as affected by grain moisture content. Transactions of ASAE 25: 1095–1099.

Haque, E., G. H. Foster, D. S. Chung, and F. S. Lai. 1978. Static pressure across a corn bed mixed with fines. Transactions of the ASAE 21: 997–1000.

Hasted, J. B. 1973. Aqueous Dielectrics, Chapman and Hall, London, U.K.

Heywood, H. 1948. Calculation of particle terminal velocity. Journal of Imperial College Chemical Engineering Society 4: 17.

Hoffman, P. C., D. Ngonyamo-Majee, and R. D. Shaver. 2010. Technical note: determination of corn hardness in diverse corn germplasm using near-infrared reflectance baseline shift as a measure of grinding resistance. Journal of Dairy Science 93: 1685–1689.

Hood, T. J. A., and G. R. Thorpe. 1992. The effects of the annisropic resistance to airflow on the design of aearation systems for bulk stored grains. Agricultural Engineering Australia 21: 18–23.

Hukill, W. V., and N. C. Ives. 1955. Radial airlow resistance of grain. Agricultural Engineering 33: 332–335.

Irvine, D. A., D. S. Jayas, N. D. G. White, and M. G. Brixton. 1992. Physical properties of flaxseed, lentils, and fababeans. Canadian Agricultural Engineering 34: 75–81.

Isik, E., and H. Unal. 2007. Moisture-dependent physical properties of white speckled red kidney bean grains. Journal of Food Engineering 82: 209–216.

Jafari, M., G. Chegini, B. Rezaeealam, and A. A. S. Akmal. 2020. Experimental determination of the dielectric constant of wheat grain and cluster straw in different moisture contents. Food Science and Nutration 8: 629–635.

Jain, R. K., and S. Bal. 1997. Properties of pearl millet. Journal of Agricultural Engineering Research 66: 85–91.

Jaky, J. 1948. Pressure in silos, pp. 103–107. In 2nd International Conference of ICSMFE, Balkema, Rotterdam.

Janssen, H. 1895. Versuche uber getreidedruck in silozellen. Zeitschrift des Vereines deutscher Ingenieure 39: 1045–1049.

Jayas, D. S. 1987. Resistance of canola (rapeseed) to airflow. PhD, University of Saskatchewan Saskatoon, Saskatchewan.

Jayas, D. S., and D. D. Mann. 1994. Presentation of airflow resistance data of seed bulks. Applied Engineering in Agriculture 10: 79–83.

Jayas, D. S., and S. Sokhansanj. 1989. Design data on resistance of airflow through canola (rapeseed). Transactions of the ASAE 32: 295–296.

Jayas, D. S., S. Sokhansanj, and N. D. G. White. 1987a. Bulk density and porosity of canola. American Society of Agricultural Engineers Microfiche collection.

Jayas, D. S., S. Sokhansanj, E. B. Moysey, and E. M. Barber. 1987b. The effect of airflow direction on the resistance of canola (rapeseed) to airflow. Canadian Agricultural Engineering 29: 189–192.

Jayas, D. S., K. Alagusundaram, and D. A. Irvine. 1991a. Resistance of airflow through bulk flax seed as affected by the moisture content, direction of airflow and foreign material. Canadian Agricultural Engineering 33: 279–285.

Jayas, D. S., S. Sokhansanj, and F. W. Sosulski. 1991b. Resistance of bulk canola seed to airflow in the presence of foreign material. Canadian Agricultural Engineering 33: 47–54.

Jayas, D. S., S. Sokhansanj, and N. D. G. White. 1989. Bulk density and porosity of two canola species. Transactions of the ASAE 32: 291–294.

Jayas, D. S., N. D. G. White, M. G. Britton, and J. T. Mills. 1992. Effect of oil used for dust control on engineering properties of stored wheat. Transactions of the ASAE 35: 659–664.

Jenike, A. W., J. R. Johanson, and J. W. Carson. 1973. Bin load – part 3: mass-flow bins. Journal of Engineering for Industry 95: 6–12.

Jian, F., and D. S. Jayas. 2018. Characterization of isotherms and thin-layer drying of red kidney beans, part I: choosing appropriate empirical and semi theoretical models. Drying Technology 36: 1696–1706.

Jian, F., D. S. Jayas, P. G. Fields, and N. D. G. White. 2017a. Water sorption and cooking time of red kidney beans: part II – mathematical models of water sorption. International Journal of Food Science & Technology 52: 2412–2421.

Jian, F., D. S. Jayas, P. G. Fields, and N. D. G. White. 2017b. Water sorption and cooking time of red kidney beans (*Phaseolus vulgaris* L.): Part I – effect of freezing and drying conditions on water sorption and cooking time. International Journal of Food Science & Technology 52: 2031–2039.

Jian, F., D. S. Jayas, and N. D. G. White. 2012. Thermal conductivity, bulk density, and germination of a canola variety with high oil content under different temperatures, moisture contents, and storage periods. Transactions of the ASABE 55: 1837–1843.

Jian, F., D. S. Jayas, and N. D. G. White. 2013. Specific heat, thermal diffusivity, and bulk density of genetically modified canola with high oil content at different moisture contents, temperatures, and storage times. Transactions of the ASABE 56: 1077–1083.

Jian, F., R. B. Narendran, and D. S. Jayas. 2019. Segregation in stored grain bulks: kinematics, dynamics, mechanisms, and minimization – a review. Journal of Stored Products Research 81: 11–21.

Jian, F., S. Yavari, R. B. Narendran, and D. S. Jayas. 2018. Physical properties of FINOLA® hemp seeds: clean and containing dockages. Applied Engineering in Agriculture 34: 1017–1026.

Jindal, V. K., N. Mohsenin, and J. V. Husted. 1974. Surface area of selected agricultural seeds and grains. Transaction of the ASAE 17: 720–725, 728.

Jones, R. N., H. E. Bussey, R. F. Little, and Metzker. 1978. Electrical characteristics of corn, wheat, and soya in the 1-200 MHz range. In National Bureau of Standards. USA Department of Commerce, Washington, DC.

Joshi, D. C., S. K. Das, and R. K. Mukherjee. 1993. Physical properties of pumpkin seeds. Journal of Agricultural and Engineering Research 54: 219–229.

Juliano, B., and D. Bechtel. 1985. The rice grain and its gross composition, pp. 17–58. In B. Juliano (ed.), Rice: Chemistry and Technology, second ed. American Association of Cereal Chemists, St Paul, Minnesota, USA.

Kaatze, U. 1989. Complex permittivity of water as a function of frequency and temperature. Journal of Chemical Engineering Data 34: 371–374.

Kaliniewicz, Z., Z. Zuk, and Z. Krzysiak. 2018. Influence of steel plate roughness on the frictional properties of cereal kernels. Sustainability 10: 1–11.

Kang, Y. S., C. K. Spillman, J. S. Steele, and D. Chung. 1995. Mechanical properties of wheat. Transactions of the ASAE 38: 573–578.

Karababa, E., and Y. Coskuner. 2007. Moisture dependent physical properties of dry sweet corn kernels. International Journal of Food Properties 10: 549–560.

Kashaninejad, M., M. Ahmadi, A. Daraei, and D. Chabra. 2008. Handling and frictional characteristics of soybean as a function of moisture content and variety. Powder Technology 188: 1–8.

Kay, R. L., C. J. Bern, and C. R. J. Hurburgh. 1989. Horizontal and vertical airflow resistance of shelled corn at various bulk densities. Transactions of the ASAE 32: 733–736.
Kengh, R. N., K. R. Kenghe, P. M. Nimkar, and S. S. Shirkole. 2013. Effect of bulk density, moisture content, and grain size on vertical airflow resistance of laghyrus (*Lathyrus sativus* L.) grains. Legume Genomics and Genetics 4: 3–21.
Kent, M. 1987. Electrical and Dielectric Properties of Food Materials, Science and Technology Publishers, Essex, England.
Kent, M., and W. Meyer. 1982. A density-independent microwave moisture meter for heterogeneous foodstuffs. Journal of Food Engineering 1: 31–42.
Kent, N. L., and A. D. Evers. 1994. 'Kent's Technology of Cereals, fourth ed. Pergamon Press Ltd., Oxford.
Kézdi, A. 1962. Erddrucktheorien, Springer, Berlin.
Khoshtaghaza, M. H., and R. Mehdizadeh. 2006. Aerodynamic properties of wheat kernel and straw materials. Agricultural Engineering International: the CIGR eJournal 13: 1–10.
Koenen, M. 1896. Berechnung des seiten- und bodendrucks in silozellen. Centralblatt der Bauverwaltung 16: 446–449.
Kolodziej, J. A. 1988. Influence of the porosity of a porous medium on the effective viscosity in Brinkman's filtration equation. Acta Mechanica 75: 241–254.
Konopka, I., M. Tanska, and S. Konopka. 2015. Differences of some chemicals and physical properties of winter wheat grain of mealy and vitreous appearance. Cereal Research Communications 43: 470–480.
Krumbein, W. C., and L. L. Sloss. 1963. Stratigraphy and Sedimentation, second ed. W. H. Freeman and Company, San Francisco.
Kudra, T., G. S. V. Raghavan, C. Akyel, R. Bosisio, and S. R. van de Voort. 1992. Electromagnetic properties of milk and its constituents at 2.45 MHz. Journal of Microwave Power 27: 199–204.
Kumar, A., and W. E. Muir. 1986. Airflow resistance of wheat and barley affected by airflow direction, filling method and dockage. Transactions of the ASAE 29: 1423–1426.
Lanzerstorfer, C. 2017. Dusts from dry off-gas cleaning: comparison of flowability determined by angle of repose and with shear cells. Granular Matter 19: 58.
Lawrence, K. C., S. O. Nelson, and A. W. Kraszewski. 1990. Temperature dependence of the dielectric properties of wheat. Transactions of the ASAE 33: 535–540.
Lee, J., and H. J. Herrmann. 1993. Angle of repose and angle of marginal stability: molecular dynamics of granular particles. Journal of Physics A: Mathematical and General 26: 373.
Li, W., and S. Sokhansanj. 1994. Generalized equation for airflow resistance of bulk grains with variable density, moisture content and fines. Drying Technology 12: 649–667.

Li, Z., J. Yang, X. Xu, X. Xu, W. Yu, X. Yue, and C. Sun. 2002. Particle shape characterization of fluidized catalytic cracking catalyst powders using the mean value and distribution of shape factors. Advanced Powder Technology 13: 249–263.

Ling, B., G. Tiwari, and S. Wang. 2014. Pest control by microwave and radio frequency energy: dielectric properties of stone fruit. Agronomy for Sustainable Development 35: 233–240.

Lukaszuk, J., M. Molenda, J. Horabik, and M. D. Montross. 2009. Variability of pressure drops in grain generated by kernel shape and bedding method. Journal of Stored Products Research 45: 112–118.

Lumay, G., F. Boschini, K. Traina, S. Bontempi, J.-C. Remy, R. Cloots, and N. Vandewalle. 2012. Measuring the flowing properties of powders and grains. Powder Technology 224: 19–27.

Makse, H. A., R. C. Ball, H. E. Stanley, and S. Warr. 1998. Dynamics of granular stratification. Physical Review E 58: 1–12.

Markowski, M., K. Żuk-Gołaszewska, and D. Kwiatkowski. 2013. Influence of variety on selected physical and mechanical properties of wheat. Industrial Crops and Products 47: 113–117.

Martin, C. R., R. Rousser, and D. L. Brabec. 1993. Development of a single-kernel wheat characterization system. Transactions of the ASAE 36: 1399–1404.

Matthies, H. J., and H. Petersen. 1974. New data for calculating the resistance to airflow of stored granular material. Transactions of the ASAE 17: 1144–1149.

Matuttis, H. G., S. Luding, and H. J. Herrmann. 2000. Discrete element simulations of dense packings and heaps made of spherical and non-spherical particles. Powder Technology 109: 278–292.

McNeill, S. G., S. A. Thompson, and M. D. Montross. 2004. Effect of moisture content and broken kernels on the bulk density and packing of corn. Applied Engineering in Agriculture 20: 475–480.

Mestres, C., F. Matencio, and A. Louis-Alexandre. 1995. Mechanical behavior of corn kernels: development of a laboratory friability test that can predict milling behavior. Cereal Chemistry 72: 652–657.

Metcalf, J. R. 1966. Angle of repose and internal friction. International Journal of Rock Mechanics and Mining Sciences & Geomechanics Abstracts 3: 155–161.

Meyer, W., and W. Schilz. 1980. A microwave method for density independent determination of the moisture content of solids. Journal of Physics D: Applied Physics 13: 1823–1830.

Meyer, W., and W. Schilz. 1981. Feasibility study of density-independent moisture measurement with microwaves. IEEE Transactions on Microwave Theory and Techniques 29: 732–739.

Milani, A. P., R. A. Bucklin, A. A. Teixeira, and H. V. Kebeli. 2000. Soybean compressibility and bulk density. Transactions of the ASAE 43: 1789–1793.

Miller, R. L., and R. J. Byrne. 1966. The angle of repose for a single grain on a fixed rough bed. Sedimentology 6: 303–314.

Misra, R. N., and J. H. Young. 1980. Numerical solution of simultaneous moisture diffusion and shrinkage during soybean drying. Transactions of the ASAE 23: 1277–1282.

Mitchell, J. K., and K. Soga. 2005. Fundamentals of Soil Behavior, third ed. John Wiley & Sons, Inc., New Jersey, NJ.

Mohsenin, N. N. 1970. Physical Properties of Plant and Animal Materials, Gordon and Breach Science Publishers, New York, NY.

Mohsenin, N. N. 1986. Physical Properties of Plant and Animal Materials, Gordon and Breach Science Publishers, New York, NY.

Molenda, M., M. D. Montross, S. G. McNeill, and J. Horabik. 2005. Airflow resistance of seeds at different bulk densities using Ergun's equation. Transactions of the ASAE 48: 1137–1145.

Morris, C. F. 2002. Puroindolines: the molecular genetic basis of wheat grain hardness. Plant Molecular Biology 48: 633–647.

Morris, C. F., and B. S. Beecher. 2012. The distal portion of the short arm of wheat (*Triticum aestivum* L.) Chromosome 5D controls endosperm vitreosity and grain hardness. Theoritical and Applied Genetics 125: 247–254.

Morris, C., and A. N. Massa. 2003. Puroindoline genotype of the U.S. National institute of standards & technology reference material 8441, wheat hardness. Cereal Chemistry 80: 674–678.

Moss, R., N. L. Stenvert, K. Kingswood, and G. Pointing. 1980. The relationship between wheat microstructure and flour milling. Scanning Electron Microscopy 3: 613.

Muir, W. E., and R. N. Sinha. 1988. Physical properties of cereal and oilseed cultivars grown in western Canada. Canadian Agricultural Engineering 30: 51–55.

Mwithiga, G., and M. M. Sifuna. 2006. Effect of moisture content on the physical properties of three varieties of sorghum seeds. Journal of Food Engineering 75: 480–486.

Nalladurai, K., K. Alagusundaram, and P. Gayathri. 2002. Airflow resistance of paddy and its byproducts. Biosystems Engineering 83: 67–75.

Neethirajan, S., D. S. Jayas, N. D. G. White, and H. Zhang. 2008. Investigation of 3D geometry of bulk wheat and pea pores using x-ray computed tomography images. Computers and Electronics in Agriculture 63: 104–111.

Neethirajan, S., C. Karunakaran, D. S. Jayas, and N. D. G. White. 2006. X-ray computed tomography image analysis to explain the airflow resistance differences in grain bulks. Biosystems Engineering 94: 545–555.

Nelson, E. 1955. Measurement of the repose angle of a tablet granulation. Journal of the American Pharmacists Association 44.

Nelson, S. O. 1965. Dielectric properties of grain and seed in the 1 to 50 –mc range. Transactions of the ASAE 8: 38–48.

Nelson, S. O. 1973a. Microwave dielectric properties of grain and seed. Transactions of the ASAE 16: 902–905.

Nelson, S. O. 1973b. Electrical properties of agricultural products – a critical review. 16 (2):. Transactions of the ASAE 16: 384–400.
Nelson, S. O. 1977a. Microwave dielectric properties of insects and grain kernels. Journal of Microwave Power 12(1): 39.
Nelson, S. O. 1977b. Use of electrical properties for grain moisture measurement. Journal of Microwave Power 12: 67–72.
Nelson, S. O. 1979a. RF and microwave dielectric properties of shelled, yellow-dent field corn. Transactions of the ASAE 22: 1451–1457.
Nelson, S. O. 1979b. Improved sample holder for q-meter dielectric measurement. Transacrions of the ASAE 22: 950–954.
Nelson, S. O. 1980. Moisture dependent kernel and bulk density relationship for wheat and corn. Transactions of the ASAE 203: 139–437.
Nelson, S. O. 1981. Review of factors influencing the dielectric properties of cereal grains. Cereal Chemistry 58: 87–492.
Nelson, S. O. 1982. Factors affecting the dielectric properties of grain. Transactions of the ASAE 25: 1045–1049.
Nelson, S. O. 1983. Observations on the density dependence of the dielectric properties of particulate materials. Journal of Microwave Power 18: 143–152.
Nelson, S. O. 1984. Density dependence of the dielectric properties of wheat and whole-wheat flour. Journal of Microwave Power 19: 55–64.
Nelson, S. O. 1991. Dielectric properties of agricultural products - measurements and applications. IEEE Transactions of Electrical Insulation 26: 845–869.
Nelson, S. O. 1992. Correlating dielectric properties of solids and particulate samples through mixture relationships. Transactions of the ASAE 35: 625–629.
Nelson, S. O. 1994. Measurement of microwave dielectric properties of particulate materials. Journal of Food Engineering 21(3): 365–384.
Nelson, S. O. 2010. Fundamentals of dielectric properties measurements and agricultural applications. Journal of Microwave Power and Electromagnetic Energy 44: 98–113.
Nelson, S. O., and A. W. Kraszewski. 1990. Grain moisture content determination by microwave measurements. Transactions of the ASAE 33: 1303–1307.
Nelson, S. O., A. W. Kraszewski, S. Trabelsi, and K. C. Lawrence. 2003. Using cereal grain permittivity for quality sensing by moisture determination. IEEE Transactions on Instrumentation and Measurement 49: 237–242.
Nelson, S. O., and D. Sc. S. Trabelsi. 2011. Sensing grain and seed moisture and density from dielectric properties. International Journal of Agricultural and Biological Engineering 4: 1–7.
Nelson, S. O., S. Trabelsi, and A. W. Kraszewski. 1998. Advances in sensing grain moisture content by microwave measurements. Transactions of the ASAE 41: 483–487.
Nield, D. A., and A. Bejan. 2013. Convection in Porous Media, fourth ed. Springer, New York.

Nimkar, P. M., and P. K. Chattopadhyay. 2002. Airflow resistance of green gram. Biosystems Engineering 82: 407–414.

Niven, R. K. 2002. Physical insight into the Ergun and Wen & Yu equations for fluid flow in packed and fluidised beds. Chemical Engineering Science 57: 527–534.

Nyfors, E., and P. Vainikainen. 1989. Industrial Microwave Sensors, Artech House, Norwood, MA.

Ohlsson, T., N. E. Bengtsson, and P. O. Risman. 1974. The frequency and temperature dependence of dielectric food data as determined by a cavity perturbation technique. Journal of Microwave Power 9: 129–145.

Pace, W. E., W. B. Westphal, and S. A. Goldblith. 1968. Dielectric properties of commercial cooking oils. Journal of Food Science 33: 30–36.

Pagano, A. M., D. E. Crozza, and S. M. Nolasco. 1998. Pressure drop through in-bulk flax seeds. Journal of the American Oil Chemists' Society 75: 1741–1747.

Pagano, A. M., D. E. Crozza, and S. M. Nolasco. 2000. Airflow resistance of oat seeds: effect of airflow direction, moisture content and foreign material. Drying Technology 18: 457–468.

Parde, S. R., A. Johal, D. S. Jayas, and N. D. G. White. 2003. Physical properties of buckwheat cultivars. Canadian Biosystems Engineering 45: 3.19–13.22.

Pasha, I., F. M. Anjum, and C. F. Morris. 2010. Grain hardness: a major determinant of wheat quality. Food Science and Technology International 0: 1–12.

Pipatpongsa, T., and S. Heng. 2010. Granular arch shapes in storage silo determined by quasi-static analysis under uniform vertical pressure. Journal of Solid Mechanics and Materials Engineering 4: 1237–1248.

Piyasena, P., C. Dussault, T. Koutchma, H. S. Ramaswamy, and G. B. Awuah. 2003. Radio frequency heating of foods: principles, applications and related properties—a review. Critical Reviews in Food Science and Nutrition 43: 587–606.

Prescott, J. K., and R. J. Hossfeld. 1994. Maintaining product uniformity and uninterrupted flow to direct-compression tableting presses. Pharmaceutical Technology 18: 99–114.

Psota, V., K. Vejrazka, O. Famera, and M. Hrcka. 2007. Relationship between grain hardness and malting quality of barley (*Hordeum vulgare* L.). Journal of the Institute of Brewing 113: 80–86.

Rackl, M., F. E. Grötsch, M. Rusch, and J. Fottner. 2017. Qualitative and quantitative assessment of 3D-scanned bulk solid heap data. Powder Technology 321: 105–118.

Ramiärez, A., G. T. Peä rez, P. D. Ribotta, and A. E. Leoä. 2003. The occurrence of friabilins in triticale and their relationship with grain hardness and baking quality. Journal of Agricultural and Food Chemistry 51: 7176–7181.

Ramli, N., M. H. F. Rahiman, L. M. Kamarudin, A. Zakaria, and L. Mohamed. 2019. A review on frequency selection in grain moisture content detection. In 5th International Conference on Man Machine Systems. IOP Publishing.

Reichert, R. D., and C. G. Youngs. 1976. Dehulling cereal grains and grain legumes for developing countries, I: quantitative comparisons between attrition and abrasive type mills. Cereal Chemistry 53: 829–839.

Riley, R. E., and H. H. Hausner. 1970. Effect of particle size distribution on the friction in a powder mass. International Journal of Powder Metallurgy 6: 17–22.

Robutti, J. L. 1995. Maize kernel hardness estimation in breeding by near infrared transmission analysis. Cereal Chemistry 72: 632–636.

Rosa-Sibakova, N., K. Poutanen, and V. Micard. 2015. How does wheat grain, bran and aleurone structure impact their nutritional and technological properties? Trends Food Science Technology 41: 118–134.

Ross, M. H. 1997. Genetic study of nonavoidance of a pyrethroid residue by German cockroaches (dictyoptera: Blattellidae). Journal of Economic Entomology 90: 1243–1246.

Rotter, J. M. 2008. Silo and hopper design for strength, pp. 99–134. In D. McGlinchey (ed.), Bulk Solids Handling: Equipment Selection and Operation Blackwell, Oxford, U.K.

Sacilik, K. 2004. Resistance of bulk poppy seeds to airflow. Biosystems Engineering 89: 435–443.

Sahoo, P. K., and A. P. Srivastava. 2002. Physical properties of okra seed. Biosystems Engineering 83: 441–448.

Samson, M. F., F. Mabille, R. Chéret, J. Abécassis, and M. H. Morel. 2005. Mechanical and physicochemical characterization of vitreous and mealy durum wheat endosperm. Cereal Chemistry 82: 81–87.

Santos, L. C. D., R. Condotta, and M. D. C. Ferreira. 2017. Flow properties of coarse and fine sugar powders. Journal of Food Process Engineering 41: 1–10.

Schulze, D. 1994. The prediction of initial stresses in hoppers. Bulk Solids Handling 14: 497–503.

Schulze, D. 2008. Powders and Bulk Solids: Behavior, Characterization, Storage and Flow, Springer, Wolfsburg, Germanay.

Schulze, D., and J. Schwedes. 1994. An examination of initial stresses in hoppers. Chemical Engineering Science 49: 2047–2058.

Seifi, M. R., and R. Alimardani. 2010. The moisture content effect on some physical and mechanical properties of corn (sc 704). Journal of Agricultural Science 2: 125–134.

Selvi, C. S., Y. Pinar, and E. Yesiloglu. 2006. Some physical properties of linseed. Biosystems Engineering 95: 607–612.

Senadhira, D., G. Gregorio, and R. Graham. 1998. Breeding iron and zinc-dense content of rice, International Workshop on Micronutrient Enhancement of Rice for Developing Countries. Rice Research and Extension Center, Stuttgart, AK.

Senthilkumar, T., F. Jian, D. S. Jayas, and R. B. Narendran. 2018. Physical properties of white and black beans (*Phaseolus vulgaris*). Applied Engineering in Agriculture 34: 749–754.

Shedd, C. K. 1953. Resistance of grains and seeds to air flow. Agricultural Engineering 34: 616–619.

Sheldon, W. H., C. W. Hall, and J. K. Wang. 1960. Resistance of shelled corn and wheat to low airflows. Trans. ASAE 3: 92–94.

Shepherd, H., and R. K. Bhardwaj. 1986. Moisture-dependent physical properties of pigeon pea. Journal of Agricultural and Engineering Research 35: 227–234.

Siebenmorgen, T. J., and V. K. Jindal. 1987. Airflow resistance of rough rice as affected by moisture content, fines concentration and bulk density. Transactions of the ASAE 30: 1138–1143.

Simmonds, D. H. 1972. The ultrastructure of the mature wheat endosperm. Cereal Chemistry 49: 212.

Simmonds, D. H., K. K. Barlow, and C. W. Wrrigley. 1973. The biochemical basis for grain hardness in wheat. Cereal Chemistry 50: 553.

Singh, S., N. Sivakugan, and S. Shukla. 2010. Can soil arching be insensitive to PHIL? International Journal of Geomechanics 10: 124–128.

Siska, J., and C. R. Hurburgh. 1994. Prediction of Wisconsin tester breakage susceptibility of corn from bulk density and NIRS measurements of composition. Transactions of the ASAE 37: 1577–1582.

Smith, E. A., D. S. Jayas, W. E. Muir, K. Alagusundaram, and V. H. Kalbande. 1992. Simulation of grain drying in bins with partially perforated floors. II. Calculation of moisture content. Trans. ASAE 35: 917–922.

Smith, D. L. O., and R. A. Lohnes. 1980. Grain bin overpressures induced by dilatancy upon unloading. In ASAE Annual Conference. ASAE, St. Joseph, Mich.

Smith, R. D., and R. L. Stroshine. 1985. Aerodynamic separation of cobs from corn harvest residues. Transactions of the ASAE 28: 893–897.

Sobieski, W., Q. Zhang, and C. Liu. 2012. Predicting tortuosity for airflow through porous beds consisting of randomly packed spherical particles. Transport in Porous Media 93: 431–451.

Sokhansanj, S., A. A. Falacinski, F. W. Sosulski, D. S. Jayas, and J. Tang. 1990. Resistance of bulk lentils to air-flow. Transactions of the ASAE 33: 1281–1285.

Sokhansanj, S., W. Li, and O. O. Fasina. 1993. Resistance of alfalfa cubes, pellets and compressed herbage to airflow. Canadian Agricultural Engineering 35: 207–213.

Sokhansanj, S., and S. O. Nelson. 1988. Transient dielectric properties of wheat associated with nonequilibrium kernel moisture conditions. Transaction of the ASAE 31: 1251–1254.

Song, H., and J. B. Litchfield. 1991. Predicting method of terminal velocity for grains. Transanctions of the ASAE 34: 225–231.

Sosa-Morales, M. E., E. S. Hernández-Gómez, J. L. Olvera-Cervantes, A. Corona-Chavez, and P. Porras-Loaiza. 2015. Dielectric properties of bean weevil, grain moth and their hosts (common bean and amaranth) using the resonant cavity technique, pp. 2358-2362. In American Society of Agricultural and Biological Engineers Annual International Meeting 2015.

Stark, B., and K. James. 1982. Airflow Characteristics of Grains and Seeds. The Institution of Engineers, Australia.

Stenvert, N. L. 1972. The measurement of wheat hardness and its effect on milling characteristics. Australian Journal of Experimental Agriculture and Animal Husbandry 12: 159.

Stenvert, N. L. J. 1974. Grinding resistance, a simple measure of wheat hardness. Flour and Animal Feed Milling 156: 24–25, 27.

Stephens, L. E., and G. H. Foster. 1976. Grain bulk properties as affected by mechanical grain spreaders. Transactions of the ASAE 19: 354–358, 363.

Stetson, L. E., and S. O. Nelson. 1972. Audiofrequency dielectric properties of grain and seed. Transacrions of the ASAE 15: 180–184,188.

Stewart, B. R. 1968. Effect of moisture content and specific weight on internal-friction properties of sorghum grain. Transactions of the ASAE 11: 260–262.

Stuchly, M. A., and S. S. Stuchly. 1980. Dielectric properties of biological substances - tabulated. Journal of Microwave Power 15: 19–26.

Sun, K., F. Jian, D. S. Jayas, N. D. G. White, and P. G. Fields. 2014. Physical properties of three varieties of high-oil canola and one variety of low-oil canola. Transactions of the ASABE: 599–608.

Szabo, B. P., E. Gyimes, A. Veha, and Z. H. Horvoath. 2016. Flour quality and kernel hardness connection in winter wheat. Acta Universitatis Sapientiae, Alimentaria 9: 33–40.

Tabak, S., and D. Wolf. 1998. Aerodynamic properties of cotton seeds. Journal of Agricultural Engineering Research 70: 257–265.

Takata, K., T. Ikeda, M. Yanaka, H. Matsunaka, M. Seki, N. Ishikawa, and H. Yamauchi. 2010. Comparison of five puroindoline alleles on grain hardness and flour properties using near isogenic wheat lines. Breeding Science 60: 228–232.

Taylor, J. R. N., and K. G. Duodu 2009. Applications for non-wheat testing methods, pp. 197–234. In S. P. Cauvain and L. S. Young (eds.), The ICC Handbook of Cereals, Flour, Dough and Product Testing. Methods and Applications. DEStech Publications, Lancaster, PA.

Teunou, E., J. Vasseur, and M. Krawczyk. 1995. Measurement and interpretation of bulk solids angle of repose for industrial process design. Powder Handling and Processing 7: 219–227.

Thompson, S. A., and I. J. Ross. 1983. Compressibility and frictional coefficients of wheat. Transactions of the ASAE 26: 1171–1176.

Tinga, W. R., and S. O. Nelson. 1973. Dielectric properties of materials for microwave processing-tabulated. Journal of Microwave Power 8: 23–65.

Trabelsi, S., A. W. Kraszewski, and S. O. Nelson. 1997. A new density-independent function for microwave moisture content determination in particulate materials. In Proceedings of IEEE Instrumentation and Measurement Technology Conference.

Trabelsi, S., and S. O. Nelson. 2012. Microwave dielectric properties of cereal grains. Transactions of the ASABE 55: 1989–1996.

Turuspekov, Y., B. Beecher, Y. Darlington, J. Bowman, T. K. Blake, and M. J. Giroux. 2008. Hardness locus sequence variation and endosperm texture in spring barley. Crop Science 48: 1007–1019.

Vadivambal, R., D. S. Jayas, and N. D. G. White. 2010. Controlling life stages of Tribolium castaneum (Coleoptera: Tenebrionidae) in stored rye using microwave energy. Canadian Entomologist 142: 369–377.

Varnamkhasti, M. G., H. Mobli, A. Jafari, A. R. Keyhani, M. H. Soltanabadi, and S. K. Rafiee. 2008. Some physical properties of rough rice (*oryza sativa* L.) grain. Journal of Cereal Science 47: 496–501.

Venkatesh, M. S., and G. S. V. Raghavan. 2005. An overview of dielectric properties measuring techniques. Canadian Biosystems Engineering 47: 7.15–17.30.

Wadell, H. 1932. Volume, shape, and roundness of rock particles. The Journal of Geology 40: 443–451.

Wehling, R. L., D. S. Jackson, D. G. Hooper, and A. R. Ghaedian. 1993. Prediction of wet-milling starch yield from corn by near-infrared spectroscopy. Cereal Chemistry 70: 720.

West, N. I. 1972. Aerodynamic force predictions. Transactions of the ASAE 15: 584–587.

Woodcock, C. R., and J. S. Mason. 1987. Bulk Solids Handling: An Introduction to the Practice and Technology Chapman and Hall, New York, NY.

Wu, Y. V., A. C. Stringfellow, and J. A. Bietz. 1990. Relation of wheat hardness to air-classification yields and flour particle size distribution. Cereal Chemistry 67: 421–427.

Yang, F., X. Liu, K. Yang, and S. Cao. 2009. Study on the angle of repose of nonuniform sediment. Journal of Hydrodynamics, Ser. B 21: 685–691.

Yang, X., and D. L. Williams. 1990. Airflow resistance of grain sorghum as affected by bulk density. Transactions of the ASAE 33: 1966–1970.

Yelcin, I., and C. Ozarslan. 2004. Physical properties of vetch seed. Biosystems Engineering 88: 507–512.

Zaalouk, A. K., and F. I. Zabady. 2009. Effect of moisture content on angle of repose and friction coefficient of wheat grain. Misr Journal of Agricultural Engineering 26: 418–427.

Zewdu, A. D., and W. K. Solomon. 2007. Moisture-dependent physical properties of tef seed. Biosystems Engineering 96: 57–63.

Zhang, Q., and X. Cheng. 2014. An intermittent arch model for predicting dynamic pressures during discharge in grain storage bins. Transactions of the ASABE 57: 1839–1844.

Zhong, W., K. Xu, X. Li, Y. Liao, G. Tao, and T. Kagawa. 2016. Determination of pressure drop for air flow through sintered metal porous media using a modified Ergun equation. Advanced Powder Technology 27: 1134–1140.

Zhou, X., and S. Wang. 2019. Recent developments in radio frequency drying of food and agricultural products: a review. Drying Technology 37: 271–286.

Chapter 2

Thermal Properties and Temperatures in Grain Bulks

Introduction

Cooking oil should not lead to a fire in the kitchen, bean sprouts should not be frizzed, flour and rice should be kept at cool and dry places – what is common in all these simple statements? All of these deal with heat transfer and require an understanding of thermal properties of food materials. Thermal properties of food materials are the necessary parameters required for heat transfer calculations involved in designing of preservation and processing equipment and facilities. Such basic information should be valuable not only to engineers but also to food scientists, manufacturers, and other scientists who may exploit these properties and find new uses. For example, an energy balance for a drying process cannot be estimated and the temperature profile within stored grain bulks cannot be determined without knowing the thermal properties of the stored bulk grain. Through precise study of these properties, the optimum operational conditions for both processing and storage can be defined. In this chapter, definitions and applications of terms related to thermal properties of bulk grains are introduced. In addition, mathematical models of these parameters are also given.

DOI: 10.1201/9781003186199-2

2.1 Thermal Properties

2.1.1 Heat Content and Heat Capacity

Inside grain kernels and processed products, water can be present in the following states: free, bound, adsorbed, and absorbed (for additional details refer to Chapter 3). Free water can have three phase transitions: crystallization on cooling (or ice melting on heating), vaporization on heating (or vapour liquefaction or condensation on cooling), and sublimation. Sublimation occurs under some conditions such as high vacuum in a freeze dryer. Ice melting and water evaporation and condensation are common phenomena in our daily life. The enthalpies of melting and vaporization of free water under atmospheric pressure are 334 J/g at 0°C and 2250 J/g at 100°C, respectively. The specific heats of ice and water are 2.05 and 4.18 $J \cdot g^{-1} \cdot K^{-1}$, respectively. The energy of sorbed water in grain kernels is different from that of free water. These differences inside the grain influence the grain thermal properties and may introduce some errors during measurements and calculations of thermal parameters and processes.

Heat content is a thermodynamic quantity, which equals to the internal energy of a system plus the product of its volume and pressure. During grain storage period, volume and pressure of the grain and associated air have negligible change. Therefore, the heat content for a grain bulk is the total energy of the dry mass of the grain material and the water. Heat content of water inside grain kernels is greater than that of free water. Similar phenomena can be found inside biomass materials with different moisture contents. This higher heat content of water is due to the bounding force between the water and the biomaterials. Therefore, additional heat must be added to remove the bound water. During water absorption by the grain kernels, additional heat will be released (isosteric heat). This released heat is not obvious because the grain kernels and the water will absorb the released heat.

Heat content of a grain bulk is usually estimated by assuming that: (1) water at 0°C has zero heat content; (2) grain kernels are composed of mixture of water and dry matter and the two components are not interrelated (Muir and Viravanichai 1972); (3) specific heat of dry mass inside the grain kernels is the heat content of the dry matter; and (4) heat content of water inside kernel is the difference of the specific heat of the grain at a moisture content and the heat content of the dry matter. Heat content of the dry matter can be measured at 0% moisture content of the material. The calculated heat content of water in wheat is larger than that of pure water (Fig. 2.1).

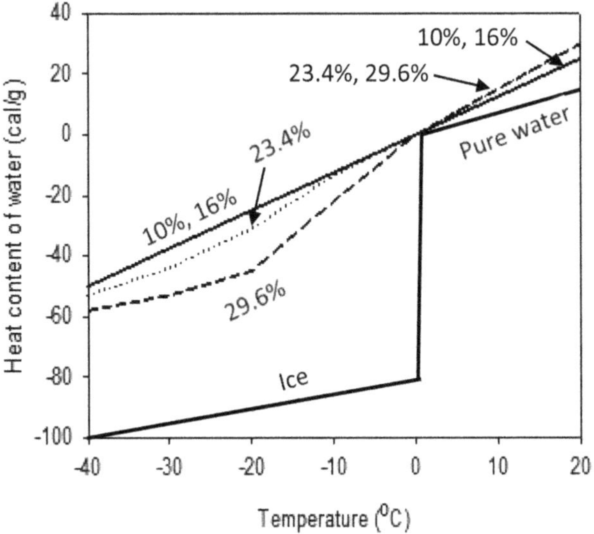

Figure 2.1 Heat content of water in wheat (cal/g = 4.186 kJ/kg at 0°C). (Reprinted with permission from Muir and Viravanichai (1972)).

Heat capacity is the ratio of an amount of heat increase to the corresponding increase in its temperature. Mathematically, heat capacity equals to a ratio of heat flow to heating rate.

$$\text{Heat capacity} = \frac{\text{Heat flow}}{\text{Heating rate}} = \frac{q/t}{\Delta T/t} = \frac{q}{\Delta T} \tag{2.1}$$

where q is the energy supplied (J), ΔT is the temperature change (°C), and t is the time (s). Heat flow and heat capacity are usually measured by differential scanning calorimeter (DSC). Heat capacity is usually used to calculate enthalpy difference between two temperatures. Heat content and heat capacity are usually calculated after other thermal properties (such as specific heat) are measured.

2.1.2 Specific Heat and Measurement

2.1.2.1 Specific Heat

Specific heat is a term used to evaluate the heat capacity of a material (such as bulk grain) because it indicates the ability of the material to hold and store the energy (heat). The value of specific heat is the amount of heat required to raise temperature by 1°C per unit mass of the material.

$$C_s = \frac{Q}{m\Delta T} \tag{2.2}$$

where C_s is the specific heat (kJ·kg^{-1}·°C^{-1}), Q is the amount of heat required (J), and m is the mass of the material (kg). This equation does not hold if a phase change is encountered, because the heat added or removed during a phase change does not change the temperature. Latent heat is related to a phase change and phase change in grain bulks is usually associated with the changes in moisture content and water phase. Therefore, the measured specific heat of grain products also usually includes the heat gain or loss associated with the phase change of water when moisture content is high and this moisture goes through phase change. At high grain moisture contents and when temperature is lower than 0°C, specific heat has a non-linear relationship with both moisture content and temperature because: (1) the free water inside the grain kernel has different physical properties from the bound water; (2) the free water will become ice or ice will become free water (phase change) at 0°C; and (3) the phase change will result in latent heat gain or release. Red spring wheat at moisture content of 23.4–29.6% releases latent heat of fusion when the wheat is cooled down to lower than 0°C (Fig. 2.2). Wheat with higher than 23% moisture content can freeze and this freezing process can kill the seed germ.

The specific heat of a grain bulk depends on the process of heat addition either at constant pressure or constant volume. Specific heats of solids and liquids do not depend on pressure much (except at extremely high pressures), and pressure changes related to agricultural materials during grain storage are usually small, therefore, specific heats of agricultural materials at a constant pressure are usually considered.

2.1.2.2 Measurement Methods of Specific Heat

Specific heat of a grain bulk can be measured using guarded plate (Mohsenin 1980), adiabatic chamber (Mohsenin 1980), comparison calorimeter (Bowers and Hanks 1962), DSC (Chakrabarti and Johnson 1972; Tang et al. 1991; Singh and Goswami 2000; Yang et al. 2002), and mixture (Ravikanth et al. 2012; Jian et al. 2013) methods. Mohsenin (1980) provides a detailed introduction for each method. The mixture method is most commonly used (Jian et al. 2013). Therefore, only the mixture method is introduced in this book because there is a slight difference from that introduced by Mohsenin (1980).

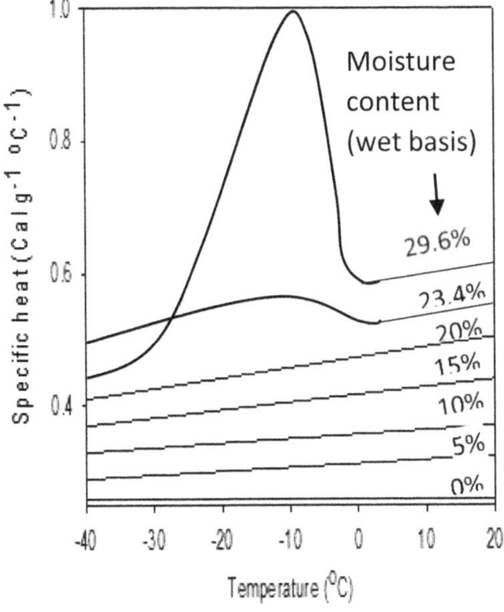

Figure 2.2 Specific heat of hard red spring wheat at different moisture contents and different temperatures (cal·g^{-1}·°C^{-1} = 4.186 kJ^{-1}·kg^{-1}·°C^{-1} at 0°C. (Reprinted with permission from Muir and Viravanichai (1972)).

To measure the specific heat of grains using the mixture method, Ravikanth et al. (2012) used a 1000-mL household Dewar flask with a rubber cork as a calorimeter (Fig. 2.3), and 4-cm-thick fibre glass to minimize the heat transfer between the flask and its surroundings. To reduce the size of the opening during the pouring of sample into the calorimeter, the rubber

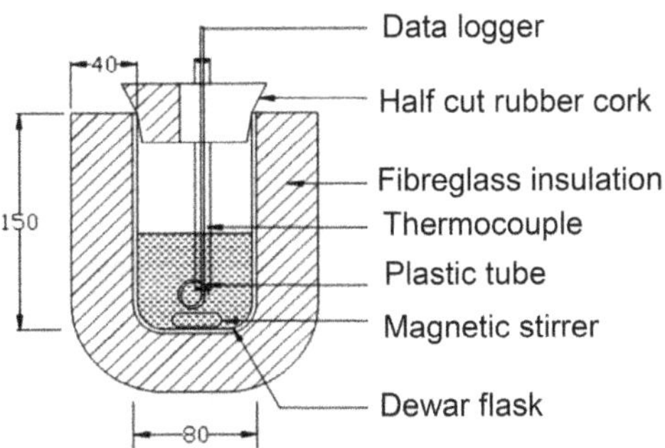

Figure 2.3 Calorimeter used to measure specific heat by using mixture method (all dimensions are in mm). (Adapted from Ravikanth et al. (2012)).

cork was cut into two equal parts, so only half was opened to introduce the sample. During the measurement, the calorimeter should be kept at a temperature similar to the temperature of the liquid inside the calorimeter, so the heat loss or gain between the calorimeter and the environment is nearly zero.

Before the specific heat of the sample is measured, the equivalent mass of water having the thermal capacity equal to that of the calorimeter system (W_e) should be determined. To determine this W_e, a desired amount of water with known temperature is poured into the calorimeter. The water is stirred continuously using the magnetic stirrer throughout the test. After the temperature inside the calorimeter is stable, W_e can be calculated using the energy conservation.

$$W_e = \frac{M_{\text{water}}\left(T_{wi} - T_{wf}\right)}{T_{ci} - T_{wf}} \tag{2.3}$$

where W_e is the mass of water having thermal capacity equal to that of the calorimeter system (kg), M_{water} is the mass of the water poured into the calorimeter (kg), T_{wi} is the initial temperature of the water (°C), T_{wf} is the finial temperature of the water and the calorimeter (°C), and T_{ci} is the initial temperature of the calorimeter (°C). Specific heat of grain samples can be measured at different temperatures and moisture contents by mixing a desired amount of water with a desired amount of grain sample at desired temperatures (Ravikanth et al. 2012; Jian et al. 2013). The specific heat can be calculated using the energy conservation.

$$C_{ss}W_s\left(T_{mf} - T_{si}\right) = C_{sw}\left(W_w + W_e\right)\left(T_{ci} - T_{mf}\right) \tag{2.4}$$

where C_{ss} is the specific heat of the sample (kJ·kg^{-1}·°C^{-1}), C_{sw} is the average specific heat of water in the measured temperature range (kJ·kg^{-1}·°C^{-1}), W_s is the mass of the sample (kg), W_w is the mass of water (kg), T_{mf} is the final temperature of the mixture (°C), and T_{si} is the initial temperature of the sample (°C). When the grain with low moisture contents is used in the mixture method, the absorption of water by the grain will generate heat of hydration (isosteric heat) because the absorbed water molecules in the grain are under constraint and have fewer degrees of freedom than free water. This generated heat will increase the rate of the water temperature rise (Sharma and Thompson 1973). However, several researchers found that dropping samples directly into water did not affect the measured specific heat value of the grain (Muir and Viravanichai 1972; Bitra et al. 2010; Ravikanth et al. 2012; Jian et al. 2013) because the grain is not at so low moisture contents.

Except water, other liquids such as toluene, N-hexane, and canola oil are also used to conduct the mixture method. The specific gravity of toluene and N-hexane is 0.86 and 0.66, respectively. The specific heat of toluene,

N-hexane, and canola oil is 0.39, 0.515, and 2.007 $kJ \cdot kg^{-1} \cdot K^{-1}$, respectively. The advantage of using other liquids is that: (1) grain kernels can easily sink in the liquid which has a smaller specific gravity than water; (2) for an equal liquid weight and the same amount of heat transferred, the other liquid with a smaller specific heat value achieves a greater temperature difference than that would be obtained with water; and (3) there is no heat generation by hydration when water is not used. For samples that float in water and also dissolve in water, the sample can be placed in capsules (Narain et al. 1978; Subramanian and Viswanathan 2003). Ice can also be used by measuring the amount of ice melted after a sample at 0°C temperature is dropped on the ice (Mohsenin 1980). However, ice melting needs time and high-level insulation of the calorimeter is required.

2.1.2.3 Prediction of Specific Heat

Specific heats of food materials depend on their composition. A quadratic equation is available to estimate the specific heats of food bulks if the components of food materials are known (Table 2.1). Experimentally determined specific heat is usually higher than the estimated value because the energy associated with the bound water, variation of component phases, and interaction of the components are not considered (Rahman 1995). Therefore, empirical models developed from experimental data are widely used (Table 2.2).

2.1.2.4 Relationship between Specific Heat and Other Physical Properties

The specific heat of a grain bulk depends on its moisture content, temperature, foreign materials, bulk density, and the factors influencing its bulk

Table 2.1 Values of Parameters in $C_c = a + bT + cT^2$ to Estimate the Specific Heat of the Component (C_c) at a Given Temperature (T). (Choi and Okos 1986; Riegel 1992)

Component	*Temperature (°C)*	*a*	*b*	*C*
Water	–40 to 0	4081.7	–5.3062	0.99516
	0–150	4176.2	–0.0909	5.4731×10^{-3}
Ice	<0	2062.3	6.0769	0
Carbohydrate	–40 to 150	1548.8	1.9625	-5.9399×10^{-3}
Protein	–40 to 150	2008.2	1.2089	-1.3129×10^{-3}
Fat	–40 to 150	1984.2	1.4373	-4.8008×10^{-3}
Ash	–40 to 150	1092.6	1.8896	-3.6817×10^{-3}

Table 2.2 Empirical Models Developed to Predict Specific Heat of Grain Bulks

Crop	*Equation*	*Source*
Canola	$C_s = (1.329 \pm 0.058) + (0.049 \pm 0.005)MC + (0.003 \pm 0.001)T$	Jian et al. (2013)
Chickpea	$C_s = -4.19 \times 10^3 + 11.9T + 2.15 \times 10^{-2} T^2 - 373MC - 0.165MC^2 + 1.38\ MC \times T$	Dutta et al. (1988)
Rough rice	$C_s = 4184(03032 + 0.00833MC)$	Morita and Singh (1979)
Soybeans	$C_s = 0.39123 + 0.0046057MC$	Alam and Shove (1973)
Wheat	$C_s = 4184(0.262 + 0.00976MC)$	Muir and Viravanichai (1972)

Abbreviations: T, temperature (°C); MC, moisture content (%, wet basis); C_s, predicted specific heat ($kJ \cdot kg^{-1} \cdot {}^\circ C^{-1}$).

density. Grain specific heat usually has a positive linear relationship with its temperature and moisture content (Figs. 2.2 and 2.4) if latent heat of fusion does not occur (Muir and Viravanichai 1972; Sharma and Thompson 1973; Subramanian and Viswanathan 2003). Dutta et al. (1988) found the specific heat of chickpea at 19–35°C and 12.4–32.4% moisture contents (d.b.) had a second-order relationship with both the temperature and moisture content. Grain quality change during storage slightly influences the specific

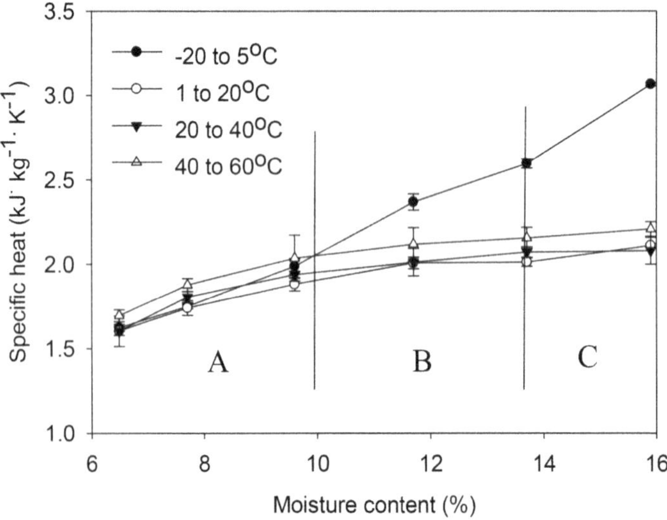

Figure 2.4 Specific heat of freshly harvested canola at different moisture contents (wb) and temperatures (Jian et al. 2013).

heat of the grain. Jian et al. (2013) found there was no significant difference between the specific heats of canola at different storage times. They concluded that the specific heat of the oilseed would not change with the spoilage of oilseed unless the kernels of the seeds were stuck together due to the growth of mould. Mohsenin (1980) found the specific heat of wheat measured by different researchers was similar if the moisture content of the wheat was the same. Subramanian and Viswanathan (2003) found the specific heats of different types of millet (finger millet, foxtail millet, poroso millet, barnyard millet, and kodo millet) were very close. The specific heat of hard red spring wheat of the same variety (cultivar) is not significantly affected by dockage or by year of harvest (Muir and Viravanichai 1972). The differences among specific heats of dusts of wheat, corn and sorghum are not statistically significant at the same moisture content (Chang and Miller 1980). Therefore, the specific heats of grain products are mainly influenced by their moisture content and temperature (Table 2.3).

2.1.3 Thermal Conductivity and Measurement

2.1.3.1 Thermal Conductivity

Thermal conductivity (k) describes the thermal conduction capacity and rate of heat transfer through a bulk food material and is the singular coefficient representing the thermal property of the materials in Fourier's law.

$$J = -k\Delta T / x \tag{2.5}$$

where J is the heat flux (W/m^2), k is the thermal conductivity (W·m^{-1}·°C^{-1}), and x is the heat transfer distance (m). Thermal conductivity of a material depends on its molecular structure at the atomic electron level (free electrons) as well as its physical lattice structure by which the molecules are held in place within the material (lattice waves). Heat transfer caused by conduction is usually interpreted either as a molecular interchange of kinetic energy or electron drift, therefore, conduction of thermal energy is influenced by these two effects and their interaction. Mobility of electrons inside a solid material is the major contribution to the thermal conductivity in metals and the effect due to lattice vibration is negligible.

Biomass with high moisture content, such as fresh apple slice or 25% MC wheat, has a high thermal conductivity because water has a high electrical conductivity. Dry food materials have low electrical conductivities because electrical conductivity of bulk porous solids depends on composition of the materials and also on many factors that affect the electric flow paths through the materials, such as void fraction (porosity), shape, size, and arrangement of void (pore) spaces, the gases contained in pores, and homogeneity of the materials and pores. The reason for a low thermal

Table 2.3 Thermal Properties of Grain Bulks

Crop (cultivar)	*P*	*MC*	*T*	C_s	*K*	α	*Sources*
Barley		6.7, 7.3	25	1392–1631	0.175–0.161	14.67–15.70	(Jangi et al. 2011)
Barley (Justo)	681–730	10–30	24	1812–2389	0.245–0.288	19.47–16.75	(Hobani and Tolba 1995)
Barley (Sahrawy)	624–697	10–30	24	1321–1671	0.193–0.239	23.06 –21.78	(Hobani and Tolba 1995)
Barley (Harrington)	673 ± 8.3	9–23	–28 to 29		0.169 –0.232		(Alagusundaram et al. 1991)
Barley (CR279)	645–689	10–30	24	1323–1872	0.210–0.285	24.33–22.72	(Hobani and Tolba 1995)
Canola	602–654	5.5–15.5	–20 to 60	1604–3065	0.072–0.112	4.99–15.74	(Jian et al. 2012, 2013)
Corn (yellow dent)	681–761	0.9–30.2	10–50	1497–2540	0.078–0.096	10.09–9.24	(Kazarian and Hall 1965)
Corn (yellow dent)	731–752	11.8–18.2	Room T		0.162–0.186		(Chang 1986)
Corn dust	240–580	8.8–17	Room T	1983–2158	0.086–0.101	800 (at ρ = 510)	(Chang and Miller 1980)
Chickpea	772–703	12.4–32.4	10–39	1464–2904	0.114–0.247	9.46–16.39	(Dutta et al. 1988)
Lentil (Laird)	790 ± 12.1	9–23	–28 to 29		0.187–0.249		(Alagusundaram et al. 1991)
Millet		10–30	100	1480–2480	0.119–0.210	16.39–21.39	(Subramanian and Viswanathan 2003)
Mung bean	774.5 ± 13.5	9.9–18.3	10–50	1630–2450	0.092–0.141	6.57–7.52	(Ravikanth et al. 2012)
Pea (Trapper)	795 ± 13.4	9–23	–28 to 29		0.187–0.257		(Alagusundaram et al. 1991)
Peanut pods (Tirupathi-2)	213–215	5.2–23.7	31.3–35.6	2100–3300	0.12–0.16	28–23	(Bitra et al. 2010)
Peanut kernels (Tirupathi-2)	635–713	5.0–30.6	31.3–35.6	1900–2800	0.15–0.19	11–10	(Bitra et al. 2010)
Peanut shells (Tirupathi-2)	58–68	3.5–28.7	31.3–35.6	2700–4100	0.11–0.18	59–67	(Bitra et al. 2010)
Rapeseed	560–730	1.0–10.0	–26 to 19	1218–2046	0.086–0.113	8.85–10.35	(Moysey et al. 1977)

(Continued)

Table 2.3 Thermal Properties of Grain Bulks (*Continued*)

Crop (cultivar)	*P*	*MC*	*T*	C_s	*K*	α	*Sources*
Rough rice (medium, long)	586–649	12–18	Room T	888–1107	0.057–0.062	10.5–8.57	(Wratten et al. 1969)
Rough rice (short grain)	632–664	11.2–23.7	Room T	1665–2025	0.113–0.127	10.97–9.0	(Morita and Singh 1979)
Rice	632–664	10.8–21.5	Room T	1666–2026	0.113–0.127	9.28–10.90	(Morita and Singh 1979)
Rice (*Oryza sativa* Linn)	500–850	10–70	50–70		0.082–0.543		(Ramesh 2000)
Soybean	719–766	0–27.5	12.0–50	1377–2343			(Alam and Shove 1973)
Sorghum	725–752	2–29	Room T	1461–2363	0.101–0.143		(Sharma and Thompson 1973)
Sorghum	728–779	11.2–18.2	Room T		0.160–0.184		(Chang 1986)
Sorghum dust	240–580	8.8–17	Room T	1961–2195	0.080–0.092	900 (at $\rho = 430$)	(Chang and Miller 1980)
Wheat (hard red spring)		1–19	–34 to 22	1339–2428			(Muir and Viravanichai 1972)
Wheat (hard red winter)	762–792	13.8–16.6	Room T		0.181–0.202		(Chang 1986)
Wheat (soft red winter)	704–730	13.5–16.7	Room T		0.167–0.190		(Chang 1986)
Wheat (soft white)	770–740	0.7–20.3	10–50	807–1214	0.065–0.077	9.26–8.0	(Kazarian and Hall 1965)
Wheat dusts	240–580	8.8–17	Room T	1927–2165	0.062–0.074	1200 (at $\rho = 280$)	(Chang and Miller 1980)

Abbreviations: ρ, bulk density of the tested grain (kg/m^3); MC, moisture content (wet basis, %); *T*, temperature (°C); C_s, specific heat (J·kg^{-1}·°C^{-1}); *k*, thermal conductivity (W·m^{-1}·°C^{-1}); α, thermal diffusivity (10^{-8} m^2/s).

conduction through grain with a high porosity is that the air inside pores in the grain bulk has low electrical/thermal conductivity as the air does not move. The thermal conductivity of grain dust at low density is about three times higher than that of air (Chang and Miller 1980). The pore structure change will influence the thermal conductivity. Therefore, thermal conductivity increases linearly with the bulk density of wheat, corn, and sorghum because contact area among seeds influences heat transfer (Chang 1986; Kara et al. 2011).

2.1.3.2 Measurement Method of Thermal Conductivity

In literature, thermal conductivity of a grain bulk (not a single kernel) is commonly reported because the practical application is for the bulk grain and it is difficult to measure thermal conductivity of a single kernel (Brown and Otten 1992). Moisture migration is the main issue when a single kernel is used, hence thermal conductivity of the single kernel can only be measured for the kernel with a low moisture content (Brown and Otten 1992). The thermal conductivity of a single soybean kernel over the moisture content of 4.5–13.6% (wb) is 0.211–0.221 $W{\cdot}m^{-1}{\cdot}°C^{-1}$ and of a single white bean over 8.0–12.5% moisture content (wb) is 0.206–0.225 $W{\cdot}m^{-1}{\cdot}°C^{-1}$ (Brown and Otten 1992). These measured data were not verified by other studies.

Both steady-state and non-steady-state (transient) methods for measuring thermal conductivity of bulk food and biological materials are documented (Mohsenin 1980; Alagusundaram et al. 1991). When the steady-state method is conducted, a time-independent heat flow passes through the sample, and boundary conditions such as surface temperature of the sample and the heat flux are monitored. The thermal conductivity of the sample is calculated by using Fourier law of heat conduction (Eq. 2.5). The heat flux (J) can be determined from the heat supplied by the source or received by the sink. The steady-state method includes the method of steady-state longitudinal heat flow, steady-state radial heat flow, and steady-state heat of vaporization. Nesvadba (1982) listed the disadvantages of the steady-state method and one of the major disadvantages of the steady-state method is that it requires a longer time than that of the transient method because it requires time to establish a linear temperature gradient in the measured materials. This method is not suitable for grain materials because moisture migration might occur during long conditioning time to establish a temperature gradient in grain (Alagusundaram et al. 1991). These disadvantages are minimized in the transient method. When a transient method is conducted, the sample is subjected to time-dependent heat flow and the temperature at one or more points within the sample or on the surface of the sample is monitored. The transient methods include Fitch method, line heat source method, frequency response method, packed bed analysis method, and thermal conductivity probe method.

Nesvadba (1982) listed the transient methods used for the measurement of thermal conductivity and thermal diffusivity of foodstuffs. DSC method is not commonly accepted due to its complexity (Pujula et al. 2016). Mohsenin (1980) provided detailed explanation on these methods and their applications. Most researchers used the line heat source method to measure thermal conductivities of different types of grains at different temperatures and moisture contents (Chandra and Muir 1971; Chang 1986; Alagusundaram et al. 1991; Denys and Hendrickx 1999). Researchers used a thermal conductivity probe, which has the same measurement principle as the line heat source method, to measure the thermal conductivities of grain and oilseeds (Choi and Okos 1986; Jian et al. 2012). The application technique of this line heat source method has been improved by these researchers. Therefore, only the line heat source method is introduced in the following section.

2.1.3.2.1 Line Heat Source Method The line heat source method is based on the principle of transient heat transfer. The sample of the grain is usually kept inside a cylindrical jar or column. Compared with the wire, which is used to produce the line heat source, the volume of the sample is large and can be assumed as an infinite homogeneous medium. This assumption simplifies the calculation and also results in a negligible error (Hooper and Lepper 1950; Chandra and Muir 1971). The line heat source (usually an electrical wire), which produces heat at a constant rate, is buried in the centre of the grain sample with a uniform initial temperature. The solution of the transient heat conduction equation in the radial direction of a cylindrical coordinate system under the assumed boundary condition gives the following equation (Hooper and Lepper 1950):

$$T_2 - T_1 = \frac{\dot{Q}}{4\pi k} \ln\left(\frac{t_1}{t_2}\right) \tag{2.6}$$

where T is the temperature (°C), and $\dot{Q}$ is the line source strength (W/m). During the measurement, the heat source is controlled (e.g., the supplied voltage and current are controlled, and the electrical resistance of the line source is measured before test) and its heat strength ($\dot{Q}$) is calculated. The temperature at the centre and mid-radius of the sample column at time t_1 and t_2 is measured as T_1 and T_2. After these measurements, the k value of the sample is calculated. The thermal conductivity probe method is same as the line heat source method, only the probe replaces the line heat source and the temperature sensors. The measured temperatures of the sample are not linear at the beginning of the measurement (less than 60 s) but gradually become linear (Jian et al. 2012). Jian et al. (2012) used the measured temperatures at 61–100 s (the linear portion of the measurement) to calculate the *k* value. Chandra and Muir (1971) used the temperatures measured at 7–9 min after heating.

Two error sources associated with Eq. 2.6 are: Only the first two terms of an infinite series are used in the derivation of this equation, and the heat source is finite in length and diameter in the actual measurement. Researchers have shown that the error caused by dropping the other terms of the infinite series is negligible (Hooper and Lepper 1950; Chandra and Muir 1971). To minimize the second error, the length of the heat source is chosen long enough to ensure radial heat flow with a negligible axial component in the test section. So, this method requires that the sample column to be long enough.

The advantage of this thermal conductivity probe method is that it eliminates moisture loss and migration within the tested materials because minimum heat is produced in a short measurement period (less than 40 min) (Jian et al. 2012). This method can be used for most food and biological materials with different moisture contents and at different pressures (Denys and Hendrickx 1999).

2.1.3.3 Prediction of Thermal Conductivity

Based on measured thermal conductivities, researchers can develop empirical models (Jian et al. 2012) (Table 2.4). For food materials, the moisture content is a predominant variable and is commonly used for estimating the thermal conductivity of the food in an empirical equation. The thermal conductivity of grain bulks has a linear relationship with moisture content (Table 2.4). At a constant moisture content, the thermal conductivity of a grain bulk also has a linear relationship with its temperature (Table 2.4) and bulk density (Chang 1986).

Theoretical models have been proposed to estimate thermal conductivity of food materials. These theoretical models include parallel model, series (perpendicular) model, Maxwell–Eucken model, and Kopelman model (Sahin and Sumnu 2006). These theoretical models are based on the models of thermal conductivity of each component inside the food materials, and usually have a lower accuracy than that of empirical models because the complexity of the chemical components of grain kernels and structural variations of kernels and grain bulks. Therefore, these theoretical models are not used widely.

2.1.3.4 Relationship between Thermal Conductivity and Other Physical Properties

The *k* value is a function of physical properties such as moisture content, bulk density, temperature, physical structure of the food product, and water binding to the starch (Ramesh 2000). The thermal conductivity of a grain bulk does not change significantly during storage period unless there is a

Table 2.4 Empirical Models Developed to Predict Thermal Conductivity of Grain Bulks

Crop	*Equation*	*Source*
Barley (Justo)	$k = (221.6 + 2.1MC)/1000$	(Hobani and Tolba 1995)
Barley (Sahrawy)	$k = (169 + 2.54MC)/1000$	(Hobani and Tolba 1995)
Barley (Cr279)	$k = (170.6 + 3.78MC)/1000$	(Hobani and Tolba 1995)
Canola	$k = (0.0671 \pm 0.0018) + (0.0017 \pm 0.0002)MC + (3.8605 \pm 0.2924) \times 10^{-5}MC \times T$	(Jian et al. 2012)
Chickpea	$k = -5.07 \times 10^{-1} + 2.55 \times 10^{-3}T - 2.13 \times 10^{-6}T^2 + 4.24 \times 10^{-3}MC - 6.56 \times 10^{-6}MC^2 + 6.48 \times 10^{-6}MC \times T$	(Dutta et al. 1988)
Rough rice	$k = 0.09999 + 0.01107MC$	(Morita and Singh 1979)
Sorghum	$k = 0.9615 \times (0.0564 + 0.000858)MC$	(Sharma and Thompson 1973)
Wheat (hard red spring)	$k = 0.000418 \times (333 + 2.86MC)$	(Chandra and Muir 1971)
Wheat (soft white)	$k = 0.9615 \times (0.0676 + 0.000854MC)$	(Kazarian and Hall 1965)

Abbreviations: T, temperature (°C); MC, moisture content (%, wet basis), k = thermal conductivity ($W{\cdot}m^{-1}{\cdot}°C^{-1}$).

significant change in the above mentioned factors. Jian et al. (2014) found the thermal conductivity of the canola seeds did not significantly decrease if the germination rate was ≥80%. The *k* value of canola was reduced by 0.0201 ± 0.0036 and 0.0269 ± 0.0047 $W{\cdot}m^{-1}{\cdot}°C^{-1}$ after 30 and 60 days of storage, respectively, due to the decrease of bulk density and germination.

Water has a higher electric conductivity than dry food materials. Thermal conductivity of food materials with enough free water has the thermal conductivity of pure water. When temperature is lower than 0°C, the *k* value of bulk food materials is influenced by the portion of the moisture crystallized to form ice (Kazarian and Hall 1965; Chandra and Muir 1971; Chang 1986; Alagusundaram et al. 1991). Thermal conductivity of food materials is between the thermal conductivity of pure water (0.614 $W{\cdot}m^{-1}{\cdot}°C^{-1}$ at 27°C) and that of air (0.026 $W{\cdot}m^{-1}{\cdot}°C^{-1}$ at 27°C). The *k* values may change by a factor of three or more at the temperature where changes like cooking, freezing, or thawing occur (when there is a phase change). Raw rice at lower moisture content had higher *k* value compared to fully cooked and dried rice (Ramesh 2000). Thermal conductivities of high moisture seeds (>18%) at ≤0°C are less than at >0°C due to the formation of ice (Alagusundaram et al. 1991). Chandra and Muir (1971) found thermal conductivity of 22.5% MC wheat had a linear relationship with temperature when temperature is >0°C, while the thermal conductivity of the 22.5% MC wheat non-linearly correlates with ≤0°C of grain temperature. Thermal conductivity of food materials only mildly depends on temperature and decreases slightly with decreasing temperature because temperature influences electric conductivities of the materials. This mild dependency is overshadowed some times by the effect of the moisture content, compaction of the material, and pore space among the grain kernels inside the grain bulk.

Thermal conductivity of food materials with high oil content is lower than of foods with low oil content (Table 2.3) because oil has a low electric conductivity than other components inside food materials. Canola with high oil content has about 23% less k value than that of rapeseed which has low oil content (Jian et al. 2012).

The difference in thermal conductivity is small between varieties of each type of grain. The moisture content and bulk density have a larger influence on the thermal conductivity than that of the varieties (Table 2.3). Hard red winter wheat has a slightly higher thermal conductivity than that of soft red winter wheat, and compaction can significantly increase thermal conductivity (Chang 1986; Jian et al. 2012). During drying of cooked rice, the k value decreased as the bulk density decreased because the surface of the material gets case hardened during drying which acts as an insulator (Ramesh 2000). The k values of rice during cooking are higher than that during drying because the surface water evaporates faster than migration of core water towards the surface during drying which causes case hardening of surface, while the space in between the rice grains is filled with water vapour (Ramesh 2000).

2.1.4 Thermal Diffusivity and Measurement

2.1.4.1 Thermal Diffusivity

The thermal diffusivity is the measurement of material's ability to conduct thermal energy relative to its ability to store thermal energy. Materials with large thermal diffusivity will respond quickly in changes of its temperature while materials with small thermal diffusivity will change its temperature more slowly. The definition in equation form is:

$$\alpha = \frac{k}{\rho C_s} \tag{2.7}$$

where α is the thermal diffusivity (m^2/s), and ρ is the bulk density (kg/m^3). A material with high k and low ρ and C_s will have a fast heat diffusion rate.

Table 2.5 Comparison of Thermal Properties of Some Materials under Grain Storage Conditions

				α		
Material	*P*	C_s	*k*	*V*	*R*	*Source*
High oil canola (9.5% MC, 25°C)	475	1853	0.092	10.5	133	(Jian et al. 2013)
Hard wheat (9.2% MC, 26.1°C)	849	860	0.084	11.5	122	(Kazarian and Hall 1965)
Wood (20°C, particleboard)	533	1730	0.164	17.9	78	(Adl-Zarrabi and Bostrom 2004)
Cast concrete (dense)	2100	840	1.400	79.4	18	(Integrated Environemnted Solutions Limited 2000)
Glass fibre quilt	12	840	0.040	396.8	4	(Integrated Environemnted Solutions Limited 2000)
Steel siding – HF-A3	7690	418	44.970	1399.0	1	(Integrated Environemnted Solutions Limited 2000)

Abbreviations: ρ, bulk density (kg/m^3); C_s, specific heat (J·kg^{-1}·°C^{-1}); *k*, thermal conductivity (W·m^{-1}·°C^{-1}); α, thermal diffusivity; *V*, value of the thermal diffusivity (10^{-8} m^2/s); *R* = diffusivity of the steel siding – HF-A3/ diffusivity of the material.

For example, the thermal conductivity of the glass fibre quilt (a common building insulation material) is about 0.04 W·m^{-1}·°C^{-1}, which is lower than wheat and canola; however, the temperature of a bin holding the glass fibre will change about 38 times faster than a bin holding canola because the glass fibre has a lower bulk density, hence higher thermal diffusivity than canola (Table 2.5). Therefore, insulation of bin walls will not influence the grain temperature at the locations which are more than 15 cm away from the bin walls. The thermal diffusivity of canola is about 1.1–1.5 times of the wheat (Tables 2.3 and 2.5), hence wheat cools faster in fall and also warms faster in spring than canola.

2.1.4.2 Measurement Method of Thermal Diffusivity

Thermal diffusivity of a material can be determined using a transient method along with the basic equations of heat transfer (Hobani and Tolba 1995).

The methods of thermal diffusivity measurement include thermal conductivity probe, cylindrical object and time-temperature data (Moysey et al. 1977, Hobani and Tolba 1995), spherical object and time-temperature data, and use of charts and graphical solution. Mohsenin (1980) provided detailed explanation of these methods and their applications. The methods used for measuring the thermal conductivity can be used to measure the diffusivity (Nesvadba 1982). The thermal diffusivity of a material is usually calculated by using measured bulk density, specific heat, and thermal conductivity of the material in most of the reported studies in literature.

2.1.4.3 Relationship between Thermal Diffusivity and Other Physical Properties

Thermal diffusivity is directly related to the bulk density, specific heat, and thermal conductivity. Therefore, any factor influencing these properties also affects the thermal diffusivity. Due to this reason, diffusivity of grain bulks has a wide range (Table 2.3) and there is no empirical equation developed to predict the thermal diffusivity instead of using Eq. 2.7. Due to this large variation of thermal diffusivity of a grain material, stored bulk grain at different locations could have different thermal diffusivity.

2.1.5 Latent Heat of Vaporization and Isosteric Heat

2.1.5.1 Definition

Moisture removal from biomaterials depends on the energy required for water evaporation, binding energy of water to the materials, and the water transfer inside the material. The binding energy of water depends on the nature of water interactions with the constituents of the materials, and is related to the isosteric heat of sorption and latent heat of vaporization of the material. Many terms related to these energies are used in literature and some terms overlap.

The latent heat of vaporization of the material is the total energy required to vaporize water from the material. The latent heat of vaporization is usually expressed as the ratio of the latent heat of vaporization of the material to the latent heat of the free water at the same temperature. When water is absorbed or bound by the solid material inside grain kernels, the kinetic, rotational, and vibrational energies of the water are all lower than those of free water. Therefore, when grain is dried (moving out the water), more energy is required. The total energy required to remove a unit mas of water from grain kernels, termed as the differential heat of sorption or as latent heat of vaporization of the material, can be partitioned into two components: namely, the latent heat of vaporization of free water (λ_{water}), and the differential heat of wetting (or net isosteric heat of sorption, Q_s). The integral heat of sorption is the heat evolved when one kilogram of the material at a given moisture content is completely wetted (kJ/kg). The differential heat of wetting is the decrease in energy of sorbed water expressed as kJ/kg of water that is sorbed at constant moisture content (Q_s is a negative value). When water is sorbed by grain, the water will release heat and it is expressed as a negative value for the material. The isosteric heat of sorption (Q_{st}) is the heat evolved when one gram of water vapour is absorbed by an infinite mass of the material at a given moisture content and temperature. The latent heat of vaporization (λ_{grain}) and isosteric heat of sorption (Q_{st}) can be expressed as:

$$\lambda_{\text{grain}} = \lambda_{\text{water}} - Q_s \tag{2.8a}$$

$$Q_{st} = Q_s + \lambda_{\text{water}} \tag{2.8b}$$

$$\lambda_{\text{grain}} - Q_{st} = -2Q_s \tag{2.8c}$$

where λ_{grain} is the latent heat of vaporization of grain sample (kJ/kg), λ_{water} is the latent heat of water (kJ/kg), Q_s is the net isosteric heat of sorption or termed as desorption specific bonding enthalpy of monolayer water (kJ/kg), and Q_{st} is the isosteric heat of sorption (kJ/kg). Therefore, latent heat of vaporization of the materials is higher than the isosteric heat of sorption because the net isosteric heat of sorption is a negative value. The latent heat of vaporization of the material is usually used to measure the total energy required to remove the water (including free, absorbed, adsorbed, and bound water) from the biomaterial, while the isosteric heat of sorption is used to describe the involved energies when water is sorbed by the material. The isosteric heat of desorption is also used to describe the total energy required to remove the water from the biomaterial. The isosteric heat of desorption is higher than the corresponding isosteric heat of sorption when the material is dry (Iglesias and Chirife 1976; Rizvi 1986; Tsami 1991). The isosteric heat of sorption and latent heat of vaporization are a good estimation of the minimum amount of heat required to remove a given amount of water from the material because both are related to the net isosteric heat of sorption. Therefore, these terms are related and used for different purposes even though an individual researcher may use one of these terms.

In general, the net isosteric heat of sorption can be viewed as the release of the kinetic energy from the molecules of the bulk phase being sorbed onto the surfaces of the adsorbent. Therefore, the net isosteric heat of sorption measures the binding energy of the forces between the water vapour molecules and the solid. It yields information for the understanding of the sorption mechanism. Moreover, it can help to detect the type of water binding that is occurring at a given moisture content.

2.1.5.2 Factors Influencing Net Isosteric Heat of Sorption

Temperature influences the net isosteric heat of sorption because a higher temperature will decrease the binding energy between molecules. Therefore, net isosteric heat of sorption also determines the temperature dependence of water activity of a biological material. Aguerre et al. (1988) found the net heat of adsorption of cereal grains at various moisture contents decreased slightly with increasing temperature, and therefore a strong dependence of the net heat of adsorption of cereal grains on temperature cannot be expected, contrary to reports for grain sorghum (Rizvi and Benado 1983).

The primary factor affecting the net isosteric heat of sorption is the water activity (relative humidity) of the material, so the moisture content of the material. Based on the assumption of the Brunauer–Emmett–Teller (BET) theory, the net isosteric heat of sorption during drying should be constant up to monolayer coverage and then suddenly increase because the energy required is high when only "bound" water is left inside the kernel (Fig. 2.4). Generally speaking, the net isosteric heat of sorption for low moisture contents is higher than the value predicted by the BET theory, and also falls more gradually with the increase of moisture content, indicating the gradual change from monolayer to capillary water (Fig. 2.4). Based on this observed net isosteric heat of sorption, a desorption isotherm can be divided into three regions (Fig. 2.4). In region C, the water molecules bind less firmly in the multiple layers of water region (free water), the net isosteric heat of sorption is lower due to the existence of capillary water. After the free water is removed, more water molecules will be evaporated from the monolayer, the vaporization enthalpy is gradually increased in region B. In region A, most water molecules are evaporated from the monolayer, hence the highest enthalpy of vaporization. Haque et al. (2007) found the net isosteric heat of desorption was higher than that for the adsorption within a moisture content range of 12–20% for all of the rice kernels. The net isosteric heat of a material at lower moisture contents is much higher than that at higher moisture contents.

The net isosteric heat of sorption might be influenced by the structural modifications that takes place during desorption (Iglesias et al. 1976; Iglesias and Chirife 1976). During desorption, binding of the sorbates through co-operative binding or to entrapment effects increase the overall energy. This is one of the reasons of the hysteresis. Therefore, the same materials with different drying and rewetting histories might have a slightly different net isosteric heat of sorption. However, this difference is negligible compared to the entire energy used during drying.

2.1.5.3 Determination of Net Isosteric Heat of Sorption

In scientific terms, the net isosteric heat of sorption is the difference between the molar enthalpy of the gaseous phase and the differential enthalpy of the adsorbed phase. Therefore, the net isosteric heat is determined using the phase information at specified temperature, pressure, and concentration or the amount adsorbed with the help of the Clausius–Clapeyron equation and Arrhenius relationship.

2.1.5.3.1 Clausius–Clapeyron Equation and Sorption Isotherm The heat of adsorption depends on the environmental conditions under which the adsorption or desorption occurs. For most applications at normal environmental temperatures and pressures, heat of adsorption can be treated as a function of humidity alone (or water activity) because the net isosteric heat of sorption is strongly related to the state of water in biomaterials. Therefore, the net isosteric heat of sorption can be calculated by Eq. 2.8.

To find Q_{st}, the Q_s should be determined first under most situations. Theoretical prediction of the net isosteric heat of sorption is not possible because of the complexity of the physical and chemical structure of biomaterials. The Clausius–Clapeyron equation (refer to Chapter 5) is the most used method to calculate the Q_s. To use Clausius–Clapeyron equation, the following conditions are assumed: (1) pure system; (2) constant moisture content during the a_w measurement across different temperatures; (3) the adsorbed phase volume is negligible as compared to the gaseous phase and it behaves as an ideal gas; and (4) the net isosteric heat of sorption is constant over the applied temperature range. However, food products are not pure systems; due to various interactions between food components and water molecules, irreversible changes can occur. Thus, it is usually suggested to determine a_w of food samples with at least 10°C interval.

After simplification and application of the associated boundary conditions, the Clausius–Clapeyron equation can be written as (Labuza et al. 1985; Lomauro et al. 1985):

$$Q_s = \frac{RT_1T_2}{T_2 - T_1} ln \frac{a_{w2}}{a_{w1}} \tag{2.9}$$

where R is the gas constant (8.31451×10^{-7} J·K^{-1}·mol^{-1}), and a_w is the water activity (decimal), and a_{w2} and a_{w1} are the water activities at temperatures T_2 and T_1 (K), respectively. The determined isotherm of a material could be used to calculate the net isosteric heat (Q_s) because the isotherm equations provide the values of a_{w1} and a_{w2} at T_1 and T_2, respectively. The BET model is the most used equation to find the a_{w1} and a_{w2} at T_1 and T_2 because BET model can be used to associate the water status of the material at different water activities.

The relationship between the isosteric heat and the differential entropy (ΔS_d) of adsorption is given by:

$$(-ln a_w)_M = \frac{Q_{st}}{RT} - \frac{\Delta S_d}{R} \tag{2.10}$$

where M is the moisture content (%, or decimal, wet basis), and ΔS_d is the differential entropy (J/K). By plotting $ln(a_w)_M$ versus 1/T, Q_{st} can be determined from the slope, and ΔS_d from the intercept.

Tsami et al. (1990) developed the following empirical equation:

$$Q_s = \Delta h_0 e^{\left(-\frac{M_e}{M}\right)} \tag{2.11}$$

where Δh_0 is the net isosteric heat of sorption of the first molecules (when $M_e = 0$) of water in the food (kJ/mol), and M_e is the equilibrium moisture content (kg/kg dry solid, decimal). Values of Δh_0 at different moisture contents (M) for several foods have been reported in the literature (Al-muhtaseb et al. 2002).

The Clausius–Clapeyron equation is based on the assumption that the material is at constant moisture content and invariant of Q_s at different temperatures, which can hold for pure systems at low temperatures, and is convenient assumption, as it allows an easy calculation of the net isosteric heat from the sorption isotherms. Irreversible changes in the binding properties at high temperatures result in temperature dependence for almost all food materials (Labuza et al. 1985).

The Q_s found using Eqs. 2.9–2.11 will be only valid under monolayer water adsorption condition. The use of the Clausius–Clapeyron equation with data from only two or three isotherms may lead to significant errors in the calculation of the sorption heat (Aguerre et al. 1988) because small errors in the interpolation of water activity values lead to very significant errors in the net isosteric heat calculation (Iglesias and Chirife 1976). Therefore, the determined Q_s has been considered more qualitative than quantitative because of measurement errors in determination of sorption isotherms and approximations in calculating the net isosteric heat of sorption from the isotherms (Mulet et al. 1999).

2.1.5.3.2 Riedel Equation and Calorimetric Technique Riedel (1977) modified Eq. 2.9 as:

$$ln\frac{a_{w2}}{a_{w1}}\frac{T_1T_2}{T_2 - T_1}\mathrm{R} = Ce^{-BM} = Q_s \tag{2.12a}$$

$$Q_{st} = Q_s + \lambda_{\text{water}} = Ce^{-BM} + \lambda_{\text{water}} \tag{2.12b}$$

where C is the constant. Another method to find the net isosteric heat is the calorimetric technique with Eq. 2.12. Calorimetric method is less cumbersome than sorption isotherm technique and requires small samples. The calorimetric technique uses a DSC meter which measures enthalpy needed to remove water and the weight loss during the heating process of the samples with different moisture contents. Any gain or loss of energy is recorded as the equipment scans the sample by heating it at a controlled rate of temperature rise over a selected temperature interval. The meter generates two thermograms: DSC enthalpy evolved and weight changed verse temperature (°C s^{-1}) (Sanchez et al. 1997). The vaporization energy per mass unit of water evaporated in the sample can be calculated by dividing the average enthalpy by the average weight loss of the water. The average enthalpy is obtained from the DSC enthalpy thermogram, and the average weight loss of the water is determined from the DSC weight thermogram. The measured average enthalpy and average weight loss are usually obtained by integration of the thermograms (Sanchez et al. 1997). This calculated Q_{st} is moisture content dependent, this calculation can be termed as:

$$\overline{Ec_i} = \frac{\int_0^M Q_{st} dM}{\int_0^M dM} \tag{2.13}$$

where $\overline{Ec_i}$ is the cumulated average vaporization enthalpy (kJ/kg). Substituting Eq. 2.12 into Eq. 2.13 and integrating, the following expression is obtained:

$$\overline{Ec_i} = \frac{C}{B}\left(\frac{1-e^{-BM}}{M}\right) + \lambda_{\text{water}} \tag{2.14}$$

where B is the constant. By using the measured $\overline{Ec_i}$ and the M, the parameter B and C can be regressed by using Marquardt procedure. Sanchez et al. (1997) found the two methods (using the isotherm equation or the DSC technique) had the similar values of the Q_s.

2.1.5.3.3 Arrhenius Relationship Another method to find the Q_s with the application of BET equation is to use the classic Arrhenius relationship (Figura and Teixeira 2007). The constant C in BET equation is temperature dependent and follows the Arrhenius relationship, which characterizes the temperature dependence of most chemical, physical, biological, and microbiological reactions. The Arrhenius relationship is:

$$lnC = lnM_m + \frac{Q_s}{R} \cdot \frac{1}{T} \tag{2.15}$$

where M_m is the constant in BET equation. The Q_s can be found by regression of the $1/T$ against the lnC after the BET equation is found. For a multilayer adsorption, extra energy is required for the second, third, and additional multiple layers. With each additional layer, less and less bounding enthalpy is needed until the adsorbed water becomes freely available to behave as free water and only the enthalpy of vaporization or condensation applies. For multilayer water adsorption, GAB model can be used to calculate the net heat of adsorption.

$$\Delta h_{s,\ \text{multi}} = \Delta h_k + \lambda_{\text{water}} \tag{2.16}$$

where $\Delta h_{s,\ \text{multi}}$ is the isosteric heat of sorption of multilayer water (kJ/kg), and Δh_k is the mean desorption specific bonding enthalpy of multilayer water ($J{\cdot}kg^{-1}$). The GAB model is most appropriate when the water activity is above 0.4 and less than 0.9 (Blahovec and Yanniotis 2010). When water activity >0.9, the general form of the GAB model can be used. However, there is no study to prove the above methods can be used to find the Δh_k.

2.1.5.3.4 Othmer Equation Othmer (1940) assumed that the heats of both sorption and condensation had the same temperature dependence. This assumption is less restrictive than that in the Clapeyron method, in which both heats are taken to be constant. Based on the Clausius–Clapeyron equation, Othmer (1940) derived the following equation by using the sorption vapour pressure:

$$\frac{d(lnP_v)}{d(lnP_0)} = \frac{Q_{st}}{h_{\text{con}}} \tag{2.17a}$$

where P_v is the vapour pressure of water over the absorbent, P_0 is the vapour pressure of pure water (Pa), and h_{con} is the heat of condensation of pure water (cal/mol). h_{con} can be calculated as:

$$h_{\text{con}} = 1.47 \times 10^4 - 22.5T_K + 4.24 \times 10^{-2} T_K^2 - 4.86 \times 10^{-5} T_K^2 \tag{2.17b}$$

where T_K is the temperature (K). After integration, Eq. 2.17a can be modified as:

$$\ln P_v = \frac{Q_{st}}{h_{\text{con}}} \ln P_0 + C, \text{ and } P_v = a_w P_0 \tag{2.18}$$

By using regression between the calculated P_0 and measured P_v, the $\frac{Q_{st}}{h_{\text{con}}}$ (slope of the regression equation) can be found. $\frac{Q_{st}}{h_{\text{con}}}$ is a constant which indicates the temperature variation of the isosteric heat and the heat of water vapour condensation are the same. The net isosteric heat of sorption can be calculated as:

$$Q_s = \left(\frac{Q_{st}}{h_{\text{con}}} - 1\right) h_{\text{con}} \tag{2.19}$$

2.1.5.3.5 Commonly Used Methods The adsorption or sorption calorimetry and application of the isotherm equation are the available methods to determine the differential heat. The BET equation combined with Clausius–Clapeyron or Arrhenius relationship can be used to find the differential heat. However, the adsorption calorimetry method requires the technique to precisely measure the small quantities of heat evolved. It is difficult to obtain a homogeneous small sample that represents the product tested. The calorimetry must be calibrated. Therefore, application of the Clausius–Clapeyron equation to sorption isotherms at different temperatures is mostly used method (Table 2.6). The second widely

Table 2.6 Net Isosteric Heat of Common Crops and Speciality Crops

	Measurement Condition					
Crop	*T (°C)*	*MC*	a_w	Q_s[a]	*Method*[b]	*Source*
Barley	20–50	5.5–21	0.11	1029	1	Gely and Pagano (2012)
Chia	15, 25, 35		0.2	1865	1	Arslan-Tontul (2020)
Malt	5–45	3.9–30		849	3	Bala (1983)
Millet	20–50	7–21%	0.11	547	1	Singh et al. (2011)
Potatoes	4–50		0.4	1050	1 and 2	Sanchez et al. (1997)
Rough rice	0–50	10–24		2100	1	Sun (1998), Öztekin and Soysal (2000)
	10–70	4–17		2090	3	Aguerre et al. (1988)
Rapeseed	5–80	5–25		700	1	Sun (1998), Öztekin and Soysal (2000)
Wheat (durum)	0–50	10–26		350	1	Öztekin and Soysal (2000), ASABE (2016)
Wheat (soft)	0–50	10–20		900	1	Öztekin and Soysal (2000), ASABE (2016)
Wheat (hard)	0–50	10–25		700	1	Öztekin and Soysal (2000), ASABE (2016)
Yellow dent corn	0–50	5–24		730	1	Öztekin and Soysal (2000), ASABE (2016)
Shelled corn	0–50	5–24		750	1	Sun (1998), Öztekin and Soysal (2000)

Abbreviations: T, temperature (°C); MC, moisture content (dry basis, %); a_w, the lowest water activity measured in the study.

[a] Maximum net isosteric heat (kJ/kg) at the lowest moisture content tested. The Q_s is estimated from the published isosteric heat minus the latent heat (2450 kJ/kg).

[b] The measurement method: 1 = Clausius–Clapeyron equation and sorption isotherm, 2 = Riedel equation and calorimetric technique, 3 = Othmer equation.

used method is the application of calorimetric techniques to Riedel equation. There is a good agreement among these two methods (Sanchez et al. 1997; Mulet et al. 1999). Aguerre et al. (1988) found Othmer equation provided the similar result as that of Clausius–Clapeyron equation. However, isotherm models at a single temperature do not predict the heat of sorption well, while multi-temperature isotherm models can, even though the single-temperature models fit the isotherm data better than multi-temperature models (Leuk et al. 2016).

2.1.5.4 Net Isosteric Heat and Latent Heat of Vaporization of Cereal Grain

Values of the net isosteric heats of adsorption and desorption of foods are available in literature. The foods include fruits, vegetables, protein foods (e.g., cheese, chicken, yoghurt, muscle, and fish protein concentrate), and spices (Iglesias et al. 1976; Iglesias and Chirife 1976; Tsami 1991; Wang and Bremmam 1991). Net isosteric heats of cereal grains are also available in literature (Aguerre et al. 1988; Öztekin and Soysal 2000; Haque et al. 2007). Net isosteric heats of desorption are higher than net isosteric heats of adsorption within 12–20% moisture contents of cereal grains, while below this moisture content, net isosteric heat of desorption increases rapidly. The differences between desorption and adsorption isosteric heats of shelled corn are larger at lower moisture contents, increasing from 2.6% at 14% moisture content to 5.1% at 5% moisture content (Öztekin and Soysal 2000). When moisture content is higher than 20%, the isosteric heats are close to the latent heat of vaporization of free water because vaporization of moisture from the grain mainly involves the removal of free water when the cereal grain moisture content is 18% or greater (Cenkowski et al. 1992). Lowering the moisture content to less than 11% increases the amount of energy required to remove water from the product, partially due to the greater resistance to moisture movement from the bound and absorbed water inside the kernels to the surface of the product. Wheat and shelled corn at 13% moisture content, the heat required for vaporization of moisture is about 1.06 times that for the vaporization of free water, while is about 1.15–1.20 times when at 9% moisture content (Johnson and Dale 1954). At 10% moisture content, isosteric heat of desorption of wheat (soft, hard, and wheat), corn (yellow dent and shelled), and rice (rough or de-hulled rice) were 3351–2789, 3391–2882, and 3346–3086 kJ/kg, respectively (Öztekin and Soysal 2000). The isosteric heat of desorption of malting barley varies from 3479 to 2393 kJ/kg (Gely and Pagano 2012). Therefore, the isosteric heat of desorption are similar among different cereal grains and their varieties under the similar water activities (Li 2012). The ratio of latent heat of vaporization of grain to latent heat of water can be described by the following equation:

$$\frac{\lambda_{\text{grain}}}{\lambda_{\text{water}}} = 1 + Be^{CM} \tag{2.20}$$

Different crop types and cultivars can have different B and C values (Cenkowski et al. 1992).

The maximum net isosteric heat of sorption of grain reported in literature is in the range of 350–2000 kJ/kg (Table 2.6). These reported values are not comparable because: (1) different isotherm models are used to estimate the Q_s and different isotherm models have different prediction accuracies; (2) different methods can be used to determine the isotherm models and different methods might have different prediction accuracies; (3) the tested materials might not be at the same water activity; and (4) the water might not be at the same status even at the same water activity due to the different structures of the materials. Different grain types or classes in the same grain type at dry moisture contents have significantly different net isosteric heats (De Oliveira et al. 2010). The maximum values are the net isosteric heat at the lowest moisture content tested and grain is usually not dried to this low moisture content, and the safe storage grain moisture content is different for different grain types. The energy associated with the irreversible processes is small compared with the overall energy changes during drying. Therefore, the net isosteric heat (including its maximum value) is usually neglected in the estimation of the heat requirements for the grain drying (Iglesias et al. 1976; Iglesias and Chirife 1976).

2.1.6 Convective Heat Transfer

When air is forced though grain bulks, convection heat transfer occurs on the surface of grain kernels. The rate of heat transfer between a solid surface and fluid is usually computed using the equation suggested by Isaac Newton in 1707.

$$q_c = h_c A (T_s - T_\infty) \tag{2.21}$$

where q_c is the energy transferred by convection (W), h_c is the convective heat transfer coefficient ($W{\cdot}m^{-2}{\cdot}K^{-1}$), A is the area (m^2), T_s is the surface temperature of solid (K), and T_∞ is the temperature of the fluid surrounding the solid (K). The convective heat transfer coefficient on surface of grain kernels is defined by using this Newton equation ($W{\cdot}m^{-2}{\cdot}K^{-1}$). For grain drying and aeration, volumetric convective heat transfer coefficient in grain bulks is usually needed because calculation of heat transfer on each single kernel is difficult due to the huge number of kernels. The volumetric convective heat transfer coefficient of a grain bulk is defined as the rate of heat transfer between the grain bulk and moving intergranular air, and is estimated based on the volume of the grain bulk (W·m^{-3}·K^{-1}). These two coefficients are a complicated function of the fluid, the thermal properties of the fluid medium and grain bulk, and physical properties of the grain kernels.

2.1.6.1 Determination of Convective Heat Transfer Coefficient

Direct measurement of convective heat transfer coefficient of grain bulk is difficult due to: (1) small surface area and irregular shape of a single kernel; (2) water evaporation and absorption by grain kernels during convection and this evaporation and absorption is influenced by this ventilation; (3) thermal conduction between grain kernels during convection and this conduction influences convection; and (4) uneven airflow distribution among kernels and at different locations of a single kernel. Therefore, the methods used in literature usually apply mathematical techniques to estimate it.

2.1.6.1.1 Schumann Method Schumann (1929) developed the following analytical model to estimate fluid and solid temperature distribution in a packed bed of material without mass exchange between the fluid and solid:

$$\frac{\partial T_s}{\partial Z} = T_a - T_s \tag{2.22a}$$

$$\frac{\partial T_a}{\partial Y} = T_s - T_a \tag{2.22b}$$

$$Y = \frac{h_c x}{h_a \varepsilon v} \tag{2.22c}$$

$$Z = \frac{h_c}{h_s(1-\varepsilon)}\left(t - \frac{x}{v}\right) \tag{2.22d}$$

where T_a is the temperature of air around grain kernel (°C), x is the distance traversed by the fluid in the bed (m), h_s is the heat capacity per unit volume of the solid (kJ·m^{-3}·K^{-1}), h_a is the heat capacity per unit volume of the fluid (kJ·m^{-3}·K^{-1}), ε is the porosity of material (fraction of total volume occupied by air and water vapor), t is the time (s), and v is the superficial velocity (m/s). Equation 2.22 is based on the following assumptions: (1) negligible conduction in the fluid itself or in the solid itself; (2) uniform temperature in the fluid and solid at any place; (3) rate of heat transfer from fluid to solid at any point is proportional to the average difference in temperature between fluid and solid at that point; (4) fluid and solid have negligible volume change; (5) constant thermal properties of the fluid and solid; and (6) no mass transfer (such as water evaporation or sorption) between the liquid and kernels. Schumann (1929) provided the solution of Eq. 2.22 as:

$$\frac{T_s}{T_0} = e^{-Y-Z} \sum_{n=1}^{\infty} Z^n M_n(YZ) \tag{2.23a}$$

$$\frac{T_a}{T_0} = e^{-Y-Z} \sum_{n=0}^{\infty} Z^n M_n(YZ) \tag{2.23b}$$

$$M_n(YZ) = \frac{\partial^n M_0(YZ)}{\partial (YZ)^n} \tag{2.23c}$$

$$M_0(YZ) = I_0\left(2\sqrt{YZ}\right) = 1 + YZ + \frac{(YZ)^2}{2!} + \frac{(YZ)^3}{3!} + \cdots \tag{2.23d}$$

where T_0 is the initial temperature of the air (K). $I_0\left(2\sqrt{YZ}\right)$ is the modified Bessel function of the first kind and order zero. $M_n(YZ)$ is the Bessel Function of the first kind, and $M_0(YZ)$ is the Bessel Function of the first kind and of order zero. The values of Y and Z can be calculated using Eq. 2.22. Schumann (1929) presented these dimensionless numbers $\left(\frac{T_s}{T_0}, \frac{T_a}{T_0}\right)$ on charts (Fig. 2.5) and suggested these can be used to estimate the convective heat transfer coefficient if the expansion of fluid is negligible. Furnas (1930) also numerically solved Eq. 2.22 and extended the temperature history curves for values of Y up to 500. Researchers used the Schumann charts to estimate the h_c for different materials such as iron ore, limestone, bituminous coal, anthracite, coke, typical blast furnace charge (Furnas 1930), granitic gravel, and rock pile (Alanis et al. 1977). Boyce (1966) studied the convective heat transfer coefficient of a layer of barley. The barley bulk was pretreated with the airstream to equilibrate with the air and then cooled the barley to a lower temperature, so there was no mass exchange between the air and barley kernels during the convection test. The temperatures of air and barley at different locations were recorded at a constant time interval, and then the values of $\frac{T_s}{T_0}$, $\frac{T_a}{T_0}$, Y and Z were calculated. These calculated values were compared with the Schumann solution (Fig. 2.5) and

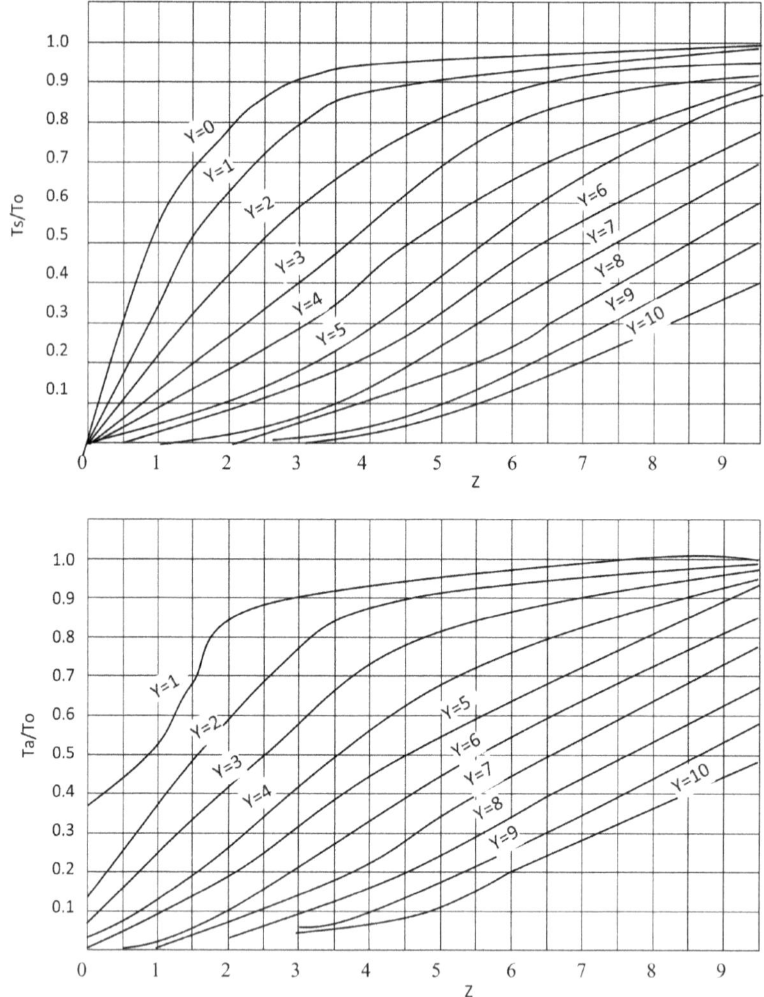

Figure 2.5 Temperature in liquid (top) and solid (bottom). (Reprinted with permission from Schumann (1929)).

the h_c was finally determined. By using this procedure and the experimental data, Boyce (1966) developed an equation to calculate the volumetric heat transfer coefficient as:

$$h_{cv} = B\left[\frac{G(T+460)}{P_{at}}\right]^{C} \tag{2.24}$$

where h_{cv} is the volume convective heat transfer coefficient (W·m^{-3}·K^{-1}), G is the mass flow rate of drying air (kg·m^{-2}·s^{-1}), T is the temperature (°C), and P_{at} is the atmospheric pressures (kPa). The constant B and C are determined by using the experimental data. Researchers found Eq. 2.24 gave an overestimation, so they modified Eq. 2.24 (Nellist 1974; Matouk 1976). Wang et al. (1979) used the same procedure as Boyce (1966) and found the heat transfer coefficient of packed bed of rice could be calculated as:

$$h_{cv} = 8.69 \times 10^4 G^{1.30} \tag{2.25a}$$

$$h_{cv} = 0.00718 G^{1.2997} \tag{2.25b}$$

Bala (1983) also used Boyce (1966) procedure and determined the volumetric heat transfer coefficient of malt as:

$$h_{cv} = 49.32 \times 10^3 G^{0.6909} \tag{2.26}$$

Bala (1983) suggested that the difference in the estimated h_{cv} between Eqs. 2.24 and 2.26 was due to the different physical properties of the crop seeds and the slightly different measurement procedure such as bed depth and the location of temperature measurement.

2.1.6.1.2 Dimensionless Analysis Gamson et al. (1943) introduced a dimensionless group (*j*-factor, j_h) representing the experimentally measured data from which heat transfer coefficient could be calculated.

$$j_h = \frac{T_1 - T_2}{\Delta T_m} \frac{S}{A} \left(\frac{C_{pa}\mu}{k} \right)^{2/3} = \frac{h_{cs}}{C_{pa}G} \left(\frac{C_{pa}\mu}{k} \right)^{2/3} \tag{2.27}$$

where j_h is the Colburn j-factor for heat transfer, ΔT_m is the mean temperature difference between the fluid and the surface (°C), S is the cross-sectional area of air flow (m^2), C_{pa} is the specific heat of the fluid (kJ·kg^{-1}·°C^{-1}), μ is the viscosity of the fluid (Pa·s), h_{cs} is the heat transfer coefficient based on surface area (W·m^{-2}·K^{-1}), and T_1 and T_2 are the temperatures at different times. Based on series of experimental data (using spherical and cylindrical catalyst carrier pellets of diameters 2.29–18.8 mm and cylindrical heights 4.78–16.9 mm under conditions of constant drying rate, at temperatures 26.7–71.1°C and at mass velocities of air 0.54–3.12 kg·s^{-1}·m^{-2}), Gamson et al. (1943) found the strong relationship between the heat transfer factor (j_h) and modified Reynolds number (d_eG/μ) as:

$$h_{cs} = 1.064 C_{pa} G \left(\frac{d_e G}{\mu} \right)^{-0.41} \left(\frac{C_{pa}\mu}{k} \right)^{-2/3} \quad Re > 350 \tag{2.28a}$$

$$h_{cs} = 18.1 C_{pa} G \left(\frac{d_e G}{\mu} \right)^{-1} \left(\frac{C_{pa}\mu}{k} \right)^{-2/3} \quad Re < 40 \tag{2.28b}$$

where d_e is the effective kernel diameter (m), and Re is the Reynolds number. Their data also showed that the ratio of Colburn j-factor for heat transfer (j_h) to Colburn *j*-factor for mass transfer (j_d) was a constant for all conditions of flow (turbulent, laminar, or transition).

$$\frac{j_h}{j_d} = 1.076 \tag{2.29}$$

where j_d is the Colburn *j*-factor for mass transfer. Based on published data and after plotting j_d against Re, Yoshida et al. (1962) found the following relationship:

$$j_d = 0.84Re^{-0.51} \quad 0.01 < Re < 50 \tag{2.30a}$$

$$j_d = 0.57Re^{-0.41} \quad 50 < Re < 1{,}000 \tag{2.30b}$$

$$Re = \frac{G}{a_v \phi \mu} \tag{2.30c}$$

$$a_v = \frac{6(1-\varepsilon)}{d_e \phi} \tag{2.30d}$$

where a_v is the area of kernel per unit volume of bed (m²), ϕ is the shape factor $\left(=\dfrac{\text{geometric mean diameter}}{\text{effective diameter}}\right)$, and ϕ is the sphericity (decimal). Equations 2.29 and 2.30 can be used to find j_h.

After summarizing published data, Yoshida et al. (1962) developed the following equations:

$$j_h = 0.61\text{Re}^{-0.41} = 0.61\left(\frac{G}{a_e \phi \mu}\right)^{-0.41} \quad 50 < Re < 1000 \tag{2.31a}$$

$$h_{cs} = 0.61C_{pa}G\left(\frac{G}{a_v \phi \mu}\right)^{-0.41}\left(\frac{C_{pa}\mu}{k}\right)^{-2/3} \tag{2.31b}$$

Bala (1983) used Eq. 2.31 and calculated the volumetric convective heat transfer coefficient as:

$$h_{cv} = 82.25 \times 10^3 G^{0.59} \text{ for malt} \tag{2.32a}$$

$$h_{cv} = 89.83 \times 10^3 G^{0.59} \text{ for barley} \tag{2.32b}$$

O'Callaghan et al. (1971) used the following equation:

$$h_{cv} = 856.8\left(\frac{GT_a}{P_{at}}\right)^{0.6011} \tag{2.32c}$$

The units of h_{cv}, G, T_a, and P_{at} in this equation are kW·m^{-3}·K^{-1}, kg·m^{-2}·s^{-1}, K, and Pa, respectively. These equations correlated the heat transfer coefficient with the air velocity. Jian and Jayas (2018) found the transient heat transfer coefficient might be smaller than the calculated by these equations because more energy transferred by convection cannot be used by grain kernels if there is not enough water in the kernel to be evaporated. Jian and Jayas (2018) suggested the Lewis number should be used to estimate the heat transfer coefficient.

2.1.6.1.3 Dimensionless Model Goyal and Tiwari (1998) suggested the following relationship between dimensionless groups:

$$\mathrm{Nu} = \frac{h_c X}{k_a} = C(GrPr)^B \tag{2.33}$$

where Nu is the Nusselt number, Pr is the Prandtl number, X is the characteristic dimension (m), k_a is the thermal conductivity of humid air (W·m^{-1}·°C^{-1}), B and C are constants, and Gr is the Grashof number. The rate of heat utilized to evaporate moisture is given as (Dunkle 1961; Tiwari and Lawrence 1991):

$$Q_e = 0.016h_c\left[P(T_g) - \gamma P(T_e)\right] \tag{2.34}$$

where Υ is the relative humidity (%, decimal), Q_e is the rate of heat utilized to evaporate moisture on surface (J·m^{-2}·s^{-1}), $P(T)$ is the partial vapour pressure at temperature T (Pa), T_g is the grain temperature (°C), and T_e is the temperature of evaporated water (°C). After substituting h_c from Eq. 2.33, Eq. 2.34 becomes:

$$Q_e = 0.016\frac{k_a}{X}C(GrPr)^B\left[P(T_g) - \gamma P(T_e)\right] \tag{2.35}$$

The evaporated moisture (water) from the bulk grain in the time interval (t) can be estimated as:

$$M_{ev} = \frac{Q_e}{\lambda}A_t t \tag{2.36}$$

where M_{ev} is the evaporated water from the grain bulk in a time period (kg). After substituting Q_e from Eq. 2.35 to Eq. 2.36 and reorganizing the equation, the following equation can be generated:

$$\ln\left[\frac{M_{ev}}{0.016\left(\frac{k_a}{X\lambda}\right)\left[P(T_g) - \gamma P(T_e)\right]A_t t}\right] = \ln(C) + B\ln(GrPr) \tag{2.37}$$

where λ is the latent heat of vaporization (kJ/kg), and A_t is the overall surface area of grain kernels in a bulk (m^2). The following parameters can be estimated from experimental data: k_a, $P(T_g)$, $P(T_e)$, γ, and A_t. The value of X can be assumed as the bed depth. The constant C and B can be found by linearly regressing the GrPr with M_{ev}. The M_{ev} is experimentally measured. After C and B are calculated, the h_c can be calculated by using Eq. 2.33.

Anwar and Tiwari (2001b) used this method to estimate the convective heat transfer coefficients during the sun drying of green chillies, green peas, Kabuli chana, onion, potato, and cauliflower. The experimental errors in terms of percent uncertainty were found to be in the range of 6 ± 16% for open sun drying and 35% for natural cooling of Kabuli chana. Akpinar (2004) used the same method and procedure to determine the convective heat transfer coefficients of mulberry, strawberry, apple, garlic, potato, pumpkin, eggplant, and onion under indoor forced convection drying conditions. The convective heat transfer coefficient of these crops was between 0.6 and 7.1 kW·m^{-2}·K^{-1}. This method is widely used by research groups in India and Turkey to determine the convective heat transfer coefficients of gooseberry (Anwar and Singh 2012), vermicelli (Sahdev et al. 2012), and other speciality crops due to its simplicity.

2.1.6.2 Selection of Determination Methods

Dimensionless analysis requires sufficient experimental data and estimation of the physical properties of both fluid and grain bed which are difficult to estimate and usually have low accuracy due to different grain storage conditions such as different dockage concentrations, filling methods, grain depths, and varieties (these factors will result in different pore size and shapes). Schumann method presupposes that the physical mechanisms are sufficiently well described in the form of partial differential equations, and the solution of these partial differential equations in the dimensionless form is compared with the experimental data to compute convective heat transfer coefficient. The final equations (for the calculation of the convective heat transfer coefficient) developed from the Schumann method are based on experimental data. Therefore, the equation will have a high accuracy. However, Schumann method requires that the grain bulk should be tested under no mass transfer condition which introduces the complexity during the test. Dimensionless model method can directly measure the convective heat transfer coefficient at any drying condition. Its calculation is also simple. Therefore, researchers usually use this method (Table 2.7). However, its accuracy might be low due to the errors caused by: (1) the estimation of the physical parameters; (2) the simplification of the dimensionless relationship (Eq. 2.33); and (3) only considering the energy used to evaporate

Table 2.7 Convective Heat Transfer Coefficient (h_c, h_{cv}) of Common Crops and Speciality Crops

	Measurement Condition					
Crop	*T (°C)*	*MC*	*v*	h_c	*Method*[a]	*Reference*
Barley	75	27.1	0.56[air]	43–59	4	Miketinac et al. (1992)
Cauliflower	S-dry	Fresh	N-C	10.0	3	Anwar and Tiwari (2001b)
Gram	80–85	30	0.4	10.9	3	Goyal and Tiwari (1998)
	25–55	30	0.4	5.2	3	Goyal and Tiwari (1998)
Green chillies	S-dry	Fresh	N-C	3.7	3	Anwar and Tiwari (2001b)
Green peas	S-dry	Fresh	N-C	8.2	3	Anwar and Tiwari (2001b)
Kabuli chana	S-dry	30	N-C	8.5	3	Anwar and Tiwari (2001b)
Malt	51–71	MC_e	0.35–0.62[m]	23.9–35.5[b]	1	Bala (1983)
	51–71	MC_e	0.35–0.62[m]	44.3–62.0[b]	2	Bala (1983
Onion flakes	S-dry	Fresh	N-C	14.0	3	Anwar and Tiwari (2001b)
Potato slices	S-dry	Fresh	N-C	26.0	3	Anwar and Tiwari (2001b)
Wheat	80–85	20	0.4	9.7	3	Goyal and Tiwari (1998)
	25–55	20	0.4	3.7	3	Goyal and Tiwari (1998)

Abbreviations: air, air velocity (m/s); fresh, fresh harvest crops; *m*, mass flow rate ($kg \cdot m^{-2}s^{-1}$); MC, moisture content (dry basis, %); MC_e, grain is tested at the equilibrium moisture content with the drying air; S-dry, fresh grain being sun dried; *T*, temperature (°C); *v*, superficial velocity (m/s),

[a] Method used to determine the convective heat transfer coefficient; 1 = Schumann method; 2 = Dimensionless analysis; 3 = Dimensionless model; and 4 = mathematical modelling estimation.

[b] Volume convective heat transfer coefficient ($kW \cdot m^{-3} \cdot K^{-1}$).

the water (Eqs. 2.35 and 2.36). The grain and air temperatures and relative humidities at different locations in a thin layer bed are different. Using different temperature in Eq. 2.37 will produce different C and B values resulting in different h_c.

Bala (1983) compared the Schumann method with the dimensionless analysis method and found the dimensionless analysis gave an estimate about 1.6–1.8 times higher than the Schuman method. Nellist (1974) also had the same conclusion. To the best of our knowledge, there is no study to compare the accuracy among the three methods. Different researchers developed or used different equations to calculate the convective heat transfer coefficient. Barker (1965) studied the relationship between j_h and Re and found that the agreement with published data was off by a factor of about 2 over Reynolds numbers ranging from 10 to 100,000 and especially in the most common range 200–4000.

The major reasons for the difficulty and difference in obtaining a single generalized experimental correlation and theoretical or semi-empirical models to evaluate the convective heat transfer coefficient in packed beds are: (1) difficulty to evaluate the properties of the fluid such as the air velocity in grain bulks; (2) errors associated with the evaluation of the properties of the grain bulks such as effective area of convection; and (3) different temperature distribution in the grain bed.

2.1.6.3 Convective Heat Transfer Coefficient of Crop Grain

It is difficult to categorize the convective heat transfer coefficient based on crop types or drying methods because of limited studies and use of different measurement methods, which have different accuracies. The convective heat transfer coefficient during grain drying and aeration should be higher than 3.7 $W{\cdot}m^{-2}{\cdot}K^{-1}$ if the drying temperature is higher than 25°C (Table 2.7). The heat transfer coefficient inside bulk grain in literatures is 14–120 $W{\cdot}m^{-2}{\cdot}K^{-1}$ and different researchers used different assumptions. Anwar and Tiwari (2001a) studied the drying of six crops (green chillies, green peas, white gram, onion flakes, potato slices, and cauliflower) and found the values of convective heat transfer coefficients varied between 1.31 and 12.80 $W{\cdot}m^{-2}{\cdot}°C^{-1}$, and between 1.25 and 10.94 $W{\cdot}m^{-2}{\cdot}°C^{-1}$ in indoor open and closed conditions, respectively. Miketinac et al. (1992) used five mathematical models simulating the process of simultaneous heat and mass transfer in a thin layer drying of barley. They found that even though the mass transfer was small and the same, the heat transfer coefficient was from 43 to 59 $W{\cdot}m^{-2}{\cdot}°C^{-1}$ depending upon the drying model used. Goyal and Tiwari (1998) reported the convective heat transfer coefficients for wheat and gram at thin layer drying condition were 9.7 and 10.9 $W{\cdot}m^{-2}{\cdot}K^{-1}$, respectively, while were 3.7 and 5.2 $W{\cdot}m^{-2}{\cdot}K^{-1}$ at natural air cooling condition, respectively.

2.2 Temperature of Grain Bulks

This section focuses on the grain temperature patterns under safe storage conditions (without heat production of grain, insects, and fungi due to their respiration) inside metal cylindrical bins, which are common grain storage structures in North America and some parts of the world (Fig. 2.6). These presented temperature patterns can be applied to a large amount of grain bulk (e.g., more than 100 tonne). These temperature patterns also affect the moisture migration in the granary during storage (refer Chapter 6).

2.2.1 Temperature Patterns

2.2.1.1 Temperature Lag

Grain temperatures at the places within less than 15 cm from the walls follow the wall temperature of metal bins, and the following factors influence the bin wall temperatures: Daily weather temperature, solar radiation, wind and wind direction, cloud cover, precipitation, shade on the walls, and the ground condition surrounding the bin (such as concrete, snow, or grassed ground). The temperature on walls following the daily ambient temperature is the combined effect of these factors. As the distance increases from the bin wall to the centre of the bin, grain temperature is gradually not influenced by the daily ambient weather temperature, while is influenced by the seasonal temperature (Fig. 2.7). The minimum average daily weather

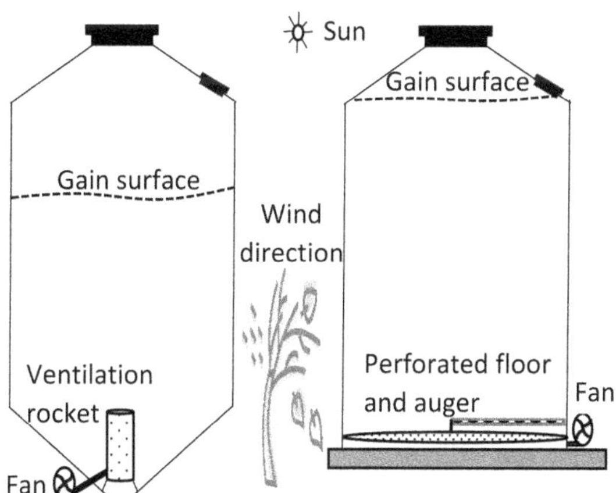

Figure 2.6 Schematic diagrams of common configurations of farm storage systems in North America.

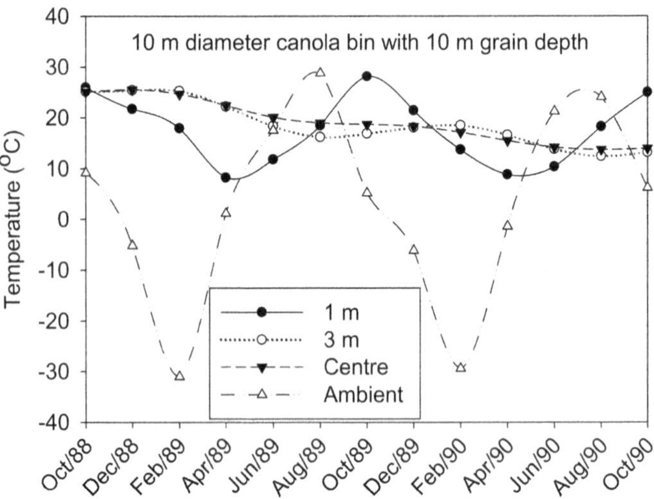

Figure 2.7 Wheat temperature at the half grain depth. Temperature was simulated by using Winnipeg weather data and the model developed by Jian et al. (2005). The initial grain condition: 25°C, 12.5% moisture content, clean, and no insect and mould infestation. In the graph, 1 m, 3 m, centre, and ambient are the grain temperatures at 1 and 3 m from the south wall, at the bin centre, and the average daily ambient temperature, respectively.

temperature occurred in February 1989 at Winnipeg, Canada, while the minimum temperature of the grain at the centre of the 10 m diameter bin occurred in August 1989. The same trend occurs between the maximum temperatures of the grain and that of the ambient weather. This time delay is termed as temperature lag (Fig. 2.7). This time delay is caused by the low thermal diffusivity of the grain, and increases with the increase of the size (diameter) of the grain bulk. Temperature lag occurs at any location and the centre has the largest temperature lag (Fig. 2.7). The time delay is near zero at the walls and increases with the increasing distance from walls towards the centre.

2.2.1.2 Temperature Gradient Patterns without Aeration

2.2.1.2.1 Initial Temperature Grain temperature at the harvest can be equal to or higher than the ambient temperature because solar radiation can increase the kernel temperature. On sunny days in August in Manitoba, kernel temperature at the head of wheat plants is 8°C higher than the ambient temperature, 3°C higher in the middle, and 1°C higher at the bottom of the plants (Prasad et al. 1978). Measured temperatures of wheat were

8°C higher in combine hoppers, 5°C higher in truck boxes, and 8°C higher in bins on sunny days (Prasad et al. 1978). Canola temperature was 5°C higher in pods on top of the plants, 2°C higher in the plant bottom, 4°C higher in combine hoppers and truck boxes, and 5°C higher in bins (Prasad et al. 1978). The grain temperatures were slightly higher than the ambient temperatures on cloudy days and the temperature increase depends on weather conditions. When canola is harvested at night, the increase is 0°C in swathes, 2°C in combine hoppers, 1°C in trucks, and 3°C in bins (Williamson 1964). The temperature increase during night is caused by friction during threshing in combine, transfer through augers, or respiration heat if grain is damp or wet. Possible cooling could occur as air is blown through the grain in combines or during grain loading and unloading. This cooling effect is usually negligible. Researchers in different countries found the same trend (Williamson 1964, Loschiavo 1985). This initial grain temperature is influenced by the grain kernel moisture content. If kernel moisture content is high, the initial temperature will be low because the part of the energy will be used to evaporate the water and the evaporative cooling will decrease the grain temperature.

Initial grain temperature is an important factor influencing the safe storage of grain, and the effect of initial temperature can influence the grain temperature in bins for more than one year in a wheat bin with 6 m diameter and 6 m grain depth (Fig. 2.8). The effect will last longer with the increase

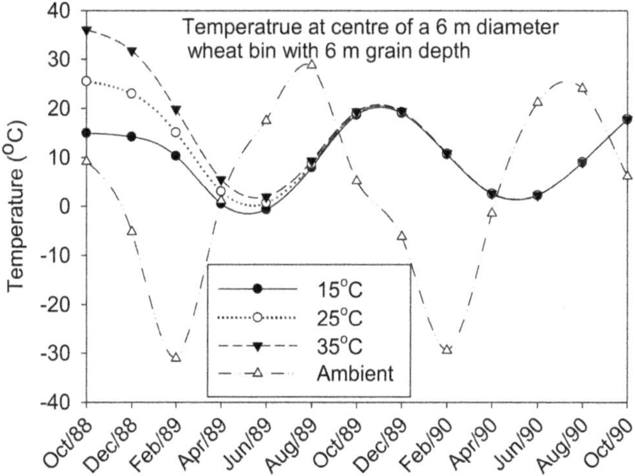

Figure 2.8 Wheat temperature at the half grain depth and the centre of a bin with 6 m diameter and 6 m grain depth. The data were generated using the same method as Fig. 2.7. Initial grain temperatures of 15, 25, and 35°C were used to conduct the simulations.

of bin diameter without aeration. Loschiavo (1985) found insect infestations were higher in grain harvested on warm sunny days than on cool days.

2.2.1.2.2 Temperature Gradients In North America such as Canada, harvesting of wheat normally occurs in late summer or early fall when weather temperature is decreasing. After grain is binned, the grain in an unventilated bin begins to cool at the boundary (walls, headspace, bottom of hopper bin, or bottom of flat-bottom bin if there is a plenum), and heat loss from centre to the boundary will be much slower than that at the walls due to the low thermal diffusivity of the grain. In flat-bottom bins without aeration floors, the grain is in contact with the concrete floor which in turn is on a soil foundation. Temperature of the grain on the floor will change more slowly than grain at the walls. Gain temperature near the bin bottoms of hopper bottom bins will be similar to those near the walls (Jian et al. 2009). Grain which is away from the boundary will have temperature lags. These temperature lags generate temperature gradients. Temperature gradients at the locations close to the centre are usually lower than that at the places close to the walls. Jian et al. (2009) found the temperature gradients developed a few days after the grain is loaded. The average temperature gradient was 5.09 ± 1.24°C/m inside a bin holding 20 tonne of wheat during 15 months of storage (Fig. 2.9). The higher temperature gradients were always at the east side of the bin with the highest being 32.4°C/m and it was located at the centre of the bin at 1.6 m height.

In fall and winter grain temperature at the centre is higher than that at the boundary (Fig. 2.10). In spring and summer, the grain temperature gradually warms up from the outside, and temperatures of grain near the walls rise above the temperature of the grain at the centre. The temperature gradient direction in the spring and summer is reversed (Fig. 2.10).

2.2.1.2.3 Free Convection Currents Temperature gradient causes convection currents because dense air at the lower temperature locations moves down and less dense air at the higher temperature locations moves up. The pressure differences (caused by gravity and buoyancy forces) between the dense and less dense air at different locations inside the bin dictates the movement direction of the convention currents. If some places have no temperature gradient, currents might also pass through the places because air might be pushed or pulled by the moving air at its adjacent locations. There will probably be some short-circuiting between the streams of cold and warm air throughout the grain bulk. In fall and winter, the dense air at the places near the walls moves down and the thin air at the places near the centre moves up (Fig. 2.10). In spring and summer, these convection currents reverse because the dense air is located at the places near the centre (Fig. 2.11). The average velocity of the currents

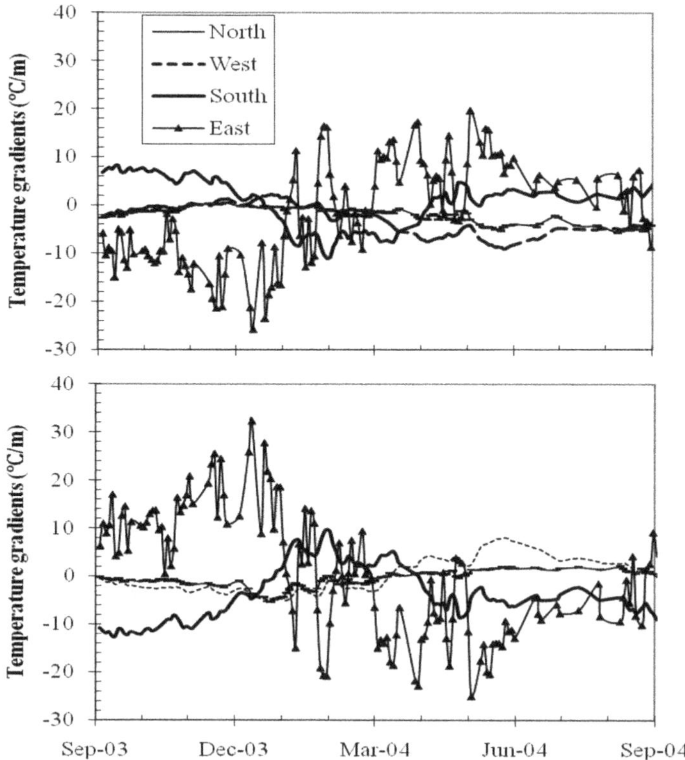

Figure 2.9 Temperature gradients in the north, west, south, and east directions inside a 3.7 m diameter bin holding 20 tonne of wheat (2.6 m grain depth) between 0.9 and 1.7 m (top); and between 0.0 m and 0.90 m (bottom) from the centre at 1.0 m depth of the grain (when temperature gradient >0, the temperature at the peripheral location was higher than that near the centre) (Jian et al. 2009).

inside a maize metal-bin under African condition is 0.8 m/hour at dawn and 0.7 m/hour at early afternoon (Gough et al. 1990). Therefore, these convection currents are very slow (about 6–15 m/day) and will be much smaller inside a bin with smaller grain temperature gradients than that inside the measured bin by Gough et al. (1990). These convection currents do not affect the grain temperature unless the Rayleigh number is large (Smith and Sokhansanj 1990). For small cereal seeds such as wheat, the Rayleigh number is not large enough for convection to influence grain temperature greatly. However, these free convection currents can result in moisture migration (refer Chapter 6).

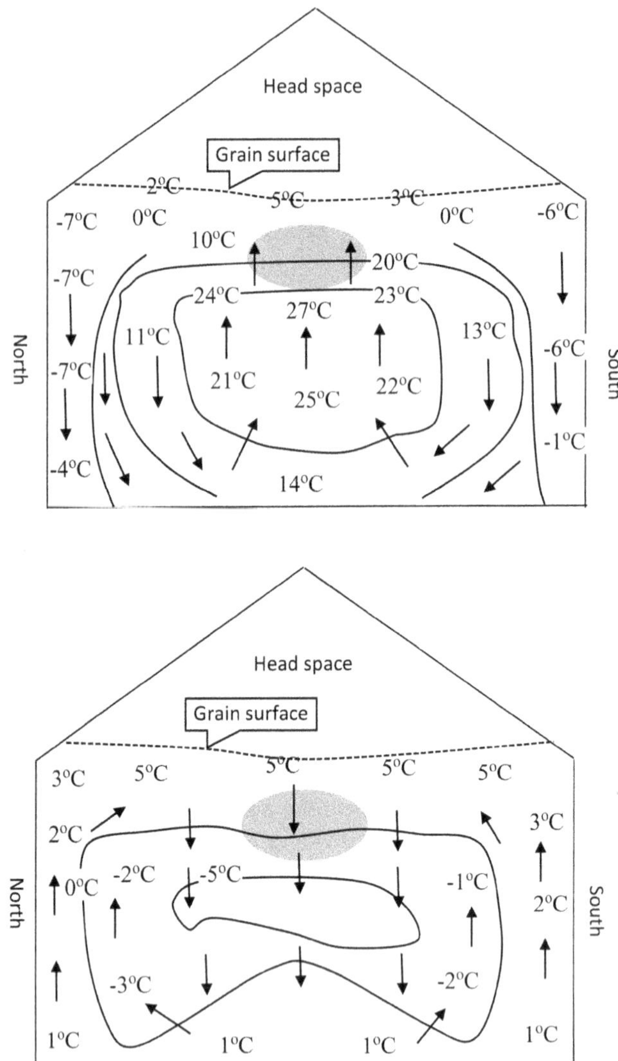

Figure 2.10 Measured temperatures and approximate isotherms of 20 tonne of wheat in late November (after 3 months of storage, top) and late April (after 8 months of storage, bottom) in a galvanized-steel bin near Winnipeg, MB, Canada. The wheat ranged in temperature from 25 to 30°C and 13.6 ±0.1% moisture content when harvested and stored to a depth of 2.6 m in the 3.7-m diameter, cylindrical bin in late August. Arrows show directions of convection currents, and shaded areas show the cumulated moisture areas. The same data source as Fig. 2.9, but unpublished by Jian et al. (2009).

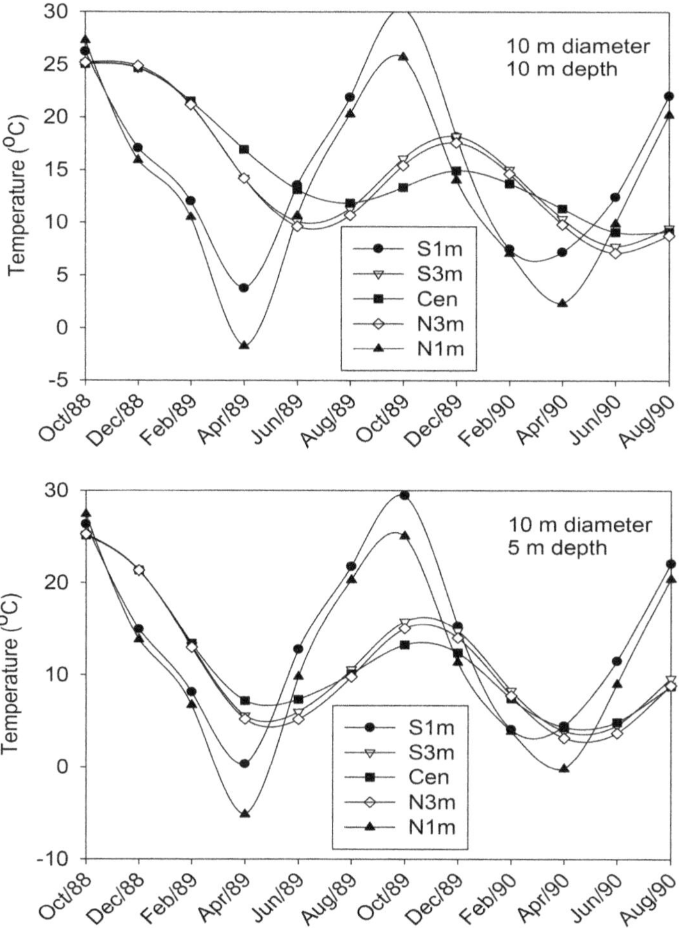

Figure 2.11 Effect of grain depth on the temperatures at the half grain depth in a galvanized bin with 10 m diameter. In the graph, S1m = 1 m from South, S3m = 3 m from South, Cen = centre, N3m = 3 m from North, N1m = 1 m from North, 5 m depth = 5 m of total grain depth, and 10 m depth = 10 m of total grain depth. The data were generated using the same method as Fig. 2.7.

2.2.2 Factors Affecting Grain Temperatures in Bins

Factors influencing the grain temperature and its distribution can be categorized into physical and biotic factors. Physical factors include bin location (geographical location; the location of the bin to sun, to other building, and to trees; snow covering, attached buildings or structures, surface slope, shading, and ground surface around the structure), weather (temperature, RH, solar radiation, precipitation, and wind), bin structure (bin shape, size, wall materials, bin foundation, openings, vents, and ventilation system), stored grain (grain type, dockage concentration, grain depth, initial temperature, and moisture content), and management (pesticide application, storage time, drying, aeration, turning, and coring). The biotic factors include the grain itself (kernel vigour, germination, dockage and foreign material in grain, and amount of damaged kernels) and activity of organisms (insects, fungi, birds, and rats). Under safe storage grain condition, the main factor influencing the grain temperature is the weather at any geographical location. If grain temperature and moisture content are higher than the recommended safe storage values, a minor factor will become the major factor affecting grain temperature and its distribution (Jian and Jayas 2014), For example, if grain moisture content is high, the heat produced by the grain and fungi will be the main factor influencing grain temperatures (Jian et al. 2014). This section focuses on main factors influencing grain temperatures inside a metal cylindrical bin holding grain at safe storage moisture contents.

2.2.2.1 Headspace, Bin Diameter, and Grain Depth

Headspace temperature is higher than or equal to the outside weather temperature due to solar radiation on roofs and walls. Jian et al. (2009) found that temperatures of the headspace of a metal bin holding 20 tonne of wheat was 2.9 ± 0.2°C higher than that of the ambient air with a maximum difference of 18.3°C and a minimum difference of 0°C. The minimum difference of temperature occurred in winter (due to snow covering the roof of the bin) and at nights. There was larger temperature fluctuation in the headspace than inside the grain mass. Inside the headspace, temperature was different at different locations especially at different heights. In a day, the average difference between the highest and lowest temperatures inside the headspace over the 15-month period was 4.3 ± 0.4°C with a maximum of 36.0°C (on October 3, 2003). The warmer temperature inside the headspace influences the temperature of the grain located at the top of the grain bulk, and this influence is affected by the bin diameter and grain depth. Larger headspace (large diameter and short grain depth) will have a significant influence on the entire grain temperature and this influence will be negligible if the headspace is small, e.g., the grain is filled to the top vent of the headspace. Vent in the headspace usually has negligible effect on the headspace temperature if the airtightness of the headspace is low. The galvanized metal bins usually have a low airtightness.

When grain depth is larger than the bin diameter, the solar radiation effect will be more obvious and the temperature difference between the south and north locations increases with the increase of grain depth (Fig. 2.11). Under this condition, the temperature of grain at the centre is mostly influenced by the heat transfer along the bin radius path because the shortest distance is the radius. Solar radiation causes the temperatures at the south and west walls to be higher than at other locations from August to March (Jian et al. 2009). Buschermohle et al. (1988) found a temperature difference of 7°C between north and south portions of corn stored in a 6.4 m diameter bin at Clemson, South Carolina. Jayas et al. (1994) found that for a 6 m diameter bin, the minimum centre temperatures increased lineally with bin height from –5°C for 3 m height to 0°C for 10 m height. If grain depth is smaller than the bin diameter, the temperature of grain at the centre is mostly influenced by the headspace temperature and the floor temperature (Fig. 2.12). In Fig. 2.11, the minimum grain temperature at the centre in the bin with 10 m diameter and 10 m grain depth was 6°C higher than that in the bin with 10 m diameter and 5 m grain depth under the same weather and configuration conditions. The temperature lag becomes smaller in the bin with short grain depth (Fig. 2.11). The lag of grain temperature at the centre of the grain bulk becomes more obvious with the increase of the grain depth and diameter (Fig. 2.12).

With the increase of bin diameter and grain depth, the centre temperature becomes more stable and changes slowly (Fig. 2.12). Small bins

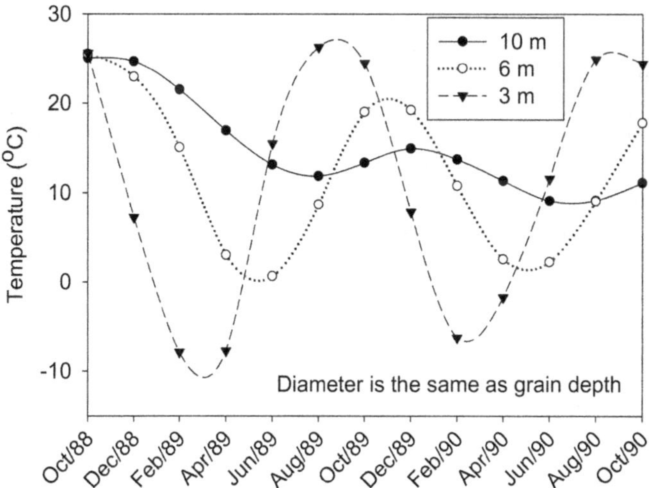

Figure 2.12 Effect of bin diameter on the temperatures at the half grain depth in a galvanized bin. The data were generated using the same method as Fig. 2.7. In the graph, 10, 6, and 3 m are the bin diameters and the grain depth in each bin is the same as the diameter.

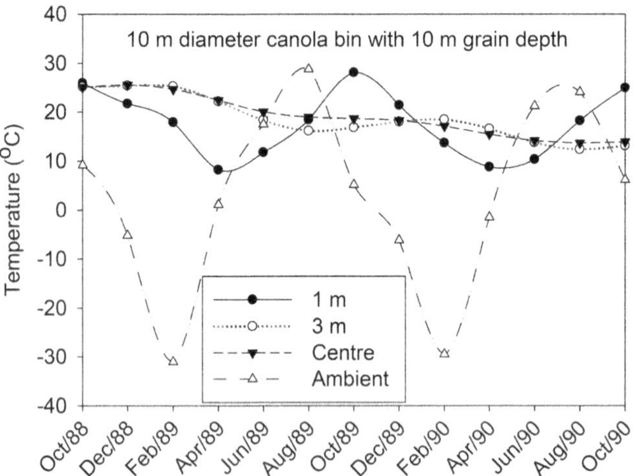

Figure 2.13 Canola temperature at the half grain depth. Temperatures were simulated using Winnipeg weather data and the model developed by Jian et al. (2005). The initial canola condition: 25°C, 9% moisture content, clean, and no insect and mould infestation. In the graph, 1 m, 3 m, centre, and ambient are the grain temperatures at 1 and 3 m from the south wall, at the bin centre, and the average daily ambient temperature, respectively.

cool most rapidly in the fall but warm up quickly in the spring. Increase in bin diameter not only causes time delays in the centre temperature, but also decreases the differences between the maximum and minimum temperatures at the same location. The effect of large diameter and grain depth on the centre grain temperature is the same as the bin, which stored the crop with small thermal diffusivity. The centre temperate of canola inside a 10 m diameter and 10 m depth bin (Fig. 2.13) has a less temperature fluctuation than that of wheat under the same weather condition and in the same bin (Fig. 2.12). Small bin has larger temperature gradients (Fig. 2.14), while has smaller temperature differences inside the bin because the difference of calculation methods (temperature gradient measures the temperature difference across a fixed length, while temperature difference measures the temperature difference at the same locations inside the bin without considering the length scale).

2.2.2.2 Bin Shape and Wall Material

The temperature of grain at centre of a grain mass is influenced mostly by the shortest heat transfer path to outside. Centre temperatures in square (6 × 6 m) and cylindrical (6 m diameter) bins containing the similar amount

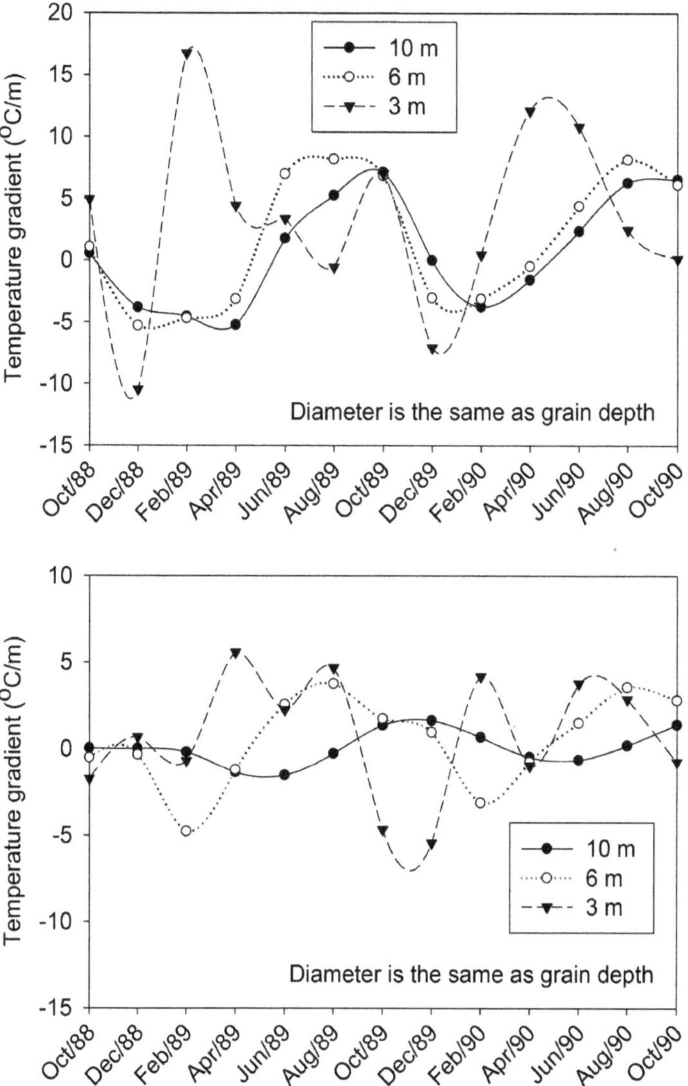

Figure 2.14 Temperature gradients at the half grain depth. The bottom graph shows the temperature gradients at the 2/3 of radius to the centre at the south side. The top graph shows the temperature gradients at south middle of the radius (1 m long). Negative temperature gradient indicates the temperature at the locations close to the walls is lower than that at the locations close to the centre. The data were generated using the same method as Fig. 2.7.

of wheat are similar (Jayas et al. 1994). The configuration of the bin walls influences the solar radiation and convection on the walls, hence affects the overall energy received by the grain mass. For the same amount of grain mass, a grain bulk with a larger surface area facing sun will have a higher overall temperature inside the bulk than the bulk that has a smaller area facing sun. Higher wind speed on walls will increase the heat loss from the bin if outside is cooler than bin wall, hence decreasing the overall grain temperature inside the bin. Jian et al. (2009) found that temperatures at the north wall were not the lowest temperatures during winter and this was inconsistent with the published model prediction. They found the low wind speed at east side of the bin was the reason causing this variance because a dike at the east direction and the identical silo at the north direction of the test silo influenced the wind speeds and directions. Montross et al. (2002) also found their pilot bins were more heavily influenced by wind than conventional sized bins.

Radiation properties of bin materials influence grain temperature patterns and gradients. Wall materials especially their emissivity and absorptivity to solar radiation are important factors influencing the overall grain temperature inside the bin. Walls with high absorptivities and low emissivities to solar radiation have high wall temperatures when exposed to the sun. The wide diurnal variations may damage the grain that is against or near the walls. An ideal material for a bin wall is one in which the heat flow into the grain is restricted (low absorptivity) while heat flow outward is rapid (high emissivity). Under the same configuration, a bin with different bin materials can have different grain temperatures (Table 2.8). A good example is

Table 2.8 Bin Wall Material Influencing the Storage Time (Days) to Cool Wheat with 14.5% Moisture Content (wb) at 25°C Initial Temperature to 20°C at the Centre of a Cylindrical Bin in Winnipeg Stored on September 1 (Yaciuk et al. 1975)

	Bin Diameter (m)			
Wall Material	*4*	*6*	*8*	*10*
Black rubber	74	134	207	301
Concrete	68	126	198	288
Galvanized steel	90	150	225	329
Insulated steel	99	161	241	356
Insulated steel, painted white	78	140	217	316
White painted steel	69	129	203	295
Wood, painted white	70	129	203	296
Wood, painted red	73	132	206	300

Table 2.9 Absorptivity and Emissivity of Bin Roof and Wall Materials (Incropera and DeWitt 1996)

Bin Material	*Absorptivity*	*Emissivity*
Black paint	0.98	0.98
Brick (red, Purdue)	0.63	0.93
Concrete	0.63	0.88–0.93
Galvanized steel (new)	0.65	0.13
Galvanized steel (weathered)	0.80	0.28
Metal plate (Black sulphide)	0.92	0.10
Metal plate (black cobalt oxide)	0.93	0.30
Metal plate (black nickel oxide)	0.92	0.08
Metal plate (black chrome)	087	009
Paint (white, acrylic)	0.26	0.90
Paint (black, Parsons)	0.98	0.98
Paint (white, zinc oxide)	0.16	0.93
Snow (fine particles, fresh)	0.13	0.82
Snow (ice granules)	0.33	0.89

the clean white paint (zinc oxide) that has an absorptivity to shortwave solar radiation of 0.16 while its emissivity for long wave radiation (surface temperatures above 25°C) is about 0.93 (Table 2.9). The centre temperature in a 6 m diameter bin holding 6 m high wheat and white painted bin can be 17.5°C lower than the same bin without paint (Fig. 2.15). The galvanized steel is the most common materials for grain silos in North America. However, it is one of the worst possible material for maintaining low grain temperatures (Tables 2.8 and 2.9, Figs. 2.6 and 2.15). Paints of other colours are not as good as white because of their solar absorptivities is less than their emissivities (except black, Table 2.9). Therefore, bins with white paint especially in the low latitudes is better for maintaining grain at low temperature. Shading may be another strategy if the bin is not painted white.

2.2.2.3 Geographical Location

Weather condition is the most important factor influencing grain temperature patterns and gradients. Different countries and geographical areas have different recommended grain moisture contents for safe storage due to the effect of weather on stored grain temperatures. At different geographical locations, the harvest times and weather conditions at the harvest period are also different. This results in different initial grain temperatures

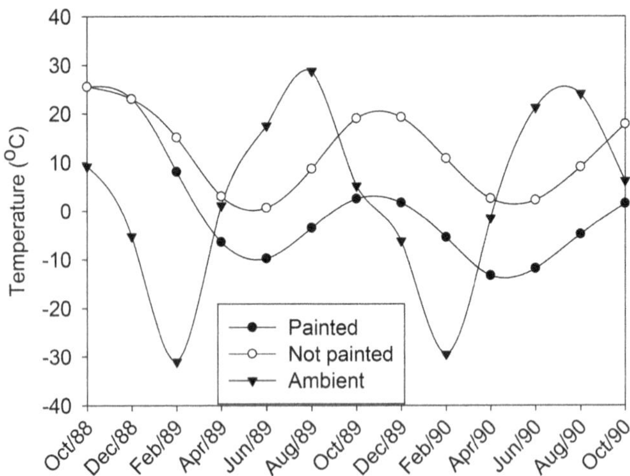

Figure 2.15 Centre temperatures of a galvanized steel cylindrical bin (6 m diameter and 6 m wheat depth) with the walls white painted or not painted. The data were generated using the same method as Fig. 2.7. Ambient is the average daily temperature.

and physical properties. Even in the same city and different areas, the same grain stored in the bin with the same configuration might have different temperature patterns and gradients due to different solar radiations, wind speeds, and precipitation. The practice and procedure of grain harvesting, transportation, and storage management at different locations might also be different. Therefore, temperature patterns and gradients in different bins will be different. Metzger and Muir (1983) simulated the storage of 100 t of wheat with 14.5% moisture content in galvanized steel bins, 6 m diameter, at four main cities (representing four climatic regions) in Canada. The predicted safe storage time of the wheat with 23.6°C initial temperature, 14% moisture content (wb) and harvested on September 1 was about: 220 days at Winnipeg, 230 days at Swift Current, 290 days at Edmonton, and 320 days at Fort St. John (Fig. 2.16).

2.3 Remark on Application of Grain Thermal Properties

The low thermal diffusivity of stored grain coupled with fluctuating daily and seasonal temperature creates temperature gradients within grain bulks. Analyses of temperature patterns indicate that many factors influence grain temperatures and grain might have different temperature

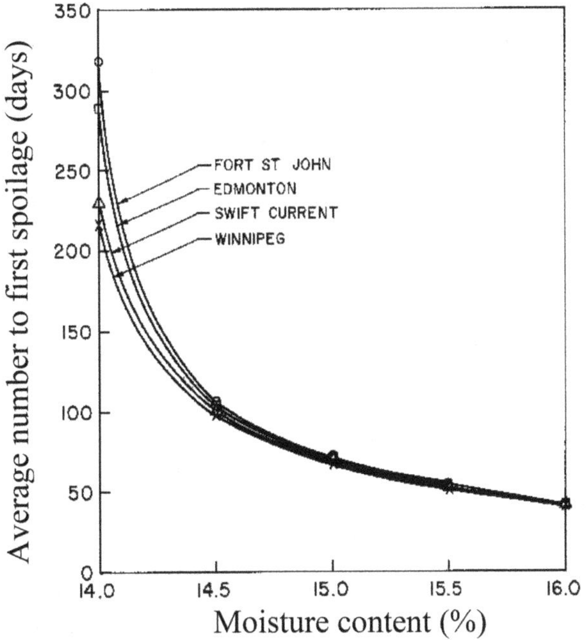

Figure 2.16 Average number of days from harvest to the first occurrence of spoilage (drop in germination or visible mould) by fungal infection. Simulations are based on 6 m diameter bin holding 100 tonne of wheat with 23.6°C initial temperature harvested and stored on September 1. (Reprinted with permission from Metzger and Muir (1983)).

distribution inside different bins. Therefore, grain storage management should take advantages of the grain thermal properties, temperature and moisture patterns, and their biological properties (such as low respiration at low temperatures and moisture contents), and at the same time should avoid side effects caused by these factors.

Grain turning and aeration are common practices taking advantages of low thermal diffusivities of stored grain. Since grain has low thermal conductivities and convective heat transfer coefficients, grain turning should provide enough surface area and time allowing the heat energy inside grain to be diffused to ambient environment, which should be at least 5°C lower than the grain temperature during turning. For a large diameter bin with deep grain depth, aeration can be conducted to cool grain at the beginning of the storage. The grain at the centre in the large bin will keep cool even for more than one year. After turning and aeration, the grain temperature gradients should be decreased. However, grain temperature at the walls will be changed again by following daily and seasonal weather temperatures. Therefore, turning or aeration should be repeated after the temperature gradient is large enough. So, long time storage is possible. For a small bin; however, turning or aeration is not needed because small bin can warm up or cool down quickly.

Considering the low thermal diffusivity of grain, insulation of grain bins is usually not recommended because daily temperatures only influences grain at locations within 15 cm from bin walls. However, insulation at the ceiling facing the head space is installed in warehouses in some countries (such as at south of China). This insulation will work for a warehouse with a large roof area and shallow grain depth because solar radiation in south of China can significantly increase headspace temperatures inside this type of warehouse. If grain temperature is controlled by an air conditioning system, insulation at the ceiling facing headspace can significantly decrease energy consumption inside this type of warehouse. Therefore, different solutions might be recommended at different conditions of storage practices, even though the same grain is stored. Grain managers should use principles of grain thermal properties to cater for their special requirements.

Any grain storage management practice cannot improve the quality of stored grain, but can maintain the quality, if proper management practices are followed. The main purpose of grain storage management is to maintain grain quality at the minimum cost. To minimize the storage cost and loss, the grain thermal properties and temperature patterns should be considered together with the stored grain ecosystem. In the ecosystem, the quality change of stored grain is mainly caused by insect infestation and fungi infection. Insects and fungi require certain temperatures and water activities to multiply. Therefore, controlling grain temperature and moisture content to less than the minimum requirement of their multiplication is usually recommended. When any recommendation is practiced, thermal properties and temperature patterns should be considered because considering these principles could decrease the management cost and loss. For example, aeration or turning with the consideration of the seasonal weather pattern and bin configuration could decrease the time of turning or aeration.

Temperature gradient patterns presented in this chapter should be used as a basic model of temperature distribution inside a metal cylindrical bin. Locations of monitoring (such as temperature and moisture cables) and sampling should be decided based on these patterns, so any abnormal change (caused by spoiled grain) can be identified by comparing with these basic patterns. This chapter used the metal cylindrical bin as the model example because extensive studies have been conducted on this type of bin. Therefore, thorough studies on other types of bins (such as warehouse holding bulk or bagged grain) are needed.

Nomenclature

α: thermal diffusivity (m^2/s);
ρ: bulk density (kg/m^3);
λ_{grain}: latent heat of vaporization of grain sample (kJ/kg);
λ_{water}: latent heat of water (kJ/kg);
ε: porosity of material (fraction of total volume occupied by air and water vapour);
μ: viscosity of the fluid (Pa·s);
ϕ: sphericity (decimal);
Φ: shape factor $\left(= \frac{\text{geometric mean diameter}}{\text{effective diameter}} \right)$;
Υ: relative humidity (%, decimal);
λ: latent heat of vaporization (kJ/kg);
A: area (m^2);
A_t: overall surface area of grain kernels in a bulk (m^2);
a_v: area of kernel per unit volume of bed (m^2);
a_w: water activity (decimal);
B: constant;
C: constant;
C_{pa}: specific heat of the fluid ($kJ \cdot kg^{-1} \cdot {}^\circ C^{-1}$);
C_s: specific heat ($kJ \cdot kg^{-1} \cdot {}^\circ C^{-1}$);
C_{ss}: specific heat of the sample ($kJ \cdot kg^{-1} \cdot {}^\circ C^{-1}$);
C_{sw}: average specific heat of water in the measured temperature range ($kJ \cdot kg^{-1} \cdot {}^\circ C^{-1}$);
d_e: effective kernel diameter (m);
$\overline{Ec_i}$: cumulated average vaporisation enthalpy (kJ/kg);
G: mass flow rate of drying air ($kg \cdot m^{-2} \cdot s^{-1}$);
Gr: Grashof number;
h_a: heat capacity per unit volume of the fluid ($kJ \cdot m^{-3} \cdot K^{-1}$);
h_c: convective heat transfer coefficient ($W \cdot m^{-2} K^{-1}$);
h_{con}: heat of condensation of pure water (cal/mol);
h_{cs}: heat transfer coefficient based on surface area ($W \cdot m^{-2} K^{-1}$);
Δh_k: mean desorption specific bonding enthalpy of multilayer water ($J \cdot kg^{-1}$);
h_s: heat capacity per unit volume of the solid ($kJ \cdot m^{-3} \cdot K^{-1}$);
$\Delta h_{s,\ multi}$: isosteric heat of sorption of multilayer water (kJ/kg);
h_{cv}: volume convective heat transfer coefficient ($W \cdot m^{-3} K^{-1}$);
Δh_0: net isosteric heat of sorption of the first molecules of water in the food (kJ/mol);
J: heat flux (W/m^2);
j_d: Colburn j-factor for mass transfer;
j_h: Colburn j-factor for heat transfer;

k: thermal conductivity ($W{\cdot}m^{-1}{\cdot}°C^{-1}$);
k_a: thermal conductivity of humid air ($W{\cdot}m^{-1}{\cdot}°C^{-1}$);
M: moisture content (%, or decimal, wet basis);
M_e: equilibrium moisture content (kg/kg dry solid, decimal);
M_{ev}: evaporated water from the grain bulk in a time period (kg);
M_m: constant in BET equation;
M_{water}: mass of the water poured into the calorimeter (kg);
m: mass of the material (kg);
Nu: Nusselt number;
P_{at}: atmospheric pressures (kPa);
Pr: Prandtl number;
$P(T)$: partial vapour pressure at temperature T (Pa);
P_v: vapour pressure of water over the absorbent (Pa);
P_0: vapour pressure of pure water (Pa);
Q: amount of heat required (J);
$\dot{Q}$: line source strength (W/m);
Q_e: rate of heat utilized to evaporate moisture on surface ($J{\cdot}m^{-2}{\cdot}s^{-1}$);
Q_s: net isosteric heat of sorption or termed as desorption specific bonding enthalpy of monolayer water (kJ/kg);
Q_{st}: isosteric heat of sorption (kJ/kg);
q: energy supplied (J);
q_c: energy transferred by convection (W);
R: gas constant (8.31451×10^{-7} $J{\cdot}K^{-1}{\cdot}mol^{-1}$);
Re: Reynolds number;
S: cross-sectional area of air flow (m^2);
ΔS_d: differential entropy (J/K);
T: temperature (°C);
ΔT: temperature change (°C);
T_a: temperature of air around grain kernel (°C);
T_e: temperature of evaporated water (°C);
Tc_i: initial temperature of the calorimeter (°C);
T_g: grain temperature (°C);
T_K: temperature (K);
ΔT_m: mean temperature difference between the fluid and the surface (°C);
T_{mf}: final temperature of the mixture (°C);
T_s: surface temperature of solid (K);
T_{si}: initial temperature of the sample (°C);
T_{wi}: initial temperature of the water (°C);
T_{wf}: finial temperature of the water and the calorimeter (°C);
T_0: initial temperature of the air (K);
T_∞: ambient air temperature (K);
t: time (s);
v: superficial velocity (m/s);

W_e: mass of water having thermal capacity equal to that of the calorimeter system (kg);
W_s: mass of the sample (kg);
W_w: mass of water (kg);
X: characteristic dimension (m);
x: distance, or distance traversed by the fluid in the bed (m).

References

Adl-Zarrabi, B., and L. Bostrom. 2004. Determination of thermal properties of wood and wood based products by using transient plane source. 8th world conference on timber engineering, Lahti, Finland. June 14–17.

Aguerre, R. J., C. Suarez, and P. E. Viollaz. 1988. The temperature dependence of isosteric heat of sorption of some cereal grains. International Journal of Food Science and Technology 23: 141–145.

Akpinar, E. K. 2004. Experimental investigation of convective heat transfer coefficient of various agriculture products under open sun drying. International Journal of Green Energy 1: 429–440.

Alagusundaram, K., D. S. Jayas, W. E. Muir, and N. D. G. White. 1991. Thermal conductivity of bulk barley, lentils, and peas. Transactions of the ASAE 34: 1784–1788.

Alam, A., and G. C. Shove. 1973. Hygroscopicity and thermal properties of soybeans. Transactions of the ASAE 16: 707–709.

Alanis, E., L. Saravia, and L. Rovetta. 1977. Measurement of rock pile heat transfer coefficients. Solar Energy 19: 571–572.

Al-muhtaseb, A. H., W. A. M. McMinn, and T. R. A. Magee. 2002. Moisture sorption isotherm characteristics of food products: a review. Food and Bioproducts Processing 80: 118–128.

Anwar, S. I., and R. D. Singh. 2012. Convective heat transfer coefficient of Indian gooseberry (*Emblica officinalis*) dried in three different forms under forced convection mode. Journal of Engineering Science and Technology 7: 635–645.

Anwar, S. I., and G. N. Tiwari. 2001a. Convective heat transfer coefficient of crops in forced convection drying — an experimental study. Energy Conversion and Management 42: 1687–1698.

Anwar, S. I., and G. N. Tiwari. 2001b. Evaluation of convective heat transfer coefficient in crop drying under open sun drying. Energy Conversion and Management 42: 627–637.

Arslan-Tontul, S. 2020. Moisture sorption isotherm, isosteric heat and adsorption surface area of whole chia seeds. Food Science and Technology 119: 1–7.

ASABE. 2016. ASAE D245.6 OCT2007. Moisture Relationships of Plant-based Agricultural Products, ASABE Standard. American Society of Agricultural and Biological Engineers, St. Joseph, MI.

Bala, B. K. 1983. Deep Bed Drying of Malt. PhD, University of Newcastle Upon Tyne, Tune and Wear, UK.

Barker, J. J. 1965. Heat transfer in packed beds. Industrial & Engineering Chemistry 57: 43–49.

Bitra, V. S. P., S. Banu, P. Ramakrishna, G. Narender, and A. R. Womac. 2010. Moisture-dependent thermal properties of peanut pods, kernels, and shells. Biosystems Engineering 106: 503–512.

Blahovec, J., and S. Yanniotis. 2010. 'GAB' generalised equation as a basis for sorption spectral analysis. Czech Journal of Food Science 28: 345–354.

Bowers, B. A., and R. J. Hanks. 1962. Specific heat capacity of soil and minerals as determined with a radiation calorimeter. Soil Science 94: 392–396.

Boyce, D. S. 1966. Heat and moisture transfer in ventilated grain. Journal of Agricultural Engineering Research 11: 255–265.

Brown, R. B., and L. Otten. 1992. Thermal conductivity and convective heat transfer coefficient for soybean and white bean seeds. Canadian Agricultural Engineering 34: 337–341.

Buschermohle, M. J., J. M. Bunn, and R. A. Spray. 1988. Moisture Migration in Stored Grain. American Society of Agricultural Engineering, St. Joseph, MI.

Cenkowski, S., D. S. Jayas, and D. Hao. 1992. Latent heat of vaporization for selected foods and crops. Canadian Agricultural Engineering 34: 281–286.

Chakrabarti, S. M., and W. H. Johnson. 1972. Specific heat of flue cured tobacco by differential scanning calorimetry. Transactions of the ASAE 15: 928–931.

Chandra, S., and W. E. Muir. 1971. Thermal conductivity of spring wheat at low temperatures. Transactions of the ASAE 14: 644–646.

Chang, C. S. 1986. Thermal conductivity of wheat, corn, and grain sorghum as affected by bulk density and moisture content. Transactions of the ASAE 29: 1447–1450.

Chang, C. S., and B. S. Miller. 1980. Thermal conductivity and specific heat of grain dust. Transactions of the ASAE 23: 1303–1306, 1312.

Choi, Y., and M. R. Okos. 1986. Effects of temperature and composition on the thermal properties of foods. In L. M. Maguer and P. Jelen (eds.), Food Engineering and Process Applications: Transport Phenomena, vol. 1. Elsevier, New York.

De Oliveira, G. H. H., P. C. Correa, E. F. Araujo, D. S. M. Valente, and F. M. Botelho. 2010. Desorption isotherms and thermodynamic properties of sweet corn cultivars (*Zea mays* L.). International Journal of Food Science and Technology 45: 546–554.

Denys, S., and M. E. Hendrickx. 1999. Measurement of the thermal conductivity of foods at high pressures. Journal of Food Science 64: 709–713.

Dunkle, R. V. 1961. Solar water distillalion: the roof type still and a multiple effect diffusion still, pp. 895–902. In International Developments in Hear Transfer. ASME, Pan V.

Dutta, S. K., V. K. Nema, and R. K. Bhardwaj. 1988. Thermal properties of gram. Journal of Agricultural Engineering Research 39: 269–275.
Figura, L. O., and A. A. Teixeira. 2007. Food Physics, Physical Properties: Measurement and Applications. Springer, Berlin Heidelberg, Germany.
Furnas, C. C. 1930. Heat transfer from a gas stream to a bed of broken solids. Transactions of the American Institute of Chemical Engineers 24: 142–169.
Gamson, B. W., G. Thodos, and O. A. Hougen. 1943. Heat, mass, and momentum transfer in the flow of gases through granular solids. Transaction of The American Institute of Chemical Engineers 39: 1–35.
Gely, M. C., and A. M. Pagano. 2012. Moisture desorption isotherms and isosteric heat of sorption characteristics of malting barley (*Hordeum distichum* L.). Latin American Applied Research 42: 237–243.
Gough, M. C., C. B. S. Uiso, and C. J. Stigter. 1990. Air convection current in metal silos storing maize grain. Tropical Science 30: 217–222.
Goyal, R. K., and G. N. Tiwari. 1998. Heat and mass transfer relations for crop drying. Drying Technology 16: 1741–1754.
Haque, M. A., N. Shimizu, T. Kimura, and B. K. Bala. 2007. Net isosteric heats of adsorption and desorption for different forms of hybrid rice. International Journal of Food Properties 10: 25–37.
Hobani, A. L., and M. H. Tolba. 1995. Bulk thermal conductivity and diffusivity of barley. Research Bulletin 53: 5–17.
Hooper, F. C., and F. R. Lepper. 1950. Transient heat flow apparatus for the determination of thermal conductivities. ASHVE Transaction 56: 309–324.
Iglesias, H. A., and J. Chirife. 1976. Isosteric heats of water vapor sorption on dehydrated foods. Part I: analysis of the differential heat curves. Lebensmittel-Wissenschaft und Technologie 9: 116–122.
Iglesias, H., J. Chirife, and P. Viollaz. 1976. Thermodynamics of water vapour sorption by sugar beet root. Journal of Food Technology 11II: 91–101.
Incropera, F. P., and D. P. DeWitt. 1996. Fundamentals of Heat and Mass Transfer, 4th ed. John Wiley & Sons, New York.
Integrated Environemnted Solutions Limited. 2000. Apache-Tables User Guide. Integrated Environemnted Solutions Limited, Glasgow, Scotland.
Jangi, A. N., S. A. Mortazavi, M. Tavakoli, A. Ghanbari, H. Tavakolipour, and G. H. Haghayegh. 2011. Comparison of mechanical and thermal properties between two varieties of barley (*Hordeum vulgare* L.) Grains. Australian Journal of Agricultural Engineering 2: 132–139.
Jayas, D. S., K. Alagusundaram, G. Shunmugam, W. E. Muir, and N. D. G. White. 1994. Simulated temperatures of stored grain bulks. Canadian Agricultural Engineering 36: 239–245.
Jian, F., and D. S. Jayas. 2014. Understanding the initiation and development of hotspots in storage-grain ecosystems. Journal of Applied Zoological Research 25: 1–10.

Jian, F., and D. S. Jayas. 2018. Characterization of isotherms and thin-layer drying of red kidney beans, part II: three dimensional finite element models to estimate transient mass and heat transfer coefficients and water diffusivity. Drying Technology 36: 1707–1718.

Jian, F., D. S. Jayas, and N. D. G. White. 2009. Temperature fluctuations and moisture migration in wheat stored for 15 months in a metal silo in Canada. Journal of Stored Products Research 45: 82–90.

Jian, F., D. S. Jayas, and N. D. G. White. 2012. Thermal conductivity, bulk density, and germination of a canola variety with high oil content under different temperatures, moisture contents, and storage periods. Transactions of the ASABE 55: 1837–1843.

Jian, F., D. S. Jayas, and N. D. G. White. 2013. Specific heat, thermal diffusivity, and bulk density of genetically modified canola with high oil content at different moisture contents, temperatures, and storage times. Transactions of the ASABE 56: 1077–1083.

Jian, F., D. S. Jayas, and N. D. G. White. 2014. Heat production of stored canola seeds under airtight and non-airtight conditions. Transactions of the ASABE 57: 1151–1162.

Jian, F., D. S. Jayas, N. D. G. White, and K. Alagusundaram. 2005. A three-dimensional, asymmetric, and transient model to predict grain temperatures in grain storage bins. Transactions of ASAE 48: 263–271.

Johnson, H. K., and A. C. Dale. 1954. Heat required to vaporize moisture. Agricultural Engineering 35: 705–709, 714.

Kara, M., L. Ozturk, S. Bastaban, and F. Kalkan. 2011. Thermal conductivity of safflower (*Carthamus tinctorius* L.) Seeds. Spanish Journal of Agricultural Research 9: 687–692.

Kazarian, E. A., and C. W. Hall. 1965. Thermal properties of grain. Transactions of the ASAE 8: 33–37, 48.

Labuza, T. P., A. Kaanane, and J. Y. Chen. 1985. Effect of temperature on the moisture sorption isotherms and water activity shift of two dehydrated foods. Journal of Food Science 50: 385–391.

Leuk, P., M. Schneeberger, U. Hirn, and W. Bauer. 2016. Heat of sorption: a comparison between isotherm models and calorimeter measurements of wood pulp. Drying Technology 34: 563–573.

Li, X. 2012. The hygroscopic properties and sorption isosteric heats of different Chinese wheat types. Journal of Food Research 1: 82–98.

Lomauro, C. J., A. S. Bakshi, and T. P. Labuza. 1985. Evaluation of food moisture sorption isotherm equations. Part I. Fruit vegetable and meat products. Lebensmittel-Wissenschaft und Technologie 18: 111–117.

Loschiavo, S. R. 1985. Post-harvest grain temperature, moisture, and insect infestation in steel granaries in Manitoba. The Canadian Entomologist 117: 7–14.

Matouk, A. M. 1976. Heat and Moisture Movements During Low Temperature Drying and Storage of Maize Grain. PhD, University of Newcastle Upon Tyne, YK.

Metzger, J. F., and W. E. Muir. 1983. Aeration of stored wheat in the Canadian Prairies. Canadian Agricultural Engineering 25(1): 127–137.

Miketinac, M. J., S. Sokhansanj, and Z. Tutek. 1992. Determination of heat and mass transfer coefficients in thin layer drying of grain. Transactions of the ASAE 35: 1853–1858.

Mohsenin, N. N. 1980. Thermal Properties of Foods and Agricultural Materials, Gordon and Breach Science, New York.

Montross, M. D., D. E. Maier, and K. Haghighi. 2002. Validation of a finite-element stored grain ecosystem model. Transactions of the ASAE 45: 1465–1474.

Morita, T., and R. P. Singh. 1979. Physical and thermal properties of short-grain rough rice. Transactions of the ASAE 22: 630–636.

Moysey, E. B., J. T. Shaw, and W. P. Lampman. 1977. The effect of temperature and moisture on the thermal properties of rapeseed. Transactions of the ASAE 20: 768–771.

Muir, W. E., and S. Viravanichai. 1972. Specific heat of wheat. Journal of Agricultural and Engineering Research 17: 338–342.

Mulet, A., J. Garcia-Reverter, R. Sanjuan, and J. Bon. 1999. Sorption isosteric heat determination by thermal analysis and sorption isotherms. Journal of food science 64: 64–68.

Narain, M., S. S. C. Bose, M. Jha, and V. K. Dwivedi. 1978. Physicothermal properties of rice bran. Journal of Food Science and Technology 15: 18–19.

Nellist, M. E. T. 1974. The Drying of Ryegrass Seeds in Deep Layers. PhD, University of Newcastle Upon Tyne, UK.

Nesvadba, P. 1982. Methods for the measurement of thermal conductivity and diffusivity of foodstuffs. Journal of Food Engineering 1: 93–113.

O'Callaghan, J. R., D. J. Menzies, and P. H. Bailey. 1971. Digital simulation of agricultural drier performance. Journal of Agriculture Engineering Research 16: 223–244.

Othmer, D. F. 1940. Correlating vapor pressure and latent heat data. Industrial and Engineering Chemistry 32: 841–856.

Öztekin, S., and Y. Soysal. 2000. Comparison of adsorption and desorption isoteric heats for some grains. The CIGR Journal of Scientific Research and Development 2: 1–17.

Prasad, D. C., W. E. Muir, and H. A. H. Wallace. 1978. Characteristics of freshly harvested wheat and rapeseed. Transactions of the ASAE 21: 782–784.

Pujula, M., D. Sanchez-Rodriguez, J. P. Lopez-Olmedo, J. Farjas, and P. Roura. 2016. Measuring thermal conductivity of powders with differential scanning calorimetry: a simplified method. Journal of Thermal Analysis and Calorimetry 125: 571–577.

Rahman, M. S. 1995. Food Properties Handbook. CRC Press, New York.

Ramesh, M. N. 2000. Effect of cooking and drying on the thermal conductivity of rice. International Journal of Food Properties 3: 77–92.

Ravikanth, L., D. S. Jayas, K. Alagusundaram, and V. Chelladurai. 2012. Measurement of thermal properties of mung bean (*Vigna radiata*). Transactions of the ASABE 55: 2245–2250.

Riedel, L. 1977. Kalorimetrische bestimmung der hydratationswarme lebensmitteln. Chem Mikrobiol Technol Lebensm 5: 97–101.

Riegel, C. A. 1992. Thermodynamics of moist air. In A. F. C. Bridger (ed.), Fundamentals of Atmospheric Dynamics and Thermodynamics. World Scientific, Singapore.

Rizvi, S. S. H. 1986. Thermodynamics of foods in dehydration, pp. 122–214. In M. A. Rao and S. Rizvi (eds.), Engineering Properties of Food. Marcel Dekker, New York.

Rizvi, S. S. H., and A. L. Benado. 1983. Thermodynamic analysis of drying foods. Drying Technology 2: 471–502.

Sahdev, R. K., N. Jain, and M. Kumar. 2012. Convective heat transfer coefficient for indoor forced convection drying of vermicelli. IOSR Journal of Engineering 2: 1282–1290.

Sahin, S., and S. G. Sumnu. 2006. Physical Properties of Foods. Springer, New York.

Sanchez, E. S., R. Sanjuan, S. Simal, and C. Rossello. 1997. Calorimetric techniques applied to determination of isosteric heat of desorption for potato. Journal of Scientific Food Agriculture 74: 57–63.

Schumann, T. E. W. 1929. Heat transfer: a liquid flowing through a porous prism. Journal of the Frankline Institure: 405–416.

Sharma, D. K., and T. L. Thompson. 1973. Specific heat and thermal conductivity of sorghum. Transactions of the ASAE 16: 114–117.

Singh, K. K., and T. K. Goswami. 2000. Thermal properties of cumin seed. Journal of Food Engineering 45: 181–187.

Singh, K. P., H. N. Mishra, and S. Saha. 2011. Sorption isotherms of barnyard millet grain and kernel. Food Bioprocess Technology 4: 788–796.

Smith, E. A., and S. Sokhansanj. 1990. Moisture transport caused by natural convection in grain stores. Journal of Agricultural and Engineering Research 47: 23–34.

Subramanian, S., and R. Viswanathan. 2003. Thermal properties of minor millet grains and flours. Biosystems Engineering 84: 289–296.

Sun, D. 1998. Selection of EMC/ERH isotherm equations for drying and storage of grain and oilseed, pp. 331–336, 13th International Congress on Agricultural Engineering, Rabat.

Tang, J., S. Sokhansanj, S. Yannacopoulos, and S. O. Kasap. 1991. Specific heat capacity of lentil seeds by differential scanning calorimetry. Transactions of the ASAE 34: 517–522.

Tiwari, G. N., and S. A. Lawrence. 1991. New heat and mass transfer relations for a solar still. Energy Conversion and Management 31: 201–203.

Tsami, E. 1991. Net isosteric heat sorption in dried fruits. Journal of Food Engineering 14: 327–335.

Tsami, E., Z. B. Maroulis, D. Morunos-Kouris, and G. D. Saravacos. 1990. Heat of sorption of water in dried fruits. International Journal of Food Science & Technology 25: 350–359.
Wang, C. Y., T. R. Runsey, and R. P. Singh. 1979. Convective Heat Transfer in Packed Bed of Rice, ASAE. ASAE.
Wang, N., and J. G. Bremmam. 1991. Moisture sorption isotherm characteristics of potatoes at four temperatures. Journal of Food Engineering 14: 269–287.
Wratten, F. T., W. D. Poole, J. L. Chesness, S. Bal, and V. Ramarao. 1969. Physical and thermal properties of rough rice. Transactions of the ASAE 12: 801–803.
Williamson, W. F. 1964. Temperature changes in grain dried and stored on farm. Journal of Agriculture Engineering Research 9: 32–47.
Yaciuk, G., W. E. Muir, and R. N. Sinha. 1975. A simulation model of temperature in stored grain. Journal of Agricultural Engineering Research 20: 245–258.
Yang, W., S. Sokhansanj, J. Tang, and P. Winter. 2002. Determination of thermal conductivity, specific heat, and thermal diffusivity of borage seeds. Biosystems Engineering 82: 169–176.
Yoshida, F., D. Ramaswami, and O. A. Hougen. 1962. Temperatures and partial pressures at the surfaces of catalyst particles. AIChE Journal 8: 5–11.

Chapter 3

Water in Biomaterials and Relationship with Its Environment

Introduction

Water is the most important factor influencing the stability of most biological materials. Water status inside biomaterials is influenced by physical and chemical properties of the material and its environment. The environmental factors include temperature and relative humidity (RH), which is calculated from the partial vapour pressure of water in a gas or a mixture of gases such as air and saturated vapour pressure of the gas or the mixture of gases at the same temperature. Biological materials sorb or desorb water from the environment. This chapter elucidates the phenomenon of water sorption and desorption by biomaterials from and to its environment and relationships that describe this phenomenon. Information about the theory of sorption isotherms, relationship between water status of biomaterials and its environment, and determination of their water properties are provided.

DOI: 10.1201/9781003186199-3

3.1 Moisture Content and Water Activity

3.1.1 Moisture Content and Measurement

3.1.1.1 Moisture Content

Moisture content of a biomaterial is the amount of water in the material and is an indicator of water status inside the biomaterial. It is determined from the mass of water in the material and can be expressed on wet or dry mass basis and either as decimal fraction or percentage. The moisture content in decimal on dry or wet basis is calculated as:

$$M_D = \frac{W}{M_g} \tag{3.1a}$$

$$M_w = \frac{W}{M_g + W} \tag{3.1b}$$

where M_D and M_w are moisture contents in dry basis and wet basis (decimal), respectively; W is the mass of water inside the material (kg), and M_g is the dry mass of the material (kg). Moisture content in percentage is calculated by multiplying 3.1a and 3.1b by 100 on the right side of the equations. Dry basis percentage moisture content can be converted to wet basis percentage moisture content, and vice versa as:

$$M_{D,\%} = \frac{100 M_{w,\%}}{100 - M_{w,\%}} \tag{3.2a}$$

$$M_{w,\%} = \frac{100 M_{D,\%}}{100 + M_{D,\%}} \tag{3.2b}$$

where $M_{D,\%}$ and $M_{w,\%}$ are the moisture contents in dry basis (%) and wet basis (%), respectively. The wet basis moisture content in percentage is mainly used by gain industry for trading commercial purpose and for describing interactions between stored-grain and moisture. When people mention moisture content of biomaterial without specification, it is typically wet basis. The dry basis moisture content in decimal is more likely used for grain drying calculations. In this book, wet basis moisture content in percentage is used unless otherwise specified.

To produce a grain sample with a desired moisture content, an amount of water can be mixed with or removed from the sample. The amount of water added or removed can be calculated as:

$$\Delta G = \frac{G(M_f - M_i)}{100 - M_f} \tag{3.3}$$

where ΔG is the mass of water added or removed (kg), G is the initial mass of the sample (kg), M_f is the final moisture content of the sample (wb, %), and M_i is the initial moisture content of the sample (wb, %). If water is added to grain kernels, a moisture gradient is established between the surface and core of the kernels. If uniform moisture content inside the kernels is required, the sample should be stored at a low temperature for a while. The common practice is to store the sample inside double layer plastic bags for at least one week at 5°C. When this sample is used for drying studies, this tempering period should be longer than one week. For large size grain kernels such as corn, soybean, and kidney bean, longer the period, the better.

The moisture content of the same grain with different moisture contents after mixing is:

$$M_{\text{mix}} = \frac{G_1 M_1 + G_2 M_2}{G_1 + G_2} \tag{3.4}$$

where M_{mix} is the moisture content of the mixture (wb, %), M_1 and M_2 are the moisture contents of sample 1 (wb, %) and 2 (wb, %), respectively; and G_1 and G_2 are the masses of samples 1 and 2 (kg), respectively. The average moisture content of the mixture will be the M_{mix}, but the moisture content of each individual kernel might not be the same because of the hysteresis of the food materials and the time is required to reach equilibrium (Pixton and Griffiths 1971).

3.1.1.2 Measurement of Moisture Content

Methods of moisture content measurement of biomaterials are classified into two categories: direct and indirect methods. When a direct method is used, water is removed by extraction, drying, distillation, or other physicochemical methods, and the mass of the materials is weighed before and after the water is removed. Direct methods include Karl Fisher titration, oven drying, microwave drying, thermos-gravimetric analysis, and freeze drying. In indirect methods, a parameter that changes with moisture content of material is measured. Indirect methods include the measurement of dielectric capacitance, alternating current (AC) and direct current (DC) conductivity, electrical resistance between electrodes, microwave absorption, microwave resonator, microwave moisture analyzer, and ultrasonic absorption. Direct methods are mostly used for laboratory analysis and usually have a high accuracy than the indirect methods. However, the direct methods usually need a longer time than indirect methods. Indirect methods are usually used for a quick measurement of moisture contents, and it is usually required to calibrate the equipment. The accuracy of the indirect methods depends on the equipment and application environment.

Methods of moisture content measurement are reviewed by many authors (Park 1996). Nelson et al. (1991) investigated several measurement methods including impedance, radio frequency, nuclear magnetic resonance, and conductance, and concluded that a measurement technique based on any of these parameters can provide moisture data with a standard error of about 0.1 percentage point.

The common method is the gravimetric oven method (or termed as oven method). The procedure of the oven method is: (1) the materials are dried by placing the samples in a gravity-convection or mechanical-convection (forced-draft) oven at a desired constant temperature for a desired time; (2) the mass of the material is measured before and after drying; and (3) moisture content of the sample is calculated by using Eq. 3.1. Oven method is a reliable method because it does not require calibration of the equipment. Oven method is also commonly used as a standard or reference method by many organizations such as American Association of Cereal Chemists and American Society of Agricultural and Biological Engineers. Balance used to weigh the samples should have a precision of 1 mg. Dishes used to hold the samples should be made of heavy gauge aluminium that do not dent readily. The standard error of the moisture content measured by the oven method is usually less than 0.1 percentage point. Therefore, this method is most used for research purpose even though it requires longer time than the other methods. For example, it requires 72 h to measure moisture content of corn and soybean at 103°C (ASABE 2019).

In the oven method, sample size to determine the grain moisture content is 5–15 g. Moisture content of single kernels can be determined by using a single kernel moisture meter which is based on the measurement of electric resistance, conductance, or hardness of each single kernel. The single kernel moisture can have about 5% standard deviation of the average moisture content, which depends on the pre-treatment of the grain. The standard deviation (wb, %) of moisture content of individual rice kernels at harvest and after drying and tempering process is from 3.46 to 5.09 and from 0.96 to 5.09, respectively (Li et al. 2003). Moisture content of single kernels in ears of corn generally decreases from the butt to the tip end with a difference up to 11 percentage points (Nelson and Lawrence 1991). The standard deviations of single-kernel moisture content of corn with similar average moisture content stored at five US locations varies from 0.6% to 1.2% (Engebretson 1994). The standard deviation of the moisture content of crossflow dried corn at an average moisture of 11.5%, decreases to 1.2% after 30 h of storage at 4.4°C in sealed plastic bags, and to approximately 0.65% after 500 h storage. However, for a crossflow dried sample with an average moisture content of 21.0 ± 3.7%, the standard deviation never decreased below 1 percentage point after drying (Montross et al. 1999). Wheat with average moisture of 14% might have individual kernels from 12 to 16%. Kernels with

16% moisture are prone to spoilage and can initiate spoilage in a bulk. The moisture movement through contact of different moisture content kernels is a slow process and moisture does not distribute fast enough to avoid initiation of spoilage by high moisture kernels. Thus, storage decisions should not be made on the basis of average moisture content of the bulk only, but also the variability of moisture content of individual kernels and location of high moisture content kernels in the bulk should be considered.

3.1.2 Water Distribution, Status, and Mobility

3.1.2.1 Distribution

Of all the resources that plants need to grow and function, water is the most required. A grain seed is part of a plant (parent of the seed) and the future of a young plant (parent of the young plant) if it is seeded. Therefore, stored grain kernels are living part of plants, and certain amount of water inside the seed is required to maintain its living ability. Processed food materials contain certain amount of water and will exchange water with its environment. Inside biomaterials, water is usually not uniformly distributed. Inside a wheat kernel, germ usually has higher moisture content than that in the region of endosperm and pericarp. Pericarp region has a low and more evenly distributed water (Ghosh et al. 2006). During drying of wheat, water is removed faster from the pericarp than from the endosperm, and the germ has the lowest water losing rate (Ghosh et al. 2006). During drying of corn, water is removed faster from the endosperm than the pericarp, and the scutellum has the lowest water losing rate (Kovacs and Nemenyi 1999). Moisture content inside seeds is also different under different mature stages and storage conditions (Garnczarska et al. 2007). Processed food products usually have more uniform distribution of water. However, this uniform distribution can be influenced by its storage environment and history of the stored product.

3.1.2.2 Status

Water inside biomaterials can be sorbed by the materials from few to several moles of water per mole of substance. Water exerts its solvent properties at water contents higher than the monolayer value. The hydration shell of the hydrated Li^+ ion is formed by four water molecules (Franks 1975), about ten water molecules are associated with phosphatidylcholine head groups (Crowe et al. 1998), urea is hydrated with seven moles of water, and guanidinium ion binds twelve moles of water (Nandel et al. 1998). Under suitable condition such as seeds with free water, the seed will sprout and microorganisms may grow on seeds, which will result in deterioration of the stored grain.

Status of water in biomaterials can be one or several of the following statuses: free water (unbound water), absorbed water which is attracted by capillary force or surface tension, adsorbed water which is attracted to the cell surfaces by intermolecular forces such as van der Waal forces, and bound water which is held by attractive force such as chemical bond force or H bond. It is usually difficult to exactly identify these water statuses inside biomaterials and it is believed that there are at least three types of interstitial water: water that is very strongly bound, weakly bound and loosely associated with polymers. The free water is loosely associated with polymers and only exists when moisture content of the biomaterial is high. The free water might be clustered. Thermodynamic and nuclear magnetic resonance (NMR) studies show that free water does not significantly exist when seed moisture content is low (Krishnan et al. 2004). Soybean seeds has strongly bound water at moisture content below 8%, weakly bound water (adsorbed and absorbed water) at moisture contents between 8% and 24%, and very loosely bound water (free water) at moisture content greater than 24% (Vertucci and Leopold 1984). For wheat kernels and flour dough, the free water occurs at about 23 or 24% moisture content.

At any moisture content, water with different statuses may coexist (Table 3.1). However, the percentage of water associated with a water status may be different at different moisture contents. For example, grain kernel

Table 3.1 Status of Water in Grain Kernels

Condition	*Description*	*Binding Type*	*Mobility*	*Heat*[a]	*Example*
Free water	Water dropping	None	Free	0	Wet surfaces
Capillary water	Coarse capillaries	Mechanical	Nearly free	0	At both inside and outside of kernels
	Fine capillaries	Physical	Nearly free	0–300	
Solute	Between molecules	Physicochemical	Decreased	0–1000	Among starches
Absorbed[b]	Sitting on surface	Physicochemical	Decreased	100–3300	Inside cells
Crystal	Crystal lattice	Chemical, stoichiometric	Non mobile	300–2200	Hydrated carbohydrate
Bound	In compounds	Chemical, stoichiometric	Non mobile	1000–6000	Bound with starch

Source: Adapted from Figura and Teixeira (2007).
[a] Heat of binding in J/g.
[b] Includes adsorbed.

at low moisture contents may have no significant amount of free water, but free water exists. The physical and chemical properties of this free water at this low water content will not show. Coexistence of water inside the lupin seeds was observed by Garnczarska et al. (2007). The same water fractions are observed in lupin seeds at both hydrating and desiccating. Lupin embryo tissues show the same water fractions at the same moisture content during both seed development and germination (Garnczarska et al. 2007). Measurement of dielectric property of wheat indicates the change of water status inside grain kernels during drying and tempering (Sokhansanj and Nelson 1988). This free water at high moisture content is important for fungi multiplication.

Skaar (1988) suggested that water has two phase sorption: monolayer or multilayer. Under monolayer condition, the attraction between the sorbent (vapour) and sorbate is larger than that between sorbent compounds at liquid condition. Under multilayer condition, the attraction between the sorbed molecules and sorbate is negligible. Two different hydration phases occur and termed as A and B shells by Lewicki (2000). The A shell water is similar to the fraction of non-freezing water. The B shell water continues to hydration level of about 1.4 g/g, which corresponds to the amount of water whose motional correlation times are different from that of bulk water in melting temperature and heat and entropies of fusion (Gregory 1998). More recent evidences show there are semi-bound water which is weakly associated with the polymer surface (Caurie 1981). This semi-bound water has no free water property. At low water content, the water status is mostly at charged and associable with the charged and polar chemicals of biomaterials. Additional water will result in water molecule clusters around polar groups at a lower affinity. More additional water will aggregate over nonpolar residues. These additional water molecules will show the physical properties of the free water. When wheat endosperm equilibrium RH is 88–99% (20–36% moisture content), water film with about 0.5–2.5 nm exists inside the endosperm (might be in capillaries) and this corresponds to approximately a four to five molecular layers of water molecules (Callaghan et al. 1979). During drying, the NMR technique show that the bound water content decreased in materials with different groups in the order: nonpolar groups, polar groups, and ionic groups (Kuntz 1971) because hydrophobic groups have very week interactions with water.

Physical and chemical properties of the bound water are different from that of free water, such as free water will become ice at 0°C while bound water will not. Some characteristics of bound water are lower vapour pressure, higher binding energy as measured during dehydration, reduced mobility, unfreezability at low temperature, and unavailability as a solvent. Bound water cannot act as free water such as solvent or reactant. Collagen containing 0.26 g water/g protein is unfreezable, but the water requires the same heat of fusion as bulk water at hydration level 0.60 g/g collagen. Bound water reduces the diffusion of water-soluble solutes in the sorbent, and decreases diffusion coefficient with decreasing moisture content. The decreased diffusion reduces drying rate because of slower diffusion of water to the surface of the drying materials. There is no single freezing temperature of water in foods. The thermodynamic freezing of water in the soybean protein occurs over a range as broad as 40 K (Johari and Sartor 1998) due to perturbations of the ice-water equilibrium by the interactions of water with protein.

At low water activities, water in monolayer is strongly absorbed by polar sites of solid materials. More layers will be sequentially added with the increase of water activity. The number of water layers, calculated by Brunauer–Emmett–Teller (BET) model in wheat, could be eight. The force acting between the polar sites and different layers of water gradually decreases with the increase of water layers, which results in different densities of water in different layers. The water in wheat flour up to the BET monolayer moisture content of 7% had a density of 1.482 g/cm^3, and the second layer moisture up to 14% has a density of 1.11 g/cm^3, while the third and higher layers had a density less than unity (0.967 g/cm^3) (Gur-Arieh et al. 1967). This conclusion was proved by other researchers in soil (Mackenzie 1958).

3.1.2.3 Water Mobility

Water molecules are in constant motion regardless of their status in liquid, gas, solid or inside food materials. The mobility of water molecules in a biomaterial is exceedingly complex. The reason for this complexity is its complex nature inside the material, the mobility during measurement period, the complexity of its determination and measurement method, and technique used to interpret the measurement result. For example, the water mobility response obtained is dependent on the physical and chemical properties of the material and measurement methods. The type and distance scale of water mobility also varies widely. In the case of molecular water mobility, the rotational and translational motions of the liquid phase of water are usually the types of mobility. Transverse relaxation times and NMR longitudinal are a measurement of the molecular rotational mobility of the water, and the NMR diffusion coefficient is a measurement the rotational and translational mobility of the water in the food matrix. The vibrational motion of the water can be measured using infrared and Raman spectroscopy. In the polymer science approach, the increase in mobility is measured as the material transfers from the glassy (least mobile) to the rubbery (more mobile) state as temperature and/or moisture content changes.

The macroscopic translational mobility of water molecules from inside to outside of the biomaterial results in the change of partial vapour pressure around the material. The water sorption by the material may not change the partial vapour pressure around the material, but the moisture content of the material may change. This mobility is mainly governed by the water gradient and effective diffusivity. The effective water diffusivity is usually used to evaluate the water mobility of the materials.

3.1.3 Vapour Pressure, Relative Humidity, and Water Activity

3.1.3.1 Vapour Pressure

Intermolecular attractive force among water molecules in gas (vapour) status is much smaller (almost negligible) than that in the liquid or solid status. Inside an enclosure containing liquid water and air space, some of the water molecules have enough kinetic energy that they can escape the attractive force of other liquid water molecules and move into the air space and become vapour. At the same time, some gas molecules move into the liquid water due to its kinetic movement and the attractive force from the liquid water molecules. A dynamic equilibrium is established between molecules leaving the liquid and entering the gaseous phase and molecules leaving the gaseous phase and entering the liquid phase. At equilibrium the movement of molecules between liquid and gas does not stop, but the number of molecules in the gaseous phase stays the same – there is always movement between phases. So, at equilibrium there is a certain concentration of molecules in the gaseous phase; the pressure produced by the gas molecules inside the air space is the vapour pressure. Pressure can be observed inside the vapour space and this measured vapour pressure is the vapour pressure at this measured temperature. If the enclosure has pure free water, this measured vapour pressure is the saturated vapour pressure at the measured temperature. This saturation vapour pressure exerts on the walls of the enclosure, and the surface of the liquid water. Inside an enclosure containing biomaterials and air space, the similar equilibrium mentioned above occurs. Therefore, the definition of the saturation vapour pressure is: the pressure exerted by vapour in thermodynamic equilibrium with its condensed phases (solid or liquid) at a given temperature in a closed system. The term saturation indicates that no more water molecules can stay inside the air space than that at the equilibrium.

In the case of a plane or flat surface of water, this saturation vapour pressure is a function only of temperature. When the water inside the enclosure is heated (so the temperature of the enclosure increase), the added heat energizes the water molecules so that more of them escape into the vapour space. This results in the increasing of the saturation vapour pressure. The saturation vapour pressure of any substance increases nonlinearly with temperature according to the Clausius-Clapeyron equation.

One of the most accurate expressions that relate the saturation vapour pressure and temperature is the Goff and Gratch (1946) equation in the range of 100°C (212°F) to –107°C (–160°F).

$$P_s = e_{st} 10^z$$

$$z = a\left(\frac{T_s}{T} - 1\right) + b\log_{10}\left(\frac{T_s}{T}\right) + \text{c}\left(10^{d\left(1-\frac{T}{T_s}\right)} - 1\right) + f\left(10^{g\left(\frac{T_s}{T}\right)} - 1\right) \quad (3.5a)$$

where $e_{st} = 101{,}324.6$, $a = -7.90298$, $b = 5.02808$, $c = -1.3816 \times 10^{-7}$, $d = 11.344$, $f = 8.1328 \times 10^{-3}$, $g = -3.49149$, P_s is the saturation vapour pressure (Pa), T is the absolute temperature (K), and T_s is the steam (boiling) point temperature ($T_s = 373.15$ K). There are simpler expressions, and the most commonly used is the equation proposed by Hunter (1987) with accuracy within ±0.3% in the range 0–60°C.

$$P_s = \frac{6 \times 10^{25}}{T_c^5} e^{\left(\frac{-6800}{T_c}\right)} \quad (3.5b)$$

where T_c is the temperature (°C). ASABE (2016a) recommends the following equation to calculate the saturation vapour pressure:

$$lnP_s = 31.9602 - \frac{6270.3605}{T_K} - 0.46057\ lnT_K \qquad 255.38 \le T_K \le 273.16$$

$$\ln\left(\frac{p_s}{H}\right) = \frac{A + BT_K + CT_K^2 + DT_K^3 + ET_K^4}{FT - GT_K^2} \qquad 273.16 \le T_K \le 533.16 \quad (3.5c)$$

where $A = -27{,}405.526$; $B = 97.5413$; $C = -0.146244$; $D = 0.12558 \times 10^{-3}$; $E = -0.48502 \times 10^{-7}$; $F = 4.34903$; $G = 0.39381 \times 10^{-2}$; $H = 22{,}105{,}649.25$; T_k is the temperature (K). Another empirical equation to calculate the saturated vapour pressure is:

$$P_s = 0.611e^{\frac{17.502T_c}{240.97+T_c}} \quad (3.5d)$$

where T_c is temperature (°C).

Vapour pressure inside a sealed chamber can be calculated by using the measured dew point temperature.

$$P_s = ae^{\frac{bT_c}{c+T_c}} \quad (3.5e)$$

where $a = 0.611$ kPa; $b = 17.502°\text{C}^{-1}$; $c = 240.97$°C, and T_c is the measured dew point temperature (°C).

Inside an enclosure (such as a grain bin), other components in a mixture also contribute to the total pressure of the enclosure. The pressure of a single component in a mixture, which contributes to the total pressure in the system, is called partial pressure. For example, air at sea level saturated with water vapour at 20°C, has partial pressure of 2.3 kPa of water vapour, and 99.9 kPa of air, totalling 102.2 kPa. Inside a grain bin, the interstitial air has the partial pressure of air and vapour. Partial pressure due to volatile components inside the interstitial air of the bin is usually negligible. During fumigation and controlled atmosphere storage, the pressure of the fumigants might be higher than the total pressure of outside air due to the contribution of applied fumigant or atmosphere gas (such as CO_2).

In an open container the molecules in the gaseous phase will just fly off and an equilibrium would not be reached, as many fewer gaseous molecules would be re-entering the liquid phase. This is a common status inside the headspace of bin with a low airtightness, and water and air inside the grain surface will exchange with the air and water inside the headspace. Under the same condition, the interstitial air inside the grain bulk might be close to equilibrium if the air flow inside grain is negligible. For example, grain moisture content will be close to its equilibrium moisture content inside the grain bulk if there is no ventilation inside the grain bulk.

3.1.3.2 Relative Humidity (RH)

The most used definition of RH of an air-vapour mixture is the ratio of the partial pressure of the water vapour in the air-vapour mixture to the saturation vapour pressure of water at the same temperature. A vacuum also has the capacity to hold water vapour as the same volume filled with air. Hence, there is a RH inside a vacuum. RH expresses the percentage of the saturation at the temperature and pressure of interest. On the psychometric chart and at a given temperature, RH of an air-vapour mixture corresponds to the moisture content (kilogram of water per kilogram of dry air) of the air. For example, the moisture content of air-vapour mixture at 60% RH and 20°C is about 0.009 kg water/kg of dry air. The moisture content of the air at 100% RH (saturation) and 20°C is about 0.015 kg water/kg of dry air. The ratio of 0.009/0.015 is 0.6 or 60%. Therefore, RH is also the measurement of the amount of water in the air at the air temperature and the given pressure, and the percentage of moisture content inside the air-mixture over the saturated moisture content at the temperature and pressure.

The vapour pressure in an air-vapour mixture of a container holding free water at 20°C is 2.34 kPa. The RH in the air-vapour mixture of the container is 100% (Fig. 3.1). The RH in the air-vapour mixture inside a container holding wheat kernels with different moisture contents will be less than 100%, and decreases with the decrease of wheat moisture contents (Fig. 3.1).

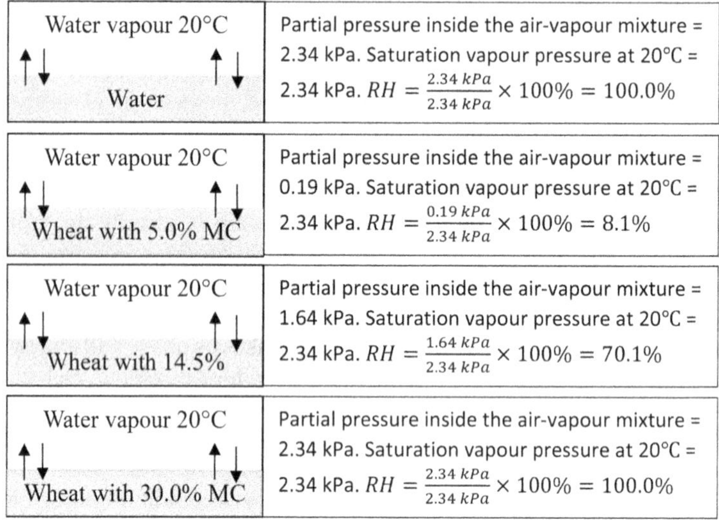

Water vapour 20°C
Water
Partial pressure inside the air-vapour mixture = 2.34 kPa. Saturation vapour pressure at 20°C = 2.34 kPa. $RH = \frac{2.34\ kPa}{2.34\ kPa} \times 100\% = 100.0\%$
Water vapour 20°C
Wheat with 5.0% MC
Partial pressure inside the air-vapour mixture = 0.19 kPa. Saturation vapour pressure at 20°C = 2.34 kPa. $RH = \frac{0.19\ kPa}{2.34\ kPa} \times 100\% = 8.1\%$
Water vapour 20°C
Wheat with 14.5%
Partial pressure inside the air-vapour mixture = 1.64 kPa. Saturation vapour pressure at 20°C = 2.34 kPa. $RH = \frac{1.64\ kPa}{2.34\ kPa} \times 100\% = 70.1\%$
Water vapour 20°C
Wheat with 30.0% MC
Partial pressure inside the air-vapour mixture = 2.34 kPa. Saturation vapour pressure at 20°C = 2.34 kPa. $RH = \frac{2.34\ kPa}{2.34\ kPa} \times 100\% = 100.0\%$

Figure 3.1 Relative humidity of air inside a container holding wheat with different moisture contents.

3.1.3.3 Measurement of RH

There are different methods to measure the RH. Here only briefly introduce the common methods used in the grain industry.

3.1.3.3.1 Psychrometric Method (Wet and Dry Bulb Method) It does not directly measure humidity, but rather measures wet and dry bulb temperatures to find RH by using the Psychrometric equations or charts. The sensing elements can be thermometers, resistance temperature detectors (RTD), or thermistors. The common one is to use thermometers to detect wet and dry bulb temperatures by swinging the wet bulb around, so air movement across the wet bulb is created. This method has good precision if the airflow is well controlled. The accuracy can be ±2 percentage points in the humidity range from 10 to 100% at 0–60°C.

3.1.3.3.2 Hygrometers, Chilled Mirror Dew Point This method is based on thermodynamic principles of dew point and do not require calibration. The system typically consists of a mirror, optical sensor, thermoelectric cooler, internal fan, and infrared thermometer. The thermoelectric cooler precisely controls the mirror temperature, the optical sensor detects droplets on the mirror, and the infrared thermometer measures the temperature on the mirror. The fan will increase the air circulation inside the system and to reduce vapour equilibrium time. The RH is calculated as:

$$\mathrm{RH} = \frac{P_s(T_d)}{P_s(T_s)} \tag{3.6}$$

where $P_s(T_d)$ is the saturated vapour pressure at dew point (Pa), and $P_s(T_s)$ is the saturated vapour pressure at the sample temperature. The value of $P_s(T_d)$ and $P_s(T_s)$ can be calculated by Eq. 3.5. The measurement range is 3–100% RH with accuracy of ±0.3 percentage points of RH and measurement time is less than 5 min. This method can only be used for non-volatile materials or use charcoal to treat the materials before measuring.

3.1.3.3.3 Electric Hygrometer The electric hygrometer consists of salt film or proprietary hygroscopic polymer film and meter to measure electrical current or voltage. After the water is absorbed by the film and the change of the electrical properties of the film is detected by the electrical meter. The disadvantages of the electric hygrometer are that each sensor should be carefully calibrated, and this calibration should be also conducted after a period of application because the electrical properties of the film might change by the environment. The reason for the change of the electrical property is that the air and film must come to thermal and vapour equilibrium for accurate RH measurement. Dust in the environment might cover the film, and the chemicals of the environment might have chemical reactions with the film. There are detailed reviews for this type of film sensors (Rahman and Sablani 2001; Rodel 2001). Mechanical and chemical filters are available to protect the sensor from contamination, but these filters usually increase the equilibrium time. This sensor is commercialized by different companies and widely used inside stored grain bins to measure RH inside stored grain bulks. The measured RH and temperature are used to estimate grain moisture contents by using the sorption isotherm equations of the grain.

The other methods to measure RH are hair or polymer hygrometer, thermocouple psychrometer, isopiestic method, vapour pressure manometer, and freezing point depression method. These methods are reviewed by Fontana (2007).

3.1.3.4 Water Activity

Scott (1957) and Salwin (Salwin 1959; Salwin and Slawson 1959) independently introduced the concept of water activity (a_w, decimal), which is defined as the ratio of free water to the total water inside a substrate (such as food materials) or microenvironment. It describes the degree of "boundness" of water and hence, its availability to participate in physical, chemical, and microbiological reactions. From the thermodynamic standpoint, a_w is a measure of the ability or tendency of water molecules to escape from the liquid phase to the vapour phase. At low pressures the difference between the ratio of fugacity and the ratio of the partial vapour pressure of water is less than 1%, so a_w is also termed as the ratio of the water-vapour pressure in the substrate to the water-vapour pressure of pure water at the same temperature: $a_w = \frac{p}{p_0} = RH$. Even though water activity of a material equals to the value of the equilibrium relative humidity (ERH) of a container holding the same material, the RH is usually used to express the percentage of saturation of air, and water activity is usually used to express the measurement of free water inside solid or liquid materials. When food materials equilibrate with its surrounding atmosphere, the water activity equals to the RH of the interstitial air inside the materials. Therefore, water activity has the same value as the RH of food materials at the equilibrium condition, and these two terms are used interchangeably sometimes.

Water activity describes the energy status or escaping tendency of the water in a sample. It indicates how tightly water is "bound," structurally or chemically, in products. Therefore, both the water content and the water activity of a sample must be specified to fully describe its water status. Water activity is closely related to the partial specific Gibbs free energy of the system. Pure water is taken as the reference or standard state from which the energy status of water in food systems is measured. The Gibbs free energy of free water is zero; thus, the water activity is 1.0 under saturation condition.

Around 0°C, the melting curves of water in high moisture foods can be used to identify the free water from bound water. The amount of free water related to the total amount of water can be calculated as:

$$w = \frac{\Delta H_m}{(1-S)\Delta H_w} \tag{3.7}$$

where $1 - S$ is the percentage of water, ΔH_m is the measured fusion enthalpy (J/kg), and ΔH_w is the fusion enthalpy of pure water (J/kg).

3.1.3.5 Relationship between Moisture Content and Water Activity

For the historical reason, a_w is not commonly used in grain storage but has several distinct advantages for specifying conditions related to seed longevity. The relationship between moisture content and longevity differs for different crop seeds, but the relationship between a_w and longevity is linear and similar from species to species (Roberts and Ellis 1989). The loss rate of seed viability increases with the increase of water potential in the range of –350 to –14 MPa. These values correspond roughly to 0.1 ad 0.9 a_w. Below –350 MPa of water potential (0.077 a_w), there is little change in longevity with decreasing moisture (Roberts and Ellis 1989). This lower water potential corresponds to seed water contents ranging from 2 to 6% of cereal grains.

The moisture content is used to represent the amount of water inside solid or liquid materials. RH and water activity depend on the temperature and pressure of the system, while the moisture content is independent on the temperature and pressure. Therefore, moisture content alone is not a reliable factor for the prediction of microbial multiplication and chemical reactions in stored products because bound water is unavailable to microbes under certain moisture content and high moisture content does not means that the product has a high water activity. Water activity is more closely related to the physical, chemical, and biological properties of biological products than its total moisture content (Table 3.2). For chemical and biochemical reaction, and multiplication and growth of microorganisms, most of them require free water to act as a solvent and for hydraulic transport of molecules across semi-permeable membranes. Lipid oxidation shows a minimum in the 0.2–0.35 a_w range and increases in rate on both

Table 3.2 General Storage and Handling Properties of Crop Seeds at Different a_w

a_w	*Water Status*	*Properties Related to Storage and Handling*
<0.2	Bound water, negligible monolayer	Over dried, glassy, brittle, hard, lipid oxidization
0.2–0.45	Monolayer water, negligible multiple layer water	Over dried, glassy, brittle, hardness, lipid oxidization with increase of storage time
0.45–0.60	Few multiple layer water	Over dried, loss of glassy and brittle, decreased hardness, exhibits rubberised properties
0.6–0.7	Some multiple layer water	Safe storage moisture content, right hardness and elasticity for storage and transportation. Negligible multiplication of microorganisms
>0.7	Multiple layer and free water	Damp, soft, decreased rubberised properties, multiplication of microorganisms

sides (Labuza and Dugan 1971). Therefore, a very dry food ($a_w < 0.2$) would most likely degrade by lipid oxidation.

3.1.3.6 Water Activity and Microorganism Multiplication

Water activity is usually used to express the water status for the multiplication of microorganisms. Walter (1924) might be the first researcher to relate relative water vapour pressure to microbial growth, the main cause of food spoilage. Many fungi in stored grain have a minimum requirement of water activity, and different species have different minimum requirements. Although general in nature, these reactions require an aqueous phase, there is a lower limit, usually at a_w between 0.2 and 0.3. Below this limit, the reactivity is 0. Above that, the reaction rate increases until reaching a maximum a_w essentially between 0.6 and 0.8, and then decreases again, reaching 0 at a_w of 1.0. The minimum requirement of most stored fungi is 0.65 a_w. No microbe can grow well at ≤0.6 a_w inside stored grain. Moulds can generally grow at lower a_w than yeast, such limits are lower than that for bacteria. Therefore, for safe storage of seeds, lower than 0.65 water activity usually eliminates the multiplication of microorganisms. However, chemical reactions and enzymatic changes may occur at considerably much wide water activities.

3.1.3.7 Surface Properties and Water Activity

Three major physical effects that lower water activity are:

1. Colligative effect. When solids dissolve in water, these interact with the water through dipole-dipole, ionic, and H-bond. This interaction is defined as the colligative effect. The water physical properties such as boiling point, freezing point, and vapour pressure will change due to this colligative effect. The chemical compounds on the surface of the food materials influence this colligative effect. Some hygroscopic compounds, such as salts, sugar, carbohydrate with short carbon chains, can form stable hydrates through colligative effect. Therefore, increasing of these chemical compounds on the surface of food materials will cause the increase of moisture content, but decrease the a_w of the food materials.
2. Capillary effect. The vapour pressure of water above a curved liquid meniscus is less than that of pure water because of change in the H-bond between water molecules as a result of surface curvature (tension). Capillary pressure inside the capillary holding pure water can be calculated as:

$$\varphi_p = -\frac{2\tau}{r} \tag{3.8}$$

where φ_p is the capillary pressure (Pa), τ – water surface tension (N/m, or Nm/m²), and r is the capillary radius (m). Equation 3.8 indicates the capillary pressure increases with the decrease of the capillary size. The pores in food materials (such as inside a wheat seed) are in the 10–300 μm range and these pores will lower the a_w due to the capillary effect which can be predicted by the Kelvin equation:

$$ln\frac{P}{P_s} = lna_w = \frac{2\tau V_l \cos\theta}{rRT_k} \text{ and } \frac{V_l}{R} = \frac{V}{nR_sM} = \frac{1}{\rho R_s} \text{ and } \rho = \frac{nM}{V} \quad (3.9)$$

where P is the vapour pressure (Pa), a_w is the water activity (decimal), V_l is the molar volume of liquid (m³/mol), θ is the wetting angle (°), R is the gas constant (8.31451×10^{-7} J·k⁻¹·mol⁻¹), V is the water volume (m³), n is the amount of water (mol), R_s is the specific gas constant (J·K⁻¹·kg⁻¹), and M is the molar mass of water (kg/mol).

The Kelvin equation predicts that the a_w is 0.989–0.999 for the 10–300 μm pores. Inside grain kernels, about 5–7% of the pore size is 0.01–0.001 μm, which can lower the a_w to 0.899–0.34. Thus, smaller capillaries have a great effect on lowering the a_w, and those capillaries are the last to be emptied by drying. The larger pores will be filled at the first when food materials have a high moisture content. These phenomena will influence the isotherm of the food materials.

For non-porous surface, mono- or multi-layer adsorption occurs. For porous surface, capillary adsorption occurs when voids in the cellular structure of seed surface are of the size which can hold the water in liquid form by forces of surface tension of the absorbed free water. The size which can hold the water by surface tension is between 1 nm to 25 nm.

3. Surface interaction. When dry food materials are exposed to an environment with certain vapour pressure, the surface of the food materials will interact with vapour of the environment. The interaction will be the water absorption and/or adsorption on the surface of the food materials through dipole-dipole, ionic bonds, van der Walls forces, and the H-bond. The water molecules progressively and reversibly combine with the solids through chemisorption, physical adsorption, and multilayer condensation. These bonded or combined water molecules on the surface of the food materials are less free to become vapour, resulting in reduced a_w. Below this moisture content at which the food materials only have monolayer water molecules, generally speaking, reaction that depends on water as a reaction phase medium, does not occur at appreciable rates. Water sorption occurs on the monolayer at the first, and

bonded water molecules mostly stay in the monolayer. When more water molecules inside the environment are available, multilayer sorption occurs and the attraction force between the surface of the food materials and the water in the multilayer is weak. Once the RH in the environment exceeds a critical value, deliquescence or dissolutions of the solid may occur. Deliquescence is a first-order phase transformation because of the high water solubility of the solid and its significant effect on the colligative properties of water. Because the water in the multilayer is weakly attracted, the a_w will increase until be close to that of the pure water.

All three effects occur over the entire moisture range of food materials and result in the characteristic moisture sorption isotherm. One effect, which may be a major factor at the certain moisture consent, may become minor factor when the moisture content of the food materials changes. Generally speaking, sorption isotherm of a material is independent of pore size distribution (Xiong et al. 1991) because the water molecules are associated with the polar groups of individual molecules of the materials and are not just adsorbed/absorbed inside pores.

3.1.3.8 Water Activity Measurement

There are several methods to measure water activity, ranging from simple laboratory procedures to specialized equipment. The method used to measure RH (such as chilled mirror dew point method) can also be used to measure the a_w. There are several extensive review on the water activity measurement and equipment (Fontana 2007; Yu et al. 2008). The methods used to measure the moisture sorption and/or desorption isotherm can be also used to measure the water activity. There are many studies that using Clausius-Clapeyron equation to estimate the isosteric heat of sorption and prediction of the a_w (Tadapaneni et al. 2017).

3.2 Sorption (Adsorption and Desorption) Isotherm

3.2.1 Sorption and Desorption

3.2.1.1 Adsorption and Absorption

People often get confused with the adsorption and absorption because the phenomena described by them are similar and connected with each other. Adsorption is the process that a substance (adsorbate) is attracted to a surface of another material (absorbent) rather than be diffused into the material. On the contrary, absorption is the process when some molecules, atoms, and/or ions enter or diffuse into other materials and/or chemically bound with the materials. For example, diffusion of an organic molecule inside a grain kernel is an absorption process, while sticking of the same molecule on the surface of grain kernels is an adsorption process. Both absorption and adsorption are sorption processes, and absorption might occur at the same time as the adsorption and it depends on the physical and chemical properties of the material, and the environmental condition such as temperature and RH. It is usually difficult to separate the absorption from the adsorption. Therefore, sorption is usually used to refer to both the adsorption and/or absorption processes.

3.2.1.2 Sorption and Desorption

Water sorption and desorption by foods is a process wherein water molecules progressively and reversibly combine with the food solids via chemisorption, physical adsorption, and multilayer condensation. The rate of sorption and desorption at the beginning of the process is governed largely by physical and chemical characteristics of the surface boundary such as temperature, RH, convection velocity, and surface structure of the grain kernels. The surface structure of the kernels includes roughness, shape, size of the kernel and the size and shape of the pores among grain kernels. A two-step process is involved in the water sorption and desorption of food materials. In the first step, water vapour must transfer across the boundary surface of the materials in the form of sorption or desorption, depending upon whether water vapour is entering or existing the material. The second step involves diffusion or adsorption of the water vapour out of or into the interior tissue structure of the material. When surface is dry and in the desorption process, water movement inside the material will be the main factor influencing the desorption process.

Surface adsorption can be molecular adsorption, corresponding to bound water due to colligative effect, or capillary adsorption and surface tension, corresponding to free water and it depends on the RH of the environment surrounding the surface. When the distance between the adsorbed water molecule and the cell wall of the grain kernel is small enough, the force of attraction might draw the water molecule into the inside of the cell wall. These water molecules might continually diffuse into the inside of the cell if the cell has enough moisture gradients. Increasing with the water adsorption, the attraction between molecules lessens and the volume of the solid-water increase. This increased volume will be close to the volume of the water adsorbed. Along with the increasing of adsorbed water, the more adsorbed water will become free water and be held by the capillary force and surface tension. The reverse process is the desorption.

The basic sorption and desorption can be expressed as:

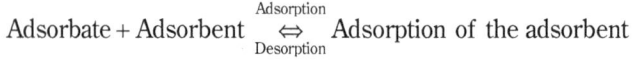

$$\text{Adsorbate} + \text{Adsorbent} \underset{\text{Desorption}}{\overset{\text{Adsorption}}{\Leftrightarrow}} \text{Adsorption of the adsorbent}$$

The direction of the dynamic equilibrium will shift in the direction where the stress can be relieved. For example, the desorption will shift to the sorption direction if the seeds are stored at higher a_w environmental conditions (Fig. 3.2).

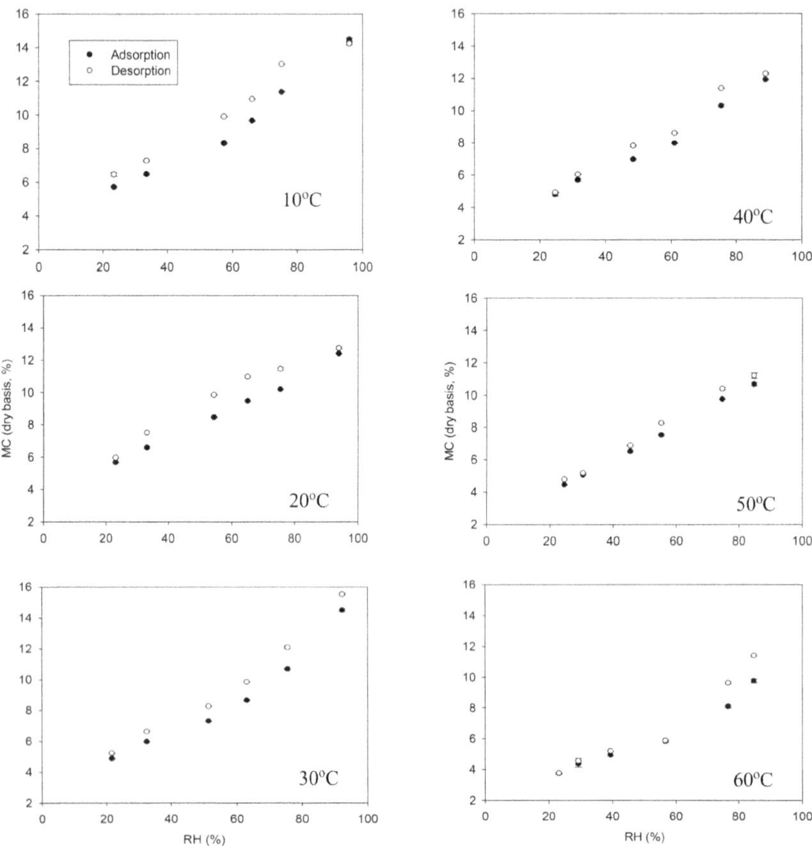

Figure 3.2 Adsorption and desorption isotherm of the dry and wet hemp seeds with 0% dockage under constant temperature and RH. Error bars do not show because all standard error ≤0.2%. (Reprinted with permission from Jian et al. (2018)).

3.2.2 Sorption Isotherm

3.2.2.1 Sorption and Desorption Isotherm

The thermodynamic relationship between equilibrium moisture content and water activity (or partial vapour pressure, RH), as a function of temperature and pressure, is commonly expressed in sorption isotherms. Usually, the pressure is the atmospheric pressure. Therefore, a plot (or mathematical equation) of the equilibrium moisture content versus air RH (water activity) at the corresponding temperatures is termed as the sorption or desorption isotherm of a material. The sorption or desorption isotherm implicates the following three information:

1. the existence of two different classes of water in the material and the environment around the materials: free water (as vapour or liquid) and not-free water (absorbed, adsorbed, and bound water) inside the material.
2. the boundary on the surface of material and this boundary separates the material from its environment.
3. thermodynamics on the boundary. Water transition between free water (as vapour or liquid) and bound water occurs inside the material and not on or outside the boundary.

For ease of interpretation, sorption isotherms are often classified as five types of isotherms (Fig. 3.3) (Brunauer 1945). Type I isotherm is the well-known Langmuir isotherm and is limited to the completion of a single monolayer of adsorbate at the adsorbent surface such as adsorption of nitrogen or hydrogen on charcoal (Fig. 3.3). Type I isotherms are typical for anticaking agents since they can hold large amounts of water at low water activities. There are a number of binding sites, but once occupied, there is little increase in moisture content at increasing relative humidifies because all the sites are filled. Type I isotherms are typical of very hygroscopic materials (Fig. 3.3).

The other four Types of sorption isotherms represent the mixture of monolayer and multilayer adsorption (Fig. 3.3). No special names have been attached to the types II and IV, and Type III is known as the Flory-Higgins isotherm. Type II and III isotherms do not exhibit a saturation limit as does Type I. Type II isotherm has the sigmoid shape and indicates an infinite multi-layer formation after completion of the monolayer and is found in adsorbents with a wide distribution of pore sizes. There are two main bends in type II, one at a_w of 0.2–0.4 and another at 0.6–0.75 (Fig. 3.3). The low a_w bend is due to build-up of monolayer and filling of small pores. The high a_w bend is due to swelling, filling of large pores, and then solute dissolution. An example is the adsorption of water vapour on carbon black at

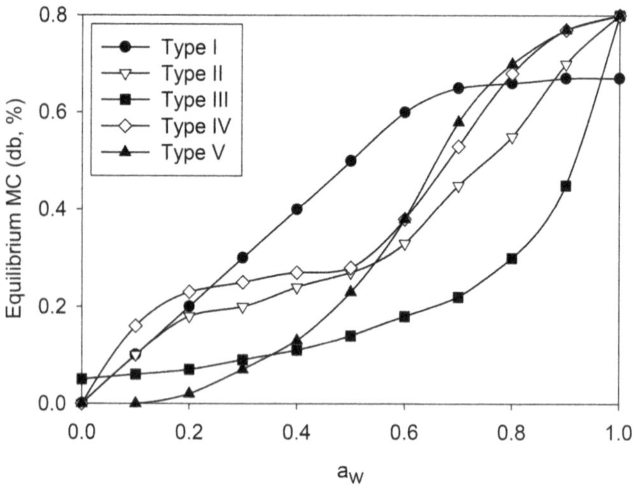

Figure 3.3 Types of adsorption isotherms. (Adapted from Brunauer (1945)).

30°C. The intermediate flat region in the Type II isotherm corresponds to monolayer formation. Most food materials have Type II isotherm, and foods that contain large amount of sugar (small soluble molecules) have J-type shape (Type III).

Type III shows there is no monolayer adsorption (Fig. 3.3). Moisture in the food is very low up to the point where the crystals begin to dissolve in the adsorbed water at the surface of the crystal, known as the deliquescent point. Example of Type III adsorption isotherm is bromine (Br_2) at 20°C on silica gel. Food materials that are composed mainly of crystalline components such as salt and sugars usually have Type III (J-shape) isotherm.

Type IV and V isotherms show phenomenon of capillary adsorption and the adsorption terminates near to a relative pressure of unity. Type IV isotherm shows monolayer adsorption at low water activities and adsorption saturation level reaches at a pressure below the saturation vapour pressure. This might be the gases condensing in the tiny capillary pores of adsorbent at pressure below the saturation pressure of the gas. Example of Type IV isotherm is the adsorption of water vapour on activated carbon at 30°C. Most food products have type IV isotherm. Type V is the BET multilayer adsorption isotherm. Example of Type V isotherm is adsorption of water (vapours) at 100°C on charcoal. Types II and III are closely related to Types IV and V, except that the maximum adsorption occurs at a pressure lower than the vapour pressure of the gas.

The adsorption and desorption isotherms of most stored biomaterials follow the sigmoidal equation (Fig. 3.3). This typical sigmoidal isotherm has been reported for starchy products (Viollaz and Rovedo 1999), crops seeds (Pixton and Warburton 1977a,b; Hunter 1987; Cenkowski et al. 1989; Mizuma et al. 2008; Sathya et al. 2009; Sun et al. 2014), fruits (Kaymak-Ertekin and Gedik 2004), vegetables (Kiranoudis et al. 1993), meats (Lomauro et al. 1985), and biomass materials (Tsami 1991).

3.2.2.2 Isotherm Shape and Hygroscopic Property

The shape of the sorption isotherm can be used to characterize food quality and storage condition as a function of water activity. The sorption isotherm can be divided into three regions (Fig. 3.4), which reflect bound, semi-bound, and free water status at the liquid-solid interface. In the region I (convex), all the water is bound and difficult to be used by microbial reactions. The water behaves as part of the solid, but is still quite exchangeable. The ERH is usually less than 35%. Most of the food material will be completely shrunken, hard, and brittle. At the same time, it is costly to dry food materials to such low moisture content.

Water activity in the region II (linear region) is about 0.25–0.3 to 0.7–0.75 (crop dependent), and most food spoilage reactions due to microorganisms cannot proceed for long time (few years), but the rate of quality loss begins to increase above 0.3 a_w for most chemical reactions. For most dry processed food, an increase in a_w by 0.1 unit in this region decreases shelf life two to three times. In this region, water is at the status of monolayer adsorption, or reaching the status of saturation of monolayer, but no free water is available. The water is slightly less mobile than bulk water. As water is added to the low-moisture end of this region, it exerts a significant plasticizing action on solutes, lowers their glass transition temperatures, and causes swelling of the solid matrix. At this a_w, the amount of water adsorbed on monolayer is enough to affect the dielectric properties and water can become solvent. The food moisture content at this status is called "intermediate moisture," and the water content at this status is sufficiently high to give a soft flexible texture to the food materials. However, microorganism usually will not develop well. Therefore, this intermediate moisture content is the recommended water activity of stored food materials.

In region III (concave), multilayer water is formed, and free water is available for chemical, biochemical, and biological activity. The water is clustered (Zhao et al. 2017) and freezable, and easily removed by drying. Water clustering is one mechanism proposed to interpret the sharp upturn in an isotherm curve. This water is available for microbial growth and enzyme activity. The multilayer molecular adsorption of water (concave) is usually completed at relative humidities greater than 70% and represents as much as 85% of the total absorbed water at relative humidities above 90%. At equilibrium RH of 90–100%, absorbed water is held loosely in capillary

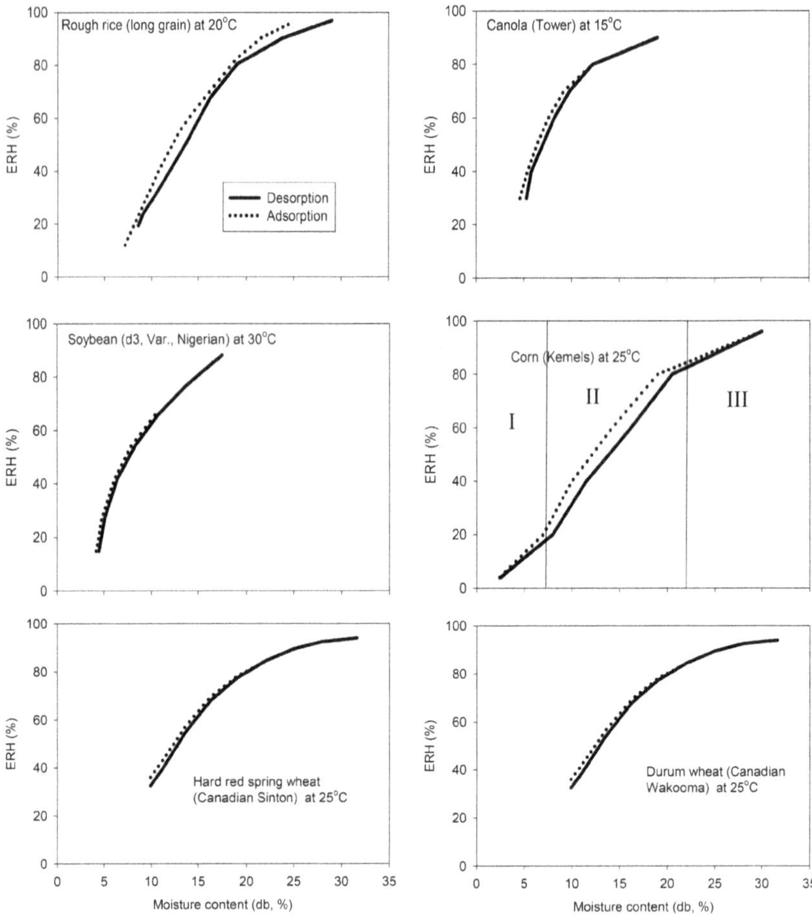

Figure 3.4 Equilibrium moisture content and equilibrium RH relationships of common crops at selected temperatures (data source: ASABE 2016c). Solid lines show the desorption, and dashed lines represent the adsorption. Graph of corn at 25°C also shows the hysteresis regions (Heldman and Lund 1992).

pores with diameters of less than 1 mm. Therefore, food materials stored at this water activity with high temperature is not recommended and should be avoided.

A material is called hygroscopic if it is able to bind water with a simultaneous lowering of vapour pressure. Molecular structure, their solubility, and the extent of reactive surface influence the hygroscopic property. Adsorption and desorption of water of stored grain shows the integrated hygroscopic properties of the complex constituents, and the interaction of food biopolymers with water. Biomaterials have complex compositions and structures, which will influence its sorption isotherms. Oil inside the oilseeds do not adsorb water. The water inside oilseeds will be absorbed by or stay at the no-oil components (such as protein, starch, and fibre). Therefore, for the same amount of dry mass and the same ERH, the non-oil component inside the oilseeds will have more water than that inside the cereal seeds. The oilseeds will have higher equilibrium RH than that of cereal grain when both have the same equilibrium moisture content. This explains why the safe storage moisture content of oilseed is lower than that of cereal seeds.

3.2.3 Factors Influencing Isotherm

3.2.3.1 Temperature and RH Effect

The equilibrium moisture content of a material is closely related to the RH. Partial pressure of free water in solids increases with the increase of temperature at roughly the same ratio as the increase in partial pressure in gases. The partial pressure of free water at different temperatures can be found from a psychrometric chart or can be calculated using equations (ASABE 2016a). The temperature dependence of the sorption isotherms inside solids with no free water is slightly different from the psychrometric chart. This difference can be seen in Figs. 3.5 and 3.6, where the RH (a_w) increases with increase of temperature at the same equilibrium moisture content, or equilibrium moisture content slightly increases with decrease of temperature at the same RH (a_w). This indicates that the food materials become less hygroscopic (holding less water) when temperature is

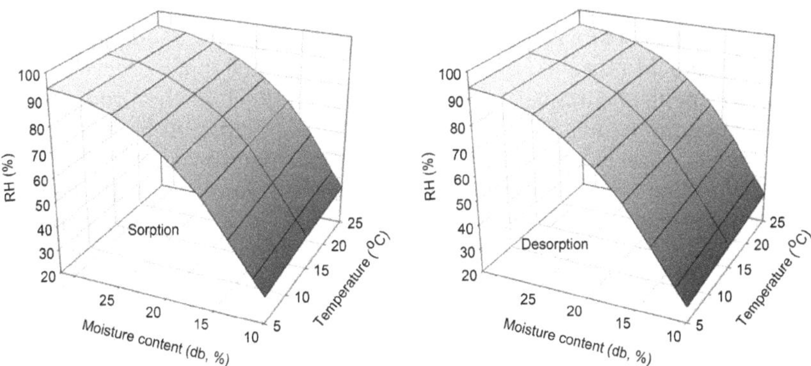

Figure 3.5 Equilibrium moisture content, equilibrium relative humidity and temperature of red spring wheat under sorption and desorption conditions (Data from ASABE 2016b).

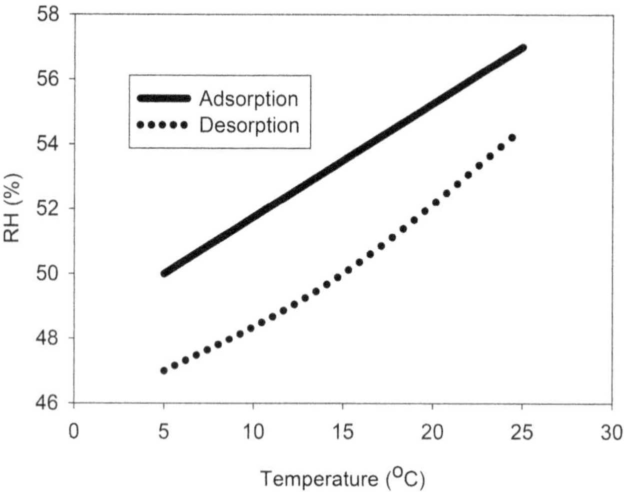

Figure 3.6 Relative humidity of red spring wheat with 13.6% initial moisture content (wet basis) under adsorption and desorption conditions (Data from ASABE 2016b).

increased due to the nature of the water binding and the increase of water vapour pressure inside the intergranular space of the material. At higher temperatures, some water molecules are activated to energy levels that allow them to break away from their sorption sites, thus decreasing the equilibrium moisture contents. The increased state of excitation of molecules at higher temperature leads to an increase in distance and corresponding decrease in attractive forces among water molecules and food materials (Mohsenin 1986; Palipane and Driscoll 1992; McLaughlin and Magee 1998). Mazza and LeMaguer (1980) suggested that increase in temperature induced physical and/or chemical changes in the product that can reduce the number of active sites for water binding. At high temperature and a_w, some new solutes may dissolve causing a crossover at high a_w (Fig. 3.7). The effect of temperature is greater at lower moisture content than that at higher moisture content (Fig. 3.5). The effect of temperature follows the Clausius-Clayperon equation, which assumes that the water activities in two different conditions are related through the net isosteric heat of sorption. The Clausius-Clayperon equation can be used to predict the isotherm values at any temperature if the corresponding excess heat of sorption is known at a constant moisture content.

$$ln\left(\frac{a_{w2}}{a_{w1}}\right) = \frac{\Delta H_c}{R}\left(\frac{1}{T_1} - \frac{1}{T_2}\right) \tag{3.10}$$

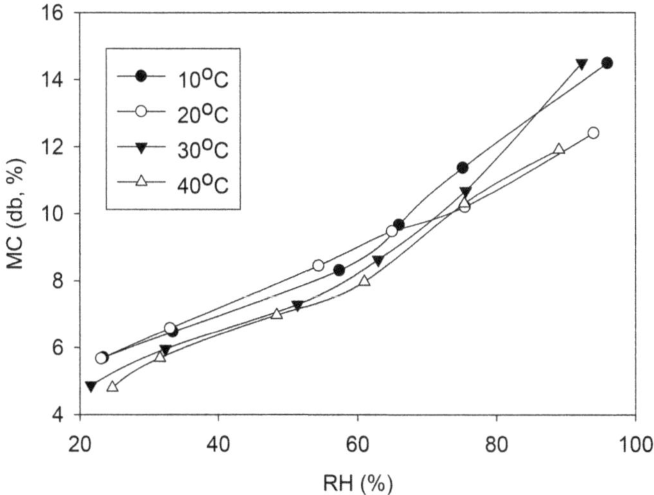

Figure 3.7 Water activity (ERH) of hemp seeds at different temperatures (Data from Jian et al. (2018)).

where a_{w2}, a_{w1} are the a_w at temperature T_2 and T_1 (K), respectively; and ΔH_c is the heat of adsorption (J/mol). Kelvin equation below can be used to estimate the isotherms at high temperatures. The vapour pressure ratio for the infilling of macropores is normally estimated by using Kelvin equation due to the lowering of vapour pressure by curved concave menisci or the increasing of vapour pressure by curved convex menisci.

$$ln\frac{P}{P_{\text{flat}}} = lnRH = -\frac{2\tau V_L \cos\theta}{rRT_k} \tag{3.11}$$

where P_{flat} is the vapour pressure on a flat surface of water droplet. Kelvin equation describes that equilibrium vapour pressure depends on droplet size. If the curvature is convex, r is positive, and $p > P_{\text{flat}}$. This indicates that when the pressure rises in a droplet, the droplet tends to evaporate more easily. If the curvature is concave, r is negative, and $p < p_{\text{sat}}$. The evaporation of the droplet will not be easy. As r increases, p decreases, and the droplets grow into bulk liquid. For a bulk with the particle size of 0.1 mm in diameter, the size of droplet among the particle is about 15 μm. The Kelvin equation predicts a value of 99.5% RH. Therefore, only very fine powders will retain bound water between the particles. There is no bound water between grain kernels, while free water exists among grain kernels at much higher grain moistures.

Certain low molecular weight food constituents such as sugars (glucose) and salt become more hygroscopic at higher temperatures due to their ability to dissolve in water. Equilibrium moisture content of some of dried fruits decreases with increase of temperature when water activity is lower than 0.55–0.7 (depending on product). However, it become opposite when water activity is higher than 0.55–0.7 (Tsami et al. 1990). This phenomenon is attributed to the fact that water is mainly at the status of the biopolymers at low a_w. Therefore, an increase of temperature would lower the moisture content. However, as a_w is raised beyond the intermediate region, water begins to be sorbed by the sugars and other low molecular weight constituents. The water held by the sugars and other low weight molecules will result in an increasing of the moisture content with the increase of temperatures. The intersection point depends on the composition of the food and the solubility of sugars (Saravacos and Stinchield 1965).

The temperature dependence of the equilibrium moisture content and RH (a_w) has an important practical meaning on chemical and microbiological reactions associated with safe storage of food materials. At the same moisture content, increasing of temperature enables a higher a_w (RH) and consequently increases the rate of microorganism multiplication which results in deterioration of the food materials. At the same temperature, increasing of grain moisture contents results a higher a_w (RH). Therefore, higher than safe storage temperature and/or grain moisture is the main reason of deterioration of stored grain products.

The ERH (a_w) is a relatively weak function of temperature but generally increases with temperature (Figs. 3.5–3.7). This fact has a significant influence on drying of food products because drying is usually conducted at higher than room temperature. At high drying temperature, a low RH will need a higher enthalpy of vaporization to remove water sorbed by hydrophilic and charged sites in proteins and polysaccharides. At high drying temperature, a phenomenon of "case-hardened" can happen. This phenomenon is usually caused by the formation of a glassy or leathery region within the dried materials. The moisture diffusivity inside the glassy or leathery region approach to zero, so the rate of drying approaches to zero. Under this drying temperature, the determined ERH (a_w) might provide wrong information because the ERH indicates the drying rate should be much larger than zero. "Case-hardened" can happen under certain combination of temperatures and moisture contents. "Case-hardened" phenomena have not been reported in the crop seed drying.

3.2.3.2 Pressure Effect

Compared with temperature, the effect of pressure on isotherm is small. Under most conditions, the pressure effect can be neglected unless elevated pressure or vacuum is used. The pressure effect is mainly on the vapour pressure, and the relationship between pressure, temperature, and water activity is:

$$ln\frac{a_{w2}}{a_{w1}} = \frac{V_L}{RT}(P_2 - P_1) \quad (3.12)$$

where a_{w2} and a_{w1} are water activity at temperature T_2 and T_1 (K), respectively. P_1 and P_2 are the pressure (Pa) at temperature T_2 and T_1. Under a high pressure such as inside an extruder, the water activity might increase.

3.2.3.3 Food Composition Effect

Compared with temperature and pressure, food composition is the most significant factor influencing the isotherm. This is the main reason that isotherm should be determined for each crop (even different classes, varieties, or cultivars) because of the difference of the composition of different crops and classes and some cultivars. For example, the equilibrium RHs of different crops are different at the same equilibrium moisture content and a constant temperature (Table 3.3). Fat, among other food components, do not contribute to moisture sorption and desorption because fats are hydrophobic. However, the moisture content is evaluated by the total mass of the material. Therefore, isotherm is also influenced by the fat. Proteins are generally associated with type II isotherms due to their easily plasticized nature, resulting in increased availability of all polar groups. Carbohydrates have more hydrogen bonds per monomer than proteins. However, plasticization and hydration are different for different internal bonding materials. For example, fibres and starches will need much more time to open up the structure to adsorption.

The isotherms of crop seeds are influenced mainly by the mixture of its components because the effect of fats, proteins, carbohydrates, and water cannot be separated. In practice, the additive rule of mixtures can be used to roughly estimate the isotherm of the mixture. The additive rule is based on the assumption of no interaction effect within each component. For a mixture with three components, the additive rule can be written as:

$$M_m = f_A M_A + f_B M_B + f_C M_C \quad (3.13a)$$

$$f_A = \frac{W_A}{W_T}, \quad f_B = \frac{W_B}{W_T}, \quad f_C = \frac{W_C}{W_T} \quad (3.13b)$$

where M_m, M_A, M_B, and M_C are the moisture content of the mixture, component A, component B and component C, respectively; W_A, W_B, and W_C are the dry masses of A, B, and C (kg), respectively; and W_T is the total mass of

Table 3.3 Equilibrium Moisture Contents of Crop Seeds at the Same Equilibrium Relative Humidity and 25°C

Crop (class or cultivar)	Equilibrium Relative Humidity (%)[a]						Reference
	70		80		90		
	Des[b]	Ads[b]	Des[b]	Ads[b]	Des[b]	Ads[b]	
Corn (dent type)	14.4	13.4	16.2	15.3	18.7	18.0	Chen (2000)
Rough rice	14.2	14.1	16.1	16.1	18.7	19.0	Chen (2000)
Wheat (soft winter)	14.3	13.9	15.9	15.5	18.0	17.6	Henderson (1987b)
(Hard red winter)	14.7		16.5		19.0		Van den Berg and Bruin (1981)
Canola (Gulle)	8.2	8.1	9.8	9.8	12.1[c]	12.4	ASABE (2016b)
(Hektor)	7.9[c]	8.3	9.8[c]	10.0	12.6	12.5	ASABE (2016b)
(Westar)	9.6	9.3	11.1	10.8	13.2	12.8	Yang and Cenkowski (1994)
Red beans	15.1	13.8	17.6	15.7	21.0	18.4	Chen (2000)
Soybean meal	19.1	11.8	26.6	13.0	38.1	14.6	Pixton and Warburton (1975)

[a] Equilibrium moisture content is calculated from Modified Henderson equation.
[b] Des = desorption, Ads = adsorption.
[c] Obtained through nonlinear regression from the original source, so some of the equilibrium moisture contents at desorption is unreasonably lower than at adsorption.

the material (kg). Equation 3.13 is rarely used because of its low prediction accuracy.

3.3 Hysteresis of Food Materials

3.3.1 Hysteresis

3.3.1.1 Mechanism of Hysteresis

The adsorption and desorption of a biomaterial are usually not the same and more water is held at the same a_w and temperature for the desorption than that for the adsorption. Therefore, when isotherm data collected from an adsorption process are plotted on the same graph as those taken from a desorption process of the same material and at the same temperature, the two isotherm curves will have different pathways (Fig. 3.4). This phenomenon is termed as the hysteresis of the biomaterial. The mechanism of hysteresis is not fully understood (Heldman and Lund 1992; McLaughlin and Magee 1998; Al-muhtaseb et al. 2002), and researchers suggested the following hypotheses:

1. The ink bottle model. This is mostly observed in porous biomaterials such as fruits and crop seeds. The open–end capillaries (pores), which are usually located on the surface of the solids, can empty differently upon desorption from that during adsorption. The pore size is usually larger when the solid is dry than that when the solid is wet. When the solids are dry and air RH around the solid is high, the water can be adsorbed at the top and middle of the pore when the pore radius is large enough (Fig. 3.8). However, the small radius at the top of the pore will govern the vapour pressure required for the process of desorption. During drying period, larger vapour pressure gradient (than the gradient required for the adsorption) at the top of the pore is required for the water to diffuse from the bottom of the pore to outside of the pore (Fig. 3.8). This gradient difference results in the different moisture contents between desorption and adsorption. One pore might connect with different sizes of pores, which results in the complicated phenomena of hysteresis.
2. Structural and phase changes. This is observed in materials with high protein and starch contents. The irreversible change may occur in the food material during the sorption/desorption process.

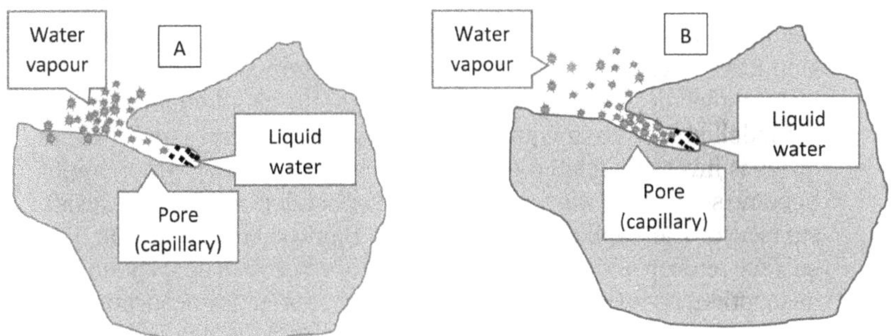

Figure 3.8 Schematic diagram shows the water adsorption (a) and desorption (b). The pore (capillary) will absorb water when the partial pressure of the vapour in air is larger than that inside the capillary (a). The capillary will desorb water when the partial pressure of the vapour in the capillary is larger than that in the air (b). Both cases occur under pressure gradients, but in opposite direction, which produces hysteresis inside the pore.

Desorption from rubbery state can reach equilibrium faster due to increase in molecular mobility, while adsorption into a glassy material can be slow due to restrictions in molecular mobility. Mohsenin (1986) proposed that the shrinkage that occurs upon drying draws chemical compounds closer enabling them to satisfy each other. For food and synthetic materials with larger shrinkage or expansion during drying or rewetting, contraction and swelling are superimposed on the drying and wetting process, which leads to varying equilibrium moisture content. Rao (1942) suggested that pores of the adsorbent become elastic and swell during adsorption. The desorption causes shrinkage and general collapse of the capillary porous structure (Ngoddy and Bakker-Arkema 1975).

3. Different energy requirement for desorption and adsorption. During drying, free water will be removed first. More bound water will be left inside the solid with the progress of drying. Extra energy is required to break the bound water molecules and adhesion force binding the monolayer to the surface boundary. When water is removed from a food product, extra heat is needed because water has to be removed against a water activity gradient and/or against increased osmotic pressure.
4. Some chemical compounds become supersaturate below their crystallization a_w during desorption, and therefore hold more water as a_w is lowered. Food with high sugar content usually hold more water as a_w is lowered due to the supersaturation effect of the sugar crystallization.
5. Incomplete wetting theory and cohesion-tension theory. When less water exists inside biomaterials, negative pressure at the surface of the water will be produced. This is caused by the cohesion and tension force between the water and the biomaterials and between the water molecules. This is the main reason of the capillary force which is modelled by Kelvin equation. The negative pressure inside capillaries is inversely related to the radius of the capillary which might become small after water sorption and/or condensation on the inside surface of the capillary (Ngoddy and Bakker-Arkema 1970). The surface tension and the wetting angle differ between adsorption and desorption, resulting in a higher moisture content for desorption.

The hysteresis might be caused by the combination of these hypotheses due to the change of hydrogen binding ability. The hysteresis prevents biomaterial from undergoing rapid changes in moisture content, such as a sudden gain of water when the RH of the ambient air increase or the material is soaked in water (Kapsalis 1981). For most stored grains, increasing of temperature decreases the total hysteresis (Fig. 3.6) (Iglesias and Chirife 1976a, b, c). The hysteresis is not found until at lower than 80°C

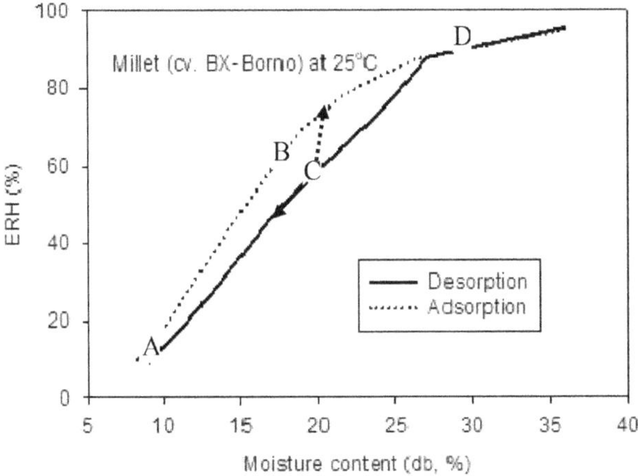

Figure 3.9 Equilibrium moisture content and equilibrium RH relationship of millet (Data source: ASABE 2016b).

(Keey 1991). The reason is unclear. This might be related to hypothesis 3 mentioned above. Isotherm measurement procedure might also be one of the reasons because equilibrium will be established in a much short time at higher temperatures than that at lower temperatures.

When biomaterials are subjected to moisture change, the isotherm direction of the food materials will depend on the drying history and states of the materials. For example, if the newly harvested millet is dried from D point to the C point in Fig. 3.9, the isotherm will switch to the adsorption when more water is added (the dash arrow direction); while it will be at the desorption direction if the millet is continually dried (the solid arrow direction). This crossover between the desorption and sorption curve is the true isotherm of the stored food products because they are usually dried to some level before storage and then subjected to either moisture gain or loss. Therefore, hysteresis has practical aspects. If the crop is dried to C point (Fig. 3.9) after harvest, small amount of water added will significantly increase a_w (ERH) of the stored grain kernels because the switch of the desorption to sorption. If the millet is dried to A point, direction of adsorption occurs when water is added to the crop.

It is not recommended to dry crop kernels to the A point because the seed kernel is brittle at this low a_w and more energy is required to dry the seeds to this low a_w. If seed kernels are over-dried, more energy is also required to mill the kernels. Under very low RH (usually less than 0.2 a_w for beans), the attraction between the material and water is so high that an "adsorption compression" may happen and this results in a net decrease in volume of the solid-water aggregate. Adsorption compression will irreversibly damage the structure of seeds which results in decrease of germination (Hobbs and Obendorf 1972). Damage during water adsorption is the greatest when initial moisture contents of soybean are lower than 13% (Hobbs and Obendorf 1972; Simon and Raja-Harum 1972). This corresponds to the region of strongest bound water (Vertucci and Leopold 1984). This compression damage can occur on other crops such as lima bean (Roos and Pollock 1971). Therefore, it is not recommended to dry grain to such low moisture content.

3.3.1.2 Types of Hysteresis

Hysteresis of biological materials can be grouped into three general categories (Al-muhtaseb et al. 2002):

1. High-sugar and high pectin foods. Hysteresis is pronounced in the lower moisture content region (Okos et al. 1992).
2. High-protein foods. Hysteresis begins at high water activity, in the capillary condensation region, and extends over the isotherm to zero water activity.
3. Starchy foods. Hysteresis is pronounced at about a_w 0.7 or within the capillary condensation region. Maximum deviation between the desorption and adsorption curves occurs (Benado and Rizvi 1985). Stored cereal grain isotherms usually fit in this category.

3.3.1.3 Hysteresis in Grain Mixture and Successive Sorption – Desorption Cycles

Mixing of food materials with different moisture contents is the conventional method in the grain industry to achieve an ideal moisture content. When the same crop grain with different initial moisture contents is mixed in a sealed package or storage bin, water exchange occurs between kernels with different moistures until an equilibrium is reached. The water activity inside the mixture will depend on the ratio of the materials with different moisture contents, the desorption and adsorption history of the materials, the hysteresis of the materials, and the slopes of the respective moisture isotherms. There are four combinations of moisture migration between the two grain sources (Fig. 3.10). Some kernels might spoil in the mixture even though the average ERH (a_w) is lower than 65% if case b occurs in Fig. 3.10 because moisture diffusion is a slow process in grain mixture (Pixton and Griffiths 1971).

Grain at or before harvesting might experience successive sorption-desorption cycles because of the fluctuated whether. Stored grain might also experience these cycles if the stored grain is aerated and dried by forcing air continuously and exposing grain to air with changing moisture at different times. To determine the isotherm of a grain type, the tested grain is tempered with one or a few cycles of sorption-desorption. This

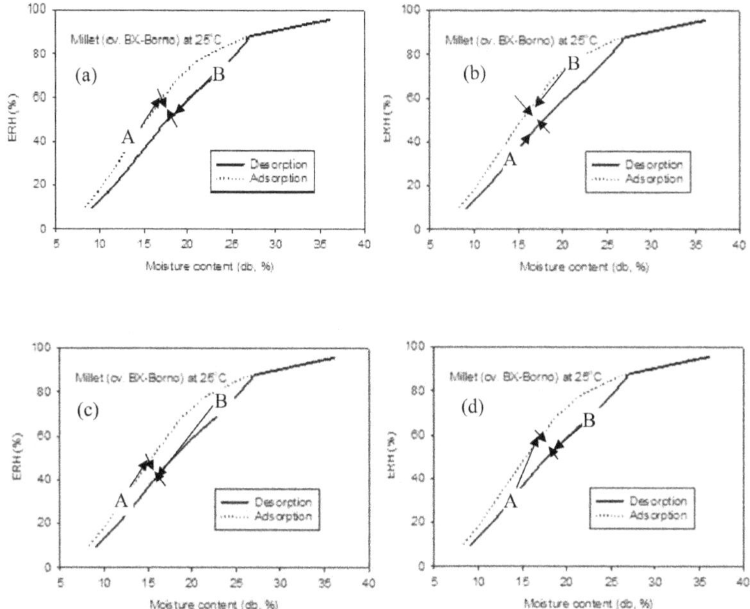

Figure 3.10 Equilibrium moisture content and equilibrium RH of materials after mixing. The two materials (A and B) before mixing are the same except they have different initial moisture contents. Arrows show the moisture trend under different combinations of grains with different initial moisture contents and different drying histories (Data source: ASABE 2016b).

sorption-desorption reduces the hygroscopicity of the stored grain, but the hysteresis exists (Yang and Cenkowski 1993). This decreased hygroscopicity will result in a higher RH at the same grain moisture content. Different grain types have different hygroscopicity reductions. The reduction for corn occurs in the intermediate to high RH (Chen and Morey 1989), while for canola it occurs in the low to intermediate RH. Therefore, the isotherm of grain in bins might be slightly different from the published isotherm.

3.3.2 Laboratory Determination of Sorption and Desorption Isotherms

3.3.2.1 General Approach

There are three general approaches to obtain the isotherm of a tested material: sorption, desorption, and working isotherms. To obtain sorption isotherms, the sample is dried to a low moisture content, and then hydrated by locating the sample in multiple environments of specified high RHs. The desorption isotherm of a crop is obtained by dehydrating the wet samples in environments of low specified RHs. A working isotherm starts with the sample "as is," or in the natural state. Laboratory determination of sorption and desorption isotherms of food materials can be generally categorized into three methods:

1. Gravimetric method: measurement of weight (moisture content) change of a food material in a closed atmosphere with constant RH and temperature at zero (Static method) or low air velocity (dynamic method). The measurements can be conducted continuously or intermittently.
2. Monomeric method: measure the vapour pressure of water surrounding the materials. To improve accuracy, oil is usually used as the fluid for the manometer. The materials are kept at constant temperatures and the material will lose water to equilibrate with the vapour space. The manometer determines the vapour pressure change. After the equilibrium status is reached, the vapour pressure will be stable. This method is not suitable for materials containing large amounts of volatiles or those undergoing a respiration process.
3. Hygrometric method: determine equilibrium RH of headspace of enclosure holding the food material at a constant moisture content and constant temperature. RH is measured by using dew-point hygrometers to detect the condensation of cooling water vapour. Electric hygrometers measure the change in conductance or capacitance of hygrometric sensors. This method requires an air-tight condition and larger amount of food sample than the other two methods. It is not practicable when the temperature and the moisture content of the food materials are high because food materials might spoil under high temperature and moisture content.

The most used methods are salt solution static method (SSS method), dynamic vapour sorption (DVS) method, and thin layer dynamic method. There are some less used methods such as dynamic Dewpoint Isotherm (DDI) method.

3.3.2.2 Salt Solution Static Method

The SSS method is recommended in the European project COST90. Most researchers refer it as the standard method. The SSS method is often conducted inside desiccator holding saturated salt solution under a constant temperature and atmospheric pressure with a constant RH. The constant

RH of different salt solutions can be found in the literature (Spiess and Wolf 1983; Kaymak-Ertekin and Sultanoglu 2001; Kaymak-Ertekin and Gedik 2004). Under this constant RH, samples will gradually lose water or gain water (depending on whether the tested sample has a higher or lower RH than that of the salt solution). When the weight of the sample stops changing (usually evaluated in three sequent measurements), it is assumed the equilibrium is reached. The accuracy of this method can be verified by using the standard chemical: microcrystalline (MCC or called Avicel) (Figura and Teixeira 2007).

The advantages of the SSS method are inexpensive and easy to use, while the disadvantages are: (1) time consuming due to lengthy period of time required to achieve equilibrium, for example, black and red kidney beans require 20–40 days to reach the equilibrium (Jian and Jayas 2018), (2) requirement of large sample units, (3) the difficulty of obtaining conclusive, accurate measurements due to the protocol of continuously removing the sample from the chamber and exposing it to the environment, which is often at a different percentages of RH, and possibly temperature, than the sample; (4) some of the saturated salt solutions, such as the commonly used $NaNO_2$ salt solution to generate 65% RH, are very hazardous and have mutagenic effects; and (5) mold growth on samples at high RH during the long equilibration times. Chemicals (mercury chloride solution, toluene), vacuum, and ultraviolet (UV) light have been studied to prevent the mold growth (Labuza 1984; Rapusas et al. 1993), but none were found to be effective at inhibiting mold and maintaining the original material properties (Yu et al. 2008; Penner and Schmidt 2013). This method is also modified by adding a small fan or mass flow controllers to decrease the measurement time. Many of the disadvantages of SSS method have been overcome by using humidity generating instruments such as DVS method and thin layer dynamic method.

The desired RH environments can also be generated by using sulfuric acid, or glycerol. The main disadvantage of using sulfuric acid solutions is corrosion. When glycerol solutions are used, the range of water activities is narrower as compared to those in the case of saturated salt and sulfuric acid solutions.

3.3.2.3 Dynamic Vapour Sorption Method

The dynamic vapour sorption (DVS) method is relatively new and uses DVS instrument to provide rapid isotherm. This method precisely controls the proportion of mixture of dry N_2 and saturated water vapour by using mass flow controllers. The measured mass change of the sample, which is exposed to the precisely controlled stream of the mixture, during the wetting or drying process is used to determine the isotherm of the sample.

The DVS instrument is usually programmed to automatically change the percentage of RH in a stepwise fashion, while reducing the measurement time and virtually eliminating sample handling by the researcher and exposure to room conditions. The sample is held at each humidity for a pre-set interval or until a steady state weight change is achieved. The operator typically chooses several humidity levels to pre-set during the set-up process. Compared with the SSS method, isotherms developed by using DVS method may have lower moisture contents during adsorption and higher moisture contents during desorption than equilibrium isotherms, resulting in higher levels of apparent hysteresis. The isotherms developed by using DVS method relies on the mass of the sample. Using lighter thinner samples leads to shorter test time, however, lighter and thinner sample might not be the representative of the tested materials. The result of DVS also depends on the operation protocol. Even though different studies had different conclusions, studies generally found that isotherms measured by the SSS method produced deviations from the DVS method (Arlabosse et al. 2003; Rahman and Al-Belushi 2006; Bingol et al. 2012; Schmidt and Lee 2012; Penner and Schmidt 2013). If the DVS generator is set at un-equilibrium protocol, DVS method will generate non-equilibrium isotherm.

3.3.2.4 Thin Layer Dynamic Method

Thin layer drying of grain seeds can also be used to determine the isotherm of the grain seeds (termed as thin layer dynamic method by (Jian et al. 2018)). This dynamic method is conducted by forcing air with constant RH and temperature through the thin layer of the tested materials. The food materials will gradually equilibrate with the supplied air. This method is faster than the SSS method. The drawback of this method is that the instrument is more expensive and must be well maintained (Penner and Schmidt 2013). This method also generates different isotherm from that of the SSS method (Jian et al. 2018).

3.3.2.5 Dynamic Dew-Point Isotherm Method

This is another method to automatically determine isotherms by using a commercial instrument, which uses the dynamic dew-point isotherm (DDI) method. The sample is exposed to air at controlled airflow rates from 10 to 1000 ml/min. Adsorption (or wetting) is measured by saturating the air with water and desorption (or drying) is measured by passing the air through a desiccant before it enters the sample chamber. The a_w is directly measured using a chilled mirror dew-point sensor, while sample weight is gravimetrically tracked by a magnetic force balance. The snap point sample weights are measured after a small change in a_w (approximately 0.015) when a short rest of the airflow is applied. During the test, the sample does not equilibrate to the supplied air; rather, the a_w and the corresponding moisture content of the sample are collected at multiple data points (often >75), over the measured a_w range and temperature. The DDI might not generate the same isotherm as the SSS and DVS methods because sample inside the generator is not required to come to equilibrium with the supplied air (Schmidt and Lee 2012). The advantage of DDI is it can be used to find the time dependent water adsorption-related properties, such as the glassy to rubbery transition, recrystallization, hydrate formation, and deliquescence.

3.3.2.6 Thermal Cell Method

This is a new method used in recent studies (Syamaladevi et al. 2016; Tadapaneni et al. 2017). The thermal cell is a small sample holder (an anodized aluminum alloy cell is usually used) and RH and temperature inside the cell are monitored during the test. The temperature of the sample inside the holder could be controlled by using a hot air oven (Syamaladevi et al. 2016) or oil bath (Syamaladevi et al. 2016). The sample moisture content is measured before the sample is loaded into the cell, and then temperature is gradually raised to a constant temperature. The equilibrium condition can be determined by monitoring the headspace RH at the constant temperature and moisture content of the material. The cell is sealed during the test period to prevent water leaking, so the equilibrium can be established. The advantage of this method is that it measures the temperature-induced changes in a_w and this is similar to the grain drying process. This technique is in the early stage of its development and it cannot measure the adsorption isotherm.

3.3.2.7 Difference Between Isotherm Methods

Different methods used to determine the isotherm of tested material usually generate different isotherm equations (Jian et al. 2018). This difference is caused by the different durations when the equilibrium is evaluated. The common method to evaluate the equilibrium between the tested material and its environment is to measure the mass of the tested material. If the mass does not change in three sequent measurements, then it is assumed the equilibrium is established. Theoretically, at equilibrium water activity of the sample is the same as that of the surrounding environment. However, in practice a true equilibrium is never attained because that would require an infinitely long time. When different methods are used, the duration among the three sequent measurements is different. For example, for the SSS method can be a week and for thin layer dynamic method or DVS method can be less than 1 h. These different durations result in the difference between the methods because different durations will result in different levels of equilibrium due to diffusion of water inside biological materials is a very slow process. A barley mixture with 7% and 23% moisture contents would have a difference of 1% moisture content after 28 days at 10°C (Henderson 1987a). Therefore, the method used to determine the desorption and sorption isotherm equation should be considered when the developed isotherm equation is used to predict grain moisture content during grain drying and storage.

Sorption and desorption isotherms of biological materials are usually not measured at very low (lower than 15%) or high (>99%) RH due to the difficulty of experimental measurement. Therefore, existing mathematical expressions usually describe the sorption or desorption data in a certain range of water activity with a high accuracy, but they usually have a low accuracy at very low or high RH. This is limitation of the extrapolation outside the range of conditions used for the experimental results (Yanniotis 1994). Therefore, the application range of the developed isotherm equation should be checked before using.

The reported isotherm data from different researches can give differences of up to 3 percentage points due to different grain source, different pre-treatment of the grain, and different procedure and conditions of desorption or sorption (Jayas et al. 1991). Jian et al. (2018) found the equilibrium moisture contents of hemp seeds determined by the SSS method were significantly lower than that measured by the thin-layer dynamic method (up to 1.8 percentage point difference at 30°C).

3.4 Soaking of Crop Seeds

3.4.1 Soaking

Soaking is usually conducted in water, and low concentration salt, or combinations of salts and alkali solutions at ambient temperature or warmer environment for a specific time. Soaking of beans may partially remove certain antinutritional chemicals of beans such as tannins and phytic acid (Deshpande and Cheryan 1983). Soaking requires from a few to 24 h at room temperature and numerous ways to shorten it have been devised. Soaked kernels are softer and more uniform in texture than unsoaked kernels (Gandhi and Bourne 1991).

Cooking time is strongly related to water sorption, which is usually tested at different temperatures in different time intervals (Sefa-Dedeh et al. 1978; Moscoso et al. 1984; Wang et al. 2003). After being submerged in water, cereal or pulse kernels absorb water with a higher rate in the first few hours and a slower sorption rate in the later sorption period (Fig. 3.11).

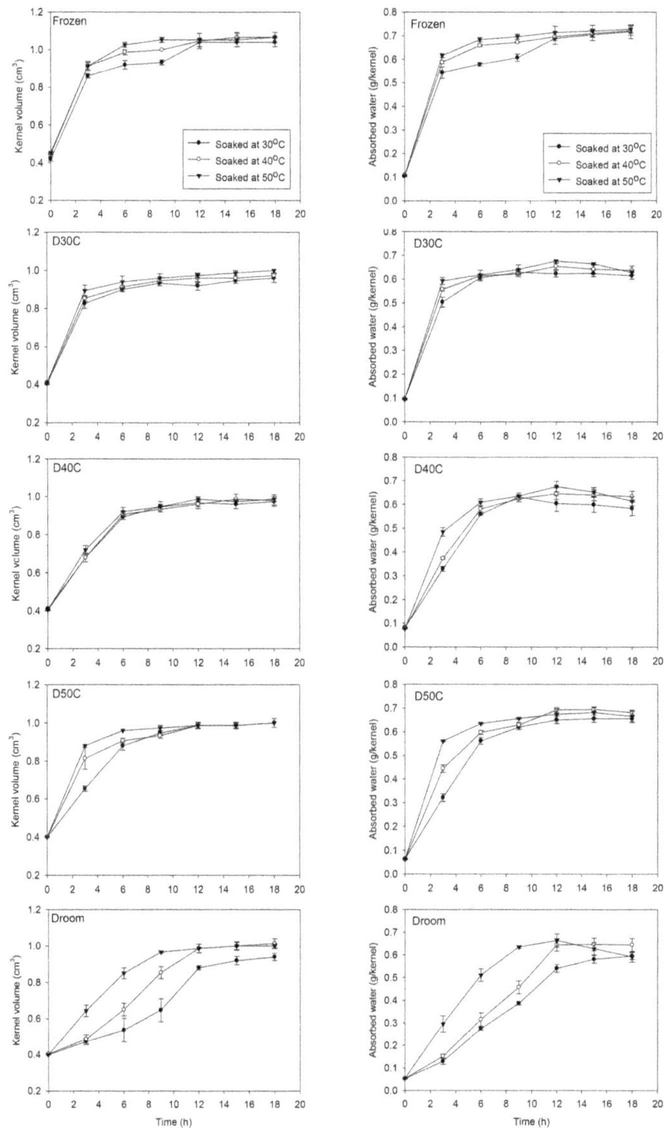

Figure 3.11 Volume of red kidney bean kernel under different soaked temperatures and times. The kidney beans were harvested and kept at −10°C and −20°C (frozen); the frozen beans were dried at room temperature (25 ± 1°C) for 4 weeks (Droom); or dried at 30°C (D30C), 40°C (D40C), or 50°C (D50C) for 7.5 h. The soaked condition was water at 30°C, 40°C, and 50°C, respectively. (Reprinted with permission from Jian et al. (2017)).

Water sorption usually reaches saturation after 12–24 h and depends on the size of kernels, initial condition of the kernels and sorption condition. Grain seeds have different water sorption capacities under different soaking conditions and the temperature is the most important factor (Shittu et al. 2012; Montanuci et al. 2013). The other factors influencing the water sorption are initial moisture content, pH of the solution, salts, soaking time, structural features of kernels (such as seed coat, thickness, seed volume, hilum sizes, true density, bulk density, and cultivar), and composition factors (such as protein content). This influence is more significant at the initial stages of the water sorption. The mode of water penetration into intact stored kernels and into fresh kernels is not substantially different (Varriano-Marston and Jackson 1981). Jian et al. (2017a) found lower initial moisture content, especially with a higher drying temperature before soaking, decreased water absorptivity, and resulted in higher percentage of uncooked kernels if the beans were not soaked before cooking.

Swelling of kernels during soaking takes place concurrently with water sorption. Soaking temperature does not influence the saturated kernel volume, while the pre-treatment condition of the beans influences the saturated kernel volume (Jian et al. 2017a, b). For example, red kidney beans dried at 50°C had a smaller saturated kernel volume than that of the fresh harvested beans. Jian et al. (2017a) found the red kidney bean decreased its sphericity during water sorption and had a larger swelling ratio in the thickness direction than in other directions. Hilum is the rate-limiting barrier in water sorption because water enters the seed at the hilum and is transported to the periphery of the cotyledon at the beginning of the water sorption (Varriano-Marston and Jackson 1981). This might partially explain the different swelling ratio in different direction of the soaked beans. During water sorption, cracks form (Grosh and Milner 1955; Mizuma et al. 2008) and the penetration of water into the kernels is due mainly to the flow of water into the cracks. So the volume increase of the grain kernel is partially equal to the weight increase divided by the fluid density (Chung et al. 1961).

3.4.2 Hard to Cook Phenomenon

Beans stored under high temperature and high humidity are susceptible to a hardening phenomenon characterized by extended cooking times. This phenomenon is termed as hard to cook (HTC) (Burr et al. 1968). If beans have the HTC, the following characterises may change: colour, moisture content, water activity, enzymatic activity, soaking characteristics, cooking quality, structure, macro-components (proteins, carbohydrates, and lipids), and micro-components (e.g., phytic acid, tannins). The HTC defect is the most important characteristic of bean quality because the cooking time required for beans to reach an acceptable texture is an important factor influencing consumer's perception. Storage conditions (such as high storage temperature and moisture) and high drying temperature (Jian et al. 2017a, b) can result in the HTC. Beans shrink during drying and the hardness of the dried beans increases. Sefa-Dedeh et al. (1978) found that stored cowpeas at 0°C and 80% RH for 9 months caused an increase in cooking time. Jian et al. (2017b) found freezing temperature at –10°C and –20°C did not result HTC defect. Studies seem to agree that at low temperatures (0–25°C) and low moisture content (5–9%) minimal changes in HTC occurrence. If temperatures are higher than 25°C and moisture contents are higher than 11%, there is no single method that could prevent HTC for a long period of storage time. However, these beans under controlled atmosphere do not have the HTC defect (Mitsuda 1980).

The HTC is thought to result from hard-shell generation or changes in the middle lamella/cell wall complex that inhibits cell separation, but the mechanisms are still not understood (Hussain et al. 1989; Liu et al. 1992). Reyes-Moreno et al. (1993) wrote an extensive review on this subject. Plhak et al. (1989) suggested two causes: HS (the seed coat is impermeable to water) and the HTC defect (the cotyledons do not soften during boiling, even though the seeds absorb water). The HS can be promoted by low humidity and high temperature, and is reversible by hydrothermal treatment, scarification, or decortication; while HTC is irreversible and accelerated by high humidity and high temperature. Some authors have suggested that the seed coat plays a significant role in the HTC (Reyes-Moreno et al. 1993). "A possible interpretation of the mechanisms participating in the HTC defect may be that both enzymatic and nonenzymatic reactions are concurrently and sequentially participating in events leading to toughness in cotyledon and seed coat in stored beans" (Reyes-Moreno et al. 1993).

3.5 Practical Considerations on the Application of Isotherm Theories

For the safe storage, control of grain moisture content and temperature is critical because this control results in the management of RH (water activity). Moisture content and temperature control is the only one tool in most situations to safely store harvested crop seeds and processed products. Therefore, understanding water status and applying the moisture sorption isotherm theories are key to extend the shelf life of products. The equilibrium isotherms for almost all crops are determined, hence the water status and equilibrium moisture content can be estimated by using these determined isotherms. However, the application of these isotherms under storage conditions is full of potholes. For example, the grain during storage and processing might never reach equilibrium because of fluctuating temperatures and relative humidities. The existence of hysteresis loops in the moisture sorption isotherms is indicative of a non-equilibrium state, no matter how reproducible the data and how accurate the isotherm equations. During grain storage practice, it is usually difficult and also too costly to evaluate the status of the equilibrium. Without knowing the status of equilibrium, the application of the equilibrium isotherms will result in errors. Therefore, application of developed theories, not only the equilibrium isotherm equations, should be considered in the practical application.

In the literature, the measurement of the relationship among materials and their environments is conducted under the controlled lab condition. These measurements under the lab condition are mostly conducted under constant temperatures, moisture contents, and/or relative humidities (water activity). During grain storage or processing, transient condition is the real situation. The theories explained in this chapter will help to understand the water status of the materials under fluctuating conditions. For example, without using the theories explained in this chapter, it might be difficult to determine whether the grain is in the desorption or sorption process during grain storage under fluctuating weather conditions. Therefore, understanding the theories and their application conditions will help to correctly apply these theories in the practice of grain storage and processing.

Nomenclature

φ_p:	capillary pressure (Pa);
τ:	water surface tension (N/m, or Nm/m^2);
θ:	wetting angle (°);
a_w:	water activity (decimal);
$a^0{}_{wj}$:	water activity of *j* component in mixture;
e_{st}:	the saturation vapoure pressure at 1 atm and 100°C (101,324.6 Pa);
G:	initial mass of the sample (kg);
ΔG:	mass of water added or removed (kg);
G_1:	mass of sample 1 (kg);
G_2:	mass of sample 2 (kg);
ΔH_c:	heat of adsorption (J/mol);
ΔH_m:	measured fusion enthalpy (J/kg);
ΔH_w:	fusion enthalpy of pure water (J/kg);
M:	molar mass of water (kg/mol);
M_A:	moisture content of the component *A*;
M_B:	moisture content of the component *B*;
M_C:	moisture content of the component *C*;
M_D:	moisture content in dry basis (decimal);
$M_{D,\%}$:	moisture content in dry basis (%);

M_f: final moisture content of the sample (%, wet basis);
M_g: dry mass of the material (kg);
M_i: initial moisture content of the sample (%, wet basis);
M_m: moisture content of the mixture (%, wet basis);
M_{mix}: moisture content of the mixture (wet basis, %);
M_w: moisture content in wet basis (decimal);
$M_{w,\%}$: moisture content in wet basis (%);
M_1: moisture content of sample 1 (wet basis, %);
M_2: moisture content of sample 2 (wet basis, %);
n: amount of water (mol);
P: vapour pressure (Pa);
P_{flat}: vapour pressure on a flat surface of water droplet;
P_s: saturation vapour pressure (Pa);
$P_s(T_d)$: saturation vapour pressure at dew point (Pa);
$P_s(T_s)$: saturation vapour pressure at the sample temperature;
R: gas constant (8.31451×10^{-7} J·k^{-1}·mol^{-1});
RH: relative humidity (%);
R_s: specific gas constant (J·K^{-1}·kg^{-1});
r: capillary radius (m);
S: percentage of solute (%);
T: temperature (K);
T_c: temperature (°C);
T_d: dew point temperature (°C);
T_k: temperature (K);
T_s: steam point temperature (°C);
V: water volume (m^3);
V_l: molar volume of liquid (m^3/mol);
W: mass of water inside the material (kg);
W_A, W_B, **and** W_C: dry masses of components A, B, and C, respectively (kg);
W_T: total weight of solids (kg).

References

Al-muhtaseb, A. H., W. A. M. McMinn, and T. R. A. Magee. 2002. Moisture sorption isotherm characteristics of food products: a review. Food and Bioproducts Processing 80: 118–128.

Arlabosse, P., E. Rodier, J. H. Ferrasse, S. Chavez, and D. Lecomte. 2003. Comparison between static and dynamic methods for sorption isotherm measurements. Drying Technology 21: 479–497.

ASABE. 2016a. ASAE D271.2 DEC92. Psychometric Data, ASABE Standard. American Society of Agricultural and Biological Engineers, St. Joseph MI.

ASABE. 2016b. ASAE D245.6 OCT2007. Moisture Relationships of Plant-based Agricultural Products, ASABE Standard. American Society of Agricultural and Biological Engineers, St. Joseph, MI.

ASABE. 2019. Moisture Measurement –Unground Grain and Seeds, ASABE Standard. American Society of Agricultural and Biological Engineers, St. Joseph, MI.

Benado, A. L., and S. S. H. Rizvi. 1985. Thermodynamic properties of water on rice as calculated from reversible and irreversible isotherms. Journal of Food Science 50: 101–105.

Bingol, G., B. Prakash, and Z. Pan. 2012. Dynamic vapor sorption isotherms of medium grain rice varieties. LWT - Food Science and Technology 48: 156–163.

Brunauer, S. 1945. The Adsorption of Gases and Vapors, Princeton University Press, Princeton, New Jersey.

Burr, H. K., S. Kon, and H. J. Morris. 1968. Cooking rates of dry beans as influenced by moisture content and temperature and time of storage. Food Technology 22: 88–90.

Callaghan, P. T., K. W. Jolley, and J. Lelievre. 1979. Diffusion of water in the endosperm tissue of wheat grains as studied by pulsed field gradient nuclear magnetic resonance. Biophysical Journal 28: 133–142.

Caurie, G. A. 1981. Derivation of full range moisture sorption isotherms, pp. 63–87. In L. B. Rockland and G. F. Stewart (eds.), Water Activity: Influences on Food Quality. Academic Press, New York.

Cenkowski, S., S. Sokhansanj, and F. W. Sosulski. 1989. Equilibrium moisture content of lentils. Canadian Agricultural Engineering 31: 159–162.

Chen, C. 2000. Factors which affect equilibrium relative humidity of agricultural products. Transactions of the ASAE 43: 673–683.

Chen, C. C., and V. Morey. 1989. Equilibrium relativity humidity (ERH) relationships for yellow-dent corn. Transactions of the ASAE 32: 999–1006.

Chung, D. S., L. T. Fan, and J. A. Shellenberger. 1961. Volume increase of wheat kernels accompanying absorption of liquid water. Journal of Biochemical and Microbiological Technology and Engineering 3: 377–393.

Crowe, J. H., J. S. Clegg, and L. M. Crowe 1998. Anhydrobiosis: the water replacement hypothesis, pp. 440–455. In D. S. Reid (ed.), The Properties of Water in Foods. ISOPOW 6. Blackie, London.

Deshpande, S. S., and M. Cheryan. 1983. Changes in phytic acid, tannins and trypsin inhibitory activity on soaking of dry beans (*Phaseolus vulgaris* L). Nutrition Reports International 27: 371–377.

Engebretson, L. R. 1994. Kernel moisture frequency distribution of U.S. corn. USDA Federal Grain Inspection Service, Kansas City, Mo.

Figura, L. O., and A. A. Teixeira. 2007. Food Physics, Physical Properties – Measurement and Applications, Springer, Berlin Heidelberg, Germany.

Fontana, J. A. J. 2007. Measurement of water activity, moisture sorption isotherms, and moisture content of foods, pp. 155–172. In G. V. Barbosa-Cánovas (ed.), Water Activity in Foods: Fundamentals and Applications. Blackwell Publishing and IFT Press, Ames, IA.

Franks, F. 1975. Water, ice and solutions of simple molecules, pp. 3–22. In R. B. Duckworth (ed.), Water Relations in Foods. Academic Press, New York.

Gandhi, A. P., and M. C. Bourne. 1991. Short communication: effect of pre-soaking on the rate of thermal softening of soybeans. International Journal of Food Science and Technology 26: 117–121.

Garnczarska, M., T. Zalewski, and M. Kempka. 2007. Changes in water status and water distribution in maturing lupin seeds studied by MR imaging and NMR spectroscopy. Journal of Experimental Botany 58: 3961–3969.

Ghosh, P. K., D. S. Jayas, M. L. H. Gruwel, and N. D. G. White. 2006. Magnetic resonance image analysis to explain moisture movement during wheat drying. Transactions of the ASABE 49: 1181–1191.

Goff, J. A., and S. Gratch. 1946. Low-pressure properties of water from –160F to 212F. Transactions of the American Society of Heating and Ventilating Engineers 52: 95–121.

Gregory, R. B. 1998. Protein hydration and glass transition, pp. 57–99. In D. S. Reid (ed.), The Properties of Water in Foods, ISOPOW 6. Blackie, London.

Grosh, G. M., and M. Milner. 1955. The penetration of moisture into the wheat kernel. Milling Productions 20: 1.

Gur-Arieh, C., A. I. Nelson, and M. P. Steinberg. 1967. Studies on the density of water adsorbed on low-protein fraction of flour. Journal of Food Science 32: 442–445.

Heldman, D. R., and D. B. Lund. 1992. Hand Book of Food Engineering, Marcel Dekker, New York.

Henderson, S. 1987a. Moisture transfer between mixed wet and dry barley grains. Journal of Agricultural and Engineering Research 37: 163–170.

Henderson, S. 1987b. A mean moisture content-equilibrium relative humidity relationship for nine varieties of wheat. Journal of Stored Products Research 23: 143–147.

Hobbs, P. R., and R. I. Obendorf. 1972. Interaction of initial seed moisture and imbibitional temperature on germination and productivity of soybean. Crop Science 13: 664–667.

Hunter, A. J. 1987. An isostere equation for some common seeds. Journal of Agricultural Engineering Research 24: 219–232.

Hussain, A., B. M. Watts, and W. Bushuk. 1989. Hard to cook phenomenon in beans: changes in protein electrophoretic patterns during storage. Journal of Food Science 54: 1367–1390.

Iglesias, H. A., and J. Chirife. 1976a. Prediction of effect of temperature on water sorption of food materials. Journal of Food Technology 11: 109–116.

Iglesias, H. A., and J. Chirife. 1976b. Isosteric heats of water vapor sorption on dehydrated foods. Part II: hysteresis and heat of sorption comparison with BET theory. Lebensmittel-Wissenschaft und Technologie 9: 123–127.

Iglesias, H. A., and J. Chirife. 1976c. Isosteric heats of water vapor sorption on dehydrated foods. Part I: analysis of the differential heat curves. Lebensmittel-Wissenschaft und Technologie 9: 116–122.

Jayas, D. S., S. Cenkowski, and S. Pabis. 1991. Review of thin-layer drying and wetting equations. Drying Technology 9: 551–588.

Jian, F., D. Divagar, J. Mhaiki, D. S. Jayas, P. G. Fields, and N. D. G. White. 2018. Static and dynamic methods to determine adsorption isotherms of hemp seed (*Cannabis sativa* l.) with different percentages of dockage. Food Science & Nutrition 6: 1629–1640.

Jian, F., and D. S. Jayas. 2018. Characterization of isotherms and thin-layer drying of red kidney beans, part I: choosing appropriate empirical and semi theoretical models. Drying Technology 36: 1696–1706.

Jian, F., D. S. Jayas, P. G. Fields, and N. D. G. White. 2017a. Water sorption and cooking time of red kidney beans (*Phaseolus vulgaris* L.): Part I – effect of freezing and drying conditions on water sorption and cooking time. International Journal of Food Science & Technology 52: 2031–2039.

Jian, F., D. S. Jayas, P. G. Fields, and N. D. G. White. 2017b. Water sorption and cooking time of red kidney beans: part II – mathematical models of water sorption. International Journal of Food Science & Technology 52: 2412–2421.

Johari, G. P., and G. Sartor 1998. Thermodynamic and kinetic features of vitrification and phase transformations of proteins and other constituents of dry and hydrated soybean, a high protein cereal, pp. 103–138. In D. S. Reid (ed.), The Properties of Water in Foods, ISOPOW 6. Blackie Academic, London.

Kapsalis, J. G. 1981. Moisture sorption hysteresis, pp. 143–178. In L. B. Rockland and G. F. Steward (eds.), Water Activity: Influency on Food Quality. Academic Press, New York.

Kaymak-Ertekin, F., and A. Gedik. 2004. Sorption isotherms and isosteric heat of sorption for grapes, apricots, apples and potatoes. LWT-Food Science and Technology 37: 429–438.

Kaymak-Ertekin, F., and M. Sultanoglu. 2001. Moisture sorption isotherm characteristics of peppers. Journal of Food Engineering 47: 225–231.

Keey, R. B. 1991. Moisture in particulate solids, pp. 29–58, Drying of Loose and Particulate Materials. Hemisphere Publishing Co., New York, Washington, Philadelphia, London.

Kiranoudis, C. T., Z. B. Maroulis, E. Tsami, and D. Marinos-Kouris. 1993. Equilibrium moisture content and heat of desorption of some vegetables. Journal of Food Engineering 20: 55–74.

Kovacs, A. J., and M. Nemenyi. 1999. Moisture gradient vector calculation as a new method for evaluating NMR images of corn (*Zea mays* L.) kernels during drying. Magnetic Resonance Imaging 17: 1077–1082.

Krishnan, P., D. K. Josh, S. Nagarajan, and A. V. Moharir. 2004. Characterisation of germinating and non-germinating wheat seeds by nuclear magnetic resonance (NMR) spectroscopy. European Biophysics Journal with Biophysics Letters 33: 76–82.

Kuntz, I. D. J. 1971. Hydration of macromolecules. 3. Hydration of polypeptides. Journal of American Chemical Society 93: 514–516.

Labuza, T. P. 1984. Moisture Sorption: Practical Aspects of Isotherm Measurement and Use. American Association of Cereal Chemists, St. Paul, MI.

Labuza, T. P., and J. Dugan. 1971. Kinetics of lipid oxidation in foods. Critical Reviews in Food Science & Nutrition 2: 355–405.

Lewicki, P. P. 2000. Raoult's law based food water sorption isotherm. Journal of Food Engineering 43: 31–40.

Li, C., J. Liu, and L. Chen. 2003. The moisture distribution of high moisture content rough rice during harvesting, storage, and drying. Drying Technology 21: 1115–1125.

Liu, K. S., K. H. McWatters, and R. D. Philipps. 1992. Protein insolubilization and thermal destabilization during storage as related to hard-to-cook defect in cowpeas. Journal of Agricultural and Food Chemistry 40: 2483–2487.

Lomauro, C. J., A. S. Bakshi, and T. P. Labuza. 1985. Evaluation of food moisture sorption isotherm equations. Part I. Fruit vegetable and meat products. Lebensmittel-Wissenschaft und Technologie 18: 111–117.

Mackenzie, R. C. 1958. Density of water sorbed on montmorillonite. Nature 181: 334–334.

Mazza, G., and M. LeMaguer. 1980. Dehydration of onion: some theoretical and practical considerations. International Journal of Food Science and Technology 15: 181–194.

McLaughlin, C. P., and T. R. A. Magee. 1998. The determination of sorption isotherm and isosteric heats of sorption for potatoes. Journal of Food Engineering 35: 267–280.

Mitsuda, H. Y. 1980. Advances in grain storage in a CO_2 atmosphere in Japan, pp. 235–246. In J. Shejbal (ed.), Controlled Atmosphere Storage of Grains. Elsevier Scientific, Amsterdam.

Mizuma, T., Y. Kiypkawa, and Y. Wakai. 2008. Water absorption characteristics and structural properties of rice for sake brewing. Journal of Bioscience and Bioengineering 106: 258–262.

Mohsenin, N. N. 1986. Physical Properties of Plant and Animal Materials, Gordon and Breach Science Publishers, New York.

Montanuci, F. D., L. M. D. M. Jorge, and R. M. M. Jorge. 2013. Kinetic, thermodynamic properties, and optimization of barley hydration. Food Science and Technology (Campinas) 33: 690–698.

Montross, M. D., F. W. Bakker-Arkema, and R. E. Hines. 1999. Moisture content variation and grain quality of corn dried in different high-temperature dryers. Transactions of the ASAE 42: 427–433.

Moscoso, W., M. C. Bourne, and L. F. Hood. 1984. Relationship between the hard-to-cook phenomenon in red kidney beans and water absorption, puncture force, pectin, phytic acid and minerals. Journal of Food Science 49: 1577–1583.

Nandel, F. S., S. Verma, B. Singh, and D. V. S. Jain. 1998. Mechanism of hydration of urea and guanidinum ion: a model study of denaturation of proteins. Pure Applied Chemistry 70: 659–664.

Nelson, S. O., and K. C. Lawrence. 1991. Kernel moisture variation on the ear in yellow-dent field corn. Transactions of the ASAE 34: 513–516.

Nelson, S. O., K. C. Lawrence, and C. V. K. Kandala. 1991. Performance comparison of RF impedance and DC conductance measurements for single-kernel moisture determination in corn. Transactions of the ASAE 34: 507–512.

Ngoddy, P., and F. W. Bakker-Arkema. 1970. A generalized theory of sorption phenomena in biological materials (part 1. the isotherm equation). Transaction of ASAE 13: 612–617.

Ngoddy, P., and F. W. Bakker-Arkema. 1975. A theory of sorption hysteresis in biological materials. Journal of Agricultural Engineering Research 20: 109–121.

Okos, M. R., G. Narsimhan, R. K. Singh, and A. C. Weitnauer 1992. Food dehydration, pp. 437–562. In D. R. Heldman and D. B. Lund (eds.), Handbook of Food Engineering. Marcel Dekker Inc, New York.

Palipane, K. B., and R. H. Driscoll. 1992. Moisture sorption characteristics of inshell macadamia nuts. Journal of Food Engineering 18: 63–76.

Park, Y. W. 1996. Determination of moisture and ash contents of food, pp. 59–92. In L. M. L. Nollet (ed.), Handbook of Food Analysis. Marcel Dekker, New York.

Penner, E. A., and S. J. Schmidt. 2013. Comparison between moisture sorption isotherms obtained using the new vapor sorption analyzer and those obtained using the standard saturated salt slurry method. Food Measure 7: 85–193.

Pixton, S. W., and H. J. Griffiths. 1971. Diffusion of moisture through grain. Journal of Stored Products Research 7: 133–152.

Pixton, S. W., and S. Warburton. 1975. The moisture content/equilibrium relative humidity relationship of soya meal. Journal of Stored Products Research 11: 249–251.

Pixton, S. W., and S. Warburton. 1977a. The moisture content/equilibrium relative humidity relationship and oil composition of rapeseed. Journal of Stored Products Research 13: 77–81.

Pixton, S. W., and S. Warburton. 1977b. The moisture content/equilibrium relative humidity relationship of a dried yeast product. Journal of Stored Products Research 13: 35–37.

Plhak, L., K. Caldwell, and D. W. Stanley. 1989. Comparison of methods used to characterize water imbibition in hard-to-cook beans. Journal of Food Science 54: 326–329.

Rahman, M. S., and R. H. Al-Belushi. 2006. Dynamic isopiestic method (dim): measuring moisture sorption isotherm of freeze-dried garlic powder and other potential uses of DIM. International Journal of Food Properties 9: 421–437.

Rahman, M. S., and S. S. Sablani 2001. Measurement of water activity by electronic sensors, pp. A2.5.1–A2.5.4. In R. E. Wrolstad (ed.), Current Protocols in Food Analytical Chemistry. John Wiley & Sons, Inc, New York.

Rao, K. S. 1942. Disappearance of the hysteresis loop. The role of elasticity of organogels in hysteresis in sorption. Sorption of water on some cereals. The Journal of Physical Chemistry 45: 531–539.

Rapusas, R. S., R. H. Driscoll, and K. A. Buckle. 1993. Moisture desorption characteristics of raw onion slices. Food Australia 5: 278–283.

Reyes-Moreno, C., O. Paredes-Lopez, and E. Gonzalez. 1993. Hard-to-cook phenomenon in common beans. A review. Critical Reviews in Food Science and Nutrition 33: 227–286.

Roberts, E. H., and R. H. Ellis. 1989. Water and seed survival. Annals of Botany 63: 39–52.

Rodel, W. 2001. Water activity and its measurement in food, pp. 375–415. In E. Kress-Rogers (ed.), Instrumentation and Sensors for the Food Industry. Woodhead Publishing Limited, Cambridge.

Roos, E. E., and B. M. Pollock. 1971. Soaking injury in lima beans. Crop Science 11: 78–81.

Salwin, H. 1959. Defining minimum moisture contents for foods. Food Technology 13: 594–595.

Salwin, H., and V. Slawson. 1959. Moisture transfer in combination of dehydrated food. Food Technology 14: 1.

Saravacos, G. D., and R. M. Stinchield. 1965. Effect of temperature and pressure on the sorption of water by freeze-dried food materials. Journal of Food Science 30: 779–786.

Sathya, G., D. S. Jayas, and N. D. G. White. 2009. Safe storage guidelines for canola as the seeds slowly dry. Canadian Biosystems Engineering 51: 29–38.

Schmidt, S. J., and J. W. Lee. 2012. Comparison between water vapor sorption isotherms obtained using the new dynamic dewpoint isotherm method and those obtained using the standard saturated salt slurry method. International Journal of Food Properties 15: 236–248.

Scott, W. J. 1957. Water relations of food spoilage microorganisms. Advances in Food Research 7: 83–127.

Sefa-Dedeh, S., D. W. Stanley, and P. W. Voisey. 1978. Effect of storage time and cooking conditions on the hard-to-cook defect in cowpeas (*Vigna unguiculata*). Journal of Food Science 44: 790–795.

Shittu, T. A., M. B. Olaniyi, A. A. Oyekanmi, and K. A. Okeleye. 2012. Physical and water absorption characteristics of some improved rice varieties. Food Bioprocess Technology 5: 298–309.

Simon, E. W., and R. M. Raja-Harum. 1972. Leakage during imbibition. Journal of Experimental Botany 23: 1076–1085.

Skaar, C. 1988. Wood-Water Relations, Springer-Verlag, Germany.

Sokhansanj, S., and S. O. Nelson. 1988. Transient dielectric properties of wheat associated with nonequilibrium kernel moisture conditions. Transactions of the ASAE 31: 1251–1254.

Spiess, W. E. L., and W. F. Wolf. 1983. The Results of the COST 90 Project on Water Activity. Physical Properties of Foods, Applied Science, London.

Sun, K., F. Jian, D. S. Jayas, and N. D. G. White. 2014. Quality changes in high and low oil content canola during storage: part I – safe storage time under constant temperatures. Journal of Stored Products Research 59: 320–327.

Syamaladevi, R. M., R. K. Tadapaneni, J. Xu, R. Villa-Rojas, J. Tang, B. Carter, and B. Marks. 2016. Water activity change at elevated temperatures and thermal resistance of salmonella in all purpose flour and peanut butter. Food Research International 81: 163–170.

Tadapaneni, R. K., R. Yang, B. Carter, and J. Tang. 2017. A new method to determine the water activity and the net isosteric heats of sorption for low moisture foods at elevated temperatures. Food Research International 102: 203–212.

Tsami, E. 1991. Net isosteric heat of sorption in dried fruits. Journal of Food Engineering 14: 327–335.

Tsami, E., Z. B. Maroulis, D. Morunos-Kouris, and G. D. Saravacos. 1990. Heat of sorption of water in dried fruits. International Journal of Food Science & Technology 25: 350–359.

Van den Berg, C., and S. Bruin 1981. Water activity and its estimation in food systems: theoretical aspects, pp. 1–43. In L. B. Rockland and G. F. Steward (eds.), Water Activity: Influence on Food Quality. Academic Press, New York.

Varriano-Marston, E. M., and G. M. Jackson. 1981. Hard-to-cook phenomenon in beans: structural changes during storage and imbibition. Journal of Food Science 46: 1379–1385.

Vertucci, C., and A. C. Leopold. 1984. Bound water in soybean seed and its relation to respiration and imbibitional damage. Plant Physiology 75: 114–117.

Viollaz, P. E., and C. O. Rovedo. 1999. Equilibrium sorption isotherms and thermodynamic properties of starch and gluten. Journal of Food Engineering 40: 287–292.

Walter, H. W. 1924. Plasmaquellung und wachstum. Zeitschrift Botanic 16: 353–417.

Wang, N., J. K. Daun, and L. J. Malcolmson. 2003. Relationship between physicochemical and cooking properties, and effects of cooking on antinutrients, of yellow field peas (*Pisum sativum*). Journal of the Science of Food and Agriculture 83: 1228–1237.

Xiong, X., G. Narsimhan, and M. R. Okos. 1991. Effect of composition and pore structure on binding energy and effective diffusivity of moisture in porous food. Journal of Food Engineering 15: 187–208.

Yang, W. H., and S. Cenkowski. 1993. Effect of successive adsorption-desorption cycles and drying temperature on hygroscopic equilibrium of canola. Canadian Agricultural Engineering 35: 119–126.
Yang, W. H., and S. Cenkowski. 1994. Effect of successive adsorption and desorption cycles and drying temperature on hygroscopic equilibrium of canola. Canadian Agricultural Engineering 35: 119–126.
Yanniotis, S. 1994. A new method for interpolating and extrapolating water activity data. Journal of Food Engineering 21: 81–96.
Yu, X., S. E. Martin, and S. J. Schmidt. 2008. Exploring the problem of mold growth and the efficacy of various mold inhibitor methods during moisture sorption isotherm measurements. Journal of Food Science 73: 69–81.
Zhao, X., H. Zhang, R. Duan, and Z. Feng. 2017. The states of water in glutinous rice flour characterized by interpreting desorption isotherm. Journal of Food Science and Technology 54: 1491–1501.

Chapter 4

Fundamental Principles of Aeration, Drying, and Rewetting

Introduction

Many drying methods are used in the food industry such as solar drying, convective drying, spray drying, lyophilisation (freeze-drying or freeze-dehydration), infrared drying, microwave drying, radiofrequency drying, osmotic dehydration, and combined processes. Different drying methods are used for different food materials based on physical property of the materials, processing requirement, and principles of drying and rewetting. Guiné (2018) provides an extensive review on this topic. Thin layer drying has been widely used in high temperature drying of harvested wet grains. Among all these drying methods, natural air drying (or termed as ambient drying) is the method that uses minimum energy and has a high quality of the dried material, if the initial moisture content is such that the drying can be completed in a reasonable time. To equilibrate temperature of the material with the air being forced through grain, aeration is usually conducted. In the grain industry, the natural air drying and aeration are conducted in bins and the grain in bins is dried, cooled, or warmed by the ambient air supplied by a fan. Ambient air always has a fluctuating temperature and relative humidity, which might result in grain drying or rewetting. The principles of thin layer drying, natural

DOI: 10.1201/9781003186199-4

air drying, aeration, and rewetting by using convection is the focus of this chapter. The mathematically modelling of these drying processes is discussed in Chapter 5.

4.1 Heat Supply and Air Properties

4.1.1 Methods of Heat Supply and Drying

To dry a material, energy is required to evaporate water from the product and the energy can be supplied by conduction, radiation, convection, or a combination of two or more of these modes. In microwave, radio frequency, and dielectric drying energy is supplied to the material to generate internal heat. In addition to energy, a medium (typically air) is required to remove water. Conduction can be done by heating solid particles (such as heated metal balls, sand, and glass balls) and mixing with the material to provide a large area for the heat transfer. The solid particles are screened out of the material after drying, reheated, and recycled through the drier. Some conduction heat transfer occurs from the walls of the dryer and from other heating surfaces if included in the dryer design. The radiation drying methods include microwave, radio frequency (RF), ultraviolet (UV), infrared, dielectric, and sun drying to supply heat. These radiation methods can be combined with other drying methods. Conduction and radiation methods are more energy efficient because the energy in the medium (solid particles for conduction method and energised waves for radiation method) can be directly transferred to the material being dried with up to 100% efficiency. However, the traditionally and mostly used method of energy supply to dry bulk grain is the convection – directly forcing air through the grain bulk continually or intermittently because even though conduction and radiation drying have a high energy efficiency, to produce the conduction and radiation energy is costly. For example, heating metal balls requires energy and extra handling facilities. For an infrared drying system, additional medium should be used to agitate the grain kernels to expose the entire surface area of grain kernels, which further increases operating cost. Energy to produce the microwaves is high, which prevents its wide-scale adoption and limiting use of the microwave drying for high-value products. Less equipment is required for the convection drying than other drying methods. A fan and a ventilation duct system are the basic components required for the convection drying. Convection drying can be designed for different drying rate and drying time. For example, high temperature drying can dry grains in less than 15 min, and natural air drying can take few weeks or even months. Convection drying is the most common method used for grain drying.

4.1.2 Energy Efficiency of Convection

The rate of moisture evaporation depends on the rate of heat supply, moisture content of the material, physical properties of the material, and differences in moisture content and temperature between grain and air. The energy efficiency of convection drying depends on these factors and material thickness (depth). There is an energy (temperature and relative humidity [RH]) difference between the material and the supplied air, so the energy transfer can be completed, but low efficiency occurs due to this energy difference if the supplied air has not enough time to equilibrate with the grain. Lower airflow usually has higher energy efficiency because the air will have enough time to equilibrate with the grain, so all the energy of the supplied air can be used in this heat and mass transfer process. In high temperature drying, the air usually does not have enough time to equilibrate with the material, so air may leave with considerable amount of unutilized heat energy.

4.1.3 Energy Consumption During Grain Drying

Farmers can harvest crops at any moisture content after the physiological maturity of the grain kernels has been reached. However, in Canada and the USA, crops are usually harvested at high moisture contents due to the following considerations: weather condition (risk of quality downgrade due to rain or frost or both for a late harvest), capacity of the drying facility, length of the harvesting time, capacity of grain handling, and time of the new crop seeding when multiples crops are harvested in a year. The harvested wet or damp cereal grains, oilseeds or pulses can then either be wet processed for feed (airtight storage or acidification) or dried before microbial deterioration occurs. To decrease field loss caused by shattering, sprouting, and the infestation of birds, insects, moulds or other pests, most farmers might harvest early even though drying of the grain would be needed. It is recommended that the high temperature drying, dryeration, or combined drying should be used if the harvested grain moisture content is 5 percentage points higher than the safe storage moisture content. If the harvested grain moisture content is less than this range, natural air drying can be conducted. When capacity of the high temperature dryer is too small for the harvested grain, aeration or natural air drying can partly dry the grain and keep it from spoiling until the dryer is available.

Drying grain is an energy intensive operation. In Ohio, USA, the corn production of an average field can require approximately 21 GJ of energy per hectare, and about 3.2 GJ of the fuel and electrical energy may be used for the conventional high temperature grain drying (Hansen et al. 2011).

In 1996, nearly 100% of the shelled corn harvested in Ohio was dried by using high temperature dryers (Hansen et al. 2011). In developed countries, energy used for drying is 12–20% of the energy requirements of national industries (Bardy et al. 2015). Therefore, finding ways to conserve energy and reduce costs associated with grain drying is a critical challenge for grain drying researchers. Natural air drying usually consumes about 25–40% of the energy used by a high temperature dryer (Hansen et al. 2011). Therefore, the natural air drying is the first recommenced method to dry grain if favourable weather conditions prevail, and farmers have enough bins equipped with correctly sized fans.

4.1.4 Air Properties Considered During Ventilation

Airflow is the key for aeration and drying. The recommended airflows for aeration, natural air drying, and high temperature drying of grains are: ≤0.15, >0.15 and ≤1.5, and >1.5 $m^3 \cdot min^{-1} \cdot m^{-3}$, respectively (Brooker et al. 1992). Airflow rate in heated air dryer is usually in the range of 15–36 $m^3 \cdot min^{-1} \cdot m^{-3}$ (Foster 1973). Regardless of the drying method, the temperature, relative humidity, airflow rate, and pressure of the supplied air are the key for the energy saving and will decide whether the grain will be dried or rewetted during aeration and drying. Under typical drying conditions during post-harvest, atmospheric pressure has a negligible change and effect. Pressure of the supplied air influence the airflow rate and air velocity distribution in bins or dryers and usually does not directly influence drying process.

Air with different temperatures and RH has different water holding capacity (Table 4.1). The air with higher temperature can hold more water and can remove more water from grain than air of lower temperatures. Air's capacity to dry depends on how much water it can hold and how much it is currently holding. Amount of water in air is usually lower in evenings but cannot take more water because the temperature in the evening is usually lower than at the noon (Table 4.1). Air with a low enough RH (typically at high temperatures) can cause an evaporative cooling effect in the material as the air has the ability to take on water from the material and this enhances the drying process. Considering the water holding capacity at <10°C is negligible (Table 4.1) and ambient air is usually lower than 30°C in northern hemisphere, it is recommended to conduct natural air drying when air temperature is higher than 10°C and RH is lower than 60%. If material has a high moisture content, the fan can be started when the ambient temperature is lower than 10°C because low grain temperature helps to prevent spoiling of the stored grain even though the drying ability of the cold air is small or negligible (Table 4.1 and Appendix A).

Table 4.1 Water Holding Capacity of Air at Different Times of a Day

Time	*Temperature (°C)*	*RH (%)*	*Water in Air*[a]	*Take Up*[b]
		70	18.8	8.2
Noon	30	50	13.1	13.9
		30	8.0	19.0
		70	5.5	2.3
Late night	10	50	3.8	4.0
		30	2.2	5.6

[a] Actual water content in the air (g/kg dry air).
[b] Amount of water took up by the air after saturation (g/kg dry air).

At different geographical locations, air usually has different drying potential in each month of a year due to different combinations of air temperature and RH. At Winnipeg, Canada, the optimum time for grain drying is from the end of March to the end of October because the average weather temperature is higher than 15°C and the average RH is lower than 70% during this time period (Fig. 4.1). During April and May, the average

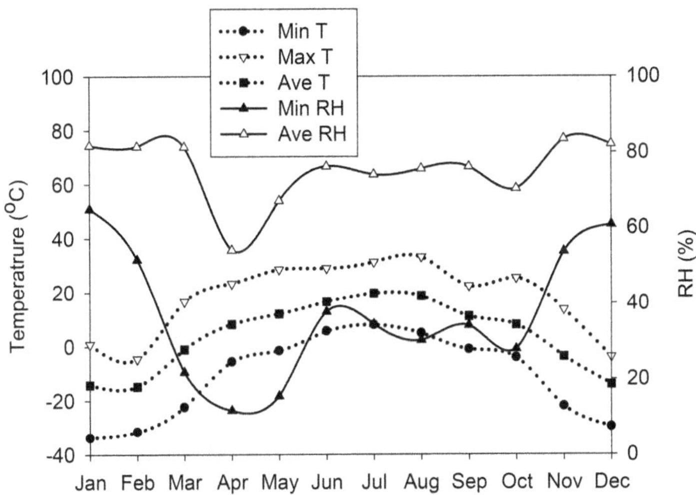

Figure 4.1 Weather temperature and RH at Winnipeg, Canada in 2010. In the graph, T = temperature (°C), RH = relative humidity (%), min = minimum, max = maximum, and ave = average.

and minimum RH are lower than 60% and 20%, respectively. This is an ideal time to conduct natural air drying.

4.1.5 Adding Heater

A heater can be added to increase the rate of natural air drying if the ambient temperature is lower than 30°C (Table 4.2). If relative humidity of the ambient air is 70%, 5°C increase in temperature will result in about 20% RH decrease in the air (Table 4.2) and 10% increase of enthalpy, hence, the increase of the capacity of water taken up because the air with low specific humidity and high enthalpy allows more water to be carried in the same volume of air (Table 4.2). For hemp seeds, ambient air at 20°C and 70% RH can dry hemp seeds to lower than 9% after 5°C is added to the ambient air (Appendix A). Increase in temperature also increases the transfer rate of moisture between the grain kernels and air because heating air makes moisture more mobile and easily released into the airflow, enhancing the potential of the air to lift moisture from the grain. Therefore, increasing temperature of air can increase the air-drying capacity. Adding heat can be conducted at ambient temperatures from 5 to 25°C. It is usually recommended that 15°C of supplied air is the minimum temperature for natural air drying. Higher temperature will result in high energy cost if heater is used, overheating can cause over drying of grain. If natural air drying is conducted at lower than 5°C by adding heat, then water freezing at the vents will occur when ambient air temperature is lower than 0°C.

Table 4.2 Effect of Adding Heat on Air's Capacity (kg Water Per Kg of Dry Air) to Dry Grain

	Without Adding Heat			*After Adding 5°C*		
Air Initial Condition (Temperature, RH)	*Water in Air (g/kg Dry Air)*	*Water Taken Up After Saturation (g/kg dry air)*	*EMC of Hemp Seed (%)*	*RH (%)*	*Water Taken Up After Saturation (g/kg Dry Air)*	*EMC of Hemp Seed (%)*
20°C, 70%	10.5	4.0	9.8	52	9.5	8.1
15°C, 70%	7.5	3.0	9.9	52	7.0	8.3
10°C, 70%	5.2	2.6	9.9	51	5.4	8.4
5°C, 70%	3.8	1.7	10.3	49	4.0	8.4
0°C, 70%	2.5	1.2	10.4	49	3.0	8.5

4.2 Principles of Aeration, Drying, and Rewetting

4.2.1 Drying Mechanism

When air (forced or unforced) moves through porous materials such as a grain bulk, both heat and mass exchanges between the grain kernels and air occur simultaneously at the surface of kernels (Fig. 4.2). To evaporate the water inside and outside kernels, heat energy must be supplied and the conservation of both energy and mass must hold during this process. Therefore, the energy can be supplied from the grain kernels such as grain cooling or from the air such as grain warming. The result of this energy exchange is that grain loses water resulting in grain drying or gains water resulting in grain rewetting. Grain kernel soaking in water is a special case of grain wetting. During these exchanges, both temperature and moisture gradients may exist, and these gradients depend on both material and air conditions.

From an energy transfer point of view, there are two main classifications for drying processes: adiabatic and non-adiabatic. Adiabatic processes are those in which the heat of evaporation is supplied by sensible heat from the drying material itself or drying gas, which carries the vapour away from the material, such as in thin layer or deep bed drying. The adiabatic drying process is a constant enthalpy process and is the same as the process occurring on a wet bulb thermometer. Non-adiabatic drying occurs when radiation and conduction (provided through or by contact with solid materials) are the main forms of heat transfer. Vacuum drying, purge drying and infrared drying are the examples of non-adiabatic drying.

Heat transfer from the surrounding environment causes internal water to be either transported to the surface of the material in liquid form and then evaporated, or evaporated internally at a liquid-vapour interface and then transported as vapour to the surface. Fortes and Okos (1980) identified six mechanisms (outlined below) to explain the water transfer in materials.

4.2.1.1 Capillary Theory

Capillary flow occurs due to pressure difference, surface tension force, gradients of capillary suction pressure, especially in powder beds and porous solids where continuous paths of liquid exist. Capillary theory (Fortes and Okos 1980) describes the water flow through capillaries, interconnecting pores by way of capillary water movement due to the adhesion forces produced by the molecular attraction between the liquid water and the solid. Flow through capillaries also depends on the permeability of the material, which depends on the pore distribution, surface tension, and liquid dynamic

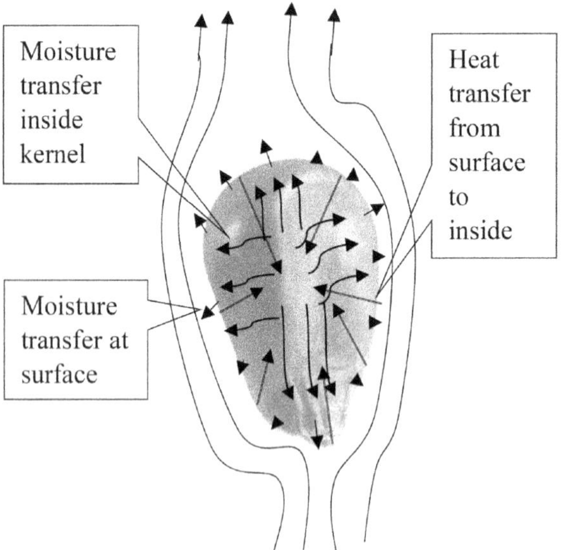

Figure 4.2 Schematic of heat and mass transfer during grain kernel drying when air is passing around a grain kernel.

viscosity. Capillary flow is observed in many situations such as fluid transfer up a wick, water migration up through soil to the ground surface, and blood flow throughout an animal body. Even though capillary flow can occur in saturated and/or unsaturated porous media, it might mostly occur in the materials (such as at the beginning of fruit drying), which have large amount of free water. Grain kernels usually do not have large amount of free water, hence, capillary movement has not often been used for modelling moisture transfer of grain kernels (Srikiatden and Roberts 2007). For the unsaturated capillary flow, the driving force is the pressure difference across the air-water interface.

4.2.1.2 Liquid Movement Due to Gravitational Effects

Hydraulic flow describes the water transfer through porous materials under a hydraulic pressure head, which can be mathematically modelled by Darcy's law. If the drying material has continuous pores filled with free water, a capillary flow may occur as a result of capillary forces caused by surface tension and or gravitational force. A modified Poiseuille equation can estimate the rate of drying. This mechanism is usually neglected in porous bodies since the pore dimension is small and gravity is negligible in a small pore. This phenomenon also does not occur during typical grain drying.

4.2.1.3 Liquid Diffusion

Liquid diffusion can occur under a concentration gradient of water between inside and surface of the material with a high moisture content. This is mostly assumed in the mathematical modelling of grain drying and predicted by using Fick's first and second laws.

4.2.1.4 Vapour Diffusion

Vapour diffusion is another mechanism describing the moisture transfer during drying of hygroscopic porous materials. It is also referred to as pore diffusion (Roman et al. 1979). Diffusion might also occur at the surface of the material if the air velocity at the surface is low (Bedane et al. 2016). Vapour diffusion could be predicted if Fick's second law equation is stated in terms of vapour pressure gradient as the driving force for diffusion. The effect of total pressure on the moisture diffusivity has been investigated in materials with different porosity such as extruded durum semolina (Litchfield and Okos 1992), dense and porous pasta (Waananen and Okos 1996) and all these studies conclude that vapour diffusion significantly contributes to internal moisture transfer during drying of porous solids (Srikiatden and Roberts 2007).

4.2.1.5 Liquid and/or Vapour Migration

Liquid or vapour or both migrations can occur at a pressure gradient induced across a porous solid caused by pressure or high temperature. This is partially explained by the evaporation-condensation theory. During drying, vapour generated in the interior regions between the surface and centre may diffuse towards the cooler centre region and re-condense. This condensed vapour releases latent heat to the material and increases the centre temperature, and at the same time the evaporation cooling keeps the temperature at the evaporation sites constant (Harmathy 1969). This assumption is consistent with the observed temperature profiles of the dough and bread samples during convective hot air drying where a constant temperature is observed at the centre of the bread samples followed by a decrease in temperature and then followed by an increase toward the drying temperature (Roberts et al. 2002). This theory has been proven in microwave drying (Tong and Lund 1993), convective hot air drying (Roberts et al. 2002), and isothermal drying (Roberts and Tong 2003). In most situations, the rate of condensation is equal to the rate of evaporation, so that liquid does not accumulate in the pores near the surface. Therefore, the internal resistance to vapour flow becomes significant. This evaporation-condensation can be the dominant mechanism during falling rate of drying (Harmathy 1969).

4.2.1.6 Effusion (Knudsen) Flow

Effusion flow occurs when the mean free path of the diffused molecules (such as water) is much larger than the diameter of pore. When Knudsen flow occurs, the diffused water molecules collide on the wall of the pore. Knudsen diffusion occurs when drying takes place at low temperatures and pressures, e.g., in freeze drying. Therefore, Knudsen diffusion is a slow process. It is important only in high vacuum drying. It is usually lumped into diffusion due to its negligible effect.

4.2.2 Combination of the Mechanisms

All the above described mechanisms might be related to bulk diffusion, Knudsen diffusion, and diffusion inside pores of molecular dimension (Krishna and Wesselingh 1997). The assumption that moisture transfer occurs only by liquid diffusion in all stages of drying is not physically realistic (Perre 2015). Two water populations with different mobility in glutinous rice flour particles exist (Zhao et al. 2017, 2019). Thus, the theories of Lewis and Sherwood which were developed by using liquid assumption are not accurate and are incomplete. The same argument may be made for any theory based on a single mechanism since there is no doubt that water or vapour transfer may occur in more than one mechanism (Waananen and Okos 1996).

The material exposed to a gas has an essentially discontinuous surface even when the material may have very fine openings. The influence of such discontinuities on the evaporative process has been analyzed through the boundary theory. These discontinuities (breaks) in a wetted surface would not influence the rate of evaporation provided they were sufficiently small compared with the boundary-layer thickness. Suzuki et al. (1972) proved that over half of the surface can be moisture-free before the evaporative rate is significantly reduced if the discontinuities are the order smaller than the thickness of the boundary layer. The thickness of boundary layer will be about 1 mm in convective drying (Keey 1991), so evaporation of water can be assumed as on the entire surface of grain kernels during convective drying.

Typical diffusivity value for vapour-phase diffusion is about 10,000 times higher than that for liquid water diffusion. Therefore, vaporization on the surface of kernels is more efficient than water diffusion inside kernels. Surface diffusion can be calculated by the surface diffusion flux.

$$J = D_s \nabla M_s \tag{4.1}$$

where J is the surface diffusion flux (kg/s), D_s is the surface diffusion coefficient (unit depends on M_s), and ∇M_s is the moisture difference between the surface and surrounding air (%, kg/kg). As drying progresses, dry material has relatively high bulk density which in turn reduces pore size (due to shrinkage), which reduces water migration. Therefore, for drying of grain with less free water, the drying is more dependent on resistance to diffusion inside the material and not the diffusion on the surface of the material.

4.3 Thin Layer Drying and Parameters

4.3.1 Thin Layer Drying

Thin layer drying is conducted by forcing conditioned air at specified temperature and RH through a thin layer of material (Fig. 4.3). A thin layer can be a single kernel deep or thickness of many kernels if the kernels in the layer are exposed fully to an airstream during drying and the properties of material in the layer are uniform. Therefore, the thickness of a thin layer can be increased with the increase of velocity of the drying air and also if the thermodynamic state of the drying air approaches the equilibrium state in heat and mass transfer with material dried in this layer (Jayas et al. 1991; Pabis et al. 1998). Drying in high temperature dryers may be analyzed using the principles of thin layer drying even though conditions of uniformity in the layer are not satisfied because high temperature causes moisture gradients across the bed thickness, which typically are 20–30 cm thick. Thin layer drying testing for high temperature dryers under laboratory condition can be done using the test set up shown in Fig. 4.3 but realizing that in a high temperature dryer grain kernels are turned or continually moved. In-bin drying (such as natural air drying) can be assumed as a stack of

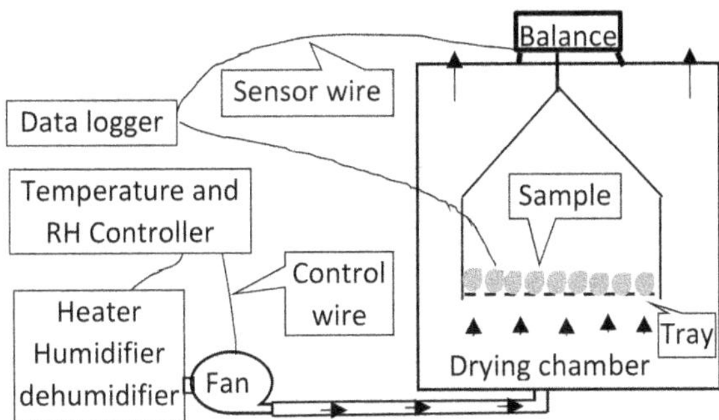

Figure 4.3 Schematic diagram of components of a lab scale thin layer drying apparatus.

multiple thin layers being dried in the bin. Therefore, parameters developed in the thin layer drying are widely used for any method of drying.

4.3.2 Drying Rate

The material gradually loses its moisture content during drying, while increases its moisture content and amount of absorbed water during water sorption or rewetting (Fig. 4.4). The rate of change of the mass (or drying rate) is different for different materials or under different conditions. The drying (wetting) rate is usually defined as the mass change over the time of drying (wetting). The drying rate can be calculated as:

$$\left(\frac{dM}{dt}\right)_n = \left(\frac{M_{n-1} - M_n}{t_n - t_{n-1}} + \frac{M_n - M_{n+1}}{t_{n+1} - t_n}\right)\Big/2 \tag{4.2a}$$

where M is the moisture content (dry basis, decimal), the $\left(\frac{dM}{dt}\right)_n$ is drying rate (kg/s) at time n, t is the time (s), and M_n, M_{n-1}, M_{n+1} are the moisture content (dry basis) at time t_n, t_{n-1}, and t_{n+1}, respectively. Equation 4.2a can be simplified as:

$$\left(\frac{dM}{dt}\right)_n = \frac{M_n - M_{n+1}}{t_{n+1} - t_n} \tag{4.2b}$$

4.3.3 Drying Phase

Evaporation takes place mainly at the surface especially when the surface is covered with a layer of free water. Capillary flow or diffusion of free water or vapour through the voids replaces the water at the surface of the material. If there is enough water to replace the water evaporated at the drying surface, the drying rate will not change. A period of constant drying rate is observed and this usually occurs at the beginning of drying when the material has a high initial moisture content (free water). The drying at a constant rate can be modelled as a wet-bulb thermometer (Fig. 4.5), which can be described as:

$$\frac{dM}{dt} = \frac{h_T A \Delta T}{\lambda} = \frac{h_m A \Delta p_v}{R_v T} \tag{4.3}$$

where h_T is the heat transfer coefficient (W·m^{-2}·K^{-1}), A is the surface area (m^2), T is the temperature (°C), λ is the heat of vaporization of water (J/kg), h_m is the convective mass transfer coefficient (kg·s^{-1}·m^{-2}), Δp_v is the difference of vapour pressure (Pa), and R_v is the gas constant of water vapour

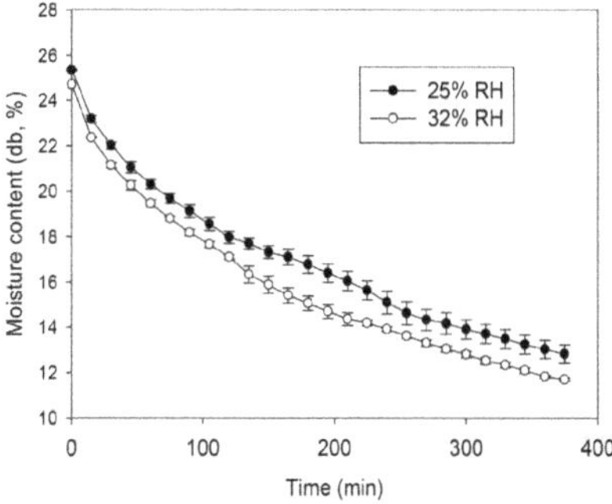

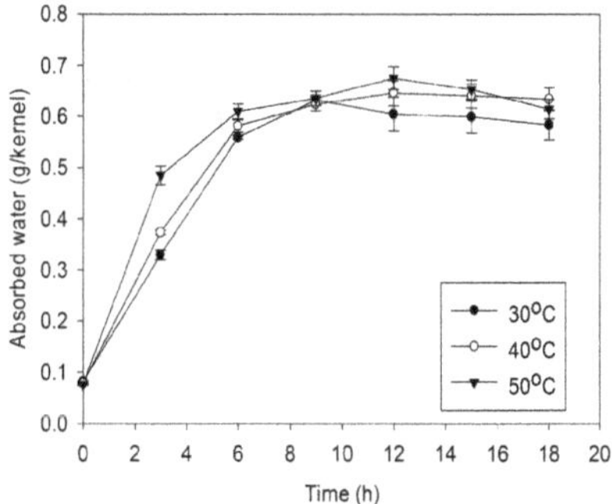

Figure 4.4 Red kidney bean drying (left) and water sorption (right). Drying condition: 50°C with 25 and 32% RH. Soaking condition: 30, 40, and 50°C in distilled water. (Reprinted with permission from Jian et al. (2017)).

(461.52 J·kg^{-1}·K^{-1}). As drying proceeds, the fraction of wet zone decreases with decreasing surface moisture content, so that the mass transfer coefficient decreases. When water passes through randomly distributed paths in a medium, there exists a percolation threshold, which usually corresponds to a critical moisture content. When the water content is less than the critical, the water is not continuous on the surface and the first falling rate period starts. Free water at the surface will evaporate at first. After the surface moisture content reaches its maximum sorptive value, no free water exists, and the surface temperature will rise rapidly because of the decrease of the evaporative cooling effect, which signals the start of the second falling rate period (Figs. 4.5 and 4.6). The reduction of wet surface to zero is the end of the first falling rate period. The first falling rate period of most biomaterials usually is very short and may not be noticed.

Only one falling rate period is usually observed for agriculture materials such as grain kernels. Therefore, drying of agricultural materials in a thin layer can be divided into constant and falling rate phases (Fig. 4.7). The moisture content separating the constant and falling rate phases is the critical moisture content (Figs. 4.6 and 4.7). Some materials such as fresh plant leaves have both phases, while other materials such as harvested grain kernels only have the falling rate phase (Figs. 4.4 and 4.7).

4.3.4 Water Activity During Drying

The moisture content will drop during drying, while a_w might remain constant or increase depending on how fast the water is removed from the kernel. The increased a_w is due to the increased temperature. The drying will continue until the equilibrium moisture content is reached and the a_w at this condition is lower than the initial a_w.

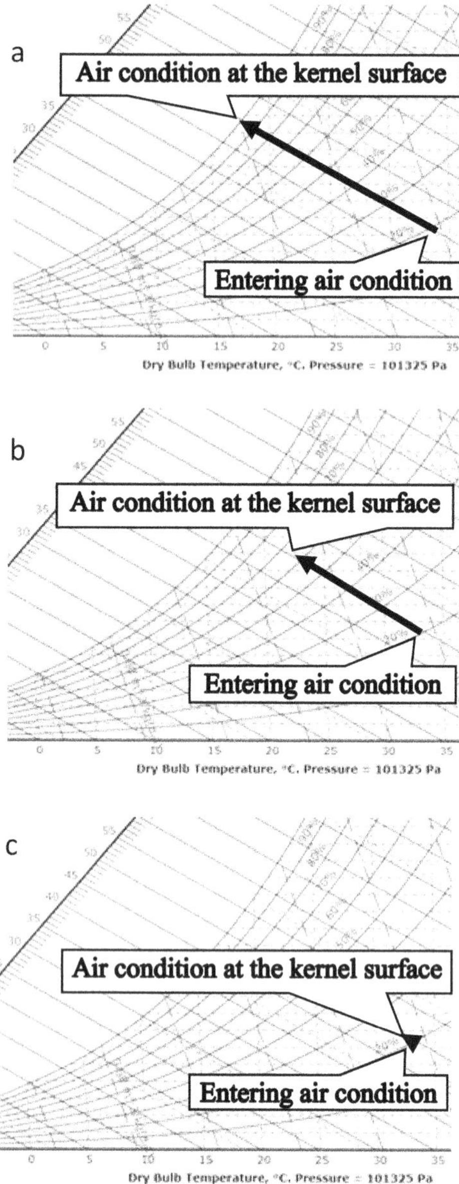

Figure 4.5 Air condition on psychrometric chart at the surface of grain kernels during constant rate period of drying (a), first falling rate period of drying (b), and second falling rate period of drying (c). Arrow shows the trend of air change from entering condition to leaving condition on the kernel surface.

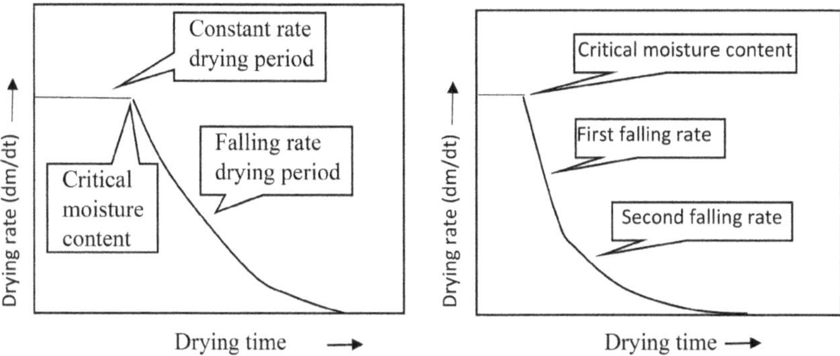

Figure 4.6 Theoretical drying rate (*dM/dt*) versus moisture content (left) and time (right).

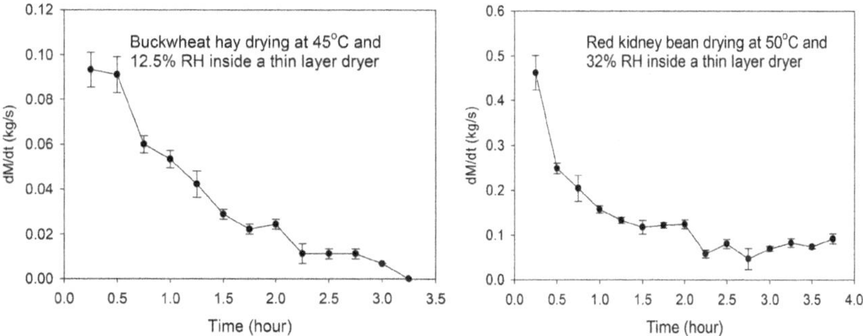

Figure 4.7 Drying of buckwheat hay (leaves) with constant rate period of drying (left) and kidney bean without constant rate period of drying (unpublished data) (right). Initial moisture contents of the buckwheat leaves and red kidney beans were 359.9 ± 0.6% and 25.0 ± 0.1% (dry basis), respectively. About 200 g buckwheat leaves and 15 g red kidney seeds were dried at the beginning of the drying.

4.3.5 Material Temperature During Drying

At the very beginning of the thin layer drying with high temperature and low RH air, the temperature at the surface of the grain kernel is the same as that inside the kernel and lower than the air temperature (Fig. 4.8). Immediately after the kernel is exposed to the high temperature air, the temperature on

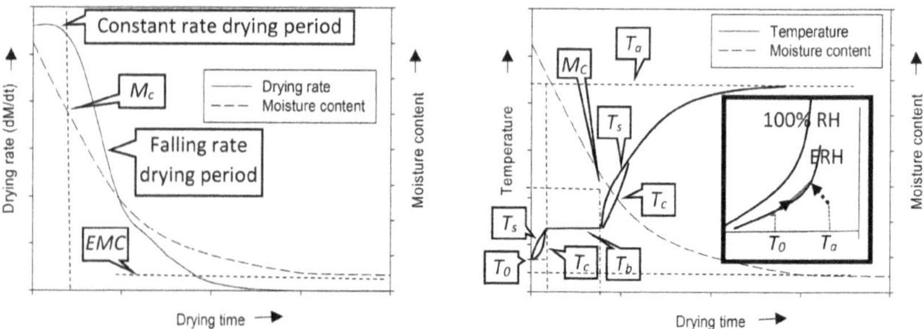

Figure 4.8 Relationship among drying rate, grain moisture content, temperature, and drying time. In the graph, T_a = air temperature, T_0 = initial grain kernel temperature, T_s = temperature on the kernel surface, T_c = temperature at the core of the kernel, T_b = wet bulb temperature, M_c = critical moisture content, EMC = equilibrium moisture content of the grain kernel, and ERH = equilibrium relative humidity of the grain. A simplified psychrometric chart is presented inside the dark frame. The dash arrow shows the change trend of the air temperature by following constant enthalpy line. The arrow shows the change trend of the kernel surface temperature.

the grain surface increases. If there is free water on the surface of the kernel, the temperature at the surface is the web bulb temperature before the critical moisture content is reached due to the evaporative cooling of the water at the surface of the grain kernel and this evaporative cooling follows a constant enthalpy line on the psychometric chart (Fig. 4.8). After the critical moisture content is reached, the temperature at the surface of the kernel will increase faster than that inside the kernel because the heat energy is transferred from the surface to inside of the kernel. There are temperature gradients between at the surface and inside of the kernel. A short time after the critical moisture content is reached, the temperature gradients will vanish (Fig. 4.8), and the temperature inside the kernel will gradually reach to the air temperature. The surface temperature of red kidney bean kernels reached the drying air temperature after 40 min (Jian and Jayas 2018b).

4.3.6 Convection Drying Energy Source

For a convection drying, energy conservation will hold, and adiabatic process occurs because the energy is only provided by the air and/or the material. When a high temperature air is used to evaporate the water at the surface of a material, the temperature of air inside the drying boundary decreases by following the constant enthalpy line on the psychrometric chart until the air reach the equilibrium relative humidity (ERH) of the grain kernel (Fig. 4.5). If the ERH is less than 100%, the temperature at the surface of the kernel will not be the wet bulb temperature, but higher than the wet bulb temperature. After the air reaches the ERH, the air will continue to decrease its temperature by following the constant ERH line on the psychrometric chart. The energy released by the air is used to evaporate the water on the surface of the material and heat up the material. At the end of this process, the air (inside the drying boundary) reaches the material temperature and no more energy from the air can be provided to dry and heat the material. This process continues and high temperature air continually replaces the air inside and close to the drying boundary. The material will increase its temperature and finally reach to the air temperature. On the psychrometric chart, the air and material conditions will change in opposite directions (Fig. 4.8).

The water on the outer layers of material evaporates much faster and more easily than that of the internal layer. This is the reason of a faster and easier drying of wheat kernels from 20% to 15% moisture content than from 15% to 10% moisture content. The time required to remove the last 10% of water is almost equivalent to the time required to remove the first 90% of water.

4.4 Factors Influencing Grain Drying and Cooling Rate

4.4.1 Physical Property of Drying Materials

Physical properties (such as kernel size, shape, porosity, different grain types, their classes and cultivars, and initial moisture content) influence drying rate (phase) and effective moisture diffusivity. The high initial moisture content could result in the material displaying short or long constant drying rate period. This is the reason that the drying constant should be determined for each type of material. Effective moisture diffusivity of rapeseed at drying temperatures of 40–60°C and RH of 30–60% varies from 1.72×10^{-11} to 3.31×10^{-11} m^2/s (Duc et al. 2011). Effective moisture diffusivity of red kidney been at drying temperatures of 30–50°C and RH of 35 and 50% varies from 8.7×10^{-11} to 33.8×10^{-11} m^2/s (Jian and Jayas 2018a). Lomauro et al. (1985) measured mass diffusion coefficient of wheat flour and raisins at 25°C and found wheat flour adsorbing to 0.75 a_w had the largest mass diffusion coefficient, 3.2×10^{-11} m^2/s; and raisins had the smallest value of 4.2×10^{-13} m^2/s. Lu and Siebenmorgen (1992) found water diffusivity of rough rice was different from the white rice, and endosperm (5.7×10^{-13} to 7.1×10^{-11} m^2/s) of both rice types had considerably higher diffusivity values than the bran (4.2×10^{-13} to 3.1×10^{-12} m^2/s) and hull (4.4×10^{-13} to 4.7×10^{-12} m^2/s).

4.4.2 Properties of Supplied Air and Initial Grain Condition

Temperature and specific humidity of the supplied air affect heat and moisture transfer rates. Of these variables, drying air temperature is the major factor influencing drying rate. Initial moisture content of grain is also an important factor. Therefore, most published studies quantify the effects of air temperature and initial moisture content of grain on drying rates. Syarief et al. (1984) found the drying air temperature (from 27 to 93°C) has the greatest effect on drying rate followed by drying air velocity and initial moisture content. Higher drying temperature with low specific humidity will result in a higher drying rate, and low drying temperature with high specific humidity will result in a low drying rate. At 20°C, an increase of relative humidity of air from 32 to 68% (specific humidity increases from 18 to 4.2 g water per kg of dry air) increases half-time of drying (reduction in moisture content to 50% of its initial value) of blue grass seeds from 1.7 to 7.3 h (Farmer et al. 1983). For most materials and most cases, the effect of initial temperature of the material on the energy consumption is negligible because the energy to remove the water is mainly supplied by the air.

4.4.3 Airflow Rate (Air Velocity)

At beginning of drying (especially in the phase of constant drying), material moisture is high and more free water is available on the surface of the material. A high air velocity is required to transfer the energy from air to the surface of the material to evaporate the water, then the maximum drying rate can be achieved. At this phase, the airflow rate (velocity) influences the drying rate. The drying rate will be decreased at the same air velocity after the critical moisture content of the material is reached because not enough free water is available for the evaporation. In the phase of falling rate drying, the internal resistance to moisture movement is much higher than the surface mass transfer resistance. The airflow rate and velocity has no effect on the drying rate during the phase of falling rate drying if the air velocity is already high enough or causes turbulence (Jian and Jayas 2018b). A higher airflow than the required will result in a higher energy consumption due to the low energy efficiency in removing of the water because the leaving air RH will be much lower than the equilibrium RH of the air. Therefore, there is an optimum airflow rate for minimizing the energy consumption. Pupinis (2008) found the optimum air velocity was 0.42 m/s to dry 20 cm thickness of barley at about 30% moisture content, and the required drying time decreased with the increase of air velocity from 0.15 to 0.42 m/s. Even though the required drying time slightly decreases with the increase of air velocity beyond 0.42 m/s the increase in energy consumption does not justify using higher than 0.42 m/s air velocity. Aguerre et al. (1982) found that the change of air velocity in the range of

12–18 m/s did not affect the drying rate of rice grain. The air velocity in the range of 0.008–3 m/s shows a dependence of the drying rate on the air velocity (Ece and Cihan 1993). When wheat with 31% is dried in a thin layer dryer, kernel surface temperature is 4.5°C lower than the air temperature for air velocities of 0.02 and 0.10 m/s, while this temperature difference disappears at 0.45 and 0.68 m/s (Pabis and Henderson 1962). Airflows from 0.1 to 0.5 m/s have little effect on the half-time drying of sunflower seed (Syarief et al. 1984). Therefore, air velocity has little effect on drying rate in certain range beyond optimum airflows (Misra and Brooker 1980; Syarief et al. 1984).

4.4.4 Tempering and Intermittent Drying

During drying, water needs time to migrate from inside grain kernels to outside kernels, while from outside to inside kernels during rewetting and soaking. If water does not migrate to the surface of grain kernels, the moisture cannot be removed from the kernels. If evaporation occurs inside grain kernels at high temperatures, the vapour still needs time to migrate to outside. Therefore, stop drying for a while and allow water to migrate to surface of kernels (termed as intermittent drying) will save energy used for drying after the drying is resumed. This is the concept used for the development of intermittent drying, combination drying, and dryeration, all of these involve a tempering step. A procedure used by the milling industry to temper grain consists of wetting the grain with warm water and allowing it to stand for 18–72 h before processing. To achieve a unit M_R value ($M_R = 1$), Sokhansanj et al. (1983) found the tempering time at 21°C for wheat, barley, canola, and corn was 32, 48, 1, and 96 h, respectively. During grain storage study, the tempering time at 5°C is at least one week for canola and hemp seeds (Sun et al. 2014; Jian et al. 2018). This tempering and intermittent drying can not only increase drying rate and save energy, but also decrease broken (fissure) rate of kernels. Li et al. (1998) found discontinuing the drying process with tempering not only decreased the stresses in the kernel, hence resulted in a low broken rate, but also decreased unit drying time. Compared with continuous drying, intermittent drying only needs 1/7 of energy used for heating of drying air (Elias et al. 2006). For deep bed drying, Franca et al. (1994) found intermittent drying led to more uniform temperature and moisture distributions, which improve grain quality.

4.5 Drying Resistance

When a grain kernel is suddenly exposed to a different temperature, resistances of heat transfer and mass transfer exist. The resistance of heat transfer includes the thermal resistance within the material which is influenced by both the size and thermal property of the material, and thermal resistance at the surface of the material. Similarly, the resistance of mass transfer includes the resistance within the solid material and at the surface of the material.

The minimum amount of energy required to evaporate pure water into air is about 2.4 MJ/kg of water. The amount of energy required during drying is more than this because the energy efficiency is less than 100% and some energy is used to energize the bound water. The rate of heat transfer from the air to the material surface is determined by the difference between the dry bulb temperature of the air and the material surface temperature (the wet bulb temperature if there is free water on kernel surface). The vapour diffusion from the material surface to the air is determined by the difference in vapour pressures. At the material surface the moisture has the vapour pressure of free water before the critical moisture content is reached. In the moving air, the vapour pressure is determined by the temperature and relative humidity of the air. After the critical moisture content is reached, free water on the surface ceases to exist and the vapour pressure at the surface begins to decrease so that the driving force for vapour diffusion into the air decreases.

4.5.1 Resistance During Constant Rate of Drying Phase Period

During the constant rate of drying phase, the rate of moisture removal from the material is mostly limited by the rates at which mass evaporates on the surface of the material. The airflow, temperature, and humidity of the air will influence these rates. Water evaporation rate at the surface of the material may be limited by low convective heat transfer rates from the air to the surface or by low mass transfer rates of moisture moving from the material surface to the air. Therefore, drying in the constant rate period is a surface-based rate governed by external conditions such as temperature difference between the drying air and wet surface, area exposed to the dry air, and external heat and mass transfer coefficients. In the constant rate drying period, there is a higher chance that the rate of energy input equals or is close to the heat loss used for the water evaporation (than that during the falling rate period).

4.5.2 Resistance in Falling Rate Period

Drying in the falling rate period is an internally controlled mechanism and the drying rate is low. In a porous material or when significant porosity is developed inside the materials such as inside a grain kernel, mass transfer may occur mainly in the vapour phase and all evaporation may occur from the interior of the material. Liquid water may transfer through capillaries. Therefore, material's water binding and porous structure influence the mechanism of moisture transfer. On the surface of the material, the heat flux from the hot air is also low because the temperature at the surface is close to the supplied air temperature. Furthermore, in order to supply the heat of vaporization to the interior, heat has to be conducted through dry solid and pore regions, both of which have low thermal conductivities. This heat transfer difficulty results in low heat transfer rate and lots of energy in the supplied air will not be used for water evaporation. During thin layer drying, this unused energy is kept inside the supplied air and moved out with the supplied air. In a deep bed drying, this unused energy can be used by the next layers of the material.

4.5.3 Resistance Due to Low Efficiency of Mass and Heat Transfer

When gases or solutes diffuse at a low moisture concentration, the molecules of the water rarely collide with each other because large spaces among diffusing molecules. The resistance of diffusion comes from collisions of diffusing molecules within the medium. In high concentration or in solids, the resistance of diffusion comes from collisions of diffusing molecules with each other, as well as with molecules of the medium because the short distance of the medium and diffusing molecules. The diffusion coefficient becomes dependent on the concentration, and the diffusion equation takes the form of the Fick's second law. When the moisture content is low and the main moisture in the kernel is bound water, the diffusion coefficient might have a weak dependence on the water concentration. Internal moisture movement is the main resistance to the moisture loss rate of sorghum in the range of 21–6% (dry basis) moisture content because the drying rate is not affected by increasing the air velocity from 8 to 10 m/s (Suarez et al. 1980). Despite the debate on which mechanism(s) contributes to the moisture transfer, diffusion has been widely considered to be the rate-limiting mechanism responsible for internal moisture transfer in drying of many food materials.

4.5.4 Parameters Used to Evaluate Drying Resistant

Drying resistance is usually evaluated by using parameters such as convective heat and mass transfer coefficients, Biot and Lewis numbers, and water diffusivity.

4.5.4.1 Convective Mass Transfer Coefficient

Convective mass transfer coefficient can be used to evaluate the drying resistance. The convective mass transfer is strongly related to convective heat transfer (refer to Chapter 2) during drying. Similar to the heat transfer coefficient, the mass transfer coefficient determines the rate of mass transfer across a medium in response to a concentration gradient of liquid or gas (kg·m^{-2}·s^{-1}, mol·m^{-2}·s^{-1}, m/s). The unit of the mass transfer coefficient depends on the unit of the driving force expressed.

The convective mass transfer coefficient is usually determined experimentally and/or estimated mathematically. Most researchers (Lu and Siebenmorgen 1992) used the Arrhenius-type function to find the mass transfer coefficient, and did not consider the drying force such as the water removal rate and moisture difference between surface of the drying material and the drying air (Jian and Jayas 2018b). Higher drying temperature will result in a higher mass transfer coefficient, therefore, Jian and Jayas (2018b) suggested the mass transfer coefficient could be calculated by using the measured mass flux of the evaporated water on the surface of the grain kernel.

$$h_{mi} = \frac{\Delta \dot{M}_i}{(M_i - M_e)\rho_a} \quad \text{and} \quad \Delta \dot{M}_i = \frac{(M_{i-1} - M_{i+1})/2}{A\beta}\rho_g V_k \tag{4.4}$$

where h_{mi} is the transient mass transfer coefficient (m/s), M_e is the equilibrium moisture content (dry basis, decimal), M_i, M_{i-1}, and M_{i+1} – moisture contents of a kernel (dry basis, decimal) at drying time i, $i - 1$, and $i + 1$, respectively; β is the drying period (s), ρ_g is the density of the material (kg/m^3), $\Delta \dot{M}_i$ is the mass flux of the evaporated water on the surface of kernel during a time period (kg·m^{-2} s^{-1}), and V_k is the volume of the material (m^3). The convective mass transfer coefficient used or reported by different researchers has large discrepancies because researchers usually assumed a number based on their model assumption or calculated by using simplified assumptions. The reported mass transfer coefficient of grain drying is from 0.18 to 1.08×10^{-6} m/s (Jian and Jayas 2018b).

4.5.4.2 Biot Number

The drying resistance in different drying phases can be evaluated by using Biot number. The Biot number (Bi) gives a ratio of the heat transfer resistances inside the solid and at the surface of the solid.

$$Bi = \frac{L_c h_T}{k} \tag{4.5}$$

where Bi is the Biot number of heat transfer, L_C is the characteristic length (m), which is commonly defined as the volume of the solid divided by the surface area of the solid; h_T is the heat transfer coefficient ($W{\cdot}m^{-2}{\cdot}K^{-1}$); and k is the thermal conductivity of the material ($W{\cdot}m^{-1}{\cdot}K^{-1}$). If the thermal resistance outside the kernel exceeds the thermal resistance inside the kernel (small $L_c h_T$, large k), the Biot number will be less than one. For a system with less than one Bio number, the interior of the material may be presumed to be at uniform temperature, although this temperature may be changing, as heat passes into the kernel from the surface. The internal temperature of a material during drying can be considered uniform if Biot number is less than 0.1 (Alzamora et al. 1979b). Negligible internal temperature gradients exist for most biomaterial during the falling rate drying (Vaccarezza et al. 1974; Vaccarezza and Lombardi 1974; Alzamora et al. 1979b). In contrast, the material might have a small thermal conductivity or large enough size, causing the Biot number to be larger than one. When the Biot number is larger than one, thermal gradients within the kernel become important.

An analogous version of the Biot number is also used in mass diffusion (Bi_m), namely Biot number of mass transfer.

$$Bi_m = \frac{L_c h_m}{D} \tag{4.6}$$

where Bi_m is the Biot number of mass transfer, h_m is the convective mass transfer coefficient ($kg{\cdot}s^{-1}{\cdot}m^{-2}$), and D is the mass (water) diffusivity (m^2/s). For drying of food stuffs, external resistance to mass transfer becomes important when a Biot number for mass transfer is less than ten (Vaccarezza et al. 1974). Kerhof (1994) proposed a modified Biot number to determine whether the drying process is controlled externally or internally.

$$Bi_m = \frac{k_{g,\text{eff}} L C_{wg}^s}{D_{\text{eff}} C_s M_{cr}} \tag{4.7}$$

where $k_{g,\text{eff}}$ is the effective mass transfer coefficient (m/s) for the change in moisture content of air at the food surface, C_{wg}^s is the saturation vapor concentration (kg/m^3), L is the dimension or thickness of the kernel (m), D_{eff} is the effective water diffusion coefficient (m^2/s), C_s is the moisture concentration of solid (kg/m^3), and M_{cr} is the critical moisture content of the kernel (dry basis, decimal).

4.5.4.3 Lewis Number

Lewis number (Le) is a dimensionless number used to characterize fluid flow where there is simultaneous heat and mass transfer. Lewis number is named after Warren K. Lewis (1882–1975) and defined as the ratio of thermal diffusivity and mass diffusivity:

$$Le = \frac{\alpha}{D} = \frac{k}{\rho_a h_m c_a} = \frac{Sc}{Pr} = \frac{St}{St_m} = \frac{h_T}{c_a h_m} \tag{4.8}$$

where *Le* is Lewis number, α is the thermal diffusivity (m^2/s), ρ_a is the density of the air (kg/m^3), C_a is the specific heat capacity of air at constant pressure ($kJ \cdot kg^{-1} \cdot K^{-1}$), *Sc* is the Schmidt number, *St* is the heat transfer Stanton number, *Pr* is the Prandtl number, and St_m is the mass transfer Stanton number.

The Lewis factor or Lewis relation (Le_f) is defined as the relative rates of heat and mass transfer in an evaporative process.

$$Le_f = \frac{St}{St_m} = \frac{H_T}{c_a h_m} = \left(\frac{Sc}{Pr}\right)^{2/3} = Le^{2/3} \tag{4.9}$$

In literature, Lewis number and Lewis relation are used interchangeably; some researchers have pointed out it as a mistake (Kloppers and Kröger 2005). Considering the Lewis number is not a unit (not equal to one) in biomass drying, the Le_f is also referred to as the Lewis number in this book.

The Lewis number is a measure of the overall mass transfer resistance to that of heat transfer. In other word, the relative rate of growth of the thermal and concentration boundary layers. When the temperature and concentration profiles coincide, $Le = 1$. In the case of wet-cooling towers the value of *Le* can be from 0.5 to 1.3 (Kloppers and Kröger 2005). Even though the value of *Le* is nearly unity for convection from a moist air stream to the surface of an adsorbent (i.e., desiccant material), it can be much greater than one if there is diffusion resistance within the materials that adds to the convective resistance (Golubovic et al. 2006).

Lewis number could be higher than one under different conditions (Chen et al. 2002; Golubovic et al. 2006; Jaturonglumlert and Kiatsiriroat 2010) because: (1) when mass transfer coefficient becomes small like during the falling rate drying, the heat transfer coefficient will also decrease, but not at the same scale because there is diffusion resistance within the desiccant that adds to the convective resistance (Jurinak and Mitchell 1984); and (2) the diffusion resistance of the water inside the drying materials during the falling rate of drying period is the main reason for this high Lewis number.

Lewis number is related to the nature of the vapour gas mixture, the nature of the boundary layer near the exchange surfaces, and the thermodynamic state of the mixture (Kloppers and Kröger 2005). The non unit of the Lewis number indicates the mass transfer is not proportional to the humidity potential, therefore, a correction term should be added (Kloppers and Kröger 2005). Researchers suggested the heat and mass transfer coefficients of biomass materials might be correlated as (Holman 1990; Miketinac et al. 1992; Chen et al. 2002; Jaturonglumlert and Kiatsiriroat 2010):

$$h_T / h_m = \rho_a c_a (Le)^{2/3} \tag{4.10}$$

Jian and Jayas (2018b) found $Le = 27$ when the kernel of kidney bean was dried at 30, 40, and 50°C with 35% and 50% RH at air velocity of 0.2 m/s. Miketinac et al. (1992) found the h_T/h_m was from 3.98×10^7 to 5.46×10^7 J m^{-3}K^{-1} for the barley drying. Haghighi and Segerlind (1988) set $h_T/h_m = 1200$ J·m^{-3}·K^{-1} and $h_T = 60$ W·m^{-2}·K^{-1} for the model of soybean drying. These prove that "mass transfer coefficient has to be reduced by a huge factor" (Perre 2015) to march with the heat transfer coefficient, and the mass transfer coefficient was much smaller than that calculated from the $Le = 1$. Therefore, the value of Le might be different under different drying conditions and periods. The physical (such as geometry of the material) and chemical (water diffusivity) properties of the material, and property (such as temperature and RH) and velocity of the air will influence the value of Le.

Young (1969) proposed a modified Lewis number (Le_m) as:

$$Le_m = \frac{k\left[\varepsilon + (1-\varepsilon)\rho_s \Upsilon\right]}{D_{\text{eff}}(1-\varepsilon)\rho_s\left(C_{ds} + C_w M + \lambda\mu\right)} \tag{4.11}$$

where Le_m is the modified Lewis number, ε is the porosity of the material (decimal), ρ_s is the mass concentration of water in solid (kg water/m^3 sample), Υ is constant, C_{ds} is the specific heat of dry solid (kJ·kg^{-1}·K^{-1}), C_w is the specific heat of moisture (kJ·kg^{-1}·K^{-1}), k is the thermal conductivity of the material (W·m^{-1}·K^{-1}), and μ is the fluid absolute viscosity (N·s/m^2).

Young (1969) found that 60 was the minimum modified Lewis number required to describe the drying curve by using the mass diffusion, and the heat and mass transfer equations had to be solved simultaneously to accurately predict drying behaviour when Le_m was smaller than 60. Most biological materials have $Le_m > 60$ and thus making drying of biological materials an internal mass transfer controlled process (Young 1969).

4.5.4.4 Water (Moisture) Diffusivity

Water diffusivity is the measurement of the rate of moisture movement in different directions in a unit time by molecular motion (m^2/s). The term of water diffusivity is interchangeable with moisture diffusivity and water diffusion coefficient. Water diffusion can be caused by moisture gradients (mass driving force), temperature gradients (thermal driving force), or pressure difference. Thermal moisture conductivity (thermal moisture diffusivity) is the coefficient of moisture diffusion within material owing to temperature gradient. This coefficient is given in terms of mass per mass-degree; hence it is usually expressed as 1/°C or 1/°F. The diffusion coefficient of moisture is the coefficient of water diffusion under a moisture gradient at a constant temperature. When different transport mechanisms occur simultaneously such as during grain drying, it is difficult to separate effect of individual mechanisms, i.e., to separate water diffusion due to moisture gradient, temperature gradient, pressure, or combination of these factors. Several other mechanisms of internal mass transfer, such as surface diffusion, Knudsen diffusion, capillary flow, vapour diffusion in air filled pores, thermal diffusion may overlap with the ordinary diffusion. None of these mechanisms can prevail throughout the total drying or rewetting process. Therefore, the rate of moisture movement is described by an effective water (moisture) diffusivity (D_{eff}) regardless of its mechanisms. The effective diffusivity is also called apparent diffusivity. Most reported water diffusivities in literature are the effective diffusivities or average of the effective diffusivities because of the reasons mentioned above.

Water diffusivity in foodstuffs has been reviewed (Zogzas et al. 1996). Water diffusivity in fruit and vegetable is from 0.72×10^{-10} to 9.89×10^{-10} (Vagenas and Marinos-Kouris 1991). Water diffusivity of foodstuffs fall between 10^{-13} and 10^{-6} m^2/s (Table 4.3), while the majority (92%) falls within 10^{-12} and 10^{-8} m^2/s (Zogzas et al. 1996). Diffusion coefficients of bulk wheat with 7.4–15.3% moisture content is 2.4×10^{-10} to 8.9×10^{-10} m^2/s at 5°C and 7.1×10^{-10} to 8.9×10^{-10} at 22.5°C, respectively.

4.5.4.4.1 Mechanism of Water Diffusion Water diffusion is the movement of water molecules by temperature, moisture and pressure gradients. At a given time, water activity equilibrium (thermodynamics) and rate of diffusion (dynamics of mass transfer) are the three main factors influencing moisture migration. Any factor influencing the water activity equilibrium and rate of diffusion will influence the moisture diffusion. The difference of chemical potential of water between inside and around material and the physical and chemical properties of the material are the main driving force of water diffusion. Mechanism of water diffusion includes capillary flow, Knudsen diffusion, transfers due to heat and pressure gradients, and external forces resulting in both vapour and liquid diffusion. When grain kernels are subjected to thermal drying, the water is transferred toward the kernel surfaces in the form of a slurry or vapour.

In thermodynamics, chemical potential of a species is a form of absorbed or released energy during a phase transition or chemical reaction. The chemical potential of a species in a mixture is the rate of change of

Table 4.3 Effective Water Diffusivity (Diffusion Coefficient) of Bulk Grain

Material	*MC (wb, %)*[a]	*Coefficient (m^2/s, $\times 10^{-10}$)*	*Source*
Chickpeas	Hydration in water	0.97–5.98 at 15–40°C	Seyhan-Gurtas et al. (2001)
Lentils	Hydration in water	3.53–13.3 at 15–40°C	Seyhan-Gurtas et al. (2001)
Red kidney bean	Hydration in water	110,000–157,000 at 22 ± 2°C	Jian et al. (2017b)
Red kidney bean	25–8.1, drying	0.04–100 at 30–50°C	Jian and Jayas (2018b)
White bean	Hydration in water	0.44–37.9 at 15–40°C	Seyhan-Gurtas et al. (2001)
Wheat	15.3–7.4, drying	2.4–8.9 at 5°C, 7.1–8.9 at 22.5°C	Pixton and Griffiths (1971)
Wheat	23–11, drying	0.069–2.77 at 20–80°C	Becker and Sallans (1957)
Wheat	Hydration in water	2290–2940 at 30–60°C	Mattioda et al. (2017)
Wheat endosperm	36–20, drying	1.8–120 at 22°C	Callaghan et al. (1979)
Sweet corn	Drying air at 9.2	0.04–0.80 at 30–90°C	Fan et al. (1963)
Popcorn	Drying air at 8.9	0.09–1.00 at 30–90°C	Fan et al. (1963)
Dent corn	Drying air at 9.9	0.020–2.30 at 30–90°C	Fan et al. (1963)
Barley	40.0 ± 0.4, drying	0.22–0.1 at 20.5 ± 0.5°C	Ghosh et al. (2009)
Sorghum	9.7–8.8, drying	0.04–0.40 at 30–90°C	Fan et al. (1963)
Sorghum	17.4–5.6, drying	0.09–0.38 at 30–50°C	Suarez et al. (1980)
Soybean	30–8.0, drying	21.54 room temperature	Deshpande et al. (1994)
Rice	33.3–4.7, drying	2–100,732 at 38–82°C	Husain et al. (1973)

[a] Moisture content of the grain or equilibrium moisture content of the supplied air.

a free energy of a thermodynamic system with respect to the change in the number of atoms or molecules of the species. Therefore, chemical potential is the partial derivative of the free energy with respect to the amount of the species, all other species concentrations in the mixture remaining constant. The molar chemical potential is also termed as partial molar free energy. When both temperature and pressure are held constant, chemical potential is the partial molar Gibbs free energy. The chemical potential of the water in a system can be calculated as:

$$\mu\left(T_k,P\right)=\mu_o\left(T_k,P\right)+RT_k ln\left(\frac{P}{p_s}\right)=\mu_o\left(T_k,P\right)+RT_k ln\left(a_w\right) \qquad (4.12)$$

where T_k is the temperature (K), P is the actual vapour pressure of water in the system (Pa), $\mu_0\ (T,P)$ is the chemical potential of pure water vapour in the system (J/mol or J/kg) at temperature (T) and pressure (P), R is the ideal gas constant (8.3144598 J·mol^{-1} K^{-1}), p_s is the vapour pressure of water at saturation (Pa), and a_w is the water activity (decimal). The a_w is less than one and the second term is therefore negative. Therefore, the chemical potential of water in a system is lower than the chemical potential of the pure water.

Theoretically, diffusion of a small molecule, such as water inside crop kernels, is controlled by the molecule's size, presence of other molecules with which water molecules may collide, and the geometry (pore size distribution and tortuous pore path inside the kernel). The smaller the pore size in the kernel, the slower is the moisture migration. In addition, all the membranes, crystals, and lipids are barriers to moisture migration. Solubility of water influences the diffusion and this can be estimated by the Guggenheim, Anderson, de Boer (GAB) equation.

Overall mass transfer is complex because it involves unsteady state diffusion as such the a_w difference decreases over time, which has an exponential dependence on moisture, while the geometric dependence is L^2 (L is the thickness or shortest minor axis of the kernel). For example, if the L is doubled, equilibrium will take four times longer.

4.5.4.4.2 Measurement of Water Diffusivity Determination of diffusion coefficient depends on the accurate determination of the equilibrium moisture content, which is always influenced by other factors and predicting equilibrium concentration in food systems is difficult (Varzakas et al. 2005). Therefore, different methods are developed to limit effects of some factors. These methods are reviewed by several researchers (Crank 1975; Saravacos and Maroulis 2001). There are direct and indirect methods for measuring the water diffusivity inside large size materials such as building materials. These methods include drying curves method, sorption kinetic method, slice dry weight method, electrical method (measure conductivity or capacitance), gamma ray attenuation, neutron radiography, nuclear magnetic resonance, computer tomography, microwave beam, thermal conductivity, thermal imaging, and evaluating from the sorption coefficient (Janz 1997).

Not all of the above-mentioned methods are used on grain kernels due to the small size of grain kernels (so the water at a point inside a grain kernel cannot be determined) and water diffusivity of grain kernels is usually evaluated at low moisture content of the kernels (so the methods used for the saturated water status cannot be used). Diffusion weighted magnetic resonance imaging (MRI) has been used to measure water diffusion in the embryo and endosperm of wheat (Callaghan et al. 1979; Gruwel et al. 2008) and hull-less barley (Ghosh et al. 2009). The water diffusivity inside different parts of grain kernels measured by MRI is significantly different. A destructive method is used by some researchers (Vaccarezza and Lombardi 1974; Tong et al. 1993). In this destructive method, the sample at different drying times is immediately frozen in liquid nitrogen to stop the moisture movement. The frozen sample is later warmed to −4°C in a freezer for cutting along its characteristic dimension, and then determining the moisture content of each section by the vacuum oven method. The most used method is the drying curves method due to the complexity of moisture transfer and measurement of effective diffusivity should be similar to the application of interest. Therefore, only this method is introduced.

4.5.4.4.2.1 Drying Curve Method This method is also referred to as quasi-stationary method, which was developed by Frank-Kamenetskii (1955). This method is based on the following over-simplified assumptions: (1) the water transfer during dehydration or sorption follows Fick's second law with negligible external resistance; (2) uniform initial moisture content and isothermal drying or water sorption; (3) the geometry of the material can be simplified as a regular shape such as infinite slab, cylinder, and sphere, so it can be simplified as one dimensional diffusion and constant diffusion coefficient; (4) negligible shrinkage or expansion; and (5) long drying time. Under these assumptions, the Fick's second law can be simplified as:

$$M_R = \frac{M_n - M_e}{M_{cr} - M_e} = A_1 \exp\left[\frac{-gD_{\text{eff}}}{L_{\text{half}}^2} t\right] \tag{4.13}$$

where M_R is the dimensionless moisture ratio, A_1 is the geometric constant, g is the another geometric constant and $g = 2.47$, 5.78, and 9.8 for slab, cylinder, sphere, respectively, M_n is the moisture content at time n (dry basis), M_{cr} is the critical moisture content (dry basis), and L_{half} is the characteristic dimension of radius or half-thickness (m).

Equation 4.13 can be modified as:

$$lnM_R = lnA_1 - \frac{gD_{\text{eff}}}{L_{\text{half}}^2} t \tag{4.14}$$

Equation 4.14 can be regressed on the experimental data (regression between M_R and time), and the D_{eff} can be calculated from the slope (Ω) as:

$$D_{\text{eff}} = -\frac{\Omega L_{\text{half}}^2}{g} \tag{4.15}$$

where Ω is the slope of the regression equation.

This method is applicable to diffusion based kinetic processes. The assumption of negligible temperature gradients during the experimental measurement is the basic requirement for the calculation of the water diffusivity. This assumption is valid during grain drying and rewetting if we accept that the resistance of diffusion inside the kernel is the main mechanism of the water diffusion. This method might not be suitable before the critical moisture content is reached, a steep moisture concentration difference at the material surface exists, and the mass transfer changes greatly with time. After the critical moisture content is reached, a continuous concentration profile is established and the mass transfer is slowed down, and the unsteady state process transforms into the quasi-stationary one, where the transfer coefficients can be assumed constant.

4.5.4.4.2.2 Determined D_{eff} by Using Different Methods Although most of the methods used to determine D_{eff} are based on Fick's law of diffusion, there are significant differences in the way of applying these laws. Equations 4.13–4.15 can also calculate a temporal variation of the D_{eff} by using determined time-moisture content data. This calculated D_{eff} decreases with the increase of drying time (Efremov and Kudra 2005). Based on finite element calculation, Jian et al. (2018) found that the water diffusivity increased when the grain moisture content was close to the M_e during grain drying because there were fewer water molecules available at the sites when moisture content was low, and the D_{eff} should be increased to transfer the water molecules out of these sites. This conclusion contradicts with some studies, which reported the increase of water diffusivity with the increase of the moisture content (Mattioda et al. 2017). Therefore, the assumption of constant water diffusivity might be held in the middle range of the drying or rewetting (before reaching M_e and after M_{cr}).

The diffusion coefficient can be measured by exposing grain bulk to dry or wet environment at a constant temperature. The wet or dry grain corresponds to adsorption or desorption diffusion coefficient of the measured grain mass, respectively. The thermal moisture diffusivity can be measured by putting grain bulk with a uniform moisture content under a temperature gradient. The effective diffusivity is usually measured by drying or rewetting the grain mass under a controlled air flow with constant or changeable relative humidity. When the diffusion coefficient is not constant, the strictly formal mathematical solutions (theoretical solutions) do not exist. Therefore, drying curve method is not suitable under this condition.

Numerical solutions from finite difference and finite element methods can be used to determine the effective diffusion coefficient. For various concentration dependent diffusion coefficients with special cases such as for non-Fickian diffusion, diffusion with chemical reaction, and simultaneous diffusion of heat and moisture, some numerical and graphical solutions can be found in the monograph by Crank (1975).

4.5.4.4.3 Factors Influencing Water Diffusivity Moisture diffusivity is mainly influenced by temperature and moisture content of materials. The pore structure and distribution of the water inside the materials also influence the water diffusivity. However, the reported diffusivity values are highly influenced by the particular analysis method, experimental condition and material composition used by the researcher due to the complexity of the water diffusivity measurement.

4.5.4.4.3.1 Temperature Effect Moisture diffusivities rise very rapidly with temperature, which is consistent with Eyring's theory which states that diffusivity is proportional to the frequency with which molecules can jump into vacant holes in the molecular lattice. This can be described by Arrhenius equation.

$$D = D_0 e^{-\frac{E_0}{RT_k}} \tag{4.16}$$

where D_0 is the Arrhenius factor (m^2/s), E_0 is the activation energy for moisture diffusion (kJ/mol).

Activation energy of food material is independent of moisture content (Vaccarezza and Lombardi 1974; Suarez et al. 1980; Aguerre et al. 1982), and Arrhenius relationship can adequately model the effect of temperature on effective moisture diffusivity (Tong and Lund 1990). Even though D_0 is important in the understanding of moisture diffusivity and D_0 is a basic constant for the energy requirement of the water migration, it is difficult to experimentally measure D_0 and different authors presents different values for the same material. The D_0 and E_0 are strongly correlated and the estimation accuracy is affected only from the combination of the values of D_0 and E_0 and not from each one separately (Kiranoudis et al. 1995). Therefore, the most used method estimating the E_0 is to conduct the regression between the experimentally determined/predicted D and the reciprocal of absolute temperature. In this regression, the Arrhenius equation should follow a straight line on the log scale. Values of E_0 of most materials are from 15 to 75 kJ g^{-1} mol^{-1} (Zogzas et al. 1996; Seyhan-Gurtas et al. 2001).

4.5.4.4.3.2 Moisture Effect The temperature dependence of moisture diffusivity in biomass materials has been verified by researchers and a general agreement to the Arrhenius equation has been achieved. Limitations of Arrhenius equation were soon realized, and few studies correlated the diffusion coefficient with both moisture content and temperature (Bruce 1985). The influence of moisture content on the estimated values of moisture diffusivity has not yet been formed into a generally accepted model even though more than 18 models were developed (Zogzas et al. 1996). Zogzas et al. (1996) plotted the water diffusivity of 61 foodstuffs against the corresponding values of moisture content and concluded a general trend that the water diffusivity slightly increased with the increase of the moisture content. This conclusion is based on the water diffusivity normally measured by the drying curve method. The drying curve method is based on the Fick's law, which might have the following relationships between the effective diffusivity and moisture content: linear, exponential, or power. All these relationships show that the D_{eff} increases with the increase of the moisture content. However, this trend is not obvious for the material in the moisture range of less than 44% and contradictory reports are published for the same materials. Researchers also reported that the diffusion coefficient of water in sorghum kernels was independent of moisture content in the approximate range of 21–6% (dry basis) moisture content (Suarez et al. 1980). This contradiction might be caused by the availability and diffusivity of the free water. During grain drying (especially at the end of grain drying), there are fewer free water molecules available at the sites with low moisture contents because most of the water should be bound at the low moisture contents, and the water diffusivity should be increased to transfer the water molecules out of these sites. The water diffusivities in starchy materials reached a maximum when the starch with low moisture was dried (Karathanos et al. 1990; Marousis et al. 1991). The water diffusivity inside beef was greatly increased as beef moisture content decreased "due probably to the opening of pores and capillaries in the structure of the meat."(Trujillo et al. 2007). Numerical models proved that the water diffusivity must increase with the decrease of moisture content (Trujillo et al. 2007; Silva et al. 2009; Jian and Jayas 2018b). Jian et al. (2018) suggested that the water diffusivity should have a larger increase at a higher drying temperature and a lower grain moisture content than that at a lower temperature and higher moisture content due to the decrease of the availability of free water at the end of drying. This conclusion is consistent with the theory related to the water status inside biomaterials and some studies (Xiong et al. 1991; Quirijns et al. 2005). The decrease in the available free water for diffusion is explained through a simple model relating the available free water to the binding energy (Xiong et al. 1991). The diffusivity decrease at the higher moisture content is due to the water clustering (Zhao et al. 2017) or water condensation which converts vapour at low RH into liquid diffusion at high RH (Srikiatden and Roberts 2007).

4.5.4.4.3.3 Effect of Physical and Chemical Properties The effective moisture diffusivity is positively related to porosity of the material (Keey 1972; Tong and Lund 1990). Structure and water soluble sugar content in starch gels influence the water diffusivity (Marousis et al. 1991). Previous processing, like thermal processing and blanching, also influence the water diffusivity (Alzamora et al. 1979). Increase of fat content in avocado increases the water diffusivity (Alzamora and Chirife 1980).

The assumption of a constant, average diffusion coefficient is valid within the moisture content range where all drying curves at a given temperature merge into one curve (regular regime curve) regardless of initial moisture content (Tong and Lund 1990). The D_{eff} is usually measured at a thin layer drying condition. The thickness of the bed influences the calculated D_{eff} and the value of the D_{eff} increased with an increase in bed depth (Tutuncu and Labuza 1996). This conclusion should be considered in the deep bed drying modelling because the deep bed drying is usually simulated by using the thin layer drying data.

4.5.4.4.3.4 Effect of Shrinkage Shrinkage can significantly influence the drying rate and water diffusivity (Perre and May 2001; Lambert et al. 2015). Different researchers got different conclusions on whether shrinkage increases or decreases the water diffusivity. Fruit, vegetable, and meat usually have more than 50% shrinkage during drying and values of D_{eff} associated with shrinkage are higher than when shrinkage is not considered (Perre and May 2001; Trujillo et al. 2007). Grain kernels usually do not have a large shrinkage during drying. When soybean with 26% initial moisture content is dried to 8%, it shrinks 8–10% and its effective water diffusivity reduces by a factor of approximately 1.5 (Misra and Young 1980). When pineapple slabs are dried at 45, 60, and 75°C, the surface area is 75–80% of the initial area and the thickness is about 30% of the initial thickness, and the calculated D_{eff} is reduced when this shrinkage is taken into account (Ramallo and Mascheroni 2013). Jian and Jayas (2018b) suggested that the effects of low moisture content and shrinkage on the D_{eff} should cancel each other out, and they used this suggestion to explain the negligible effect of the shrinkage on the water diffusivity of the red kidney bean during low temperature drying.

Drying materials shrink after losing moisture content, thus making it difficult to measure and estimate the diffusion coefficient using the drying curve method because this method is based on the assumption of a constant thickness of the drying materials. A model taking the shrinkage effect into consideration will more accurately predict the drying process. Some methods such as moving frame of reference (Crank 1975) and arbitrary volume change (Gekas and Lamberg 1991) have been introduced to solve this shrinkage issue. Simple linear relationship between volumetric or bulk shrinkage and moisture content during drying of vegetables and fruits is usually assumed (Suzuki et al. 1976; Balaban and Pigott 1988; Rahman and Poturi 1990). Other models such as analogy of thermal expansion and material balance equation considering dry fibre, water and air phases have been proposed (Rahman and Poturi 1990). Finite element and finite difference methods have been used to solve the differential equations of mass transfer in a moving boundary problem.

4.5.4.4.3.5 Effect of Bed Configuration and Desorption and Sorption Diffusivity of a single kernel is different from the bulk of the same kernels at the same environmental condition (Tutuncu and Labuza 1996) if there is no airflow in the bulk. The effective diffusivity increases with the increase of the thickness of the material bulk. The reason for this increase is the captured air in the bulk of the materials has a higher diffusivity than that inside the kernels (Tutuncu and Labuza 1996). Generally, the effective diffusivity of bulk shows a thickness squared dependence. This explains the fact that the adsorption and desorption rates and time to reach equilibrium and the associated water diffusivities are not only affected by the increase in bed depth but also by other contrary factors such as shrinkage of the bed and kernels.

Water diffusivity is different at desorption and sorption conditions. The red kidney bean has the water diffusivity of 10^{-3}–10^{-7} m^2/s at water soaking condition, while it is $0.04–100 \times 10^{-10}$ m^2/s at drying temperature of 30–50°C (Jian et al. 2017b; Jian and Jayas 2018b).

4.6 Glass Transition

4.6.1 Principle of Glass Transition

Glass transition or glass-liquid transition is the gradual and reversible transition of amorphous materials (or in amorphous regions within semicrystalline materials) from a hard and relatively brittle "glassy" state into a viscous or rubbery state as the temperature of the material is increased, or a reverse transition as the temperature is decreased. After temperature is further increased, the material in rubbery state will be melted and become liquid. The glass state can be viewed as a super cooled liquid and a non-equilibrium and disordered solid state. The glass transition temperature (T_g) of a material is the range of temperatures over which this glass transition occurs. Glass transition temperature is always lower than the melting temperature of the material if one exists. For example, the T_g of hard plastics like polystyrene is well above room temperature, so polystyrene is hard and in its glassy state at room temperature. Rubber elastomers like polyisoprene and polyisobutylene are usually used above their T_g, which means the elastomers are in the rubbery state and soft and flexible at room temperature.

The existence of glassy or rubbery state is related to the molecular motion inside the material. Glass is rigid and brittle below its T_g since the molecular chains do not have enough energy to move around below T_g. The molecules are locked into a rigid amorphous structure due to short chain length, and molecular groups branch off the chain and interlock with each other due to a rigid backbone structure. Although molecular mobility is reduced by orders of magnitude below T_g, there may still be sufficient mobility below T_g to enable degradation over a long time period from months to years. When heat is applied, the molecules gain some energy to move around. At some point the heat energy is enough to change the amorphous rigid structure to a flexible structure, and finally move freely around each other. This transition point is the glass transition temperature. The glassy structure does not relax below the transition temperature range. If slower cooling rates are used to cool liquid, structural relaxation (or intermolecular rearrangement) may result in a glass state. This transition occurs with no change in order or structural reorganization of the liquid and, therefore, is not a thermodynamic first-order process since there is no change in entropy, enthalpy, or volume, while it is a thermodynamic second-order phase transition with a jump in the heat capacity or expansivity of the sample that occurs over a temperature range.

4.6.2 Measurement of Glass Transition

Glass transition is not a phase transition and it is a phenomenon characterized by a change in rate of volume expansion with temperature, a discontinuity (a smooth step) in thermal expansion coefficient, and a discontinuity in heat capacity, constant cooling rate, and a viscosity threshold. When the temperature of a material is gradually increased, it undergoes a change in structure (from rigid to flexible) which results in a change in the heat capacity (rapid increase) of the material (Fig. 4.9). Therefore, the most

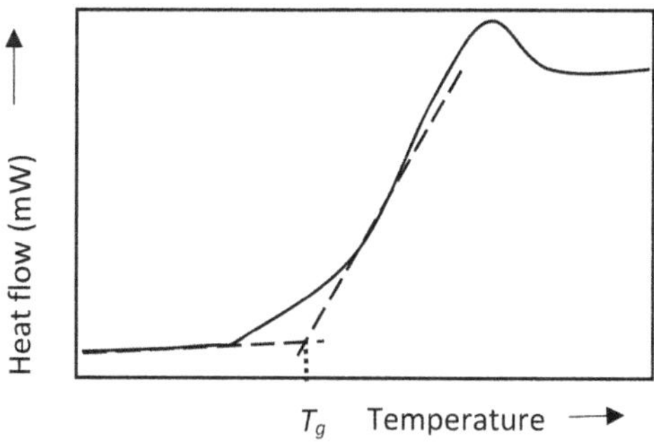

Figure 4.9 Measurement of T_g by scanning calorimetry.

used method to measure the T_g is based on this principle (by measuring the change of heat capacity). Several methods can be used to measure the T_g, including differential scanning calorimetry (DSC), thermally stimulated current, high resolution nuclear magnetic resonance (NMR), and thermomechanical analysis (TMA). The most widely used method is DSC, and commercial DSC instrument is available. Therefore, only this method is introduced.

4.6.2.1 DSC Method

After the sample (usually a thin slice with less than 100 mg) is loaded in DSC calorimetry (such as Pyris 1, Perkin-Elmer, Norwalk, CT), the sample is hermetically sealed and cooled to a low temperature (–50 to –80°C) with 10°C/min (different researchers used different rates), and then heated with the same rate to a high temperature (80–250°C). During this heating period, scanning of the sample temperature is conducted by measuring the heat flow and temperature of the sample (Fig. 4.9). The produced graph is termed as thermogram (Fig. 4.9). The T_g from each thermogram is determined by identifying the transition corresponding to a slope change in the heat capacity of the sample (Fig. 4.9) (Perdon et al. 2000). The base line of the thermogram is obtained by scanning empty pans under the same test conditions as that used for the sample (Sablani et al. 2007).

4.6.3 Factors Influencing Glass Transition

The most important factors influencing glass transition are the physical and chemical properties of the material, temperature, and moisture content, and can be predicted by the rigidity theory. Cooling or compression can result in glass transition. This gradual transition process can result in an increase of viscosity of a material by as much as 17 orders of magnitude within a temperature range of 500 K without any pronounced change in material structure. The consequence of this viscosity increase is the exhibition of solid-like glass mechanical properties. For many materials, rapid cooling will skip this transition and slow cooling results in a glass transition at some lower temperatures. Many polymers easily form glasses even upon very slow cooling or compression.

4.6.4 Glass Transition in Grain Drying

Solids in food materials are either completely amorphous or partially crystalline. Amorphous materials can exist in a glassy or rubbery state. Food material at storage conditions are within the temperature range of

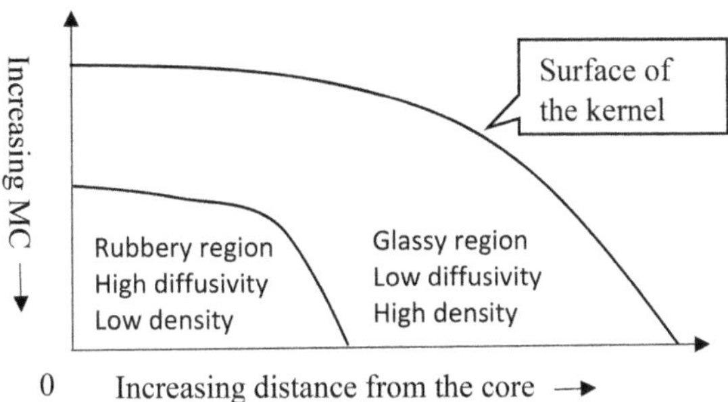

Figure 4.10 Glass transition relationship of a grain kernel with different moisture contents at different locations and at a constant temperature. The origin of the coordinate axis is the core of the kernel. In the Y-Axis label, MC = moisture content.

glass transition (Bhatt and Nagaraju 2009). Grain drying temperatures are within the temperature range of glass transition (Perdon et al. 2000; Cnossen et al. 2002). Different moisture contents at different places of a grain kernel will result different glass transition temperatures (Fig. 4.10). The same material with a lower moisture content will have a higher glass transition temperature than that with a higher moisture content. If the kernel temperature increases above the T_g, the starch transitions from a glassy to a rubbery state with higher expansion coefficients, specific volume, and diffusivity (White and Cakebread 1966). The higher expansion coefficient and specific volume will result in the changes in volumetric expansion and specific volume, which will have an effect on kernel fissuring and drying rate (Cnossen et al. 2002). Many physical properties such as isosteric heat, specific volume, viscoelasticity will be influenced by this change of physical status.

During drying, temperature gradient usually is negligible, while the moisture gradient exists. The material at the surface of the kernel where there is a low moisture content might be at glassy state, while the core of the grain kernel where there is a high moisture content might be at rubbery state (Figs. 4.8, 4.10, and 4.11). Water diffusivity inside glassy material is low (Cnossen et al. 2002). This partially explains the low drying rate after the grain kernel surface is dried to the equilibrium moisture content (EMC) of the supplied air. Therefore, if the supplied air has a low RH (hence a low moisture content) and to avoid the glassy state at the surface, the drying air temperature should be increased (Figs. 4.8 and 4.10).

Whether a rice kernel is above or below T_g significantly affects several drying parameters, including drying rate and fissure initiation in the rice kernel (Cnossen et al. 2002). When grain kernel temperature is below T_g, the starch granules are at glassy state and water associated with the

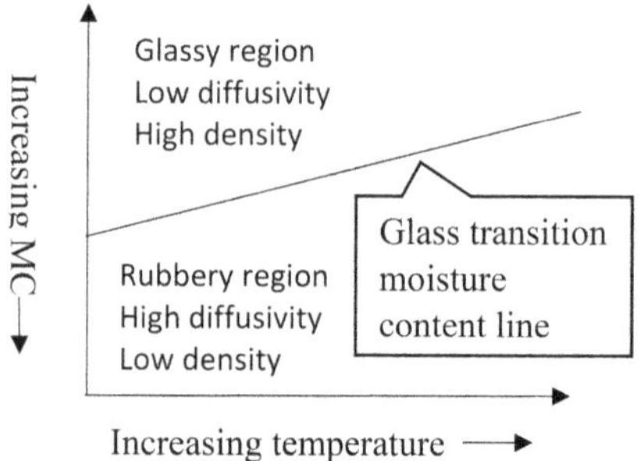

Figure 4.11 Glass transition relationship of a grain kernel with combinations of different moisture contents and temperatures. The origin of the coordinate axis is the lowest temperature and highest moisture content. In the Y-Axis label, MC = moisture content.

starch is relatively immobile, hence a low water diffusivity which results in a low drying rate. Siebenmorgen et al. (2004) found the T_g of brown rice kernels decreased with increasing moisture content and followed a linear relationship.

4.7 Collapse, Breakage, Volume Change, and Pore Formation

During drying, biomaterials undergo volume changes either by shrinkage due to water loss or by expansion due to porosity increase. Shrinkage and expansion of kernels might not be uniform because water loss or gain at different parts of kernels is not the same. Residue of these forces (tension and compression) and uneven shrinkage and expansion cause the kernel to pull itself apart (cracking of kernels) especially when kernels with high initial moisture content are exposed to a high temperature (Kunze 1995). Both intrinsic and extrinsic factors influence these results. The intrinsic factors are the change of the chemical composition (due to water loss) and physical properties (such as structure) during drying. The extrinsic factors are temperature, pressure, relative humidity, gas atmosphere (such as velocity and flow direction), and the electromagnetic radiation applied in the process. The physical mechanisms that control the shrinkage, collapse, breakage, and pore formation are pore pressure, glass transition, mechanism of moisture transport, mechanical strength of the matrix (such as surface tension of the water in capillaries, tension and contraction), environmental pressure, surface electrical charge, and gravitational force. Even though many factors can contribute to this collapse, the main factors proved in literatures are surface tension, glass transition, and unbalanced internal pressures. Using a single concept (factor) cannot always explain the physics of pore formation, collapse, and shrinkage.

Crack can occur in the field before harvest, in the combine, transport truck, storage bins, and in dryers. The most reported crops with a high crack rate are corn (Gunasekaran et al. 1985), rice (Hayashi et al. 2015), soybean (Toledo et al. 2010), and wheat (Kunze 1995). At field, when a rice kernel is dried to below 15% MC during the day and then adsorbs moisture at night, the kernel may fissure (Kunze 1995). Fissured rice kernels usually break during milling.

4.7.1 Forces Influencing Volume Change and Pore Formation

4.7.1.1 Surface Tension

Surface tension exists in capillaries when liquid water is removed from pores inside the material. Higher the surface tension, higher the collapse when the structure of the material cannot hold against this surface tension and the water in the capillaries is removed. It is proven that the capillary collapse can be reduced by lowering surface tension such as application of surfactants or organic solvents (Salas and Labuza 1968), and by flashing off the water instantaneously such as at high drying temperature because the increased vapour pressure inside the pore will counterbalance the collapse force (Genskow 1996).

Rahman (2001) conducted an extensive review on the pore formation during processing and suggested that capillary force is the main force responsible for collapse, while many factors such as surface tension, pore pressure, structure, environmental pressure, and mechanisms of moisture transport play important roles in explaining the shrinkage and formation of pores during drying. The counterbalancing of these forces causes formation of pores and lower shrinkage. The counterbalancing forces are generated by internal pressure due to vaporization of water or other solvents which are influenced and varied in moisture transport mechanism. The following factors might result in the strength of solid matrix: ice formation, case hardening, permeability of water through crust, change in tertiary and quaternary structure of polymers, presence or absence of crystalline, amorphous, and viscoelastic nature of solids, matrix reinforcement, and residence time.

4.7.1.2 Glass Transition

Material can be at glass state when at low temperatures and rubbery states when at high temperatures. Negligible collapse (more pores) in material is observed if it is dried below glass transition because the material is in glass state. Higher the drying temperature, higher the collapse because the material is in rubbery state. Fissures are produced when the rubbery region is rapidly transitioned to glassy region (Cnossen et al. 2002). The result between the freeze-drying and hot-air drying proves this hypothesis (Achanta and Okos 1996). Shrinkage or collapse is stopped when processing temperature is lower than the glass temperature (Fan et al. 1996). The ice sublimation might create pores during freeze-drying, the walls of the pores may shrink due to surface forces or gravity (Karathanos et al. 1993). However, many experimental results do not always support this glass transition theory (Rahman and Sablani 2001). Rahman (2001) concluded that the glass transition theory to explain the shrinkage and formation of pores does not hold true for all products or processes. Temperature is the critical factor influencing this glass transition and crust at the surface of the material.

4.7.1.3 Unbalanced Internal Forces

4.7.1.3.1 Expansion and Puffing At a high drying rate and temperature, the rapid decrease of moisture lets the surface of processed foodstuffs such as a bread loaf become stiff (case hardening phenomenon), which might limit subsequent shrinkage. After the crust is produced, the impermeable crust coupled with extreme drying conditions might result in rapid moisture vaporization causing large internal pressures to build-up, resulting in product expansion/puffing/cracking (Gogoi et al. 2000; Rahman 2001). Electromagnetic energy-based drying (such as microwave drying, radiofrequency drying) process will result in vapour pressure inside the material, and this drying method may produce materials with a desired structure such as less collapse or formation of high pores (Nijhuis et al. 1998).

Vacuum and air drying or frying causes a rapid boiling and evaporation of internal water due to volumetric expansions or tiny explosions of bound water within the raw material into escaping vapour by heating. This produces a puffing in many products, such as popped corn, potato chips, and extruded products. This puffing effect is temperature dependent and higher the temperature, higher the expansion. For example, good volumetric expansion of potato chips is obtained only from frying dried raw chips at their optimal moisture content by creating internal pores and a superficial crust when heat is applied at high rate by convection from hot oil (Soekarto 1993). Popped corn is obtained by rapid boiling of internal water.

Using nitrogen gas to dry food materials can produce a greater number of open pores connected to the exterior surface of the material as compared to air and vacuum drying. Carbon dioxide at high pressure has the potential in controlling the pore formation of extruded products at low temperature followed by drying (Bruce et al. 1996; Gogoi et al. 2000).

4.7.1.3.2 Fissure and Breakage of Grain Kernels During and After Drying

Fissure is the large internal fracture found to be perpendicular to the long axis of the grain kernels. However, in literature, fissure can also refer to any crack or partial breakage of a kernel. In this book, crack and fissure are exchangeable. The direct result of these fissuring damages is the germination loss of grain kernels. Grain kernel can be easily shattered after any stress crack happens. Cracking can occur in the field, in bins during drying and storage, and in dryers. Crack can occur during grain moisture desorption (drying) and moisture adsorption (wetting). For desorption, most of the cracks are irregular. For adsorption, the cracks are straight. During and after drying, many apparently sound kernels contain internal fractures which originate at the centre and develop along the minor axis toward the outside surface (Kunze 1979).

4.7.1.3.2.1 Crack During Drying During drying, moisture and temperature gradients exist (Yang et al. 2002) and these gradients might not be evenly distributed inside the kernel (Jian and Jayas 2018b). These gradients and their uneven distribution generate regions with different mechanical properties, which result in a complex stress pattern. Researchers agreed that the moisture gradients and their uneven distribution are the main reasons of fissuring (Kunze 1979). Low moisture grains adsorb moisture in a deep bed dryer until the drying front reaches them. Moisture adsorbed by low-moisture grains in freshly harvested field rice causes these grains to fissure (Kunze and Prasad 1978). Crack can occur during and after grain drying, and uneven compression and/or tension force caused mainly by the moisture migration within the grain after drying is believed to cause these fissures to develop (Kunze 1995). Rapid drying produces a steep moisture gradient in kernels. During and after grain drying, a grain kernel can change from glass state to rubbery state and different regions inside a kernel can be at different states (Zhao et al. 2020). The hardness of any specified point within an endosperm increased or decreased linearly according to the increase or decrease of its moisture content (Nagato and Ebata 1964). This complex pattern of mechanical properties, which depends on the rate and amount of moisture removed, may cause the different patterns of stress and tension (Kunze 1979; Sharma and Kunze 1982; Chen et al. 1997). Based on the hardness distribution, Nagato and Ebata (1964) concluded that crack formation during drying was a consequence of the unequal shrinking of the endosperm which results from uneven dehydration of the kernel.

These residuals and non-uniform distributions will become more significant for grain kernels with a small sphericity (irregular shape such as corn) and when kernels collide during handling after drying (Gunasekaran et al. 1985). Crack is mostly caused by moisture and thermal stress, and not probably caused by temperature gradients (Ekstron et al. 1966). Therefore, drying rate is the most significant factor in stress crack development (Thompson and Foster 1963). Total stress crack counts are 92–98% in corn dried in crossflow dryers from 20 to 30% initial moisture content to 14% final moisture content at air temperatures of 60 and 145°C, respectively (Thompson and Foster 1963).

4.7.1.3.2.2 Crack after Drying Moisture gradients exist inside grain kernels after heated air drying because the core of grain kernels usually has a higher moisture content than that at the outer layer of the kernel. Time is required for these moisture gradients to recline. After drying and as the gradient reclines, the external cells expand while adsorbing moisture from the central portion of the kernel or from environment. The cells in the central portion of the kernel contract because they are diffusing moisture to the surface cells (Kunze 1979; Sharma and Kunze 1982). The result is expansion and compression in the outer layer and tension and contraction in the central portions of the grain kernel (Fig. 4.12). When the compressive stresses at the outer layer exceed the tensile strength of its interior, the kernel pulls apart or fissures (Kunze 1979). Fissuring continues until the moisture gradient in the grain has reclined. Delayed cooling effectively reduces possible breakage susceptibility and results in improved test weights over conventional drying and cooling within the dryer (Gustafson and Morey 1979). Researchers found most fissures occur after high temperature drying and during the cooling period. Rice fissuring occurs mostly in 24 h after drying (during cooling period) especially when the grain is dried at a high temperature and cooled immediately (Yang et al. 2002).

Few kernels fissure while drying, but most of the fissuring occur within 48 h thereafter (Sharma and Kunze 1982). Cracking rate might be

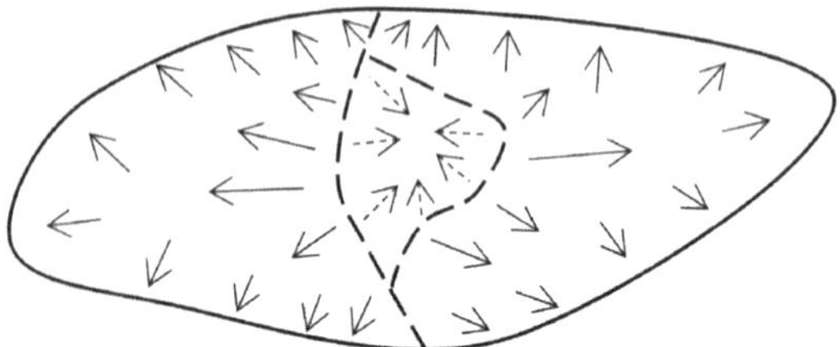

Figure 4.12 A hypothetical distribution of residual force within a rice kernel: compaction and expansion at the surface due to water sorption from the ambient and centre of the kernel (the solid arrows), tension and contraction at the centre due to water loss after drying (the dash arrows). The dash lines show the fissures because the compaction and tension forces are in the opposite directions at the fissuring locations.

different between the sample milled immediately after heated air drying and that milled several days later (Sharma and Kunze 1982).

4.7.1.3.2.3 Crack Due to Swelling Swelling due to water absorption can also result in fissured kernels. Expansion of kernels might not be uniform because water absorption at different parts of kernels is not the same (Jian et al. 2017a). Moisture adsorbed through grain surfaces results in compressive stresses in most parts of the kernel and tension stress in some parts of the kernel due to the uneven expansion (Fig. 1.6). Initial moisture content (especially grain kernels with very low moisture content) is the important factor for the crack during swelling. Temperature and water sorption duration also influence crack rate. Crack rate did not change if the grain is repeatedly rewetted and dried (Hayashi et al. 2015). Damage during water adsorption is the greatest when initial moisture content of soybean is lower than 13% (Toledo et al. 2010). To evaluate cracks of rice, Hayashi et al. (2015) recommend to test it at the following conditions: 12% initial moisture content (wb), 25°C, and 5 h sorption duration on the wet filter paper.

4.7.2 Determination of Stress-Cracked Kernels

Determination of cracked kernels is usually done by manual inspection. The most used method is to obverse kernels under a microscope, use naked eyes, or candle kernels against a bright light background. Although these methods are subjective, their results correlate well with the breakage susceptibility as determined by the Wisconsin type breakage susceptibility tester (Paulsen and Hill 1985). Image processing studies have been conducted to evaluate cracked kernels (Gunasekaran et al. 1987).

4.8 Effect of High Temperature and Safe Drying Temperature

4.8.1 Effect of High Temperature

Although drying is an alternative to extend the shelf life of food and also facilitating storage and transportation, the quality of dehydrated food dried at high temperatures is usually reduced as compared to that of the original foodstuff or the material dried at a low temperature. High temperature drying is considered as a highly destructive operation, as indicated by membrane damage, increased electrolyte and sugar leakage, enzyme and protein denaturation, starch reduction and hydrolysis, internal cracks, split seed coats, and discoloration (Seyedin et al. 1984; Zhao et al. 2020). There are extensive reviews on effects of drying on physical, chemical, sensorial, and nutritional properties of food materials (Guiné 2018). The main effects of high drying temperature are:

1. Physical changes: Drying may result in physical changes such as shrinkage, puffing, crystallization, colour change, glass transition, and fissures. These physical changes at low level usually will not result in damage.
2. Enzymatic inactivation: Higher temperature causes degradation of grain quality due to increased enzymatic inactivation. Too high temperature of wheat kernels can cause denaturation of the gluten proteins, which adversely affects the baking quality. For seeding, seed vigour after drying is one of the most considered factors for selecting the drying temperature. Seedling vigour as judged by shoot and root dry weight is also significantly reduced (Seyedin et al. 1984).
3. Membrane damage: High drying temperature results in reduction in the number and size of starch grains in the embryonic axis of corn kernels. The embryo might be the primary source of leachate sugars (Seyedin et al. 1984). High drying temperature significantly reduces the ability of the embryo to germinate (Seyedin et al. 1984). Membrane damage is particularly problematic for thermally sensitive materials such as fruits and vegetables.
4. Change of properties and components of chemical substances: Malumba et al. (2008) found solubilities of the albumin, globulin, and zein fractions of corn decreased 63.9, 76.9, and 52.4%, respectively, with an increase in the drying temperature from 54 to 130°C. At the same time, high drying temperatures resulted in the disappearance of some water-salt-soluble polypeptides. Timm et al. (2020) showed that reduction in the protein solubility and lipase activity was associated with an increase in the drying temperature and the drying method. Nutritional and energetic value is decreased due to these changes of properties and components of chemical substances. Ethanol yield is 389.83, 378.29, and 366.40 L/t of corn dried at 80, 100, and 120°C, respectively (Coradi et al. 2016).
5. Influence of industrial performance of end products: The increase in the drying temperature increases stiffness of starch granules, affects starch isolation, increases thermal resistance, changes colour, and reduces swelling capacity, water binding capacity, and the water solubility (Malumba et al. 2008; Malumba et al. 2009; Ziegler et al. 2020). The result of these influences is the reduced digestibility of the dried food and feed (Kaczmarek et al. 2014). Malumba et al. (2009) reported that the increase in

drying temperature from 54 to 130°C reduced the extraction yield of starch by 28.18% and increased the residual protein content by 84.06%.

6. Fissure and breakage: Fissure, crack, and/or breakage occur during and after high temperature drying. When subjected to a high drying potential (high temperature and low RH), high moisture kernels will have a higher potential of fissuring than low moisture rice kernels (Sharma and Kunze 1982). Corn dried under natural air drying conditions, dryeration dried at temperatures between 60 and 100°C, and crossflow dried at 45–80°C has the stress crack of 1.4%, between 10.0–16.9% and 57.5–63.5%, respectively (Brown et al. 1979). Older rice showed more fissured kernels than the recently harvested rice after drying (Kunze 1979). Corn dried at 37.8°C and 71.1°C has an average stress crack of 77 and 99%, respectively (Westerman et al. 1973).

4.8.2 Safe Drying Temperature

Increase in drying temperature increases drying rate but possibly decreases grain quality (Malumba et al. 2008). The consideration for the safe drying temperature is that when drying temperature is higher than a certain range, the processing (milling) and baking qualities of the grain will be reduced, and storage ability and germination will also decline if the grain will be used as seeds. Safe drying temperature is usually defined as the temperature at which the grain can be dried without any significant loss in its quality and quantity. Therefore, end use for different types of grain is the first considered factor for the selection of drying temperature (Table 4.4). The safe drying temperature is selected based on the following considerations:

1. End usage: Safe drying temperature for seedling and malting will be much lower than that for bread wheat. The recommended drying grain temperature range is about 38–43°C to maintain seed viability (Hall 1980). Over drying will decline seed quality such as seed coat splitting and changes of chemical compositions.
2. Crop type: Different crop types and varieties have different tolerances under the same drying condition.
3. Initial moisture content: Safe drying temperature is influenced by the initial moisture content of the grain and the residence time in the dryer. Therefore, safe drying temperature and residence time of bulk grain with different initial moisture contents should be tested in lab or industrial trials.
4. Drying method: Natural air drying with or without heater usually has a lower temperature than the temperature used in a high

Table 4.4 Recommended Safe Drying Grain Temperature (°C) for Different Grain Types[a]

Grain	*MC (wb, %)*	*Seed or Malting*	*Commercial Use*	*Feed*
Wheat	Below 24	49	66	74–85
	Above 24	43	60	74–85
Oats	Below 24	50	60	63–79
	Above 24	43	60	63–79
Barley	Below 24	49	55	74–85
	Above 24	41	55	74–85
Rice	Below 20	44	60	
	Above 20	40	60	
Rye	Below 24	45	60	80–100
	Above 24	43	60	80–100
Corn	Below 24	45	60	71–82
	Above 24	43	60	71–82
Popcorn		38	38	
Flax		38	49	
Rapeseed		41	49	
Peas		45	66	66
Mustard		45	45	
Sorghum		43	60	
Soybean		38	49	
Sunflower		45	60–77	77–91
Buckwheat		45	45	

[a] Data show grain temperatures. Drying air temperature can be higher than the grain temperature.

Source: Maier and Bakker-Arkema (2002); Jayas and Ghosh (2006); Canadian Grain Commission (2018).

temperature dryer. In high temperature drying, the ambient air is heated by 45–200°C using external sources of energy.

5. Dryer type and residence time: Depending on the initial grain condition, a short residence time usually needs high drying rate, high temperature increase, and high moisture stresses in the kernels. The potential danger during drying is the exposure of very wet seeds to high temperatures or to a critical range of temperatures for long period. Grain mixing during drying is one of the methods to avoid the same kernels from being kept at a high temperature location for a long time. Evaporative cooling can also decrease the

temperature on surface of grain kernels. This explains why the maximum air temperature for drying wheat is about 50°C in cross-flow dryers but 175–200°C in concurrent-flow dryers (Maier and Bakker-Arkema 2002). Very low relative humidity of the drying air and mechanical damage during the drying are also harmful condition for seed quality. Different dryers can be used for drying of different types of seeds, but one type of dryer under the same drying condition cannot be used for drying all crop seeds.

Based on drying temperature, drying of grains can be divided into four broad categories (Maier and Bakker-Arkema 2002): (1) natural air drying, which uses unheated air or air heated by up to 8°C, (2) medium temperature drying with heated air, which keeps grain temperatures below 45°C for seed grains and below 60°C for processed grain, (3) high temperature drying with heated air which keeps air temperatures up to 275°C for animal feed (but usually lower than 82°C), and (4) combination (dryeration) drying, which uses both low and high temperature drying. All these categories are used on-farm in North America except the drying system is smaller on farms compared with dryers at commercial elevators. Dryers located at commercial elevators are usually continuous-flow systems with a high drying capacity (at least 12.5 wet metric tonne/h).

4.9 Deep Bed Drying, Zones and Fronts

4.9.1 Deep Bed Drying

Deep bed drying or fixed bed drying is the drying of grain in a structure with a grain depth of higher than thin layer of grain. Deep bed could consist of two or more thin layers. Deep bed drying is mostly conducted inside bins or in-bin dryers, and the grain is dried mainly by convection heat transfer and in which gradients of temperature and moisture exist. In-bin drying method usually applies a lower temperature than that in a high temperature drier. For theoretical understanding of principles of deep bed drying, deep bed drying is usually assumed as a combination of multiple layers of thin layer drying. The outgoing air from the previous layer passes through the current layer and then passes through the next layer (Fig. 4.13). In each layer, the fundamental principle for a single thin layer drying holds. The outgoing air might not equilibrate with the grain kernels inside each layer and the percentage of the equilibrium will depend on the air velocity, thickness of the layer, and physical and chemical properties of the grain bulk. Most studies indicate that equilibrium exists when airflow is low such as during aeration (Jian et al. 2019).

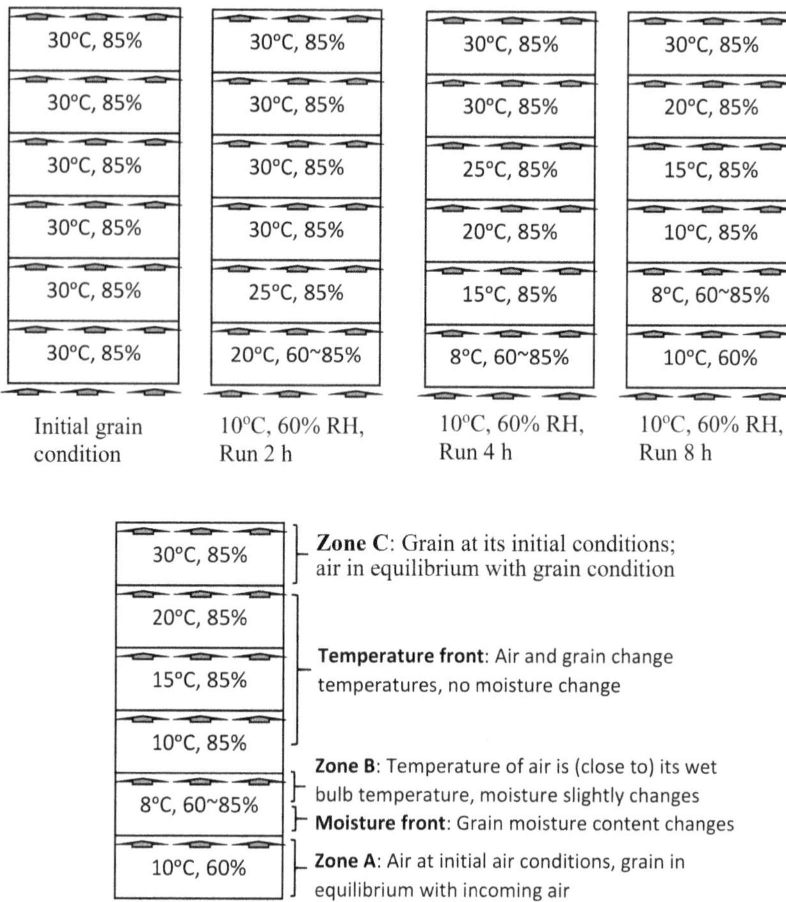

Figure 4.13 Diagrammatic representation of a deep bed drying and the development of zones and fronts (the thickness of the layers is not scaled). Arrows show air moving direction.

4.9.2 Zones and Fronts

As air moving thorough the deep bed of grain, zones and fronts are developed and move in the direction of the airflow (Fig. 4.13), and these zones and fronts can be presented on a psychrometric chart (Fig. 4.14). In zone A, where the air enters the grain bulk, the grain is in equilibrium with the conditions of the entering air and the grain temperature and the air temperature are equal (Figs. 4.13 and 4.14). The grain moisture content can be

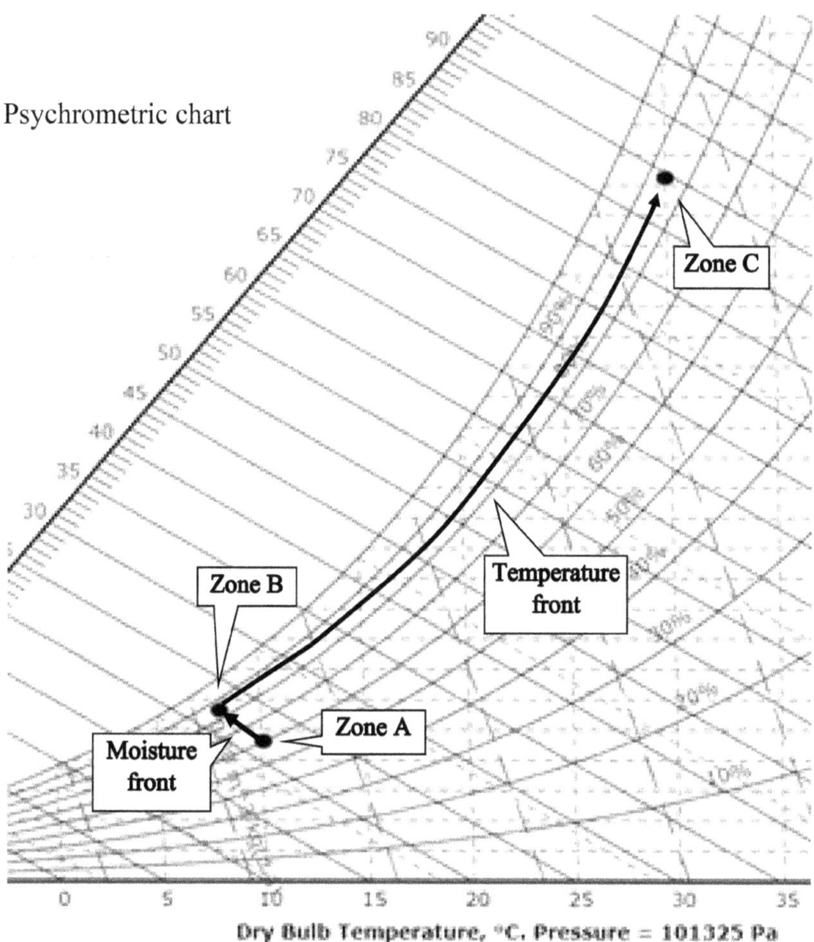

Figure 4.14 Initial condition of grain and air in Fig. 4.13 presented on a psychrometric chart. During the aeration, zone A gradually moves up of the grain bulk, and the grain condition in zone C gradually changes to the condition of zone B and zone A.

estimated from the equation of desorption (for drying, Fig. 4.14) or sorption (for wetting) isotherm of the grain at the incoming air condition.

In the moisture (adiabatic) front, the air and grain come into an intermediate equilibrium. If the grain is dried in this front, the heat to evaporate the water from the grain is supplied by the air (Fig. 4.15a) or from grain (Figs. 4.14 and 4.15b). For case a in Fig. 4.15, this heat continues to be available as long as the air is flowing through the grain. For case b in Fig. 4.15,

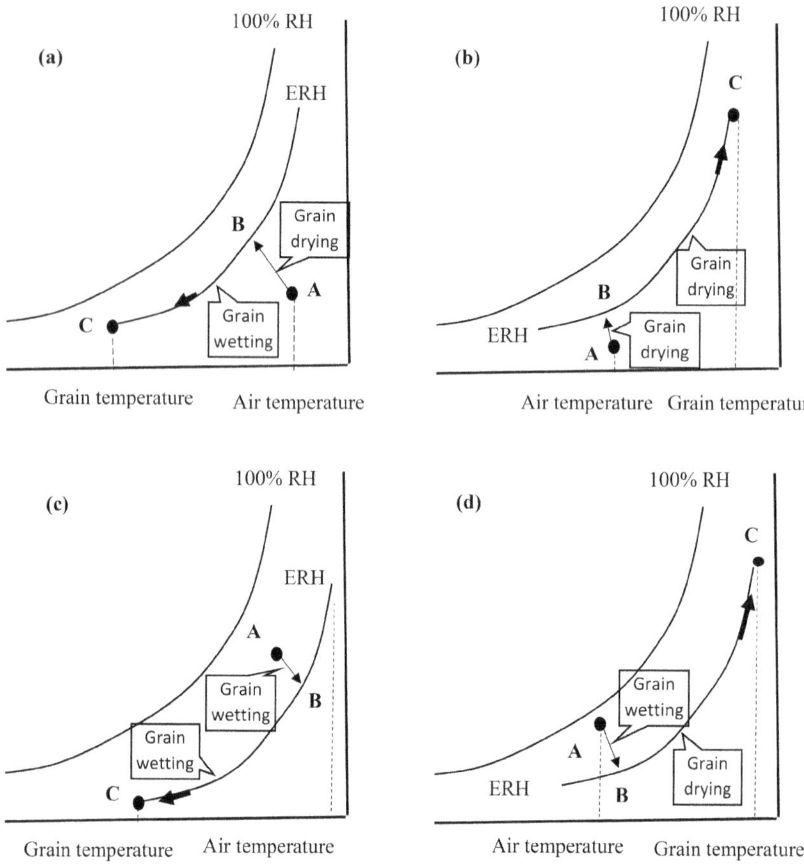

Figure 4.15 Possible psychrometric processes during drying, rewetting, cooling, and warming when supplied air is warmer and drier (a), cooler and drier (b), warmer and wetter (c), and cooler and wetter (d) than the grain.

this heat continues to be available as long as the grain temperature is higher than air. If the grain is wetted (Figs. 4.15c and d), the air increases its temperature and the heat to increase the air temperature come from the grain. Thus, it is an adiabatic or constant wet-bulb process and can be shown on a psychrometric chart as a constant enthalpy line from point A to B (Fig. 4.15).

The air leaving the zone B is close to the ERH of the grain. The location of B on the constant enthalpy line is determined by the equilibrium moisture content of the grain at that point. If the moisture content of the grain is high and grain is dried by this convection process, its ERH is near 100%, and the temperature in zone B will be or close to the wet bulb temperature of the air entering grain. Otherwise, the temperature in zone B will be between the dry and wet bulb temperatures of the entering air (Figs. 4.14 and 4.15). If the grain is wetted, the air RH in zone B will decrease along the line of the constant enthalpy (Fig. 4.15c and d).

In a faster moving temperature front, the major effect in zone B is a change in temperature from the initial grain temperature towards the air temperature and an associated small change in moisture content. In a slower moving moisture front the major effect is a change in moisture after grain temperature change.

In zone C, grain is at its initial temperature and moisture content. On the psychrometric chart, the air in zone C is in equilibrium with the initial grain conditions (Figs. 4.14 and 4.15). In the temperature front (from B to C on the psychrometric chart), the psychrometric process approximately follows the constant RH of the grain except some drying or rewetting of the grain occurs as the moisture content of the air increases or decreases from B to C (Fig. 4.15). The increased or decreased moisture content of air will influence the grain moisture content in this front, and grain spoilage might occur in the temperature front when grain is rewetted and initial grain moisture content and supplied air temperature is high (Fig. 4.15a and c).

The temperature front moves through the bed more rapidly than the moisture front because the energy in the grain is finite and less energy is required to change the grain and air temperature than that to remove water from the grain after the grain temperature has been changed. Only the moisture front and zones A and B exist in most time of deep bed drying.

For deep bed drying, airflow influences the movement of zones and fronts (Fig. 4.13). The reason for this is that some energy in the air is not used by the grain in the lower layers of the grain bed, while this energy can be used in the adjacent next up-layers. Therefore, higher airflow will move the zones and fronts quickly up and pass through the grain bed, and the temperature and ERH of the grain will quickly change to the temperature and ERH of the supplied air.

4.9.3 Possible Psychrometric Processes During Grain Drying, Wetting and Aeration

Grain bulk might be wetted or dried and cooled or warmed by forcing air through a grain bulk regardless of the grain depth. There are four possible combinations of psychrometric processes during this process (Fig. 4.15). For any combination, the grain moisture content will equilibrate with the supplied air if the grain layer becomes zone A (Figs. 4.13–4.16). The grain is wetted in the temperature front in the cases a and c in Fig. 4.15. If grain is below 0°C, the water holding capacity of the air is low at this low grain

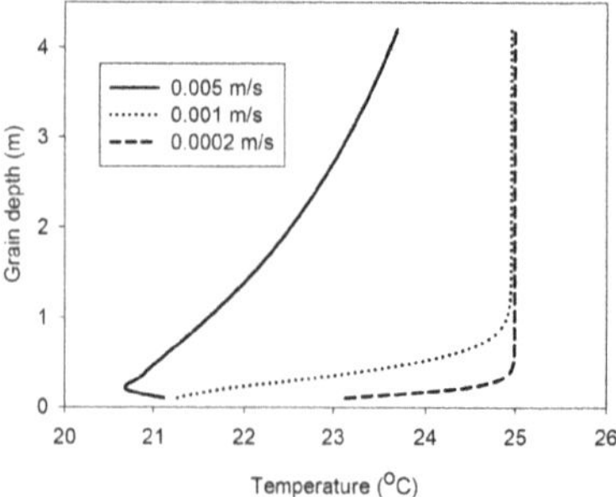

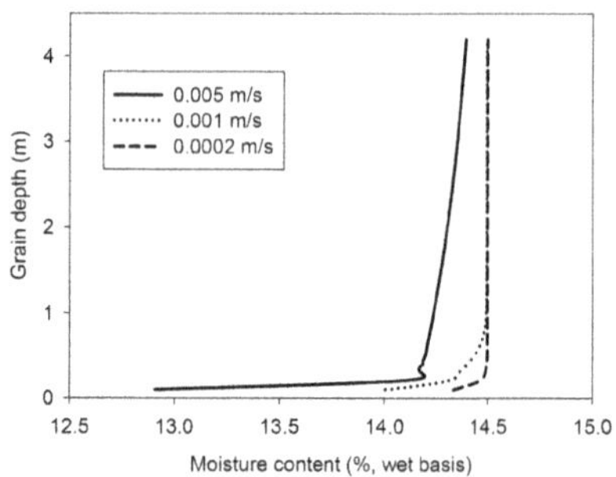

Figure 4.16 Simulated temperature (left) and moisture content (right) of canola filled to 4.3 m in a bin with different airflow rates. The initial grain temperature is 25°C and moisture content is 14.5% (wet basis). The canola is aerated by air at 25°C and 65% RH (equilibrium moisture content is about 9.1%) and at different superficial air velocities (0.005, 0.001, and 0.0002 m/s) and 20 h aeration. (Data were generated by using the model developed by Jian et al. (2019)).

temperature, the air may be cooled to its dew point temperature and the condensed water might be frozen on the surface of the grain kernels. The kernels can be frozen together which reduces or prevents airflow through grain. This can happen in the spring when a cold bulk of grain is ventilated with warm air.

Two combinations result in grain wetting in the moisture front (Fig. 4.15c and d) and another two combinations result in grain wetting in the temperature front (Fig. 4.15a and c). Grain can be dried in the moisture front in two combinations (Fig. 4.15a and b). During grain drying or aeration, the ambient air usually has fluctuating temperature and RH. Therefore, more zones and fronts can be produced and grain might be re-wetted or re-dried at different times. Over the full ventilation period some fronts and zones pass out of the grain bulk while others stay in the grain bulk.

4.9.4 Example of Aeration and Drying

4.9.4.1 Aeration

Jian et al. (2019) conducted an aeration study by aerating canola with air at 0.00275 m/s superficial velocity, 20°C, and 75 ± 10% RH. The canola was held in a 0.2 m diameter pipe to a depth of 1.5 m. The initial temperature and moisture content of the canola was 30°C and 11.8%, respectively. The ERH of the canola before the aeration was 88%, and the canola EMC

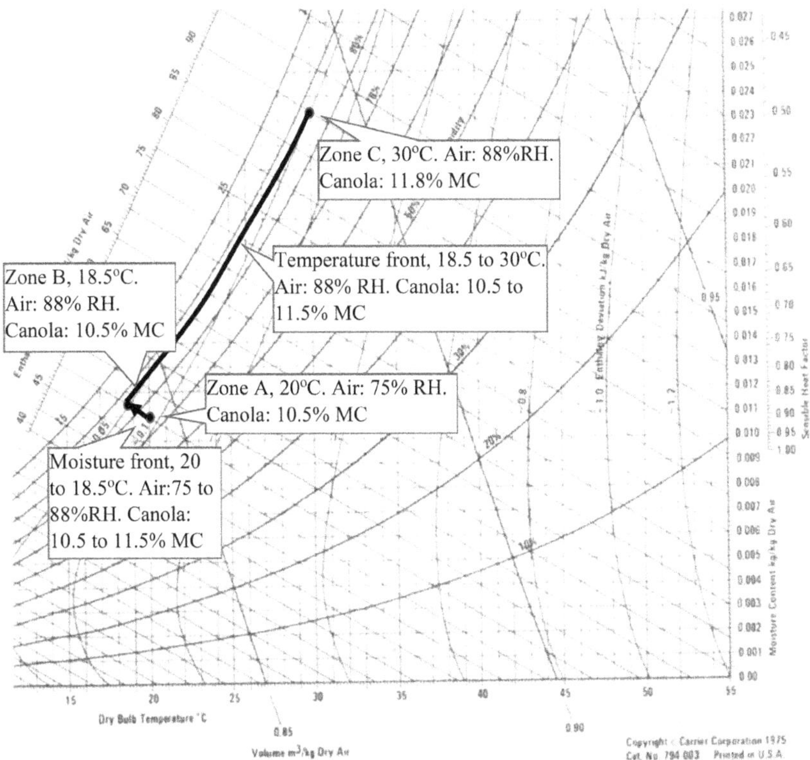

Figure 4.17 Initial canola and air conditions in zones and fronts when canola is aerated with cold air at 20°C and 75% RH. The initial canola condition is 30°C and 11.8% MC. During the aeration, Zone A gradually moves up of the grain bulk, and the grain condition in Zone C gradually changes to the condition of Zone B and Zone A.

at the aeration condition was 10.5% (Appendix A). The grain and air conditions in zones and fronts are presented in Fig. 4.17. During the continuous aeration for 10 h, the temperature of the canola was gradually cooled down and temperature front gradually moved up and moved out of the canola after 10 h (Fig. 4.18). At the end of aeration, bottom of the canola (about 0.3 m) was dried to the air condition (Fig. 4.18). The canola moisture content in the zone B and moisture front was about 11.4%. The canola moisture content after the temperature front was about 11.5%. Therefore, the aeration removed about 0.4 (from 11.8% to 11.4%) percentage point moisture content.

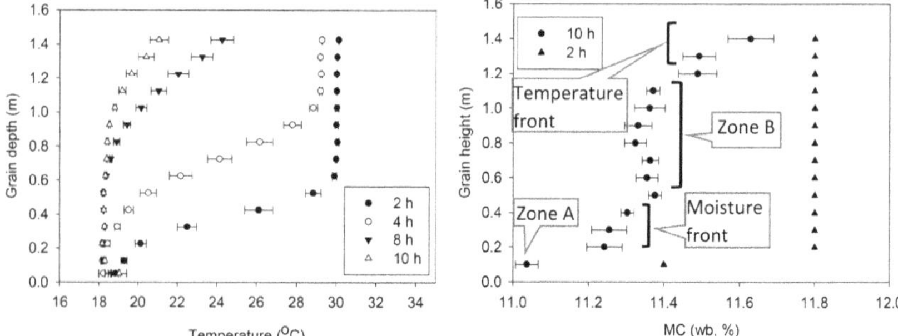

Figure 4.18 Canola temperatures and moisture contents at different grain depths after 2–10 h aeration (modified from Jian et al. 2019 with permission). The initial moisture content and temperature of the canola are 11.8 ± 0.1% (wb) and 30 ± 0.5°C, respectively. The condition of the aeration is 20°C and 70% RH.

4.9.4.2 Natural Air Drying

Muir and Sinha (1986) conducted a natural air drying study in a cylindrical, galvanized steel bin, 4.3 m diameter, and equipped with a fully perforated floor and a 5.4 kW axial fan. The bin held 26 tonnes of freshly harvested rapeseed (cv. Regent) at initial moisture content of 14.1% near the bottom of the bin, 14.8% near the centre depth, and 14.6% near the top levelled surface (Fig. 4.19). Airflow rate was 17.7 (L/s)/m^3, and the ambient air was warmed about 3.8°C by the fan, hence the relative humidity of the entering air was about 20% below that of the ambient air. The rapeseed was harvested on September 27, 1980, and the bin, located at about 20 km South of Winnipeg, Canada, was continually ventilated for 21 days.

After 14 days of drying, a drying front had developed and located near the centre depth of the bin (Fig. 4.19). Above the drying front, the rapeseed remained almost at its initial moisture content (only reduced about 0.4 percentage points after the temperature and zone B moved out of the bin). Below the drying front, the grain bulk dried to the EMC corresponding to the relative humidity of the entering air. The oilseeds were over dried and were much lower than the recommended "dry" storage moisture content (Appendix A) due to the fan warming. The drying front was not evenly moved up due to uneven airflow (Fig. 4.19).

After continuous drying for another one week (21 days), the drying front passed through the rapeseed bulk, and the bottom was re-wetted due to high ambient RH in that year. This re-wetting is beneficial because the saleable mass at the bottom was increased and the moisture content was still lower than the recommended "dry" storage moisture content for selling

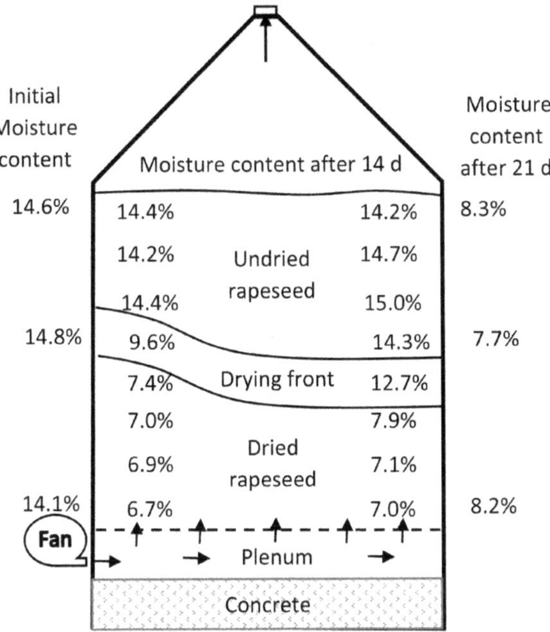

Figure 4.19 Natural air drying of 26 tonnes of rapeseed at 17.7 (L/s)/m^3. The drying was started on September 27, 1980 and continued until October 18, 1980. Data from Muir and Sinha (1986).

rapeseed. However, the rapeseeds at the middle depth of the bulk were over dried after the bottom was re-wetted because the moisture in the ambient air was adsorbed by the lower layers of rapeseeds and then the dried air picked up moisture from the rapeseed located at middle depth of the bulk. Therefore, it is difficult to dry all grain in a bin to a uniform moisture content by forcing air vertically because all grain cannot be dried simultaneously. Forcing air horizontally dries grain more uniformly (Chelladurai et al. 2015).

During natural air drying, the grain above the drying front remains near its initial moisture content until the drying front passes through that grain. The ambient air temperature used for natural air drying is usually higher than 25°C. Fan warming can raise the temperature of the entering air to about 30°C. Thus, the grain above the drying front stays moist and warm. This warm and moist grain can spoil due to mould multiplication before it is dried. This requires that the airflow rate through the grain must be sufficient to move the drying front through all the grain before the last grain has unacceptable spoilage. The cost of the fan and the energy to operate the fan increases with airflow rate, so does the fan size. Therefore, fan should be properly sized and the ventilation system should be well designed. To ensure successful natural air drying during all years, the fan sizing is usually based on the worst weather at the geographical location and the highest initial grain moisture content expected to be dried. Using such system, natural air drying will always be completed in all other years and for grain at initial moisture content less than the design moisture content; however, more overdrying will occur in such cases. During natural air drying, it is also recommended not to stop the fan during rainy days because of the reason mentioned above.

4.9.4.3 Dryeration and Combination Drying

Dryeration is a multi-stage and intermittent drying process developed for drying corn at Purdue University in the late 1960s. Dryeration is usually conducted as a three-stage process (Fig. 4.20). Stage 1: grain is dried using hot air dryer until its moisture content is about 2–5% above the target moisture content. Stage 2: grain is tempered for 4–10 h, which allows moisture migration from the centre of kernels to outside. Stage 3, grain is natural air dried, typically in a bin with fully perforated floor, to the target moisture content using properly designed airflow rate. Aeration may be conducted after natural air drying. For example, the procedure of corn dryeration can be: the corn at 43–54°C is transferred from a hot air dryer into a natural air drying bin at a moisture content of 2–3 percentage points higher than desired. After tempering for 8–12 h, the corn is ventilated with ambient air at 6–8 $L{\cdot}s^{-1}{\cdot}m^{-3}$ (0.6–0.8 $m^3{\cdot}min^{-1}{\cdot}t^{-1}$) (Morey et al. 1981) and the sensible heat stored in the grain evaporates moisture from the grain. Dryeration can be conducted by using two (Fig. 4.20), three, and four bin systems. The extra bins are usually used to conduct the natural air drying at the same time.

The advantages of the dryeration are: (1) increased drying capacity because lesser moisture is removed in the hot air dryer, so drying air temperature can be increased, drying time can be reduced; (2) converting the

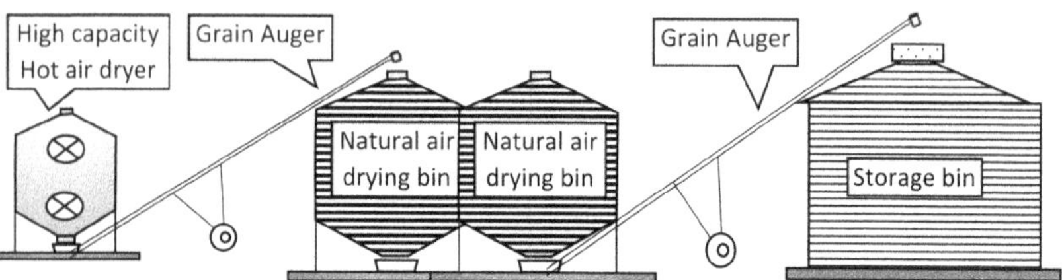

Figure 4.20 A three bin dryeration system.

cooling section of the dryer to heated air section or eliminating the cooling time will also increase drying capacity and decrease drying time; (3) energy consumption is reduced because less thermal energy and fan energy are required in the hot air dryer to remove the last 2–3% of moisture, and sensible heat in the grain in stage 3 evaporates this last 2–3% moisture more efficiently after moisture migration; (4) improved grain quality (maintain germination and reduce cracks) because: (a) the grain is dried in the hot air dryer for a shorter time and is discharged at a higher moisture content than traditional hot air drying; (b) the moisture migration during stage 2 decreases the unbalanced residual force inside kernels, transferring from dryer to bin and from bin to bin also mixes grain well; and (c) stage 3 also takes the advantages of natural air drying. Thompson and Foster (1967) reported the dryeration process could reduce breakage attributable to heated air drying by 80% since 63% of the kernels did not present stress cracks. Drying capacity can increase 50–70% and drying energy efficiency can increase 15–30% when dryeration is used (Montross and Maier 2000; Shouse et al. 2011; Santos et al. 2012).

Combination drying is a similar process as the dryeration, and the combination can be hot air drying + another hot air drying (drying air temperatures inside different dryers are different, also termed as multi-staged hot air drying), hot air drying + natural air drying + aeration, hot air drying + natural air drying with heater + aeration, and hot air drying + aeration. For example, a combination drying system uses a high-temperature, high-speed dryer for the initial stage of drying. After the corn is dried to 20–22% moisture content, it is discharged to another dryer (usually an in-bin dryer) for a second stage drying until the drying is completed. The first and second stages are usually completed within 2–24 h and 4–8 weeks after harvest (Morey et al. 1981). For a combination drying, partial drying in the hot air dryer can be about 5% above target moisture content and then natural air drying is conducted. This process is usually used for grain harvested at high moisture content and one cycle of hot air drying cannot dry grain to target moisture content due to high energy consumption, low efficiency, low drying capacity, and low grain quality. Generally speaking, dryeration and combination drying have a higher energy efficiency than only hot air drying because natural air drying is less expensive than hot air drying, but more facility, more bins and handling equipment, including larger fans to produce a higher airflow rate than used in a conventional storage bin for natural air drying, is required. The same principle is used for in-storage cooling and intermittent drying. In-storage cooling eliminates the extra handling step that is sometimes a disadvantage of dryeration, which makes in-storage cooling a feasible alternative. After harvest, grain dryer is often the bottleneck that limits harvest rate. Dryeration and combination drying can narrow down this limitation.

Dryeration and combination drying have been widely accepted and extensively used in the world (Winfield 1969). In Canada and the USA (Morey et al. 1981), farmers conduct this process every year. In France, corn is dried to 17–19% moisture content and transferred at 50–60°C to a natural air drying bin. After 12–14 h tempering, the grain is dried at 10–15 $L \cdot s^{-1} \cdot m^{-3}$ airflow rate (Nellist and Bruce 1995). Morey et al. (1981) conducted field experiment of combination drying for four years by using a centralized drying system including a 10 m^3 automatic batch dryer supplying about 8 $L \cdot s^{-1} \cdot m^{-3}$ (80 $m^3 \cdot min^{-1} \cdot t^{-1}$) air flow rate, a 2 m^3/min vertical elevator, a 1800 m^3 bin with drying floors and 1.1 kW fans providing about 0.1 $L \cdot s^{-1} \cdot m^{-3}$ (1.1 $m^3 \cdot min^{-1} \cdot t^{-1}$) airflow rate, and 110 m^3 storage bins with 0.1 kW aeration fan. They concluded that combination drying resulted in a lower energy requirement compared to conventional high temperature drying with in-dryer cooling. Adding heat to the in-storage stage drying provided a modest increase in drying reliability. The authors also concluded that by lowering the drying temperature and increasing the tempering period resulted in less breakage.

4.10 Remark on Application of Drying Theories

In this chapter, the basic concept and parameters related to drying, wetting, and aeration are introduced. Drying, wetting and aeration are complex processes which involve complex heat and mass transfer and changes of chemical and physical properties. Even though lots of principles explained in this chapter can be applied to these processes, different drying methods might have different drying principles. Only one concept cannot be used to explain mechanisms and physics of most drying processes. Therefore, it is the scientists' real challenge to develop a universal principle or model based on all the mechanisms and physics of the drying processes. An experienced engineer should select the right principles to conduct a correct design. Before selecting the drying method for the design, it is important for the engineer to properly evaluate the drying quantities and parameters in an adequately correct manner with an uncertainty analysis. The following parameters should be considered in this evaluation: energy source, drying air source (possibly other types of gases or fluids or ambient air), pressure, velocity, relative humidity, shape and type of drying materials, retention time, drying capacity, flow configuration, drying medium conditions, initial and operation cost, weather condition (regardless of whether ambient air will be used), etc. Evaluation of all of these parameters is a challenging task. Therefore, using the principles discussed in this book or developing a mathematical model to optimize all of these parameters is the first step of designing grain drying systems. Mathematical modelling will be discussed in the next chapter.

Nomenclature

α: thermal diffusivity (m^2/s);
λ: heat of vaporization of water (J/kg);
β: drying period (s);
ρ_a: density of the air (kg/m^3);
ρ_g: density of the material (kg/m^3);
ρ_s: mass concentration of water in solid (kg water/m^3 sample);
ε: porosity of the material (decimal);
Υ: constant;
μ: fluid absolute viscosity ($N{\cdot}s/m^2$);
$\mu_0\ (T, P)$: chemical potential of pure water vapour in the system (J/mol or J/kg) at temperature (T) and pressure (P);
Ω: slope of the regression equation;
A: surface area (m^2);
A_1: geometric constant;
a_w: water activity (decimal);
Bi: Biot number of heat transfer;
Bi_m: Biot number of mass transfer;
C_a: specific heat capacity of air at constant pressure ($kJ{\cdot}kg^{-1}{\cdot}K^{-1}$);
C_{ds}: specific heat of dry solid ($kJ{\cdot}kg^{-1}{\cdot}K^{-1}$);
C_s: moisture concentration of solid (kg/m^3);
C_w: specific heat of moisture ($kJ{\cdot}kg^{-1}{\cdot}K^{-1}$);
C_{wg}^s: saturation vapour concentration (kg/m^3);
D: mass (water) diffusivity (m^2/s);
D_{eff}: effective water diffusion coefficient (m^2/s);
D_s: surface diffusion coefficient (unit depends on M_s);
D_0: Arrhenius factor (m^2/s);
ERH: equilibrium relative humidity (%);
EMC: equilibrium moisture content (wet basis, %);
E_0: activation energy for moisture diffusion (kJ/mol);
g: geometric constant and g = 2.47, 5.78, and 9.8 for slab, cylinder, sphere, respectively;
h_m: convective mass transfer coefficient ($kg{\cdot}s^{-1}{\cdot}m^{-2}$);
h_{mi}: the transient mass transfer coefficient (m/s);
h_T: heat transfer coefficient ($W{\cdot}m^{-2}{\cdot}K^{-1}$);
J: surface diffusion flux (kg/s);
k: thermal conductivity of the material ($W{\cdot}m^{-1}{\cdot}K^{-1}$);
$k_{g,\text{eff}}$: effective mass transfer coefficient (m/s) for the change in moisture content of air at the food surface;
L: dimension or thickness of the kernel (m);
L_C: characteristic length (m), which is commonly defined as the volume of the solid divided by the surface area of the solid;

Le:	Lewis number;
Le_f:	Lewis factor or Lewis relation;
Le_m:	modified Lewis number;
L_{half}:	characteristic dimension of radius or half-thickness (m);
M:	moisture content (dry basis, decimal);
M_e:	equilibrium moisture content (dry basis, decimal);
M_i, M_{i-1}, and M_{i+1}:	moisture contents of a kernel (dry basis, decimal) at drying time i, $i-1$, and $i+1$, respectively;
M_{cr}:	critical moisture content of the kernel (dry basis, decimal);
M_R:	dimensionless moisture ratio;
∇M_s:	moisture difference between the surface and surrounding air (%, kg/kg);
M_n, M_{n-1}, M_{n+1}:	moisture content (dry basis) at time t_n, t_{n-1}, and t_{n+1}, respectively;
P:	actual vapour pressure of water in the system (Pa);
Pr:	Prandtl number;
p_s:	vapour pressure of water at saturation (Pa);
Δp_v:	$p_{vwb} - p_{va}$;
p_{va}:	vapour pressure at air temperature (Pa);
p_{vwb}:	vapour pressure at wet bulb temperature (Pa);
R:	Ideal gas constant (8.3144598×10^{-3} kJ·mol^{-1} K^{-1});
RH:	relative humidity (%);
R_v:	gas constant of water vapour (461.52 J·kg^{-1}·k^{-1});
Sc:	Schmidt number;
St:	heat transfer Stanton number;
St_m:	mass transfer Stanton number;
T:	temperature (°C);
T_g:	range of temperatures over which glass transition occurs;
T_k:	temperature (K);
t:	time (s);
V_k:	volume of the material (m^3).

References

Achanta, S., and M. R. Okos. 1996. Predicting the quality of dehydrated foods and biopolymers-research needs and opportunities. Drying Technology 14: 1329–1368.

Aguerre, R., C. Suarez, and P. E. Viollaz. 1982. Drying kinetics of rough rice. Journal of Food Technology 17: 679–686.

Alzamora, S., and J. Chirife. 1980. Some factors controlling the kinetics of moisture movement during avocado dehydration. Journal of Food Science 45: 1649–1651.

Alzamora, S. M., J. Chirife, P. Viollaz, and L. M. Vaccarezza. 1979. Heat and mass transfer during air drying of avqcado. In A. S. Mujumdar (ed.), Developments in Drying. Science Press, NJ.

Balaban, M., and G. M. Pigott. 1988. Mathematical model of simultaneous heat and mass transfer in foods with dimensional changes and variable transport parameters. Journal of Food Science 53: 935–939.

Bardy, E., M. Hamdi, M. Havet, and O. Rouaud. 2015. Transient exergetic efficiency and moisture loss analysis of forced convection drying with and without electrohydrodynamic enhancement. Energy 89: 519–527.

Becker, H. A., and H. R. Sallans. 1957. A theoretical study of the mechanism of moisture diffusion in wheat. Cereal Chemistry 34: 395–409.

Bedane, A. H., M. Eic, M. Farmahini-Farahani, and H. Xiao. 2016. Theoretical modeling of water vapor transport in cellulose-based materials. Cellulose 23: 1537–1552.

Bhatt, C. M., and J. Nagaraju. 2009. Studies on glass transition and starch re-crystallization in wheat bread during staling using electrical impedance spectroscopy. Innovative Food Science and Emerging Technologies 10: 241–245.

Brooker, D. B., F. W. Bakker–Arkema, and C. W. Hall. 1992. Drying and Storage of Cereal Grains and Oilseeds, Springer US, New York.

Brown, R. B., G. N. Fulford, T. B. Daynard, A. G. Meiering, and L. Otten. 1979. Effect of drying method on grain corn quality. Cereal Chemistry 56: 529–532.

Bruce, D. M. 1985. Exposed-layer barley drying: three models fitted to new data up to 150°C. Journal of Agricultural and Engineering Research 32: 337–347.

Bruce, H. L., F. H. Wolfe, S. D. M. Jones, and M. A. Price. 1996. Porosity in cooked beef from controlled atmosphere packaging is caused by rapid CO_2 gas evolution. Food Research International 29: 189–193.

Callaghan, P. T., K. W. Jolley, and J. Lelievre. 1979. Diffusion of water in the endosperm tissue of wheat grains as studied by pulsed field gradient nuclear magnetic resonance. Biophysical Journal 28: 133–142.

Canadian Grain Commission. 2018. Drying guidelines. Canadian Grain Commision.

Chelladurai, V., V. R. Parker, D. S. Jayas, and N. D. G. White. 2015. Evaluation of a horizontal air flow in-bin grain drying system. Applied Engineering in Agriculture 31: 793–798.

Chen, X. D., S. X. Q. Lin, and G. Chen. 2002. On the ratio of heat to mass transfer coefficient for water evaporation and its impact upon drying modeling. International Journal of Heat and Mass Transfer 45: 4369–4372.

Chen, H., T. J. Siebenmorgen, and B. P. Marks. 1997. Relating drying rate constant to head rice yield reduction of long-grain rice. Transactions of ASAE 40: 1133–1139.

Cnossen, A. G., T. J. Siebenmorgen, and W. Yang. 2002. The glass transition temperature concept in rice drying and tempering: effect on drying rate. Transactions of the ASAE 45: 759–766.

Coradi, P. C., L. V. Milane, L. J. Camilo, and M. G. O. Andrade. 2016. Drying and storage of corn grains for ethanol production in Brazil. Bioscience Journal 32: 1175–1190.

Crank, J. 1975. The Mathematics of Diffusion, Oxford University Press Inc., New York.

Deshpande, S. D., S. Bal, and T. P. Ojha. 1994. A study on diffusion of water by the soybean grain during cold water soaking. Journal of Food Engineering 23: 121–127.

Duc, L. A., J. W. Han, and D. H. Keum. 2011. Thin layer drying characteristics of rapeseed (*Brassica napus* L.). Journal of Stored Products Research 47: 32–38.

Ece, M. C., and A. Cihan. 1993. A liquid diffusion model for drying rough rice. Transactions of the ASAE 36: 837–840.

Efremov, G., and T. Kudra. 2005. Model-based estimate for time-dependent apparent diffusivity. Drying Technology 23: 2513–2522.

Ekstron, G. A., J. B. Liljedahl, and R. M. Peart. 1966. Thermal expansion and tensile properties of corn kernels and their relationship to cracking during drying. Transactions of the ASAE 9: 555–561.

Elias, M. C., F. da F. Barbosa, J. C. da Rocha, F. M. das Neves, G. Cella, and A. R. G. Dias 2006. Grain quality and energy consumption by evaluation intermittent methods of rice drying, pp. 1043–1052. In I. Lorini, B. Bacaltchuk, H. Beckel, D. Deckers, E. Sundfeld, J. P. dos Santos, J. D. Biagi, C. Celaro, d' Faroni, L. d, F. Bortolini, M. R. Sartori, C. Elias, C. Guedes, G. da Fonseca, and V. M. Scussel (eds.), Proceedings of the 9th International Working Conference on Stored-Product Protection. Brazilian Post-harvest Association ABRAPOS, Campinas, Brazil.

Fan, L. T., P. S. Chu, and J. A. Shellenberger. 1963. Diffusion of water in kernels of corn and sorghum. Cereal Chemistry 40: 303–313.

Fan, J., J. R. Mitchell, and J. M. V. Blanshard. 1996. The effect of sugars on the extrusion of maize grits: I. The role of the glass transition in determining product density and shape. Internation Journal of Food Science Technology 31: 55–65.

Farmer, G. S., G. H. Brusewitz, and R. W. Whitney. 1983. Drying properties of bluestem grass seed. Transaction of ASAE 26: 234–237.

Fortes, M., and M. R. Okos. 1980. Drying theories: their bases and limitations as applied to foods and grains, pp. 119–154. In A. S. Mujumdar (ed.), Advances in Drying, vol. 1. Hemisphere Publishing Corp., Washington, DC.

Foster, G. H. 1973. Heated-air grain drying, pp. 189–208. In R. A. Sinha and R. E. Muir (eds.), Grain Storage: Part of a System. Avi Publ. Co. Inc, Westport, CT.

Franca, A. S., M. Fortes, and K. Haghighl. 1994. Numerical simulation of intermittent and continuous deep-bed drying of biological materials. Drying Technology 12: 1537–1360.

Frank-Kamenetskii, D. A. 1955. Diffusion and Heat Exchange in Chemical Kinetics, Princeton University Press, Princeton, NJ.

Gekas, V., and I. Lamberg. 1991. Determination of diffusion coefficients in volume-changing systems—application in the case of potato drying. Journal of Food Engineering 14: 317–326.

Genskow, L. R. 1996. Considerations in drying consumer products, pp. 38–45. In A. S. Mujumdar and M. A. Roques (eds.), Drying '89. Hemisphere Publishing Corp, Bristol, PA

Ghosh, P. K., D. S. Jayas, and L. H. Gruwel. 2009. Measurement of water diffusivities in barley components using diffusion weighted imaging and validation with a drying model. Drying Technology 27: 382–392.

Gogoi, B. K., S. H. Alavi, and S. S. H. Rizvi. 2000. Mechanical properties of protein-stabilized starch-based supercritical fluid extrudates. International Journal of Food Properties 3: 37–58.

Golubovic, M. N., H. D. M. Hettiarachchi, W. Belding, and W. M. Worek. 2006. A new method for the experimental determination of Lewis' relation. International Communications in Heat and Mass Transfer 33: 929–935.

Gruwel, M. L. H., P. K. Ghosh, P. Latta, and D. S. Jayas. 2008. On the diffusion constant of water in wheat. Journal of Agricultural and Food Chemistry 56: 59–62.

Guiné, R. P. F. 2018. The drying of foods and its effect on the physical-chemical, sensorial and nutritional properties. International Journal of Food Engineering 4: 93–100.

Gunasekaran, S., T. M. Cooper, A. G. Berlage, and P. Krishnan. 1987. Image processing for stress cracks in corn kernels. Transactions of the ASAE 30: 266–271.

Gunasekaran, G., S. S. Deshpande, M. R. Paulsen, and G. C. Shove. 1985. Size characterization of stress cracks in corn kernels. Transactions of the ASAE 228: 1668–1672.

Gustafson, R. J., and R. V. Morey. 1979. Study of factors affecting quality changes during high-temperature drying. Transactions of the ASAE 22: 926–932.

Haghighi, K., and L. J. Segerlind. 1988. Modeling simultaneous heat and mass transfer in an isotropic sphere-a finite element approach. Transactions of the ASAE 31: 629–637.

Hall, C. W. 1980. Drying and Storage of Agricultural Crops, AVI Publishing Company Inc., Westport, Connecticut.

Hansen, R. C., R. J. Keener, and Gustafson. 2011. Natural air grain drying in Ohio, pp. 1–4, Extension Factsheet. The Ohio State University, Columbus, Ohio.

Harmathy, T. Z. 1969. Simultaneous moisture and heat transfer in porous systems with particular reference to drying. Industrial & Engineering Chemistry Fundamentals 8: 92–103.

Hayashi, T., A. Kobayashi, K. Tomita, and T. Shimizu. 2015. A new method for evaluation of the resistance to rice kernel cracking based on moisture absorption in brown rice under controlled conditions. Breeding Science 65: 381–387.

Holman, J. P. 1990. Heat Transfer, McGraw-Hill, New York.

Husain, A., C. S. Chen, and J. T. Clayton. 1973. Simultaneous heat and mass diffusion in biological materials. Journal of Agricultural and Engineering Research 18: 343–354.

Janz, M. 1997. Meathods of Measuring the Moisture Diffusivity at High Moisture Levels, pp. 73. Lund University, Lund, Sweden.

Jaturonglumlert, S., and T. Kiatsiriroat. 2010. Heat and mass transfer in combined convective and far-infrared drying of fruit leather. Journal of Food Engineering 100: 254–260.

Jayas, D. S., S. Cenkowski, and S. Pabis. 1991. Review of thin-layer drying and wetting equations. Drying Technology 9: 551–588.

Jayas, D. S., and P. K. Ghosh. 2006. Preserving quality during grain drying and techniques for measuring grain quality, pp. 969–982. In I. Lorini, B. Bacaltchuk, H. Beckel, D. Deckers, E. Sundfeld, J. P. dos Santos, J. D. Biagi, J. C. Celaro, L. R. d'A Faroni, L. O. F. Bortolini, M. R. Sartori, M. C. Elias, R. N. C. Guedes, R. G. da Fonseca and V. M. Scussel (eds.), Proceedings of the 9th International Working Conference on Stored-Product Protection. Brazilian Post-harvest Association ABRAPOS, Campinas, Brazil.

Jian, F., D. Divagar, J. Mhaiki, D. S. Jayas, P. G. Fields, and N. D. G. White. 2018. Static and dynamic methods to determine adsorption isotherms of hemp seed (*Cannabis sativa* l.) with different percentages of dockage. Food Science & Nutrition 6: 1629–1640.

Jian, F., and D. S. Jayas. 2018a. Characterization of isotherms and thin-layer drying of red kidney beans, part I: choosing appropriate empirical and semi theoretical models. Drying Technology 36: 1696–1706.

Jian, F., and D. S. Jayas. 2018b. Characterization of isotherms and thin-layer drying of red kidney beans, part II: three dimensional finite element models to estimate transient mass and heat transfer coefficients and water diffusivity. Drying Technology 36: 1707–1718.

Jian, F., D. S. Jayas, P. G. Fields, and N. D. G. White. 2017a. Water sorption and cooking time of red kidney beans (*Phaseolus vulgaris* L.): Part I – effect of freezing and drying conditions on water sorption and cooking time. International Journal of Food Science & Technology 52: 2031–2039.

Jian, F., D. S. Jayas, P. G. Fields, and N. D. G. White. 2017b. Water sorption and cooking time of red kidney beans: part II – mathematical models of water sorption. International Journal of Food Science & Technology 52: 2412–2421.

Jian, F., J. Liu, and D. S. Jayas. 2019. A new mathematical model to simulate sorption, desorption and hysteresis of stored canola during aeration. Drying Technology 38: 2190–2201.

Jurinak, J. J., and J. W. Mitchell. 1984. Effect of matrix properties on the performance of a counterflow rotary dehumidifier. Journal of Heat Transfer: 638–645.

Kaczmarek, S. A., A. J. Cowieson, D. Józefiak, and A. Rutkowski. 2014. Effect of maize endosperm hardness, drying temperature and microbial enzyme supplementation on the performance of broiler chickens. Animal Production Science 54: 956–965.

Karathanos, V., S. Anglea, and M. Karel. 1993. Collapse of structure during drying of celery. Drying Technology 11: 1005–1023.

Karathanos, V. T., G. Villalobos, and G. D. Saravacos. 1990. Comparison of two methods of estimation of the effective moisture diffusivity from drying data. Jouranl of Food Science 55: 218–231.

Keey, R. B. 1972. Drying Principles and Practice, Pergamon Press, New York.

Keey, R. B. 1991. Moisture in gases, pp. 117–162, Drying of Loose and Particulate Materials. Hemisphere Publishing Co., New York, Washington, Philadelphia, London.

Kerhof, P. J. A. M. 1994. A test of lumped-parameter methods for the drying rate in fluidized bed driers for bioproducts. Drying Technology 13: 1099–1111.

Kiranoudis, C. T., Z. B. Maroulis, and D. Marinos-Kouris. 1995. Heat and mass transfer model building in drying with multiresponse data. International Journal of Heat and Mass Transfer 38: 463–480.

Kloppers, J. C., and D. G. Kröger. 2005. The Lewis factor and its influence on the performance prediction of wet-cooling towers. International Journal of Thermal Sciences 44: 879–884.

Krishna, R., and J. A. Wesselingh. 1997. The Maxwell-Stefan approach to mass transfer. Chemical Engineering Science 52: 861–911.

Kunze, O. R. 1979. Fissuring of the rice grain after heated air drying. Transactions of ASAE 22: 1197–1201.

Kunze, O. R. 1995. Effect of drying on grain quality – moisture readsorption causes fissured rice grains, International Conference on Grain Drying East Asia, Bangkok, Thailand.

Kunze, O. R., and S. Prasad. 1978. Grain fissuring potentials in harvesting and drying of rice. Transactions of the ASAE 21: 361–366.

Lambert, C., H. Romdhana, and F. Courtois. 2015. Reverse methodology to identify moisture diffusivity during air-drying of foodstuffs. Drying Technology 33: 1076–1085.

Li, Y. B., C. W. Cao, and J. Li .1998. Study on rice fissuring during intermittent drying, pp. 1660–1664. *In* Z. Jin, Q. Liang, Y. Liang, X. Tan and L. Guan (eds.), Proceedings of the 7th International Working Conference on Stored-Product Protection. Sichuan Publishing House of Science and Technology, Beijing, China.

Litchfield, J. B., and M. R. Okos. 1992. Moisture diffusivity in pasta during drying. Journal of Food Engineering 17: 117–142.

Lomauro, C. J., A. S. Bakshi, and T. P. Labuza. 1985. Moisture transfer properties of dry and semimoist foods. Journal of Food Science 50: 397–400.

Lu, R., and T. J. Siebenmorgen. 1992. Moisture diffusivity of long-grain rice components. Transactions of the ASAE 35: 1955–1961.

Maier, D. E., and F. W. Bakker-Arkema. 2002. Grain drying systems, pp. 1–53, Facility Design Conference of the Grain Elevator and Processing Society, St. Charles, Illinois.

Malumba, P., C. Massaux, C. Deroanne, T. Masimango, and F. Béra. 2009. Influence of drying temperature on functional properties of wet-milled starch granules. Carbohydrate Polymers 75: 299–306.

Malumba, P., C. Vanderghem, C. Deroanne, and F. Béra. 2008. Influence of drying temperature on the solubility, the purity of isolates and the electrophoretic patterns of corn proteins. Food Chemistry 111: 564–572.

Marousis, S. N., V. T. Karathanos, and G. D. Saravacos. 1991. Effect of physical structure of starch materials on water diffusivity. Journal of Food Processing and Preservation 15: 183–195.

Mattioda, F., L. M. M. Jorge, and R. M. M. Jorge. 2017. Evaluation of water diffusivity in wheat hydration (Triticum spp): isothermal and periodic operation. Journal of Food Process Engineering 41: 1–8.

Miketinac, M. J., S. Sokhansanj, and Z. Tutek. 1992. Determination of heat and mass transfer coefficients in thin layer drying of grain. Transactions of the ASAE 35: 1853–1858.

Misra, M. K., and D. B. Brooker. 1980. Thin-layer drying and rewetting equations for shelled yellow corn. Transactions of ASAE 23: 1254–1260.

Misra, R. N., and J. H. Young. 1980. Numerical solution of simultaneous moisture diffusion and shrinkage during soybean drying. Transactions of the ASAE 23: 1277–1282.

Montross, M. D., and D. E. Maier. 2000. Simulated performance of conventional high-temperature drying, dryeration, and combination drying of shelled corn with automatic conditioning. Transacrions of the ASAE 43: 691–699.

Morey, R. V., R. J. Gustafson, and H. A. Cloud. 1981. Combination high-temperature, ambient-air drying. Transactions of the ASAE 24: 509–512.

Muir, W. E., and R. N. Sinha. 1986. Theoretical rates of flow of air at near-ambient conditions required to dry rapeseed. Canadian Agricultural Engineering 28: 45–49.

Nagato, K., and M. I. Ebata. 1964. On the formation of cracks in rice kernels during wetting and drying of paddies. Nippon Sakumotsu Gakkai Kiji 33: 82–89.

Nellist, M. E., and D. M. Bruce 1995. Heated-air grain drying, pp. 609–659. In D. S. Jayas, G. White and W. E. Muir (eds.), Stored-Grain Ecosystems. Marcel Dekker Inc, New York.

Nijhuis, H. H., H. M. Torringa, S. Muresan, D. Yuksel, C. Leguijt, and W. Kloek. 1998. Approaches to improving the quality of dried fruit and vegetables. Trends Food Science Technology 9: 13–20.

Pabis, S., and S. M. Henderson. 1962. Grain drying theory, III. The air/grain temperature relationship. Journal of Agricultural Engineering Research 6: 21–26.

Pabis, S., D. S. Jayas, and S. Cenkowski. 1998. Grain Drying: Theory and Practice, John Wiley & Son, Inc., New York.

Paulsen, M. R., and L. D. Hill. 1985. Corn quality factors affecting dry milling performance. Journal of Agricultural Engining Research 31: 255–263.

Perdon, A. A., T. J. Siebenmorgen, and A. Mauromoustakos. 2000. Glassy state transition and rice drying: development of a brown rice state diagram. Cereal Chemistry 77: 708–713.

Perre, P. 2015. The proper use of mass diffusion equations in drying modeling: introducing the drying intensity number. Drying Technology 33: 1949–1962.

Perre, P., and B. K. May. 2001. A numerical drying model that accounts for the coupling between transfers and solid mechanics: case of highly deformable products. Drying Technology 19: 1629–1643.

Pixton, S. W., and H. J. Griffiths. 1971. Diffusion of moisture through grain. Journal of Stored Products Research 7: 133–152.

Pupinis, G. 2008. Grain drying by use of changeable air flow method. Agronomy Research 6: 55–65.

Quirijns, E. J., A. J. B. Van Boxtel, W. K. Van Loon, and G. Van Straten. 2005. Sorption isotherms, GAB parameters and isosteric heat of sorption. Journal of the Science of Food and Agriculture 85: 1805–1814.

Rahman, M. S. 2001. Toward prediction of porosity in foods during drying: a brief review. Drying Technology 19: 1–13.

Rahman, M. S., and P. L. Poturi. 1990. Shrinkage and density of squid flesh drying. Journal of Food Engineering 58: 133–143.

Rahman, M. S., and S. S. Sablani 2001. Measurement of water activity by electronic sensors, pp. A2.5.1–A2.5.4. In R. E. Wrolstad (ed.), Current Protocols in Food Analytical Chemistry. John Wiley & Sons, Inc, New York.

Ramallo, L. A., and R. H. Mascheroni. 2013. Effect of shrinkage on prediction accuracy of the water diffusion model for pineapple drying. Journal of Food Process Engineering 36: 66–76.

Roberts, J. S., and C. H. Tong. 2003. The development of an isothermal drying apparatus and the evaluation of the diffusion model on hygroscopic porous material. International Journal of Food Properties 6: 165–180.

Roberts, J. S., C. H. Tong, and D. B. Lund. 2002. Drying kinetics and time-temperature distribution of pregelatinized bread. Journal of Food Sciences 67: 1080–1087.

Roman, G. N., E. Rotstein, and M. J. Urbicain. 1979. Kinetics of water vapor desorption from apples. Journal of Food Science 44: 193–197.

Sablani, S. S., S. Kasapis, and M. S. Rahman. 2007. Evaluating water activity and glass transition concepts for food stability. Journal of food engineering 78: 266–271.

Salas, F., and T. P. Labuza. 1968. Surface active agents effects on drying characteristics of model food systems. Food Technology 22: 80–84.

Santos, G. T., M. Fortes, R. P. Amantea, and W. R. Ferreira. 2012. Optimization aspects of dryeration based on drying simulation analysis of a single corn kernel, ASABE Annual Meeting. ASABE, Dallas, Texas.

Saravacos, G. D., and Z. B. Maroulis. 2001. Transport Properties of Foods, Marcel Dekker, Inc., New York.

Seyedin, N., J. S. Burris, and T. E. Flynn. 1984. Physiological studies on the effects of drying temperatures on corn seed quality. Canadian Journal of Plant Science 64.

Seyhan-Gurtas, F., M. M. Ak, and E. O. Evranuz. 2001. Water diffusion coefficients of selected legumes grown in turkey as affected by temperature and variety. Turkey Jouranl of Agriculture 25: 297–304.

Sharma, A. D., and O. R. Kunze. 1982. Post-drying fissure developments in rough rice. Transactions of the ASAE 25: 465–468.

Shouse, S., M. Hanna, and D. Petersen. 2011. Dryeration and Combination Drying for Increased Capacity and Efficiency, University Extension. Iowa State University, Ames, Iowa.

Siebenmorgen, T. J., W. Yang, and Z. Sun. 2004. Glass transition temperature of rice kernels determined by dynamic mechanical thermal analysis. Transactions of the ASAE 47: 835–839.

Silva, W. P., C. M. D. P. S. Silva, P. L. Nascimento, D. D. P. S. Silva, and C. D. P. D. Silva. 2009. Influence of the geometry on the numerical simulation of isothermal drying kinetics of bananas. World Applied Sciences Journal 7: 1818–4952.

Soekarto, S. T. 1993. Kinetics of volumetric expansion of indonesian chips (krupuk) during deep frying, pp. 736–744. In O. B. Liang, A. Buchanan and D. Fardiaz (eds.), Development of Food Science and Technology in South East Asia. IPB Press, Bogor, Indonesia.

Sokhansanj, S., W. P. Lampman, and J. D. MacAulay. 1983. Investigation of grain tempering on drying tests. Transacrions of the ASAE: 293–296.

Srikiatden, J., and J. S. Roberts. 2007. Moisture transfer in solid food materials: a review of mechanisms, models, and measurements. International Journal of Food Properties 10: 739–777.

Suarez, C., P. Viollaz, and J. Chirife. 1980. Diffusional analysis of air drying of grain sorghum. Journal of Food Technology 15: 523–531.

Sun, K., F. Jian, D. S. Jayas, N. D. G. White, and P. G. Fields. 2014. Physical properties of three varieties of high-oil canola and one variety of low-oil canola. Transactions of the ASABE: 599–608.

Suzuki, M., A. Dndo, S. Ohtani, and S. Maeda. 1972. Mass transfer from a discontinuous source, pp. 267–276. In Ist Pacific Chemical Engineering Congress PACHEC, Kyoto.

Suzuki, K., K. Kubota, T. Hasegawa, and H. Hosaka. 1976. Shrinkage in dehydration of root vegetables. Journal of Food Science 41: 1189–1193.

Syarief, A. M., R. V. Morey, and R. J. Gustafson. 1984. Thin layer drying rates of sunflower seed. Transactions of the ASAE 27: 195–200.

Thompson, R. A., and G. H. Foster. 1963. Stress Cracks and Breakage in Artificially Dried Corn. USDA, Washington, DC.

Thompson, R. A., and G. H. Foster. 1967. Dryeration: high speed drying with delayed aeration cooling, ASAE Annual Meeting. ASAE, Detroit.

Timm, N. S., G. H. Lang, A. H. Ramos, R. S. Pohndorf, C. D. Ferreira, and M. Oliveira. 2020. Effects of drying methods and temperatures on protein, pasting, and thermal properties of white floury corn. Journal of Food Processing and Preservation 44: e14767.

Toledo, M. Z., C. Cavariani, J. D. B. França-neto, and J. Nakagawa. 2010. Imbibition damage in soybean seeds as affected by initial moisture content, cultivar and production location. Seed Science and Technology 38: 399–408.

Tong, C. H., and D. B. Lund. 1990. Effective moisture diffusivity in porous materials as a function of temperature and moisture content. Biotechnology Progress 6: 67–75.

Tong, C. H., and D. B. Lund. 1993. Microwave heating of baked dough products with simultaneous heat and moisture transfer. Journal of Food Engineering 19: 319–339.

Tong, C. H., A. Parent, and D. B. Lund. 1993. Temperature and moisture distributions in porous food materials during microwave heating, Sixth International Congress on Engineering and Food.

Trujillo, F. J., C. Wiangkaew, and Q. T. Pham. 2007. Drying modeling and water diffusivity in beef meat. Journal of Food Engineering 78: 74–85.

Tutuncu, M. A., and T. P. Labuza. 1996. Effect of geometry on the effective moisture transfer diffusion coefficient. Journal of Food Engineering 30: 433–447.

Vaccarezza, L. M., and J. L. C. Lombardi. 1974. Kinetics of moisture movement during air drying of sugar beet root. Journal of Food Technology 9: 317–327.

Vaccarezza, L. M., J. L. Lombardi, and J. Chirife. 1974. Heat transfer effects on drying rate of food dehydration. Canadian Journal of Chemical Engineering 52: 576–579.

Vagenas, G. K., and D. Marinos-Kouris. 1991. Finite element simulation of drying of agricultural products with volumetric changes. Applied Mathematical Modelling 15: 475–482.

Varzakas, T. H., G. C. Leach, C. J. Israilides, and D. Arapoglou. 2005. Theoretical and experimental approaches towards the determination of solute effective diffusivities in foods. Enzyme Microbial Technology 37: 29–41.

Waananen, K. M., and M. R. Okos. 1996. Effect of porosity on moisture diffusion during drying of pasta. Journal of Food Engineering 28: 121–137.

Westerman, P. W., G. M. White, and I. J. Ross. 1973. Relative humidity effect on the high temperature drying of shelled corn. Transactions of the ASAE 16: 1136–1141.

White, G. W., and S. H. Cakebread. 1966. The glassy state in certain sugar–containing food products. Journal of Food Technology 1: 73–82.

Winfield, R. G. 1969. Dryeration experience with grain corn in Ontario. Canadian Agricultural Engineering 11: 9–11.

Xiong, X., G. Narsimhan, and M. R. Okos. 1991. Effect of composition and pore structure on binding energy and effective diffusivity of moisture in porous food. Journal of Food Engineering 15: 187–208.

Yang, W., C. C. Jia, T. J. Siebenmorgen, T. A. Howell, and A. G. Cnossen. 2002. Intra-kernel moisture responses of rice to drying and tempering treatments by finite element simulation. Transactions of ASAE 45: 1037–1044.

Young, J. H. 1969. Simultaneous heat and mass transfer in a porous hygroscopic solid. Transactions of the ASAE 12: 720–725.

Zhao, L., J. Yang, S. Wang, and Z. Wu. 2020. Investigation of glass transition behavior in a rice kernel drying process by mathematical modeling. Drying Technology 38: 1092–1105.

Zhao, X., W. Li, H. Zhang, X. Li, and W. Fang. 2019. Reaction–diffusion approach to modeling water diffusion in glutinous rice flour particles during dynamic vapor adsorption. Journal of Food Science and Technology 56: 4605–4615.

Zhao, X., H. Zhang, R. Duan, and Z. Feng. 2017. The states of water in glutinous rice flour characterized by interpreting desorption isotherm. Journal of Food Science and Technology 54: 1491–1501.

Ziegler, V., N. S. Timm, C. D. Ferreira, J. T. Goebel, R. S. Pohndorf, and M. Oliveira. 2020. Effects of drying temperature of red popcorn grains on the morphology, technological, and digestibility properties of starch. International Journal of Biological Macromolecules 145: 568–574.

Zogzas, N. P., Z. B. Maroulis, and D. Marinos-Kouris. 1996. Moisture diffusivity data compilation in foodstuffs. Drying Technology 14: 2225–2253.

Chapter 5

Mathematical Modelling of Isotherm, Drying, and Wetting

Introduction

The amount and kinematics of moisture diffusion highly depend on the achievement of the thermodynamic equilibrium, because moisture loss or gain from one region to another region inside the material will occur continually in order to reach thermodynamic equilibrium with the surrounding components and environment. Since the introduction of the terms "water activity" and "isotherm," water sorption isotherm is used as the fundamental characteristic of biomass materials because it influences almost every aspect of the dehydration or hydration process and the storage stability of the product. Mathematical modelling is an important tool to understand these complex systems, and understanding of isotherm, drying, and wetting processes is advanced with the development of the mathematical models. More than 270 theoretical, semi-theoretical, and empirical equilibrium isotherm equations are developed (Labuza et al. 1985; Lomauro et al. 1985a; Labuza and Altunakar 2007). Theoretical, semi-theoretical, and empirical models of drying and wetting are also developed. The origin, development, range of applicability (to both food type and water activity), and applications in grain storage and drying of some of the widely used mathematical models are presented in this chapter in alphabetical order under each

DOI: 10.1201/9781003186199-5

topic. The aim of this critical compilation is to provide useful information to researchers interested in the mathematical description of the water sorption isotherms, drying, wetting, and soaking of biomass materials.

5.1 Sorption (Equilibrium) Isotherm Equations (Models)

5.1.1 Theoretical and Semi-Theoretical Models

Water sorption equilibrium isotherm is one special case of surface sorption isotherm. Various models of sorption equilibrium isotherm used in gas and liquid sorption in the chemistry study have been applied to water sorption. Various models have been developed to relate moisture content with water activity (relative humidity) over a wide range of sorption equilibrium isotherm (0 to 1 a_w). Most of the theoretical models are developed by using physical mechanisms such as molecular sorption or capillary sorption or the combination of these two mechanisms. Most of the theoretical and semi-theoretical models assume that there are different statuses of water such as water is in monolayer or multilayer, bound, clustered, or free water. These assumptions are theoretically supported and are proven by experimental data (Hansen 1976). The reaction – diffusion approach assumes that a certain amount of water is adsorbed and temporarily immobilized to adsorption sites. This fraction of water is available to absorption or desorption (Zhao et al. 2019) and there is no chemical reaction. Several researches suggested that at least two water populations (groups) with different mobility exist in solid materials (Labuza et al. 1985; Lomauro et al. 1985b; Farroni and Buera 2014). All of these assumptions are based on the fact that water present in biomaterials behaves as if it is in solution and shows a lowered vapour pressure.

5.1.1.1 Brunauer-Emmett-Teller (BET) Equation

Brunauer-Emmett-Teller (BET) equation is named after Stephen Brunauer, Paul Hugh Emmett, and Edward Teller (Brunauer et al. 1938). The BET equation describes a multilayer sorption isotherm by assuming that: (1) sites, which are available for the water sorption, are randomly distributed; (2) in the water sorption process, the sites can be covered by one monolayer of water, then two and then multilayers; (3) the first monolayer of water evenly covers the sites (polymer surface of solid materials) and is very tightly bound in the monolayer; (4) subsequent layers of water have little interaction with the polymer surface; (5) these subsequent layers of water have bulk water properties such as becoming ice at <0°C; and (6) once multilayer adsorption occurs, amounts of absorbed water will begin to increase dramatically, which produces Type II–IV adsorption isotherm (refer to Chapter 3, Section 3.2.2 for the definition of Type I–V) and can be modelled as:

$$M = \frac{CM_m a_w}{(1-a_w)(1+(C-1)a_w)} \tag{5.1a}$$

where M is the moisture content of solids (kg/kg of the solid), M_m is the monolayer moisture content (kg/kg), a_w is the water activity (decimal), and C is the constant. This equation can be rearranged to yield the linear relationship.

$$\frac{a_w}{M(1-a_w)} = \frac{1}{M_m C} + \frac{C-1}{M_m C} a_w \tag{5.1b}$$

The BET constant C is related to surface net heat of sorption. The C value decreases with the increase of multilayer water and depends on the adsorption characteristics of the adsorbing sites. For a Type II isotherm, C varies from 2 to 50 (Kumar and Balasubrahmanyam 1986). For Type III isotherm, C is less than two. The monolayer moisture content (M_m) is a parameter of interest in dried foods.

If the data of a_w and moisture content of a material at a given temperature are available, linear regressing the a_w against $\frac{a_w}{M(1-a_w)}$ can determine the constants C and m_m of the BET equation from the slope and intercept.

BET equation is originally developed on the thermodynamic principle for the physical adsorption of non-polar gases on homogenous metal surfaces without chemical reaction. It assumes the sorption process is regulated by two mechanisms: Langmuir kinetics at the monolayer and condensation in subsequent layers. Langmuir kinetics specifies that gas molecules are directly absorbed on selected sites at the solid surface until the latter is covered by a single layer of gas molecule. Its main feature is the monolayer concept; hence it is now widely used for determining the optimum moisture conditions of biomaterials, especially of dehydrated foods.

BET model represents a fundamental milestone in the interpretation of multilayer sorption isotherms, particularly Types II and III. It is also an effective method for estimating the amount of bound water in specific polar sites of dehydrated food systems. BET equation is one of the most widely used models and gives good fit for a variety (cultivar) of food materials over the range $0.05 < a_w < 0.45$ (Timmermann et al. 2001). BET model is valid for a wider range of water activity than that of the Langmuir and Freundlich model. It is a useful tool to estimate the area of the monolayer on the surface boundary of materials with porous surfaces. For example, the monolayer area of a surface can be calculated from the sorbed amount of nitrogen molecules and using the area of each nitrogen molecule (1.62×10^{-19} m^2).

Harlwood and Horrobin (1946) developed a more general equation, and BET equation is one of the special cases of the Hailwood and Horrobin equation (referred to as H-H equation). The H-H equation assumes that the absorbed water has two states: water in simple solution and combined water with a definite unit of the material molecules. The dissolved water molecules in the solid phase are mobile. The water in the solid phase is: dissolved water, hydrated molecules (having its polar groups saturated with water), and anhydrous polymer. In the solid material, the following equilibrium can be established:

$$\text{Anhydrous polymer} + \text{Dissolved water} \leftrightarrows \text{Hydrated molecules}$$

This water status at the equilibrium condition can be kinetically expressed as:

$$\frac{\partial N_1}{\partial t} = \alpha\left(\frac{N_o}{N_o + N_1 + n}\right)n - \beta N_1 \tag{5.2}$$

where N_o, N_1, and n are the number of anhydrous groups, monohydrate groups, and water molecules in the solid solution, respectively; t is the time (s); and α and β are the parameters. At equilibrium, $\frac{\partial N_1}{\partial t} = 0$, and the mass action law can be expressed as:

$$\frac{N_o n}{N_1} = \frac{\beta}{\alpha}(N_o + N_1 + n) \tag{5.3}$$

The solid solution and water vapour will, therefore, be bivariant. The vapour pressure of the water at the ideal condition of the solid solution is:

$$P = P_s \frac{n}{N_o + N_1 + n} \tag{5.4}$$

where P and P_s are the total (vapour) pressure and saturation vapour pressure (Pa), respectively. The water in the solid solution is "non-normal liquid water" but is somehow associated with the substrate. Therefore, the vapour pressure was modified by Harlwood and Horrobin (1946) as:

$$\frac{P}{P_s} = a_w = \gamma \frac{n}{N_o + N_1 + n} \tag{5.5}$$

where γ is a constant. For the materials with pure water, $\gamma \approx 1$. For the material at low water activity, $\gamma > 1$, which indicates the interstitial water is more structured (has less entropy) than normal water.

If $C = \frac{n+N_1}{N_o+N_1}$ and $m = \frac{M_a}{M_s}C$, then Eq. 5.5 can be written as:

$$\frac{C}{a_w} = \frac{1}{\gamma - a_w} + \frac{1}{a_w + \gamma k} \tag{5.6}$$

where M_a is the molecular weight of water, M_s is an average value of the molecular weight of a polar unit, m is the mass of the adsorbent or mass of the material (kg), and k is the constant. If $\gamma = 1$, Eq. 5.6 is mathematically identical to the BET equation.

The maximum value of the water activity in the H-H equation is:

$$a_{w\max} = \frac{1}{2}\gamma(1-k) \tag{5.7}$$

where $a_{w\max}$ is the maximum value of water activity to which the material has been subjected (decimal). Although the H-H equation may fit sorption data for almost any type of food, it satisfies thermodynamic requirements (i.e., prediction of the temperature dependence) only for proteins and starchy foods (Fontan et al. 1982).

Researchers modified the BET equation and the most used modified BET equation (Brunauer 1945) is:

$$M = \frac{M_m C a_w}{1-a_w} \frac{1-(L+1)a_w^L + n a_w^{L+1}}{1+(C-1)a_w - C a_w^{L+1}} \tag{5.8}$$

where L is the number of layers. The term a_w^L and a_w^{L+1} are the water activity in layer L and $L + 1$, respectively. The modified BET equation proposes that the radius of the capillary defines the upper limit for the number of layers of water that can be built up within the capillary inside biomaterials. The modified BET equation can give a reasonably good representation of the wheat flour isotherm up to 0.8 a_w. There are other modified BET equations (Dincer and Esin 1996), but these modified equations are criticized by other researchers (Furmaniak et al. 2007).

The following assumptions that were used to develop the BET equation have been questioned: (1) the rate of condensation on the first layer is equal to the rate of evaporation from the second layer; (2) binding energy of all of the adsorbate on the first layer is same; and (3) binding energy of the other layers is equal to that of pure adsorbate (water). The assumptions of uniform adsorbent surface and absence of lateral interactions between adsorbed molecules are incorrect in the view of the heterogeneous food surface interactions. Some researchers suggested the original BET model is thermodynamically unrealistic (Furmaniak et al. 2007). Despite the theoretical limitations, the BET monolayer concept is reasonable to characterise various aspects of dried foods (Iglesias and Chirife 1982; Timmermann et al. 2001).

Different researchers suggested different equations to calculate the M and C values. The Freundlich (1926) equation is:

$$M = Ca_w^{1/n} \tag{5.9}$$

The Freundlich equation has no theoretical foundation and is empirical in nature. McGavack and Patrick (1920) equation is equivalent to Freundlich equation.

Staudt et al. (2013) suggested the following equation to calculate the C.

$$C = C_o e^{\left(\frac{Q}{RT_k}\right)} \tag{5.10}$$

where C_o is a constant, Q is the sorption energy (J/mol), R is the gas constant (8.31451×10^{-7} J·K^{-1}·mol^{-1}), and T_k is the temperature (K).

5.1.1.2 Bradley Equation

Bradley (1936) assumed that the first layer of water is sorbed because of strongly induced dipoles and these dipoles in turn polarize the second layer that in turn polarizes the next layer.

$$\ln\left(\frac{1}{a_w}\right) - K_3 = K_2 K_1^M \tag{5.11}$$

where K_1, K_2, and K_3 are constants. Both K_1 and K_2 are temperature dependent. K_1 is related to the dipole moment of the sorbed vapour and the characteristic of the sorbed molecules on the sorptive sites. K_2 is related to the function of the sorptive polar groups. Bradley equation fits well the isotherm of protein products in the a_w range of 0.05–0.95 (Chirife and Iglesias 1978). However, some researchers reported the deviation from experimental data above water activities of 0.70–0.85 depending on the material (Hansen 1976). The following equations are equivalent to Bradley equation: Hoover and Mellon (1950) equation, simplified Chen and Clayton (1971) equation, Chung and Pfost equation (Chung and Pfost 1967a,b), and de Boer and Zwikker (1929) equation.

5.1.1.3 Clausius–Clapeyron Equation

The Clausius–Clapeyron equation was published in 1834 and 1850 and derived from the Clapeyron equation with assumptions of ideal gas behaviour for vapour phase and negligible liquid molar volume when compared to the vapour molar volume. The equation is derived from the equality of Gibbs free energy for two phases of a pure substance in equilibrium. The Clausius–Clapeyron equation states a relation between temperature and saturation pressure through the latent heat of vaporization. For example, from one temperature to another temperature, the water saturation pressure of the material at the two temperatures is related through the latent heat of vaporization.

$$\frac{dP_s}{P_s} = \frac{\Delta H_{\text{vap}}}{R}\frac{dT_k}{T_k} \tag{5.12a}$$

where H_{vap} is the latent heat of vaporization (J/mol). Another format of the Clausius–Clapeyron equation is:

$$\frac{d\ln P_s}{dT_k} = \frac{\Delta H_{\text{vap}}}{RT_k^2} \tag{5.12b}$$

If all previous Clausius–Clapeyron assumptions hold and only water changes phase, then a simple equation can be derived as:

$$\frac{d\ln P_w}{dT_k} = \frac{\Delta H}{RT_k^2} \tag{5.13}$$

where P_w is the partial pressure of water (Pa). This is equivalent to just replacing the saturation pressure by the water partial pressure and the latent heat of vaporization by the enthalpy change because the Gibbs partial molar energy should hold under previous assumptions.

After introducing the water activity and $\Delta H = Q_s + \Delta H_{\text{vap}}$, the Clausius–Clapeyron equation can be written as:

$$\frac{d\ln a_w}{dT_k} = \frac{Q_s}{RT_k^2} \tag{5.14a}$$

where Q_s is the net isosteric heat of sorption (J/mol). Another format of this equation is:

$$\frac{da_w}{dT_k} = a_w\frac{Q_s}{RT_k^2} \tag{5.14b}$$

This equation indicates that the degree of dependence of the water activity on the temperature is defined by the isosteric heat of sorption. Therefore, Q_s depends strongly on temperature. Staudt et al. (2013) suggested the following equation to calculate the Q_s:

$$Q_s = \frac{Qa_w^2 - 2Qa_w + Q}{\left(C_0e^{Q/RT_k} - 1\right)a_w^2 + 1} \tag{5.15}$$

The net isosteric heat of sorption is expressed as a function of the BET model parameters. The sorption energy Q can be calculated by using Eq. 5.10.

The Clausius–Clapeyron equation is not used to evaluate the isotherm of materials, but is the starting point of several developed models such as Poyet equation, receding front theory, and Whitaker model. Poyet equation (Poyet 2009) was developed on two assumptions of the Clausius–Clapeyron equation: The absorbate (water vapour) is ideal gas and the volume of the absorbed phase (water vapour) is negligible. At a given moisture content of a material, the Clausius–Clapeyron equation can be expressed as:

$$Q_s = -R\left[\frac{\partial \ln(P)}{\partial\left(\frac{1}{T_k}\right)}\right]_M \tag{5.16}$$

The subscript M indicates that the derivation operation must be carried out for a constant amount of adsorbed water. If the Q_s is constant between two equilibrium states (at T_r and T_1) and the material is at the constant water content, the above Clausius–Clapeyron equation can be integrated and produce the Poyet equation:

$$a_{w1} = a_{wr}\frac{P_s^r}{P_s^1}e^{\left[\frac{Q_s}{R}\left(\frac{T_r - T_1}{T_r T_1}\right)\right]} \tag{5.17}$$

The subscript and superscript 1 and r refer to the state 1 and reference state, respectively.

The Poyet equation (Eq. 5.17) indicates that the sorption isotherm of a material can be found if the sorption isotherm at the reference condition is found and if the isosteric heat of sorption is constant (Poyet and Charles 2009). Poyet and Charles (2009) used the GAB (Guggenheim, Anderson, de Boer) model to calculate the Q_s after the isotherms at two states were found.

The main reason for developing the Poyet equation is to add the temperature effect to the isotherm equation. Poyet and Charles (2009) used the equation to describe isotherms of concrete at two different temperatures, and proved it was possible to estimate accurately the desorption isotherm at any other temperature. This model has been criticised because of the assumption of constant isosteric heat of sorption. Since the isosteric heat is temperature dependent, the temperature interval between T_r and T_1 must be kept small for the heat to be sensibly constant. It is also extremely difficult in practice to measure the change in the RH necessary to maintain a constant water content in the material as the temperature is changed. Therefore, Poyet equation is not widely used.

5.1.1.4 D'Arcy and Watt Equation

The D'Arcy and Watt (1970) model assumes that there is a fixed number of water binding sites with different discrete binding energies. The D'Arcy and Watt equation has three sorption terms and the first one is the Langmuir isotherm describing the monolayer adsorption onto strong sorption sites. The second term has the form of the Henry law and describes monolayer adsorption onto weak sorption sites. The third term has the form of Raoult law and describes the multilayer sorption and it is limited by the properties of the substrate (D'Arcy and Watt 1970).

$$M = \frac{M_L K_L a_w}{1 + K_L a_w} + K_H a_w + \frac{M_b K_b a_w}{1 - K_b a_w} \tag{5.18}$$

where M_L is the Langmuir capacity constant, K_L is the Langmuir adsorption equilibrium constant, K_H is the solubility coefficient of the Henry's law, K_b is the adsorption equilibrium constants, and M_b is the maximum water adsorption at secondary sites.

Each constant in the D'Arcy and Watt equation has the physical meaning. M_L is the Langmuir capacity constant representing the number of strong water-binding sites, multiplied by the molecular weight of water and divided by Avogadro's number (6.023×10^{23}). K_L is the Langmuir adsorption equilibrium constant representing the strength of the attraction of the strong water binding sites for water. K_H is the solubility coefficient of the Henry's law representing the measurement of the affinity and the number of weak binding sites. M_b is the maximum water adsorption at secondary sites. K_b is adsorption equilibrium constant for secondary sites related to the water activity. The number of strong, weak and multi-molecular water binding sites are $M_L N_A / M_a$, $K_H N_A / (M_a P_s)$, and $M_b N_A / M_a$, respectively. Where N_A and M_a are Avogadro's number (6.023×10^{23}) and molecular weight of water, respectively.

Furmaniak et al. (2007) modified the D'Arcy and Watt equation after incorporating the following assumptions: not all water molecules adsorbed on primary sites can be the secondary centres; and one water molecule attached to primary site can create more than one secondary adsorption site. The modified D'Arcy and Watt equation is:

$$M = \frac{m_{mp} K a_w}{1 + K a_w} \frac{1 - k(1 - w) a_w}{1 - k a_w} \tag{5.19}$$

where m_{mp} is the maximum sorption on primary sites (%), and w is the ratio of water molecules adsorbed on the primary sites converted into the secondary adsorption sites. Constant k is the kinetic constant associated with the secondary adsorption sites (of Dubinin–Serpinsky type). The coefficient w is the ratio of water molecules adsorbed on the primary sites, which is converted into the secondary adsorption sites. The value of adsorption on primary sites (a_{prim}) is:

$$a_{\text{prim}} = \frac{a_{mL} K_L P}{\frac{1}{P_s} + K_L P} \tag{5.20}$$

where a_{prim} is the value of adsorption on primary sites, and a_{mL} is the surface concentration of Langmuir-type primary adsorption sites (mol/g). Liu et al. (2011) introduced the temperature factor into the D'Arcy and Watt equation as:

$$M = \frac{m_{mp}\left(\frac{K}{T_c}\right)a_w}{1+\frac{K}{T_c}a_w} \frac{1-k(1-w)a_w}{1-ka_w} \tag{5.21}$$

D'Arcy and Watt equation has been used widely for the analysis of desiccation tolerant and intolerant plant tissues and is valid as the GAB model. The assumption of the three types of water binding sites is the advantage of the D'Arcy and Watt equation than the GAB equation. Furmaniak et al. (2007) found the modified D'Arcy and Watt equation could successfully describe the set of sorption isotherms measured on pineapple, macaroni, sardine, and pistachio nut paste in the $a_w \geq 0.1$. Liu et al. (2011) found the modified D'Arcy and Watt equation had the same performance as the modified Peleg and GAB equations to fit the moisture sorption data of a typical broiler feed.

5.1.1.5 Freundlich Adsorption Equation

The Freundlich adsorption isotherm is the earliest semi-theoretical model, which was proposed by Freundlich in 1909 and termed as Freundlich adsorption isotherm, Freundlich adsorption equation, or Freundlich isotherm. Freundlich adsorption equation can be used to calculate the mass (monolayer adsorbed) on the heterogeneous surface of the materials under different pressures.

$$M = kP^{\frac{1}{\lambda}} \tag{5.22}$$

where λ is the constant. The k value depends upon adsorbent and gas (water vapour) at a particular temperature. Freundlich isotherm is also written as:

$$\log M = \log k + \frac{1}{\lambda}\log P \tag{5.23}$$

Freundlich isotherm had a low prediction accuracy when pressure is close to saturation because Freundlich equation assumes the linear relationship and there is no saturation of adsorption. This equation has a higher accuracy when monolayer adsorption occurs. The value of k and λ can be regressed by using experimental data of the P and M.

5.1.1.6 GAB (Guggenheim, Anderson, de Boer) Equation

The GAB equation, derived independently by Guggenheim, Anderson, and de Boer in 1966, 1946, and 1953, respectively, is a semi-theoretical, multi-molecular, localized, and homogeneous adsorption model. The failure of the BET model at higher water activity levels had led to the development of the GAB model with three parameters having physical meanings. Therefore, GAB model has the similar assumptions as the BET equation, and the Langmuir and BET theories are further refined by the GAB model theory. This model postulates that the state of sorbate molecules in the second layer is identical to the one in superior layers, but different from those of the liquid state.

$$M = \frac{M_m C k a_w}{(1 - k a_w)(1 - k a_w + C k a_w)} \tag{5.24a}$$

$$\frac{M}{M_m} = \frac{(C-1) k a_w}{1 - k a_w + C k a_w} + \frac{k a_w}{1 - k a_w} \tag{5.24b}$$

Constant C and k are dimensionless GAB parameters, which are related to the heat of adsorption of monolayer and multilayer regions, respectively. C is the ratio of the partition function of the first molecule sorbed on a site and the partition function of molecules sorbed beyond the first molecule in the multilayer, while k is the ratio of the partition function of molecules in bulk liquid and the partition function of molecules sorbed in the multilayer. k is also called a correction factor, since it corrects the properties of the multilayer molecules related to the bulk liquid. The first term of the GAB equation (Eq. 5.24b) in the right side describes the classical mono-molecular layer expression in Langmuir's adsorption isotherm, and the second term describes the multilayer adsorption corresponding to Raoult's law.

The value of C can be calculated by Eq. 5.10 and k can be calculated using Arrhenius type equation as:

$$k = K_0 e^{\frac{\Delta H_k}{RT_k}} \tag{5.25}$$

where K_0 is a constant, and H_k is the heat of condensation of water vapour (J/mol).

In Eqs. 5.10 and 5.25, constant C_o and K_0 are entropic accommodation factors; Q and ΔH_k are functions describing the heat of adsorption and condensation of the water vapour (J/mol), respectively.

$$Q = \Delta H_{s,\text{mono}} - \Delta H_{\text{vap}} \quad (5.26a)$$

$$\Delta H_k = \Delta H_l - \Delta H_{\text{vap}} \quad (5.26b)$$

where $H_{s,\text{mono}}$ is the total heat of adsorption of mono layer (J/mol), and H_l is the heat of condensation of pure water (J/mol). Another form of the GAB model is:

$$\frac{a_w}{M} = a + Ba_w + ca_w^2 \quad (5.27a)$$

$$a = \frac{1}{M_m Ck} \quad (5.27b)$$

$$B = \frac{(C-2)}{M_m C} \quad (5.27c)$$

$$c = \frac{k(1-C)}{M_m C} \quad (5.27d)$$

The constant a, B, and c are calculated from a regression analysis of the Eq. 5.27a with the experimental data points. The GAB parameters M_m, k, and C are then obtained by solving the combination of Eqs. 5.27a–5.27d.

Both GAB and BET equations discriminate between monolayer and condensate properties of molecules sorbed on the first layer of water at the site. The GAB model represents a refined extension of the BET equation, postulating that the state of the sorbate molecules in the second and higher layers is equal, but different from that in the liquid-like state. Compared with BET model, this assumption introduces an additional constant k by which the GAB model gains its greater versatility (Timmermann et al. 2001). Introducing k assumes that multilayer molecules have interactions with the sorbent, and this interaction is termed as the second stage. This allows the GAB isotherm to be successful up to high water activities (i.e., $a_w \approx 0.9$) (Timmermann et al. 2001). $k = 1$ indicates there is no distinction between multilayer molecules and liquid molecules of water, and the GAB equation became the BET equation. $k > 1$ indicates that the adsorption isotherm is infinite, which is physically flawed.

The GAB model can be also used to determine the amount of water absorbed in monolayer and multilayers inside stored crop seeds. The k value is mostly influenced by temperature, while values of C and M_m are slightly dependent on temperature. For example, k value of freeze-dried lactoglobulin increases by 32% in the temperature range 25–40°C, while C varies only 1% (Fontan et al. 1982). When $C \geq 2$, the GAB equation gives a sigmoidal shape curve with point of inflection; when $0 < C < 2$, the isotherm curve has no point of inflection. Both BET and GAB equation can be used to find the amount of monolayer water (M_m) and the energy constant related to surface net heat of sorption (C). However, the M_m value found by BET equation is lower than that found by GAB, while the value of C found by BET is larger than that by GAB. These values found by GAB are more representative than the corresponding BET values (Timmermann et al. 2001).

The GAB model underestimates the water content values at high water activities ($a_w > 0.93$). The recommended application range of the GAB model was up to 0.9 of water activity (Blahovec and Yanniotis 2008). Therefore, the GAB model was refined for higher water activities by some researchers (Timmermann and Chirife 1991; Viollaz and Rovedo 1999). Timmermann and Chirife (1991) suggested that the second adsorption stage introduced by the GAB model may be limited to a certain number of layers and that the true liquid-like properties (a third stage) of the sorbed water, as postulated by the original BET model, might exist. Timmermann and Chirife (1991) introduced a fourth dimensionless parameter H to describe the so-called third adsorption stage and modified the GAB model as (termed as Timmermann GAB model):

$$\frac{M}{M_m} = \frac{Cka_w HH'}{(1-ka_w)\left[1+(CH-1)ka_w\right]} \tag{5.28a}$$

$$H = 1 + \frac{1-k}{k}\frac{(ka_w)^h}{1-a_w} \tag{5.28b}$$

$$H' = 1 + \frac{H-1}{H}\left(\frac{1-ka_w}{1-a_w}\right)\left[h+(1-h)a_w\right] \tag{5.28c}$$

where H' is the parameter, and h is a dimensionless constant. When $h \to \infty$, Eq. 5.28 becomes the GAB equation.

To obtain a good fitting for higher a_w, Viollaz and Rovedo (1999) modified the GAB model by adding an additional dimensionless parameter. This modified empirical equation is termed as Viollaz GAB model:

$$\frac{M}{M_m} = \frac{Cka_w}{(1-ka_w)\left[1+(C-1)ka_w\right]} + \frac{CkK_2a_w^2}{(1-ka_w)(1-a_w)} \tag{5.29}$$

When $K_2 \to 0$, Viollaz GAB model becomes the GAB model. K_2 has a very low weight for low values of a_w, so the values of M_m, C, and k are not substantially affected by the addition of this new term (K_2). The weight of K_2 increases with increasing of a_w.

The saturated salt solution method usually does not afford sufficient information to get a complete adsorption curve (Basu et al. 2006; Blahovec and Yanniotis 2008; Jian and Jayas 2018b). The GAB model and most of its modifications as well as the Peleg's four parameter model, have the same weakness, they predict a finite adsorption at water activity 1. Jian and Jayas (2018b) found there was no relationship between the temperature and the monolayer moisture content or the heat of condensation of the water vapour when the saturated salt solution method was used. They modified Eqs. 5.10 and 5.25 as:

$$c = c_0 e^{\frac{\Delta H}{RT}} \tag{5.30a}$$

$$k = K_0 e^{\frac{\Delta H}{RT}} \tag{5.30b}$$

where ΔH is the enthalpy change (J/mol).

The GAB equation has become the most versatile adsorption model available in the literature and has been adopted by a group of West European food researchers (Bizot 1983). It was formally utilized by the European project group in 1980s (COST 90). Boquet et al. (1979) reported on the excellent fitting abilities of the Hailwood-Horrobin equation, which is mathematically identical to the GAB equation (Boquet et al. 1980).

GAB equation is used and modified for different grain types to fit the collected lab data under constant RH and temperature conditions (Chirife et al. 1991; Timmermann et al. 2001). The modified equation transfers the five-parameter GAB model to three-parameter GAB model. Jayas and Mazza (1993) reported the modified GAB equation was the second best in describing the equilibrium, moisture content (EMC) data of oats. Jian and Jayas (2018b) found the modified GAB equation was the best equation to describe isotherms of hemp seeds under both adsorption and desorption conditions. The GAB equation is one of the six equations recommended by ASABE (2016) to estimate grain isotherms.

5.1.1.7 Harkins–Jura Equation

Harkins–Jura equation, developed in 1940s, is restricted to regions in which the adsorbed molecules form a condensed film, and it assumes the possibility of multilayer adsorption on the surface of absorbents having heterogeneous pore distribution. Harkins and Jura proved experimentally that the adsorptive film was of the condensed type and hence its surface pressure, the difference of the surface tensions of the bare surface and film covered surface, was a linear function of the area per molecule.

$$\ln a_w = \beta - A/M^2 \tag{5.31a}$$

$$\beta = A/V_{ad}^2 \tag{5.31b}$$

where A is the constant related to surface area, and V_{ad} is the adsorbed volume at saturation (m^3).

Harkins–Jura equation fits Type II isotherms and is in good agreement with the experimental data up to water activity of 0.33 (Sakaki 1953). Foo and Hameed (2010) reported that the Harkin–Jura isotherm model showed a better fit to the chemical adsorptive data than the Freundlich, Halsey, and Temkin models. Labuza et al. (1972) proved that the approximation of this equation was not very good for most food materials, and usually it did not hold above a water activity of 0.40–0.50 (depending on food material).

5.1.1.8 Langmuir Isotherm Equation

Langmuir isotherm equation proposed in 1916 is based on the existence of dynamic equilibrium of monolayer adsorption (physi-adsorption or chemi-adsorption). The basic assumptions of the Langmuir adsorption are: (1) the surface containing the adsorbing sites is a perfectly flat homogeneous plane with no corrugations; (2) the adsorbed molecules are at immobile state; (3) all available sites are equivalent and each site can hold at most one molecule (monolayer coverage only); and (4) there are no interactions between adsorbate molecules on adjacent sites. Therefore, Langmuir equation depicts a relationship between the number of active sites of the surface undergoing monolayer adsorption and vapour pressure. This is the region in the adsorption isotherm where the curve tends to flatten out. The assumption of this model indicates that the rates of adsorption k and desorption k' must equal when adsorption and desorption reach equilibrium. Based on this assumption, the total available sites of adsorption are fixed, so the maximum of adsorption exists (Langmuir 1916).

$$kP(1-m) = k'm, \text{ or} \tag{5.32a}$$

$$\theta = \frac{K_{eq}P}{1+K_{eq}P}, \text{ or} \tag{5.32b}$$

$$m = \frac{kP}{kP+k'}, \text{ or} \tag{5.32c}$$

$$m = m_{\max} \frac{P}{P+b}, \text{ or} \tag{5.32d}$$

$$\frac{1}{m} = \frac{1}{m_{\max}} + \frac{b}{m_{\max}} \frac{1}{P} \tag{5.32e}$$

where k' is the constant, K_{eq} is the equilibrium constant, $m_{\max}$ is the maximum mass of the adsorbent (kg), and b is the Langmuir constant (Pa).

The model parameters b and $m_{\max}$ can be determined by regressing the experimental data using the linear equation (Eq. 5.32e). When there are two distinct adsorbates, e.g., two species A and B that compete for the same adsorption site, the Langmuir adsorption equation (competitive adsorption) is:

$$\theta_A = \frac{K_{eq}^A P_A}{1 + K_{eq}^A P_A + K_{eq}^B P_B} \tag{5.33a}$$

$$\theta_B = \frac{K_{eq}^B P_B}{1 + K_{eq}^A P_A + K_{eq}^B P_B} \tag{5.33b}$$

The subscript and superscript A and B indicate that the parameter is related to the species A and B, respectively. The assumptions of this competitive adsorption equation are that each site can hold at most one molecule of A or one molecule of B, but not both; and other assumptions for the single species adsorption still hold. Based on the assumption of identical, independent adsorption sites, Langmuir (1918) proposed the following equation to relate water activity with moisture content:

$$a_w \left(\frac{1}{M} - \frac{1}{M_m} \right) = \frac{1}{CM_m} \tag{5.34}$$

This equation described the Type I isotherm. The Langmuir model is generally not applicable to most food materials because it only gives a fairly good prediction within a_w of 0–0.3, and has a limited application when $\theta \to 1$ (close to saturation) and $\theta \to kP$ (very low pressure). This model also does not account for the effect of multilayer adsorption, surface roughness, and interaction among adsorbates. However, Langmuir's isotherm is the most critical equation among the theoretical models because it is based on the force acting between the product surface and the water condensed from the vapour as a monomolecular layer. The Langmuir approach is the theoretical underpinning for the BET and GAB isotherm theories.

5.1.1.9 Lewicki Model

Based on Raoult's law, Lewicki (2000) assumed water in food materials has two statuses: free water with properties of the bulk water and as water of hydration. Hydrated molecules are considered as new entities with molecular weights larger than those of non-hydrated molecules. Hydration reduces the free concentration of bulk water and thus affects water activity in solution. According to this model, molecules after hydration form a complex with properties different than those of molecules and the bulk water.

$$M = a\left(\frac{1}{a_w} - 1\right)^{B-1} \tag{5.35}$$

Lewicki equation is mathematically similar to Oswin equation, but extends the water activity to 1.0. Lewicki (2000) used this equation to model the water adsorption isotherms of 38 products and 31 model mechanical mixtures, and concluded that it had better fit of adsorption isotherms than that the GAB model. It predicts infinite adsorption at the 1.0 water activity.

5.1.1.10 Park Equation

As the D'Arcy and Watt (MDW) equation, Park (1986) equation also has the three terms: Langmuir isotherm, Henry's law, and Raoult's law. Different from the MDW model, the Park model hypothesizes that the water molecules adsorbed according to the Henry's law make clusters, and the clustering reaction of water can be expressed by equilibrium $nH_2O \leftrightharpoons H_2O$.

$$M = \frac{M_L K_L a_w}{1 + K_L a_w} + K_H a_w + K_{eq} a_w^N \tag{5.36}$$

where N is the mean number of water molecules per cluster.

The equilibrium constant K_{eq} is the clustering reaction constant, and N is the mean number of water molecules per cluster.

Zhao et al. (2017) found the Park equation was a more realistic and mechanism-based approach for describing water desorption of glutinous rice flour than the modified D'Arcy and Watt and Hailwood–Horrobin equations.

5.1.1.11 Peleg Equation

The Peleg semi-theoretical model used for water soaking (Peleg 1988) is:

$$M_t = M_i \pm \frac{t}{K_1 + K_2 t} \tag{5.37}$$

where M_i is the initial moisture content (dry basis, %), and M_t is the moisture content (dry basis, %) at time t.

If the process is absorption, the "±" becomes "+", otherwise, it is "–". K_1 is the Peleg rate constant and K_2 is the Peleg capacity constant. Sopade and Obekpa (1990) reported constant K_1 was temperature dependent, while K_2 was not.

Peleg (1993) suggested an empirical four parameter equation to model isotherms.

$$M = K_1 a_w^{n_1} + K_2 a_w^{n_2} \tag{5.38a}$$

where n_1, n_2 are the constant. The constant K_1 and n_1 are always the dominant at the lower range of the water activity. At higher water activities either term or both terms can be dominant. Peleg (1993) claimed Eq. 5.38a had the same mechanism as the BET and GAB models in the range of water activity from 0.0 to up to about 0.95 because both models can be written as:

$$M = \frac{Ka_w}{B + a_w} + \frac{Ca_w}{D - a_w} \tag{5.38b}$$

The constant K, B, C, and D can be mathematically related to parameters in the BET and GAB models. The first term in Eq. 5.38a represents sorption with a decreasing rate with respect to a_w, and the second term with an ever increasing rate of a_w. Therefore, Eq. 5.38a has the same or better fit than GAB and BET models (Peleg 1993), and the model could also be used as a basis for understanding hydration kinetics (Peleg 1993).

In literature, both Eqs. 5.37 and 5.38a are named as Peleg equation (model), but they have different applications and theoretical backgrounds. Equation 5.37 is widely used to model water soaking (Sopade and Obekpa 1990; Abu-Ghannam and McKenna 1997; Turhan et al. 2002; Johnny et al. 2015). Sopade et al. (1992) used the Peleg equation to model water sorption by maize, millet, and sorghum at 10–50°C. Equation 5.38a has been extensively used to model isotherms of cereal and legume seeds. Peleg (1993) showed that the model had the same or better fit than the GAB model in the water activity of up to 0.95 for the equilibrium moisture sorption of agar-agar, carrageenan, gelatin, low methoxyl pectin, and wheat bran. Al-Muhtaseb et al. (2004) found Peleg equation was the best to describe the isotherm of high amylose starch powders in the water activity 0.05–0.95. Bingol et al. (2012) found the four parameter Peleg model gave the best fit for all forms of rice than the following evaluated models: Halsey, Oswin, and GAB. After evaluating GAB, BET, Henderson, Iglesias and Chirife, Oswin, Peleg, Smith and Caurie models, Arslan and Togrul (2006) found the Peleg four parameter equation was the most suitable model to describe the relationship between equilibrium moisture content and water activity for the whole range of temperature and relative humidity of tea. Peleg model has a good fit to equilibrium isotherm of food products such as: corn snacks (Palou et al. 1997), corn starch (Peng et al. 2007), mint leaves (Dalgic et al. 2012), freeze-dried strawberry (Ciurzyńska and Lenart 2009), and soy protein (Pan 2003).

5.1.1.12 Raoult's Law and Henry's Law

Raoult's law is a thermodynamics law established by French chemist François-Marie Raoult in 1886. For ideal water solution, Raoult's law states that the water activity in the solution is equal to molar fraction of water (X_w). For non-ideal solutions, the activity coefficient λ corrects for deviation from nonideality (Raoult 1886).

$$a_w = \lambda X_w \tag{5.39}$$

where X_w is the mole fraction of water.

Raoult's law is based on the principle that the particles inside the biological materials reduce the escaping tendency (fugacity) of the water. The ideal condition is the substance does not interact with water and the particles of the material are approximately the same size as the water molecule. These two assumptions are not always true and this is the reason for introducing λ. Norrish (1966) proposed the following semi-empirical correction to Raoult's law:

$$a_w = KX_w X_s^2 \tag{5.40}$$

where X_s is the mole fraction of solute.

For a multicomponent system, Ross (1975) assumed that the activity coefficient of individual components with water were equal to their activity coefficients in binary mixtures in water, and extended the Norrish equation (Eq. 5.40) for multicomponent systems as:

$$\ln(a_w) = \sum_{j=1}^{k} \ln(a_{wj}) \tag{5.41}$$

where a_{wj} is the water activity of the j component in the binary mixture (decimal).

The Ross equation can be used to find the water activity of biological materials (Chirife et al. 1985; Herman et al. 1999), and the water activities of non-solute parts can also be evaluated with the Henderson equation by using the data for starch, protein, and fibre (Roman et al. 2004).

Raoult's law is normally applied to the higher water activities and is better than BET equation in some cases. The limitation on the number of layers of water that can build up around solute particles in solution is two.

Henry's law was formed by the English physicist and chemist William Henry in 1803. Henry's law states that at very low water activity, pressure, or low adsorptions, the solubility of a gas in a liquid or the amount of water vapour adsorbed on a solid is directly proportional to its pressure of water activity.

$$M = ka_w \tag{5.42}$$

Henry's law is normally applied to low water activities and is a special case of Raoult's law at high water activity. Raoult's law and Henry's law are not used to model isotherm of products because they only predict one status of the water sorption, but are fundamental to other isotherm models.

5.1.1.13 Smith Equation

Peirce (1929) developed an equation that is equivalent to Smith equation. Smith model describes the final curved portion of the water adsorption isotherm of high molecular weight bio-polymers (Smith 1947).

$$M = a + B\log(1 - a_w) \tag{5.43}$$

Constant a is the quantity of water in the first sorbed fraction, and B is the quantity of water in the multilayer moisture fraction. Smith equation is based on the following assumptions: there are two fractions of water sorbed onto a dry surface; the first exhibits a higher than normal heat of vaporization and would be expected to follow the Langmuir model. The second fraction, which consists of multilayers of condensed water molecules, can form only after the first fraction has been sorbed, this effectively prevents any possible evaporation of the initial layer. The moisture content in the second fraction is proportional to the logarithm of the difference between the a_w of the sample and pure water.

Smith model was widely used for food components in the water activity range from 0.5 to 0.96. Becker and Sallans (1956) found that Smith equation could be used to describe adequately the desorption isotherm of wheat between 0.5 and 0.95 water activity. Smith equation was also successfully used for sucrose starch products (Chinachoti 1990) and peanuts with water activity of higher than 0.30 (Young 1976).

5.1.1.14 Young and Nelson Equation

Young and Nelson (1967) equation considers the effect of hysteresis and assumes that the water is held in the food by three mechanisms: (1) water in unimolecular layer and bound to the surface of the cells; (2) water in multimolecular layers, and (3) water in cells. The hysteresis is caused by the resistance of water movement through the cell wall.

$$M_{\text{sob}} = C(\theta + \alpha) + B\varphi \tag{5.44a}$$

$$M_{de} = C(\theta + \alpha) + B\theta a_{w\max} \tag{5.44b}$$

$$\theta = \frac{a_w}{a_w + (1 - a_w)E} \tag{5.44c}$$

$$\varphi = a_w \theta \tag{5.44d}$$

$$E = e^{-\frac{q - q_L}{\beta T_K}} \tag{5.44e}$$

$$\alpha = -\frac{Ea_w}{E - (E-1)a_w} + \frac{E^2}{E-1}\ln\left(\frac{E - (E-1)a_w}{E}\right) - (E+1)\ln(1 - a_w) \tag{5.44f}$$

where M_{sob} is the equilibrium moisture content in sorption process (dry basis, %), M_{de} is the equilibrium moisture content in desorption process (dry basis, %), θ is the fractional occupancy of adsorption sites (fraction of surface already covered by a layer of bound moisture), φ is the fraction of surface covered by two or more layers of molecules, q is the sorption energy (W), q_L is the heat of condensation (J/kg or J/mol), and β is the Boltzmann's constant ($1.38064852 \times 10^{-23}$ J/K).

Young and Nelson equation is unable to fit the data for all food groups. Young and Nelson equation fit the isotherms of fruits and starchy foods better than BET equation and Chen equation (Boquet et al. 1979) and had the same level of the fitness to the isotherm data of protein products (Boquet et al. 1979).

5.1.2 Empirical Models

Most researchers tried to mathematically model the full range of water activity for different types of food products. Empirical models are developed on purely mathematical manipulation to cope with the water status and interaction among surface of the products and water in different layers. The empirical models presented in Table 5.1 are the equations which are not widely used but are assembled for the completeness in a single source, while the equations presented in the following text are the most used equations.

5.1.2.1 Chen Equation

Chen equation is developed based on the diffusion principle mode of mass transfer during drying (Chen and Clayton 1971).

$$a_w = e^{\left(k + ae^{BM}\right)} \tag{5.45a}$$

Table 5.1 Some Sporadically Used Empirical Models to Predict Equilibrium Isotherm of Food Products

Model	*Equation*	*Development Background*	*Example of Application*
Caurie (1970)	$M = e^{(k+Ca_w)}$	To predict safe moisture levels of dehydrated food products	Could fit the water activity from 0 to 0.85 for most foods (Caurie 1970). The best equation to predict isotherm of pistachio shells (Soleimanifard and Hamdami 2018)
Fontan et al. (1982)	$\ln\left(\frac{\gamma}{a_w}\right) = \alpha M^{-r}$	To predict equilibrium isotherm of various types of foods	Could fit the isotherm from 0.1 to 0.9 a_w. It has a slight lower prediction accuracy than that of BET equation (Iglesias and Chirife 1995)
Iglesias and Chirife (1976b)	$\ln(M + \sqrt{M^2 + M_{0.5}} = Ba_w + a$	Based on a multilayer adsorption equation, originally for non-uniform surfaces	Has a fairly good fit for fruits and other high sugar products (Kaymak-Ertekin and Gedik 2004)
Linear	$M = a + Ba_w$	Simple	Successfully used for tea up to 0.55 a_w (Labuza et al. 1972), cabbage at 0.07–0.25 a_w (Mizralii et al. 1970), wheat at 0.12–0.65 a_w (Becker and Sallans 1956)
Strohman and Yoerger (1967)	$a_w = e^{\left[K_1 e^{(-K_2 M)} \ln P_s - K_3 e^{(-K_4 M)}\right]}$	Based on the linear form of the Othmer plots of equilibrium moisture content	The best to describe the isotherms of rice in the tested four equations using 763 experimental data points extracted from 18 published papers (Sun 1999)
White and Eyring (1947)	$M = \frac{1}{\frac{1}{M_m} - \frac{k}{M_m} a_w}$	Model the sorption of materials with long chain chemicals such as fibres	Fugassi and Ostapchenko (1959) equation is equivalent

This equation can be written as:

$$M = \frac{1}{B}\ln\left(\frac{1}{a}(\ln a_w - k)\right) \tag{5.45b}$$

Parameters k, a, and B are temperature dependent. The de Boer and Zwikker (1929) equation is equivalent to Chen equation with three parameters. Chen and Clayton (1971) applied this equation to various crop grains and reported the good fit. Chen and Clayton (1971) modified Chen equation to two parameter equation because they found the k was close to unity.

$$a_w = e^{\left(-ae^{-bM}\right)} \tag{5.46}$$

This two parameter equation is mathematically equivalent to Bradley equation. Chen and Clayton (1971) found the three parameter equation had a better prediction accuracy than the two parameter equation, which is better than Henderson's equation.

To predict the temperature effect, Chen and Clayton (1971) empirically modified Chen equation to four-parameter equation as:

$$a_w = e^{\left(-K_1 T_k^{a1} e^{\left(-K_2 T_k^{a2} M\right)}\right)} \tag{5.47}$$

where a_1 and a_2 are constants.

The four parameter equation describes adequately the temperature dependency of corn isotherms between 4.4 and 60°C and better than Henderson four parameter equation (Chen and Clayton 1971). Chen's equation and Chen's modified equation were mostly tested with experimental sorption data in cereal grains and other field crops (Boquet et al. 1978).

5.1.2.2 Halsey Equation

The equation developed by Halsey (1948) can describe condensation of multilayers at a relatively large distance from the surface. The assumption of Halsey model is that the potential energy of a molecule varies inversely as the nth power of its distance from the surface.

$$a_w = \exp\left(-\frac{aM^{-c}}{RT_k}\right) \tag{5.48}$$

Halsey developed the equation to theoretically account for the criticism of the BET assumptions. The magnitude of the C characterizes the type of interaction between the vapour and the solid. If C is large, the attraction of the solid for the vapour is very specific and does not extend far from the surface; when C is small the forces are more typical of van der Waals and are able to act at a greater distance. Therefore, Halsey equation is a semi-theoretical model.

Halsey (1948) showed this equation was a good representation of adsorption for Type I, II, or IV isotherms. Halsey model fits the chemical absorptive data well (Ayawei et al. 2017). By using literature data of 69 different food materials, Iglesias and Chirife (1976b) concluded that Halsey equation had a better fit than Henderson equation in most cases. Halsey model was modified by Iglesias and Chirife (1976b) and was termed as Modified Halsey equation:

$$a_w = \exp\left[-\frac{\exp(a + BT_c)}{M^C}\right] \tag{5.49}$$

where T_c is the temperature (°C).

The use of temperature term does not eliminate the temperature dependence of constant a, B, and C. The Modified Halsey equation has a wider application than the Halsey equation. Iglesias et al. (1975) and Iglesias and Chirife (1976c) reported that the Modified Halsey equation could be used to describe the adsorption isotherms of 69 foods in the water activity range of 0.1–0.8. Kaymak-Ertekin and Gedik (2004) reported that the Halsey equation gave the best fit to the isotherms of grapes, apricots, apples, and potatoes within water activity of 0.113–0.90 at 30–60°C among 11 tested models including BET, GAB, Chung-Pfost, Modified Chung-Pfost, Chen, Modified Halsey, Henderson, Modified Henderson, Iglesias-Chirife, and Oswin equations. Sun and Byrne (1998) found the Modified Halsey equation was the best equation for describing the isotherms of rapeseed among four commonly cited equations. The Modified Halsey equation is a simpler equation than the usual four-parameter equations reported in the literature (Day and Nelson 1965; Chen and Clayton 1971; Iglesias and Chirife 1976a).

5.1.2.3 Henderson Equation

Henderson (1952) proposed an equation of the following form:

$$a_w = 1 - \exp\left[-aT_cM^B\right] \tag{5.50}$$

The term T_c does not eliminate the temperature dependence of the constants a and B. To eliminate the temperature dependence, Day and Nelson (1965) modified the Henderson equation to four parameter equation as:

$$a_w = 1 - \exp\left[-K_1T^{a_1}M^{K_2T^{a_2}}\right] \tag{5.51}$$

where T is the temperature (K or °C).

Thompson et al. (1968) added another constant to the temperature term and named it as a Modified Henderson equation.

$$a_w = 1 - \exp\left[-a(T_c + C)M^B\right] \tag{5.52}$$

Modified Henderson equation is one of the most used isotherm equations, which can fit many kinds of food materials (Singh and Ojha 1974; Chirife and Iglesias 1978).

5.1.2.4 Modified Chung–Pfost Equation

The equation proposed by Chung and Pfost (Chung and Pfost 1967a,b) is:

$$a_w = \exp\left[-\frac{a}{RT_c}\exp(-BM)\right] \tag{5.53}$$

This equation was modified by Pfost et al. (1976) as:

$$a_w = \exp\left[-\frac{a}{T_c + C}\exp(-BM)\right] \tag{5.54}$$

This Modified Chung–Pfost equation is one of the most used equation to model isotherm of cereal grain seeds. The Modified Chung–Pfost equation is the most appropriate equation for rough rice than other three parameter equations: the modified Henderson, modified Oswin and modified Halsey equations (Basunia and Abe 2001). The Modified Chung–Pfost equation was identified as the best equation for wheat through comparison over 1000 experimental data points extracted from 33 publications (Sun and Woods 1994).

5.1.2.5 Oswin Equation

Oswin (1946) developed an empirical equation for the sigmoid-shaped isotherm:

$$M = K\left(\frac{a_w}{1 - a_w}\right)^C \tag{5.55}$$

Oswin equation has been used to relate the water activity of up to 0.5 for wide range of food products. Mousa et al. (2014) found Oswin equation was the best model to describe isotherm of paddy rice in the water activity range of 0.113–0.976 among the following tested models: GAB, polynomial, Halsey, Smith, Chung-Pfost, and Henderson models. The parameter K is a linear function of temperature. Therefore, the Oswin equation was modified as (Chen 1988):

$$a_w = \left[\left(\frac{a + BT_c}{M} \right)^2 + 1 \right]^{-1} \tag{5.56}$$

After comparing 591 experimental data points extracted from 20 publications, Sun (1998) concluded the Modified Chung–Pfost and Modified Oswin equations were the most appropriate equations for describing corn equilibrium isotherm among three commonly cited isotherm equations (Modified-Henderson, Modified-Chung–Pfost and Modified-Oswin equations), and the modified Oswin equation had slightly better fit than the modified Chung–Pfost equation.

5.1.3 Model Selection and Most Used Models in Grain Storage

5.1.3.1 Temperature Effect on Model Selection

Even though most models use the temperature as one of the independent parameters, all the models are developed based on the data collected at constant temperatures (this is the meaning of isotherm) because it is difficult to reach equilibrium at transient temperatures, and the effect of the temperature is interpolated from the isotherms at different temperatures with the following limitations:

1. Temperatures may affect the sorption isotherm through irreversible changes in the materials. This partially explains that the isotherm validated at low temperature might not be applicable at higher temperatures, and might not explain the difference between sorption and desorption isotherms (hysteresis). The heat changes involved in irreversible processes might be small compared to the overall energy changes for some food products, and therefore could be neglected in a general qualitative description (Iglesias and Chirife 1976a).
2. Information about a_w below freezing is limited. This field has been in some uncertainty due to contradictory reports. Some reported that vapour pressures of animal tissues over the temperature range −26 to −1°C ranged from 13 to 20% lower than those of pure ice at the same temperatures (Dyer et al. 1966; Hill and Sunderland 1967). However, other study showed the same vapour pressure as the ice when temperature was lower than 0°C (Storey and Stainsby 1970). It is not known whether the validated isotherm equations could be extended to temperatures below freezing.

3. An increase in the temperature decreases their mutual distance and attractive forces (including bound water) among chemicals, and hence reduces the formation of hydrogen bonds. Consequently, an increase in temperature decreases the binding energy and induces a reduction in the equilibrium moisture content at a specific water activity. Most developed models can handle this temperature effect with the help of Kelvin equation or Clausius–Clapeyron equation. The isotherms at high temperature can be estimated by differentiating the Kelvin equation. Kelvin equation mathematically models the equilibrium of water in capillary which has meniscus. Therefore, it only can handle the condensed water and will provide a guide for >60% RH. Kelvin equation predicts that the average change in relative humidity (water activity) is 0.5 percentage points per degree Celsius change over the temperature range from 20 to 100°C at 60% RH. This is consistent with the fact that the equilibrium moisture content changes between 0.5 and 1% of percentage points per degree Celsius. The reason for such a temperature dependence is not known. Kelvin equation is not widely used because of the difficulty to quantify the surface tension (σ) and density of the condensed water.

The thermodynamic function used to express the temperature dependence of vapour pressure is the Clausius–Clapeyron equation, which requires the isosteric heat of sorption (Q_s). Different models have been developed by using Clausius–Clapeyron equation. However, measuring the Q_s limits its application because Q_s is temperature and material dependent, e.g., different crop seeds have different Q_s at the same temperature, or the same crop seeds have different Q_s at different temperatures. The hysteresis of materials also influences the value of the Q_s. According to La Mer (1967), the Clausius–Clapeyron equation could be used only to compare the heat of sorption calculated from the equation, with the corresponding enthalpy values obtained by direct calorimetry.

5.1.3.2 Selection of Composite or Single Component Models

The difficulty of having a unique mathematical model for describing accurately the sorption isotherm in the whole (or most of it) range of water activity and for different type of biomaterials, is mainly due to: (1) multifactor influence the isotherm and different factor may be predominant in a given range of water activity; (2) the material has multi-components and the sorption isotherms represent the integrated hygroscopic properties of the numerous constituents and their sorption properties may change as a consequence of physical and/or chemical interactions induced by water sorption or desorption; (3) the material usually undergoes changes of constitution, dimensions, and physical and chemical properties during water sorption and desorption; (4) swelling and shrinking are the examples of dimension change, and common adsorption theories ignore this change in the substrate surface; and (5) water sorption may also lead to phase transformations of some materials such as food with high concentration of sugars (Iglesias et al. 1975). These also partially explain why the isotherm models are usually not able to extrapolate the isotherm of the materials.

Water at low water activity will be absorbed by solid material in a monolayer. This equilibrium isotherm at low relative humidity can be well-correlated with the commonly used monolayer models, such as Langmuir and Freundlich. Monolayer adsorption is gradually predominated by capillary and/or multilayer condensation with the increase of water activity (Liu et al. 2013). Liu et al. (2013) divided the total pore volume of a porous adsorbent into three parts: micro-pores where capillary condensation may not take place, macro-pores where capillary condensation may take place, and the region next to the macro-pore walls. They used the Langmuir and Freundlich models to describe both the micro-pore adsorption and the surface adsorption on the macro-pore walls, and used the Kelvin theory to describe water vapour condensation in the remnant pore core volume of the macro-pores under the assumption that the effective pore size follows a Gaussian distribution without considering the interferences between vapour condensation and surface adsorption. Liu et al. (2013) found the fitting accuracy can be further improved by using a more suitable expression for the effective pore size distribution. Many developed composite models used this approach (Boquet et al. 1979; Liu et al. 2011; Zhao et al. 2017). The most used composite models are Modified D'Arcy and Watt equation, Park equation, and Hailwood-Horrobin equation. A composite model usually can handle a wider range of water activity than that of single component model.

5.1.3.3 Selection of Models With More or Less Parameters

No single model adequately fits the experimental data for all foods tested (Boquet et al. 1978, 1979). For different classes or cultivars of a crop, some models might have a higher prediction accuracy than that of other models. For the same model, the values of the constants in the model are different for different types of crops. The value of these constants in the adsorption equation are also different from the desorption equation for the same crop. In many cases, the model that is suitable for certain food products is not suitable for a different one (Chen and Clayton 1971). It has been frequently observed, a curve may be described by many different equations. This proves that the agreement of the experimental sorption data with the isotherm equation does not provide a proof of the correctness of the assumption of the model.

More than 270 models with two to five parameters have been published. An equation with fewer parameters (two to four) cannot be expected to describe the sorption process with the same precision as that of the equation with five parameters because few parameter equation might not take account the overall sorption process. With the spread and ease of computer aided curve fitting techniques, the traditional empirical models and the model with less parameter have lost some of their advantages. Introducing more parameters typically increases the quality of the fitting, while making the calculation more complicated and occasionally less accurate in their predictive ability. Some of the simpler two-parameter equations also give fits of comparable or even better accuracy than the three parameter equations (Boquet et al. 1979). Composite models considering multiple phenomena usually have more parameters than that of single component model only considering a single phenomenon. One cautionary note to use multi-parameter models is to have enough data covering full water activity range at multiple temperatures.

5.1.3.4 The Most Used Equations

Boquet et al. (1979) recommended the following four equations to fit the water sorption data of probably any food: Hailwood & Horrobin, Halsey, Oswin, and Iglesias & Chirife (Chirife and Iglesias 1978). The most widely reported equations in the literature are: GAB, Oswin, Henderson, Peleg, BET, Halsey, Caurie, and Lewicki.

The empirical equations recommended by the ASABE standard (ASABE 2016) for grains (cereal grains, oil seeds and pulses) are: Modified Henderson equation, Modified Chung-Pfost equation, Modified Halsey equation, and Modified Oswin equation. The theoretical or semi-theoretical equation recommended be the ASABE standard (ASABE 2016) is the GAB. Chen and Morey (1989) found the Modified-Henderson and Chung–Pfost equations had a lower relative percentage deviation and standard error of the estimated value for starchy grains and fibrous materials. The Modified-Halsey equation had a higher prediction accuracy for high oil and protein products. The Modified-Oswin equation was more suitable for popcorn, corncobs, whole pods of peanut and some varieties or cultivars of corn. Boquet et al. (1978) considered the Oswin equation to be the best one for describing the isotherms of starchy food, and a reasonably good fit for meat and vegetables. Lomauro et al. (1985a) found the GAB equation gave the best fit for more than 50% of the fruits, meats and vegetables analysed. Al-Muhtaseb et al. (2002) reported that GAB equation gave a good fit for over 75% of the food isotherms (starchy foods, fruits, vegetables and meat products).

5.2 Drying and Wetting Models

5.2.1 Theories of Water Movement in Solids

Moyne and Perre (1991) and (Perre and Moyne 1991) summarised the history of the mathematical model development, developed the equations (three partial derivative equations) of mass and energy balances in porous medium, and demonstrated the 2-D computer codes which were used to solve these equations associated with the high temperature wood drying, microwave drying of light concrete, unsaturated to saturated porous media drying, and saturated to unsaturated media drying. They also highlighted three particular points: the importance of the air mass balance, the right way to write the energy equation (and the associated boundary conditions), and the importance of taking into account of the adsorbed water. Interested reader can refer these articles to find the history and development process of mathematical models associated with drying. Therefore, only the most important models of drying are presented in alphabetic order in this section.

5.2.1.1 Fick's Law of Diffusion and Effective Diffusivity

5.2.1.1.1 Fick's Law Diffusion can occur in gaseous, liquid and solid materials, and the diffused water can be in vapour or liquid state or in both states. The general diffusion of water (moisture content) inside a crop kernel in one direction is governed by the Fick's first law:

$$J = -D_f \frac{dM}{dx} \tag{5.57}$$

where J is the flux of moisture flow rate (kg·m·s^{-1}), D_f is the diffusion coefficient (m^2/s), and x is the distance (m). The flux (or quantity of moisture flow rate, J) is defined as amount of water exchanged per unit of time per unit area. If the unit of the moisture content (M) is dry basis (%), this Fick's first law can be written as:

$$J = -D_f \rho_g \frac{dM}{dx} \tag{5.58}$$

where J is the density of moisture flow rate (kg·s^{-1}·m^{-2}). Fick's first law can be applied to steady state system when the concentration is kept constant. The local concentration and diffusion flux through the unit area $\dot{A}$

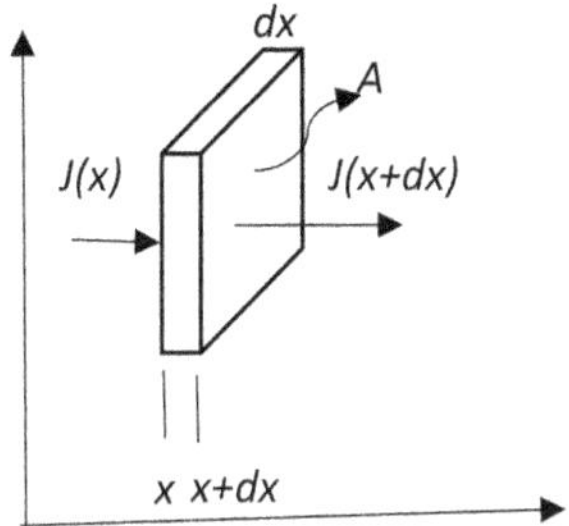

Figure 5.1 Schematic of diffusion in one dimensional transient condition.

at position x are $M(x,t)$ and $J(x)$, respectively (Fig. 5.1). The first law at position dx is:

$$dM(x,t) = \frac{[J(x) - J(x+dx)]dt\dot{A}}{\dot{A}dx} \tag{5.59}$$

where $\dot{A}$ is the surface area (m^2). Equation 5.59 can be written as:

$$\frac{dM(x,t)}{dt} = -\frac{dJ}{dx} \tag{5.60}$$

From the first law,

$$J = -\frac{\partial[D_x M(x,t)]}{dx} \tag{5.61a}$$

where D_x is the diffusion coefficient in x direction (m^2/s) and is varying with x such as in heterogeneous materials. If D_x is constant with x during the diffusing period, the flux can be simplified as:

$$J = -\frac{D_x \partial[M(x,t)]}{dx} \tag{5.61b}$$

And the non-constant diffusion can be approximated as:

$$\begin{aligned} \frac{dM(x,t)}{dt} &= -\frac{\partial^2}{\partial x^2}\left(D_f(x,t)M(x,t)\right) \\ &= -\frac{\partial}{\partial x}\left(D_f(x,t)\frac{\partial M(x,t)}{dx} + M(x,t)\frac{\partial D_f(x,t)}{\partial x}\right) \end{aligned} \tag{5.62}$$

For the diffusion in inhomogeneous materials, the equation (Eq. 5.62) is termed as Fokker-Planck diffusivity law, which includes a drift with velocity equal to $\partial D_f(x,t)/\partial x$ (van Milligen et al. 2005). Fokker–Planck diffusivity law is widely ignored due to its complexity (introducing $\partial D_f(x,t)/\partial x$ and its second derivative makes it impossible to solve mathematically for most cases).

Fick's law assumed $J = -D_f(x)\dfrac{\partial M(x,t)}{dx}$, and

$$\frac{dM(x,t)}{dt} = -D_x\frac{\partial}{\partial x}\left(\frac{\partial M(x,t)}{dx}\right) \tag{5.63}$$

This is the Fick's second law. By assuming the D_f is constant in the t time period, then,

$$\frac{dM}{dt} = -D_{\text{eff}}\frac{\partial^2 M}{\partial x^2} \quad \text{One dimension} \tag{5.64a}$$

$$\frac{dM}{dt} = D_{\text{eff}}\left(\frac{\partial^2 M}{\partial x^2}+\frac{\partial^2 M}{\partial y^2}+\frac{\partial^2 M}{\partial z^2}\right) \text{An isotropic diffusion} \tag{5.64b}$$

$$\frac{dM}{dt} = D_x\frac{\partial^2 M}{\partial x^2}+D_y\frac{\partial^2 M}{\partial y^2}+D_z\frac{\partial^2 M}{\partial z^2} \text{ An anisotropic diffusion} \tag{5.64c}$$

where D_{eff} is the effective diffusivity (m^2/s), and D_y and D_z are the diffusion coefficients in y and z directions, respectively (m^2/s).

Fick's law can also be developed by using capillary theory to describe the drying. The effective diffusion coefficient (D_{eff}) is a function of time, moisture content, temperature, material type, and drying history. In solving the diffusion equation, these parameters are usually taken as constant or in the form of linear, exponential, or polynomial functions of moisture content or position or both. When D_f has varying functions, the model cannot be solved analytically and it is assumed constant over short time intervals and over small areas during numerical solution process.

The diffusion theory supports the diffusion mechanism of water in liquid or vapour during drying or wetting. The diffusion can be the rate-limiting mechanism during the falling rate drying period regardless of the water status inside the drying material. During the constant rate drying period, the surface of the materials can be assumed as a wet body at the wet bulb temperature. Drying grains usually does not exhibit the constant rate drying period. Therefore, most developed models for grain drying use the diffusion theory. Under the falling rate of drying, the diffusion equation may be applicable provided one uses a variable diffusivity. Meinders and van Vliet (2009) found sorption and desorption of bread crust could be best described by the Fickian diffusion model for low RH (lower than 0.5 a_w for adsorption, and 0.7 for desorption) and the exponential model for large RH due to the relaxation effects of the polymer matrix.

Isothermal conditions exits during drying by convective hot air for either a non-porous or a porous material having a very small characteristic dimension (Waananen and Okos 1996; Jian and Jayas 2018a). To obtain an isothermal condition for a large material such as bred, the use of volumetric heating, such as microwave heating, along with surface heating, such as convective air, is usually used. It has been proven that drying (especially during the falling rate phase) is controlled by internal mass transfer (Waananen and Okos 1996). Therefore, Fick's law is usually used to model water diffusion inside drying materials.

Fick's law of diffusion was used by researchers before 1940s such as Lewis (1921), Ceaglske and Hougen (1937), and (Sherwood 1929) to develop the liquid-diffusion theory based on the main assumption that the movement of moisture is by liquid diffusion (Fortes and Okos 1980). However, around the same time researchers pointed out that the moisture distribution cannot be calculated correctly only from Eq. 5.63 (Hougen et al. 1939; Vu and Tsotsas 2018). This theory was subjected to severe criticism and called misleading because the prediction was different from the experiment results. Even though diffusion theory has this disadvantage and the diffusion model can misinterpret the moisture distribution or drying behaviour, this model is still widely used as a simple way to describe drying.

Fick's law is usually used to describe the diffusion of single species (like water molecule) in porous materials (binary system). Maxwell–Stefan approach should be used to describe the diffusion of multi-components such as multiple chemical compounds diffusion in porous materials (Krishna and Wesselingh 1997; Leonardi and Angeli 2010). Liquid and vapour water, and free and bound water might co-exist during drying and reaction-diffusion model, which divides water into two or more groups with individual diffusivity, is an alternative approach and Langmuir-diffusion model can be used. Maxwell–Stefan and Langmuir-diffusion models are beyond the scope of this book.

5.2.1.1.2 Effective Diffusivity In the absence of internal temperature gradients, when water vapour pressure or moisture gradient is the only driving force for diffusion (and may be used interchangeably), and if a linear water sorption isotherm exists, the effective diffusivity can be calculated (Suarez et al. 1980) as:

$$D_{\text{eff}} = \frac{MD'P_w^0}{\rho_s RT_k}\left(\frac{P}{P-P_w}\right)\left(\frac{\partial a_w}{\partial M}\right)_{T_k} \tag{5.65}$$

where D' is the effective vapor space diffusion coefficient (m²/s), and P_w^0 is the vapor pressure of water (Pa). The term $(\partial a_w/\partial \text{M})_{T_k}$ can be linear or non-linear depending whether the effective diffusivity can be considered constant or dependent on moisture content. If the relationship between vapour pressure and moisture content is not linear as seen in most food isotherms, this term is nonlinear and the effective diffusivity varies with moisture content. Constant effective diffusivity has been assumed for grains and processed products (Whitaker and Young 1972; Suarez et al. 1980).

The above equation can be further modified if the heat consumed to vaporize the moisture and vapour mass diffusion is considered (King 1968) as:

$$D_{\text{eff}} = \frac{MD'P_w^0}{\rho_s RT}\left(\frac{P}{P-P_w}\right)\left(\frac{\partial a_w}{\partial M}\right)_T\left(\frac{\beta}{1+\beta}\right) \tag{5.66}$$

where ρ_s is the mass concentration of solid (kg/m^3). The term $\frac{\beta}{1+\beta}$ can be explained as:

$$\frac{\beta}{1+\beta} = \frac{\text{Heat transfer}}{\text{Heat transfer} + \text{Heat transfer by mass transfer}}$$

If the thermal conductivity (k_T) is significantly smaller than D', $\beta \ll 1$, and $\frac{\beta}{1+\beta} = \beta$. This indicates that heat conduction controls the drying process. If k_T is significantly larger than D', $\beta \gg 1$ and $\frac{\beta}{1+\beta} = 1$. This indicates that mass transfer controls the drying process, and Eq. 5.66 is simplified as Eq. 5.65.

Equation 5.66 holds under the following assumptions: (1) a local equilibrium exists between the vapour phase and the sorbed moisture; (2) the heat of desorption is constant and considerably greater than sensible heat of the medium; and (3) the vapour diffused out of the material is constantly replenished by the sorbed moisture. Even though the D_{eff} is mathematically expressed by Eqs. 5.66 and 5.65, it is difficult to be evaluated because it is difficult to evaluate P_w^0, P_w and $\left(\frac{\partial a_w}{\partial M}\right)_T$.

For most cases, the D_{eff} is not a constant because the moisture movement in a solid is not only due to diffusion but is also due to other mechanisms such as capillarity, convection, vaporization-condensation, and pressure. Moisture movement through liquid diffusion cannot denied. More sophisticated theories take Fick's law as representation of liquid and vapour movement (diffusion). The driving force in diffusion can be pressure or concentration gradients or both. A few models were developed by calculating the diffusion of liquid water and vapour separately (Dietl et al. 1995; Thorvaldsson and Janestad 1999). After this assumption, the meaning of diffusivity (diffusion coefficient) is either lost or interpreted as a lumping of all simultaneous effects, besides being dependent on concentration and temperature. Therefore, the Fickian description of the water transfer during desorption and sorption is conceptual in nature, even though the resultant fitness is good (van Milligen et al. 2005).

5.2.1.2 Capillary Theory

Biomaterials are basically arrays of cells. Various cells exist in nature, the tissue can differ widely in porosity and bulk properties; so can differ in the intercellular spaces and voids. These voids (pores) can be the channels (capillaries) for water movement. The capillary theory considers cellular membrane as perfect semi-permeable membrane and may act as capillary paths for both bound and free water to move. Capillary theory assumes there is free water inside capillaries in the wet place of a solid, the pores (voids) provide capillary paths for the free water to flow, and capillary potential (tension or pressure gradient or both) drive the saturated and unsaturated capillary flow. Kelvin equitation governs the capillary flow.

When the capillary is filled with water, the pressure gradient and gravity inside the capillary drive the water movement. As evaporation proceeds, the water surface recedes into the pore located inside the material which has a high moisture content. A suction potential is developed in the capillary. The pores do not have uniform diameter and the suction force and surface tension will change when the water recedes from a place with large pore size to a place with a small pore size or vice versa. As drying proceeds, the continuity of liquid in the capillary will be broken up due to different suction force and capillary potentials. This breaking up will interrupt the free flow of liquid by capillary action. The liquid flux can be calculated as:

$$J_l = -\rho_w \frac{K_h}{\mu}\left(\nabla P_g - \nabla P_l - \rho_w g\right) \tag{5.67}$$

where J_l is the liquid flux (kg·s^{-1}m^{-2}), ρ_w is the water density (kg/m^3), K_h is the permeability (m^2), μ is the dynamic viscosity (Pa·s), P_g and P_l are pressure of gas (Pa/m) and liquid (Pa/m), respectively, and g is the gravitational acceleration constant (9.806 m/s^2).

The assumptions of this equation are:

1. The flux is driven by the gravity and the pressure of gas and water inside capillary;
2. the material is macroscopically homogeneous;
3. the flow in the capillary paths is laminar; and
4. there is no significant temperature gradient.

If the effect of the gas pressure and gravity force are negligible, Eq. 5.67 can be simplified as:

$$J_l = \rho_w \frac{K_h}{\mu} \nabla P_l = \rho_w \frac{K_r K_s}{\mu} \nabla P_l \tag{5.68}$$

where K_r is the relative permeability of porous material (decimal), and K_s is the single phase permeability of porous material (m^2).

Even though different researchers have different explanations on whether the ∇P_l is independent of water concentration, capillary flow has been accepted as one of the fundamental mechanisms of water movement especially during the period of high moisture drying. In homogeneous media with high moisture content and negligible gravity, tension is proportional to moisture content (Chen and Pei 1989). Therefore, water flux can be calculated as:

$$J_l = -\rho_L K_c \nabla M \tag{5.69}$$

where K_c is the capillary (liquid) conductivity (m^2/s).

The water flux can be described by using the Hagen–Poiseuille law:

$$J_l = \frac{1}{\dot{A}} \int_{r_m}^{r_{max}} \rho_w \frac{\acute{r}^2}{8\mu\tau} \nabla P_l d\dot{A} \tag{5.70}$$

where r_m and r_{max} are the minimum and maximum radius of capillaries, respectively (m).

From Eqs. 5.68 and 5.70, K_h can be calculated as (Chen and Pei 1989):

$$K_h = K_r K_s = \int_{r_m}^{r_{max}} \frac{\acute{r}^2}{8\tau} \int(\acute{r}) d\acute{r} \tag{5.71}$$

where $\dot{r}$ is the radius (m), τ is the tortuosity of capillary paths, and $\int(\acute{r})$ is the pore volume density function.

Empirical equations are suggested to calculate the $\int(\acute{r})$ and K_h (Chen and Pei 1989).

5.2.1.3 Receding Front Theory and Modelling

5.2.1.3.1 Theory After the development of diffusion and capillary theory, different receding front models were developed in order to obtain a better understanding and describe the influence of other than diffusion such as capillarity, gravity, or external forces in gradients of pressure and temperature on the motion of water during drying (Berger and Pei 1973; Chen and Pei 1989). The receding models describe the existence of evaporation front that gradually moves into the interior of the solid body when the falling rate period starts and continues.

The simple theory of water vapour diffusion in porous media under temperature and moisture gradients neglected the interaction of vapour, liquid and solid phases, and the difference between average temperature gradient in the air-filled pores and in the solids as a whole. With these factors taken into account, the receding front theory is developed. The receding evaporation front often appears in the second falling rate drying period and this front divides the material into two regions (zones), the wet region (zone) and the dry (sorption) region (zone). The dry zone is the sorption zone due to the adsorptive nature of moisture retention. In the dry zone, the main mechanism of moisture transfer is vapour flow, and the movement of adsorbed water may also play an important role. In the wet zone, the water is in capillary and the water movement is continuous in laminar flow and follows Darcy law and Hagen–Poiseuille law (Chen and Pei 1989), even though discontinuities in the liquid phase typically do exist in drying of capillary-porous bodies (Chen and Schmidt 1990). In the wet zone, the vapour pressure is determined by Clausius–Clapeyron equation; while in the sorption region no free water exists and the vapour pressure is determined by the sorption isotherms. Evaporation takes place at the front as well as in the whole sorption region. Vapour flows through the sorption region to the surface.

5.2.1.3.2 Mathematical Modelling In the wet zone, capillary movement is considered to be the dominant mechanism, and the moisture movement (liquid water and vapour) and heat transfer can be modelled as:

$$\rho_0 \frac{\partial M_w}{\partial t} = \rho_0 \nabla\left(D_L \nabla M_w\right) - M_{ev} \tag{5.72}$$

$$\frac{\partial\left(\varepsilon_g \rho_v\right)}{\partial t} = \nabla\left(\frac{D_{vw} M_w}{RT} \nabla \mathrm{P}\right) + M_{ev} \tag{5.73}$$

$$\rho C_p \frac{\partial T_w}{\partial t} = \nabla\left(K_{\mathrm{eff}} \nabla T_w\right) + q_w \tag{5.74}$$

where ρ, ρ_0, and ρ_v are density of the material, initial density of the material, and density of vapour (kg/m^3), respectively; M_{ev} is the vapour content (kg/kg); M_w is the moisture content in the wet zone (kg/kg); D_L is the capillary conductivity (m^2/s); D_{vw} is the vapour transfer coefficient in wet zone (m^2/s); ε_g is the porosity occupied by vapour; C_p is the specific heat of the material (J·kg^{-1}·K^{-1}); T_w is the temperature of the wet material (°C); K_{eff} is the effective thermal conductivity (W·m^{-1}·K^{-1}); and Q_w is the liquid water transfer in bulk flow (kg·m^{-2}·s^{-1}).

In the dry or sorption zone, combination of vapour diffusion and bound liquid movement are assumed to predominate, and the main mechanisms of moisture transport are movement of bound water and vapour transfer. In general, the temperature and mass of water in this zone vary with time and are not known until the solution is obtained. Therefore, an additional equation is required to relate the moving velocity of this receding front. The assumption of constant temperature on this front can simplify this calculation even though the temperature might not be constant. After these assumptions, the moisture and heat transfer can be modelled as:

$$\rho_0 \frac{\partial M_{\text{sorb}}}{\partial t} = \rho_0 \nabla \left(D_{\text{sorb}} \nabla M_{\text{sorb}} \right) - M_{ev} \tag{5.75}$$

$$\frac{\partial(\varepsilon \rho_v)}{\partial t} = \nabla \left(\frac{D_{v\text{sorb}} M_w}{R T_{\text{sorb}}} \nabla P \right) + M_{ev} \tag{5.76}$$

$$\rho C_p \frac{\partial T_{\text{sorb}}}{\partial t} = \nabla \left(k_{\text{sorb}} \nabla T_{\text{sorb}} \right) - M_{ev} \Delta h_v + q_{\text{sorb}} \tag{5.77}$$

where M_{sorb}, is the adsorbed water content (kg/kg), D_{sorb} is the bound water conductivity (m^2/s), ε is the porosity of material (fraction of total volume occupied by air and water vapour), $D_{v\text{sorb}}$ is the vapour transfer coefficient in sorb zone (kg·m^{-2}·s^{-1}·Pa^{-1}), T_{sorb} is the temperature of the dry zone (K), h_v is the convective heat transfer coefficient of vapour (W·m^{-2}·K^{-1}), and q_{sorb} is the heat production (heat source) in the dry zone (W/m^3).

Different moisture transfer coefficients in the two zones may exist (Luikov 1975). The vapour transfer coefficient covers the contribution of both convective and diffusive flows. The equations at the surface of the material are:

$$\frac{D_{v\text{sorb}} M_w}{RT} \frac{\partial P}{\partial \acute{r}} + \rho_0 D_{\text{sorb}} \frac{\partial M_{\text{sorb}}}{\partial \acute{r}} = \frac{h_m M_w}{RT} (P_{va} - P) \tag{5.78}$$

$$K_{\text{sorb}} \frac{\partial T_{\text{sorb}}}{\partial \acute{r}} = h_h \left(T_a - T_{\text{sorb}} \right) + \rho_0 D_{\text{sorb}} \frac{\partial M_{\text{sorb}}}{\partial \acute{r}} \Delta h_v \tag{5.79}$$

where h_m is the mass transfer coefficient (kg·m^{-2}·s^{-1}), P_{va} is the vapour pressure in the air (Pa), K_{sorb} is the thermal conductivity in the dry zone (W·m^{-1}·K^{-1}), h_h is the convective heat transfer coefficient (W·m^{-2}·K^{-1}), and T_a is the supplied air temperature (K).

The mass and heat transfer at the moving front (between the dry and wet zones) must fulfil the following conditions:

$$\rho_0 D_L \frac{\partial M_w}{\partial r'} = \rho_0 D_{\text{sorb}} \frac{\partial M_{\text{sorb}}}{\partial r'} + \frac{D_{v\text{sorb}} M_w}{RT} \frac{\partial P}{\partial r'} \tag{5.80}$$

$$k_{\text{eff}} \frac{\partial T_w}{\partial r'} = k_{\text{sorb}} \frac{\partial T_{\text{sorb}}}{\partial r'} + \Delta h_v \frac{D_{v\text{sorb}} M_w}{RT} \frac{\partial P}{\partial r'} \tag{5.81}$$

$$T_w = T_{\text{sorb}}; M_w = M_{\text{sorb}} = M_{ms} \tag{5.82}$$

Different research groups modified the above mentioned equations (Berger and Pei 1973; Luikov 1975; Chen and Pei 1989), but the basic principle of the receding front theory is the same. Luikov (1966a) developed similar models but included the effect of gradient in the total pressure. Researchers (Luikov 1975; Chen and Pei 1989) developed empirical equations to calculate the parameters such as: flux of capillary flow of free water (kg m^{-2} s^{-1}), velocity of flow in capillary (m/s), relationship between the pore saturation and the pore density function, liquid permeability (m^2), capillary conductivity (m^2/s), bound water conductivity (m^2/s), flux of bound water (kg m^{-2} s^{-1}), flux of vapour (kg m^{-2} s^{-1}), and heat and mass transfer coefficients. The liquid water transfer in bulk flow (Q_w, kg m^{-2} s^{-1}) is described by Darcy law:

$$Q_w = -\rho_w \frac{K_a}{\mu} (\nabla P_w - \nabla \psi_w) \tag{5.83}$$

where K_a is the absolute permeability (m^2), and ψ_w is the liquid phase potential (Pa).

By expressing the term ∇P_w as a function of moisture and temperature, the liquid water flux can be written as a combination of three components due to the moisture gradient, temperature gradient, and gravity:

$$Q_w = -\delta_{wm} \nabla M - \delta_{wT} \nabla T + \nabla \psi_w \tag{5.84}$$

$$\delta_{wm} = \rho_w \frac{K_w}{\mu} \left(\frac{\partial P}{\partial M} \right) \tag{5.85}$$

$$\delta_{wT} = \rho_w \frac{K_w}{\mu} \left(\frac{\partial P}{\partial T} \right) \tag{5.86}$$

where δ_{wm} is the isothermal diffusivity of water (m^2/s), δ_{wT} is the thermal diffusivity of water (m^2/s), and K_w is the permeability of liquid (m^2).

The water vapour transfer is described by Fick's first law with the assumption of steady diffusion in a closed system between an evaporation source and a condensation sink.

$$m_v = -\delta_{vm}\nabla M - \delta_{vT}\nabla T \quad (5.87)$$

where M_v is the water vapour concentration (dry basis, %), δ_{vm} is the isothermal diffusivity of vapour (m^2/s), and δ_{vT} is the thermal diffusivity of vapour (m^2/s).

The mass and heat conservation equations can be developed by assuming that gradient of gas pressure and convective energy are negligible as:

$$\frac{\partial M}{\partial t} = \nabla\cdot(\delta_T \nabla T) + \nabla\cdot(\delta_m \nabla M) + \nabla\cdot\left(\frac{Kk_w}{\mu}\nabla\psi_w\right) \quad (5.88)$$

$$\rho C_P \frac{\partial T}{\partial t} = \nabla\cdot(K_T \nabla T) + \Delta h_v \nabla\cdot(\delta_{vm}\nabla M) \quad (5.89)$$

where δ_m is the overall isothermal mass diffusivity ($\delta m = \delta_{wm} + \delta_{vm}$) (m^2/s), δ_T is the overall thermal mass diffusivity ($\delta_T = \delta_{wT} + \delta_{vT}$) (m^2/s), K_T is the thermal conductivity (W·m^{-1}·K^{-1}). The receding front model presents a theoretical prediction of drying of hygroscopic porous materials which can be observed in the drying studies and be reasonably explained from the physical point of view (Berger and Pei 1973; Chen and Pei 1989). A drawback of the receding front model is that the front tracking scheme itself are generally complicated to implement. It is difficult to evaluate the moving evaporation front and the coefficients for heat and mass transfer because experimental methods used to distinguish between liquid and vapour transfer do not exist. In many grain drying situations, the initial moisture content of the material is equal to or less than the maximum sorption moisture content and no constant rate period will appear. In that case, the model of the sorption region only should be used.

5.2.2 Types of Mathematical Models and Assumptions

5.2.2.1 Types of Mathematical Models

Heat and mass transfer simultaneously occur during dehydration and water sorption, hence effect of heat transfer should be taken into account along with mass transfer when developing a drying or wetting model. Modelling of the drying or wetting mechanisms involve making suitable assumptions, developing the heat and mass transfer equations correctly, solving the coupled equations using appropriate method, verifying the developed model, and then calibrating with experimental moisture and temperature profiles.

Compared to the low temperature drying, mathematical modelling for high temperature drying is relatively simple because: (1) high temperature drying of wet material with free water at beginning of drying is usually in the constant rate drying period, and the constant rate period is amenable to a simple equation of wet bulb thermometer; (2) the ambient air condition has a less influence on the high temperature drying than that of the low temperature drying; (3) grain is usually in a thin layer in the high temperature dryer, while the low temperature drying is usually in a deep bed; and (4) high temperature usually have a simple grain drying strategy (just dry the material in a short time period), while the low temperature drying usually involves management strategy requiring a balance between proper drying without grain spoilage and with minimum energy use.

Reviews on the drying mechanism and the developed models can be found in literature (Keey 1980; Sharp 1982; Jayas et al. 1991; Cenkowski et al. 1993). Drying models can be classified into different types by using different drying principles. Based on the equilibrium of the moisture content with the environment, the published models are classified as: non-equilibrium (partial differential equations [PDEs]), equilibrium, or logarithmic types. Based on the solving method of the equations, the models can be classified as analytical and numerical models. Based on implications of assumptions and complexity of the developed models, the models can be classified as: (1) empirical equations for simplified processes; (2) simultaneous heat and mass transfer equations; and (3) coupled heat and mass transfer equations. The last type may involve more than one mechanism of mass transport and thus is more sophisticated than the first two types. Based on the drying process, the models can be classified as: (1) models of characteristic drying curves in different drying stages; (2) distributed parameter models which use coupled heat and mass diffusion equations; and (3) empirical models obtained entirely by simple or multivariable regression methods. Based on the drying mechanism, the models can be classified as: (1) diffusion models based on the diffusional transport of water and/or vapour; (2) receding front models based mostly on capillary transport; and (3) coupled models based on the complete conservation equations.

The available mathematical models have all been derived for one particular type of drying – fixed bed thin layer drying. Such models (used as a basic unit) are modified for different conditions such as deep bed drying, concurrent flow drying, countercurrent flow drying, crossflow drying, spray drying, and microwave drying.

5.2.2.2 Model Assumptions

Many models have been developed for drying based on either lumping the moisture and heat transfers to one average value (lumped model), separating liquid diffusion and vapour diffusion, or separating heat and mass transfer at different locations inside the drying material (distributed model). After applying assumptions, lumped equations can be derived from distributed equations. The levels of model complexity can be increased by considering shrinkage, non-isothermal conditions, and dependence of diffusion coefficient on moisture content. Two major boundary conditions distinguishing these models are moving boundary condition (shrinkage is taken into consideration) and equilibrium at the surface (no external resistance).

Different assumptions have been made in the literature, so different models have been developed. The major differences among these assumptions deal with how the water inside the drying material is transferred. Water can be transferred in liquid or vapour phase through mass flow in capillaries or diffusion. Viscous saturated flow and capillary transport are rare unless grain kernels have lots of free water (much high moisture contents). The viscous saturated flow is often governed by Darcy's law. The capillary transport is driven by the pressure difference and can be described by Darcy's law or Poiseuille law. It is generally assumed that water transfer inside grain kernels is controlled by water molecular diffusion and the driving force is the vapour pressure difference at isothermal conditions (moisture gradients) or water thermal energy difference under temperature gradients. Diffusion due to temperature gradients is usually negligible because the small size of grain kernel (negligible temperature gradient inside a kernel). The most accepted assumption is the diffusion of water from the interior of the material to evaporate from its outside surface regardless of the water phase. The concept of liquid diffusion being the only mechanism for moisture movement is not accepted by researchers, and diffusion through capillaries and/or vapour-phase movement of diffusion in porous materials especially for a single grain kernel are widely accepted.

5.2.3 Mathematical Modelling of Drying and Wetting

5.2.3.1 Theoretical Model

5.2.3.1.1 Mass and Energy on Surface of Drying Materials

Mass balance on the surface of a water drop (or wet material covered by water) is:

$$\frac{dM}{dt} = -h_m f \dot{A}\left(Y_{\text{sat}} - Y_\infty\right) \tag{5.90}$$

where f is parameter, Y_{sat} is the saturation humidity at the droplet surface under the surface temperature (%), and Y_∞ is the surrounding gas humidity (%).

If the surface of the material is not covered by water, the mass balance can be described as:

$$\frac{dM}{dt} = -h_m f \dot{A}\left(Y_s - Y_\infty\right) \tag{5.91}$$

where Y_s is the humidity at the surface of the material at the surface temperature (%).

Chen and Xie (1997) suggested the following relationship:

$$Y_s = \Phi Y_{\text{sat}} \tag{5.92}$$

where ϕ is the fractionality relative to the saturation vapour concentration at the interface.

The fractionality relative to the saturation vapour concentration (Φ) at the interface is ≤1, and can be estimated as (Chen and Xie 1997):

$$\Phi = \exp\left(\frac{\Delta E_v}{RT}\right) \tag{5.93}$$

where E_v is the correction factor in apparent activation energy for drying due to the increasing difficulty of removing water at low water content levels. The correction factor (ΔE_v) is influenced by water moisture content, shrinkage of the material, and decreases to zero when liquid water fully covers the solid and increases to a large value when the water content becomes minimal.

The parameter f and M_R have the following relationship (Langrish et al. 1991):

$$f = M_R^i \tag{5.94}$$

$$M_R = \frac{M_t - M_e}{M_i - M_e} \tag{5.95}$$

where M_e is the equilibrium moisture content at the drying condition (dry basis, %), M_R is the dimensionless moisture ratio, and M_t and M_i are the moisture contents at time t and beginning of the drying, respectively.

For nonconductive materials (such as the moisture movement is hindered by a relatively impermeable outer shell), $i = 1$ in Eq. 5.94. If the drying is controlled by a fraction of the surface that is wetted and the material has a high conductivity, $i = 2/3$. A value of 0.6 for i describes the thorough-circulation drying of thin mats of loose wool fibres (Keey 1991). When $\Phi > M_{cr}$, $i = C$. When $\Phi < M_{cr}$, $f = M_{cr}^{C-D}\varphi^D$, and C and D are fitting parameters (Van Meel 1958).

The general mass balance equation (Eq. 5.90) can be simplified as (Chen and Xie 1997):

$$\frac{dM}{dt} = -h_m \dot{A}\left(Y_{\text{sat}} - Y_\infty\right) \tag{5.96}$$

Energy balance at the surface of the material can be estimated as:

$$MC_p \frac{dT}{dt} = h_h \dot{A}(T_\infty - T) + L_v \frac{dM}{dt} \tag{5.97}$$

where T_∞ is the temperature of bulk drying air (K), and L_v is the latent heat of water vaporization (J/kg).

5.2.3.1.2 Logarithmic Model Logarithmic model assumes that heat energy lost by the air provides solely the latent heat of vaporisation required to dry the material and neglects sensible heating of the material. Based on this assumption, the following equation holds:

$$G_a C_a \frac{\partial T}{\partial z} = \rho_g L_v \frac{\partial M}{\partial t} \tag{5.98}$$

where G_a is the air flow rate ($kg{\cdot}s^{-1}{\cdot}m^{-2}$), C_a is the specific heat capacity of the air ($W{\cdot}kg^{-1}{\cdot}K^{-1}$).

After application of boundary condition and further simplification, Barre et al. (1971) obtained the analogous expression as:

$$M_R = \frac{e^a}{e^a + e^t - 1} \tag{5.99}$$

$$T_R = \frac{T - T_e}{T_0 - T_e} = \frac{e^a}{e^a + e^t - 1} \tag{5.100}$$

where T_e is the temperature at equilibrium (°C), T_0 is the initial temperature of the material (°C), T_R is the dimensionless temperature ratio.

Logarithmic models are only acceptable in low airflow, low temperature, and short drying time application due to the limitation of its assumption. Logarithmic models are rarely used now due to its low accuracy of prediction.

5.2.3.1.3 Evaporation-Condensation Model The basic assumption of the evaporation-condensation model is that the moisture (vapour) is held in the pores (capillaries) and can migrate by creeping along the capillary walls or by successive evaporation and condensation between unconnected liquid columns in capillaries. Two regions (zones) exist: sorption and wet regions. In the sorption region, liquid moisture transfer still exists and is a strong function of free water content (Whitaker and Chou 1983). In the dry regions and when the adsorbed water molecules receive enough energy to break the sorptive bonds, they may leave and migrate until captured at other sites by condensation. The flow can occur via capillaries or through cellular membranes. This results in sorption diffusion and this sorption diffusion is repeated during drying (Bramhall 1979). In the wet regain, this sorption diffusion might occur, but it is usually negligible compared to the capillary flow. In the sorption region, this sorption diffusion is significant and can be described by diffusion theory.

An evaporation-condensation model was proposed by Henry (1948) under the following assumptions: (1) the moisture concentration, water vapour concentration and temperature can be expressed in a linear moisture sorption isotherm equation; (2) diffusion coefficient, over-all heat conductivity, porosity, specific heat, and density are constant (independent of moisture concentration and temperature); (3) the latent heat equals to that of free water; (4) no hysteresis; (5) vapour phase is the main mechanism of moisture diffusion in the pores with continuous network in the solid; and (6) a local equilibrium exists between the solid and pore spaces. The linear relationship is (Henry 1948):

$$M = B + CM_v - \omega T \tag{5.101}$$

where ω is the frequency of precession (radians/s).

The energy and mass balance equations of the evaporation-condensation model are:

$$\frac{\partial T}{\partial t} = D_T \frac{\partial^2 T}{\partial x^2} + \frac{\Delta H_{\text{vap}}}{C_p} \frac{\partial M}{\partial t} \tag{5.102}$$

$$\frac{\partial C_w}{\partial t} = D_{AB} \frac{\partial^2 C_w}{\partial x^2} - \frac{\rho_s (1-\varepsilon)}{\varepsilon} \frac{\partial M}{\partial t} \tag{5.103}$$

where D_{AB} is the diffusivity of the water and vapour (m^2/s), D_T is the thermal diffusivity (m^2/s), and C_W is the water concentration (%).

Henry's model was modified and tested by different research groups (de Vries et al. 1989; Thorvaldsson and Janestad 1999) and found the predicted results agreed reasonably well with the experimental data. However, the first assumption rarely exists in food materials. Therefore, it is rarely used in the literature.

5.2.3.1.4 Luikov's Model Luikov's model considers the combination of vapour and liquid movement by assuming water moves by both water diffusion and capillary flow. Irreversible thermodynamics is used to express the relations between vapour and liquid fluxes in terms of temperature and concentration gradients (Luikov 1966a; Luikov 1966b; Luikov 1975). DeVries

(1958) developed the similar model. Luikov's model assumes the vapour flux can be calculated as:

$$J_v = -\tau_e V_v D_{atm} \frac{P}{P - P_v} \nabla \rho_v \tag{5.104}$$

where J_v is vapour flux (kg·s^{-1}·m^{-2}), τ_e is the tortuosity factor allowing for extra path length, V_v is the volumetric air content of medium (m^3/m^3), and D_{atm} is the molecular diffusivity of water vapor in air (m^2/s).

After introducing the moisture transfer potential, the liquid potential can be calculated as (Luikov 1966a).

$$dP_m = \frac{\partial P_m}{\partial M} dM + \frac{\partial P_m}{\partial T} dT \tag{5.105}$$

where P_m is the moisture transfer potential (Pa/m).

This model indicates that the mass transfer occurs at opposing concentration and temperature gradients. Chen and Johnson (1969) and Hussain et al. (1972) presented analyses of Luikov models. This model has been criticized because the P_m lumps together a number of effects and thus tends to mask the actual physical processes involved (Chen and Pei 1989). The difficulty to determine the tortuosity factor allowing for extra path length (τ_e) limits the application of this model.

Several models that consider moisture movement as a combination of movement of vapour by diffusion and liquid by capillary action have been developed. In these developed models, both terms are expressed as functions of the temperature and moisture content gradients. Different approaches are used to develop these models. One approach is to consider the changes in moisture content of the liquid and vapour phases separately (De Vries 1958; Sharma et al. 2017). Another approach is to assume the liquid continuity within the pores and capillaries (Philip and De Vries 1957; Luikov 1966a,b, 1975). The most important models developed are based on the receding front theory and the Whitaker's models. Phillips (1989) provided a detailed review on the development of these models.

5.2.3.1.5 Whitaker's Model Whitaker (1977) showed that diffusion theories of drying could not describe the complete spectrum of moisture transport mechanisms that occur during the drying of a granular porous medium. In particular, a gas phase convective transport caused by a pressure gradient must exist when the saturation falls below the critical saturation. Not like the receding front theory, Whitaker and Chou (1983) proposed three regions in the drying materials. In region I, the saturation was greater than the irreducible saturation and the liquid phase in capillaries was continuous. In region II, the liquid phase became discontinuous and liquid transport was negligible. In region III, the water vapour pressure was no longer tied to the temperature by the Clausius–Clapeyron equation and pressure gradient resulted in the vapour transport. Whitaker and Chou (1983) concluded that the gas phase momentum equation must be included in any comprehensive theory of drying granular materials.

By using the volume averaging method, the conservation equations of water in the liquid and gas phases, air in the gas phase, and energy are developed. Whitaker and Chou (1983) incorporated almost all mechanisms of heat and mass transfer: liquid flow due to capillary forces, vapour and gas flow due to convection and diffusion, and internal evaporation of moisture and heat transfer by convection, diffusion, and conduction. The equations developed by Whitaker and Chou (1983) are similar to that presented by Luikov (1966a,b).

The conservation equation for water in both liquid and gas phases is (Truscott and Turner 2005):

$$\frac{\partial}{\partial t}\left(\rho M+\rho_v\varepsilon_g\right)+\nabla\cdot\left(\rho_w V_w-\rho D_b\nabla M_{\text{bound}}\right)=\nabla\cdot\left[\rho_{\text{gas}}D_v\nabla\omega_v\right] \quad (5.106)$$

where V_W is the liquid phase velocity (m/s), D_b is the diffusivity of bound liquid (m^2/s), M_{bound} is the moisture content of bound water (kg/kg), ρ_{gas} is the gas density (kg/m^3), D_v is the diffusivity of vapour (m^2/s), and ω_v is the vapour mass fraction (kg/kg).

The conservation equation for air in the gas phase is:

$$\frac{\partial}{\partial t}\left(\rho_{\text{gas}}\varepsilon_g\right)+\nabla\cdot\left(\rho_{\text{gas}}V_g\right)=\nabla\cdot\left[\rho_v D_v\nabla\left(\frac{\rho_{\text{gas}}}{\rho_v}\right)\right] \quad (5.107)$$

where V_g is the gas volume (m^3).

The conservation equation of energy is (Truscott and Turner 2005):

$$\frac{\partial}{\partial t}\left[\rho\left(H_s+MH_w\right)+\varepsilon_g\left(\rho_v H_{va}+\rho_{\text{gas}}H_a\right)-\int_0^{\rho M_b}\Delta H_e d\rho\right]$$
$$+\nabla\cdot\left(\rho_w H_e V_w-H_b\rho D_b\nabla M_b\right)=\nabla\cdot\left[\rho_g\left(H_{va}-H_a\right)D_v\nabla\omega_v+K_{\text{eff}}\nabla T\right] \quad (5.108)$$

where H_s is the enthalpy of the solid (J/mol); H_w is the differential heat of sorption (J/kg), H_{va} is the enthalpy of water vapour (J/mol), H_a is the enthalpy of air (J/mol), H_e is the enthalpy of liquid water (J/mol), and H_b is the enthalpy of the bound water (J/mol).

In the above equations, the velocity of water can be calculated using the equation of motion of the liquid phase and Darcy law as:

$$V_w=-K_{absw}\frac{K_{rw}}{\mu_w}\nabla\psi_w,\quad \nabla\psi_w=\nabla P_w-\rho_w g\nabla\chi \quad (5.109)$$

where K_{absw} is the absolute (intrinsic) liquid permeability (m²), K_{rw} is the relative liquid permeability (decimal), μ_w is the liquid dynamic viscosity (Pa·s), g is the gravitational acceleration constant (9.806 m/s²), and χ is the depth scalar (m).

The velocity of air can be calculated using the equation of motion of the gas phase as:

$$V_g = -\frac{K_{\text{absw}} K_g}{\mu} \nabla P_g \tag{5.110}$$

where V_g is the velocity of the air (m/s), P_g is the pressure of gas (Pa/m), and K_g is the relative permeability tensor of gas (decimal).

At the boundary (the surface of the drying material), the fluxes of mass and heat are described as convective drying by using the boundary layer theory with Stefan correction as:

$$J_w \cdot \hat{n} = m_v = h_m \frac{\rho_{\text{gas}} M_v}{RT} \ln\left(\frac{P_g - P_{v,\infty}}{P_g - P_v}\right) \tag{5.111a}$$

$$J_e \cdot \hat{n} = q + \Delta h_v m_v = h_h \left(T - T_\infty\right) + \Delta h_v h_m \frac{P_g M_v}{RT} \ln\left(\frac{P_g - P_{v,\infty}}{P_g - P_v}\right) \tag{5.111b}$$

where J_e is the flux of heat (W/m²), J_w is the flux of water (kg·s⁻¹·m⁻²), $\hat{n}$ is the outward-pointing normal vector at the boundary surface, and $\rho_{v,\infty}$ is the density of vapour at boundary (kg/m³). The internal pressure on the liquid and gaseous migration is not considered in Whitaker's model. Therefore, these equations cannot handle the drying at boiling conditions. The enthalpy difference between vapour and liquid can account for the latent heat of vaporisation, which enables the heat transfer to couple with mass transfer. Even though Whitaker's model offers a good representation of the physical phenomena occurring in porous media during drying, the difficulty in determining its complicated transport coefficients is obvious. Numerical solutions have been studied by different research groups (Schrader and Litchfield 1992; Ketelaars et al. 1995; Li and Kobayashi 2005; Srikiatden and Roberts 2005) and the Whitaker's model is usually consistent with the experimental data (Vu and Tsotsas 2018).

Different research groups also simplified the Whitaker's model by assuming constant gas pressure and negligible gas momentum and quasi-steady state (Whitaker and Chou 1983). Nasrallah and Perre (1988) added the effect of the bound water into the Whitaker's model. Based on Whitaker's model, Perre (1997) developed a heterogeneous drying model for wood. The effect of the gaseous pressure at boiling conditions was considered by Nasrallah and Perre (1988). The variation of the material properties such as capillary pressure and absolute permeability were taken into account. Based on 2D model developed by Perre and Turner (2002), Truscott and Turner (2005) developed a 3D heterogeneous drying model for wood.

5.2.3.2 Semi-Theoretical and Empirical Model

5.2.3.2.1 Development of Semi-Theoretical and Empirical Model Theoretical (analytical) solution of Eq. 5.64 exists for constant or variable D_{eff} for materials with simplified geometry such as sphere, infinite slab, and cylinder. Many of these solutions are given by Crank (1970). For the variable diffusion coefficients, Crank (1970) provided the solution only for a small interval over which the drying rate does not change. For variable diffusivities with irregular shapes, researchers use the numerical modelling method or semi-theoretical and empirical models.

The semi-theoretical models are generally developed by simplifying the Fick's second law or from Newton's law of cooling by using the drying or rewetting curve of the materials in a thin layer. To model this drying or rewetting process, assumptions to simplify this process (hence the developed model) have been made. The assumptions include diffusion in the one direction with a constant diffusivity, the same moisture at the surface of the materials and the air, negligible temperature gradient inside the drying materials, and that the main restriction of the drying or rewetting is the water diffusion inside the grain kernel. After these simplifications, integration of Eq. 5.64 gives the following solution for different shapes of the material:

$$M_R = \frac{M_t - M_e}{M_{cr} - M_e} = \sum_{n=1}^{\infty} \beta_n \exp\left(-\eta^2 F_0\right) \tag{5.112}$$

$$F_0 = \frac{D_{\text{eff}}}{X_\Theta^2} t \tag{5.113}$$

where M_{cr} is the critical moisture content (dry basis, %), β_n and η are the factors dependent on the geometry of a body and the Biot number, F_0 is the Fourier number, and X_θ is the semi-thickness of the drying material (m).

The critical moisture content is usually considered as the value beyond that free water on the material surface is removed by surrounding gas (unhindered process) or where a crust may be formed which starts to retard water removal. The initial water content is used to replace M_{cr} by Langrish et al. (1991) because there was no constant rate period in their experiments on drying solids. This approach provides a rescaling parameter and was successfully used by many researchers (Sharp 1982; Jayas et al. 1991; Kucuk et al. 2014).

The general form of Eq. 5.112 can be written for each special shape (Ece and Cihan 1993):

$$\text{Infinite slab: } M_R = \frac{8}{\pi^2}\sum_{i=1}^{\infty}\frac{1}{(2i+1)^2}\exp\left(-\frac{(2i+1)^2\pi^2}{4X_\theta^2}D_{\text{eff}}t\right) \tag{5.114a}$$

$$\text{Cylinder: } M_R = \frac{8}{X_\theta^2}\sum_{i=1}^{\infty}\sum_{j=1}^{\infty}\frac{1}{\alpha_i^2\beta_j^2}e^{\left[-(\alpha_i^2+\beta_j^2)D_{\text{eff}}t/X_\theta^2\right]} \tag{5.114b}$$

$$\text{Infinite cylinder: } M_R = \sum_{i=1}^{\infty}\frac{4}{\alpha_i^2}e^{\left[-\alpha_i^2 D_{\text{eff}}t/X_\theta^2\right]} \tag{5.114c}$$

$$\text{Sphere: } M_R = \frac{6}{\pi^2}\sum_{i=1}^{\infty}\frac{1}{i^2}\exp\left(-\frac{i^2\pi^2}{X_\theta^2}D_{\text{eff}}t\right) \tag{5.114d}$$

α_i (i = 1, 2, ...) is the roots of the Bessel function of the first kind and zero order ($\alpha_1 = 2.408$), and β_j is defined as:

$$\beta_j = \frac{(2j-1)X_\theta\pi}{2L_h},\ j = 1, 2...... \tag{5.115}$$

where L_h is the half length of material (m).

α_i (i = 1, 2, ...) has been tabulated by Abramowitz and Stegun (1972), and for a small i, the α_i can be calculated as:

$$\alpha_i = \frac{\alpha}{4}+\frac{1}{2\alpha}-\frac{31}{2\alpha^3}+\frac{3779}{30\alpha^5},\quad \text{and}\quad \alpha = \pi(4i-1) \tag{5.116}$$

The first three terms of the series of Eq. 5.114 are:

Infinite slab:

$$M_R = \frac{8}{\pi^2}\left[\frac{1}{9}\exp\left(-9\left(\frac{\pi}{2}\right)^2 N_{Fi}\right)+\frac{1}{25}\exp\left(-25\left(\frac{\pi}{2}\right)^2 N_{Fi}\right)+\frac{1}{49}\exp\left(-49\left(\frac{\pi}{2}\right)^2 N_{Fi}\right)\right] \tag{5.117a}$$

where N_{Fi} is the Fick number ($=D_{\text{eff}}t/X_\theta^2$).

Cylinder:

$$M_R = \frac{L_h^2}{X_\theta^4}\left[0.5676\exp\left(-\left(5.7180+2.4649\frac{X_\theta^2}{L_h^2}\right)N_{Fi}\right)\right.$$
$$+0.0119\exp\left(-\left(30.4297+22.1841\frac{X_\theta^2}{L_h^2}\right)N_{Fi}\right)$$
$$\left.+0.0017\exp\left(-\left(74.8070+61.6265\frac{X_\theta^2}{L_h^2}\right)N_{Fi}\right)\right] \quad (5.117b)$$

Infinite cylinder:

$$M_R = 0.6996\exp(-5.7180N_{Fi})+0.1315\exp(-30.4297N_{Fi})$$
$$+0.0535\exp(-74.8070N_{Fi}) \quad (5.117c)$$

Sphere:

$$M_R = 0.6085\exp(-9.8596N_{Fi})+0.1521\exp(-39.4384N_{Fi})$$
$$+0.0676\exp(-88.7364N_{Fi}) \quad (5.117d)$$

For calculation of the D_{eff} by using Eq. 5.114a with an error of 1%, $i = 20$. When $i = 1$ and 5, the error is 19% and 4%; respectively (Efremov and Kudra 2004, 2005). For a sphere and $i = 1$, 15, and 58, the error is 39%, 4%, and 1%, respectively.

Equation 5.112 could also be solved for prolate and oblate spheroid, cardioid, hexagon, corrugated, and epitrochoid shapes (Kahveci and Cihan 2008). For a long drying time where the Fick number is greater than 0.1 and $M_R < 0.6$, the first term of the series dominates. When $i = 1$, Eq. 5.112 can be simplified as:

$$M_R = A_1\exp\left(\frac{G_c}{A_2}D_{\text{eff}}t\right) \quad (5.118)$$

where G_c, A_1, and A_2 are the geometric constants.

The A_1 value of infinite slab, sphere, 3-dimensional finite slab, and infinite cylinder should be close to $\frac{8}{\pi^2}, \frac{6}{\pi^2}, \left(\frac{8}{\pi^2}\right)^3$, and 1, respectively. The geometric constant A_1 is usually determined by using the drying kinetic data. The geometric constant $G_c = \pi^2$ for slab or sphere, and $=\alpha_i^2$ for cylinder shape. $A_2 = 4X_\theta^2$ for one dimensional slab and sphere. For the 3-dimensional finite slab, $A_2 = \frac{1}{L_{hx}^2 + L_{hy}^2 + L_{hz}^2}$. For multidimensional geometrics such as 3-dimensional slab, the Newman's rule can be applied.

The rate of change of the mean moisture content with time becomes:

$$\frac{d(M_t - M_e)}{dt} = -\frac{G_c}{A_2} D_{\text{eff}}(M_t - M_e) \tag{5.119}$$

Which corresponds to a linear falling-rate period of drying. Therefore, for a material with only falling rate of drying, this simplification is acceptable. This equation also indicates that the surface moisture content will fall progressively rather than instantaneously to the equilibrium moisture content. For a first-order rate process, it will be an exponential decline:

$$M_t = M_0\left[1 - \exp\frac{-\beta_j t}{X_\theta}\right] \tag{5.120}$$

Which corresponds to the boundary condition:

$$-D_{\text{eff}}\left(\frac{\partial M}{\partial X_\theta}\right)_{X_\theta} = h_m(M_{\text{suf}} - M_e) \tag{5.121}$$

where M_{suf} is the moisture at the surface of the material (kg/kg).

The effective mass transfer coefficient (h_m) is based on the assumption that the rate of moisture loss is directly proportional to the excess moisture content above equilibrium.

The parameter A_1, A_2, and G_c are influenced by the shape and size of the drying material. For a infinite slab, Eq. 5.114 can be further simplified as (Efremov and Kudra 2005):

$$M_R = \frac{8}{\pi^2}\exp\left(-\pi^2 F_0\right) \tag{5.122a}$$

The above simplified equation indicated that $M_R = 0.81$ at $t = 0$. Therefore, this simplified equation can only be used for approximate calculations, and it is more suitable for a long drying time. After this simplification and $\frac{8}{\pi^2} \approx 1$, then

$$M_R = \exp\left(-\pi^2 F_0\right) \tag{5.122b}$$

This simplification gives significant deviations of experiment results from the model predictions in a regular regime ($F_0 > 0.04$), reaching 23.4%

for a plate and 64.5% for a sphere (Efremov and Kudra 2005). Therefore, the following equation was proposed by Efremov and Kudra (2005):

$$M_R = \exp\left(-\pi^2 F_0^a\right) \tag{5.123}$$

where a is a correction factor, determined from a condition of minimum deviation from exact solution given by Fick's law. For a sphere and plane sheet, $a = 0.83$ and 0.91, respectively, with the maximum relative deviations of ±17% and ±12% (Efremov and Kudra 2005).

5.2.3.2.2 Semi-Theoretical and Empirical Equations If the loss of the accuracy is acceptable, Eq. 5.114 is further simplified to different equations such as Lewis (Newton), two-term, and Page models (Table 5.1). For example, Lewis model, which is analogous to the Newton's law of cooling, is developed based on the diffusion theory and assumes that the resistance to diffusion occurs mainly in the thin outer layer of the drying material. This model assumes the rate of drying is directly proportional to the difference between the moisture content of the material and its equilibrium moisture content.

$$\frac{dM}{dt} = -K\left(M - M_e\right) \tag{5.124}$$

Lewis model have been modified by different researchers by replacing the K with different coefficients.

$$\int_{M_0}^{M_t} \frac{dM}{\left(M - M_e\right)} = -k\int_0^t dt \tag{5.125}$$

$$\ln\left(\frac{M_t - M_i}{\left(M - M_e\right)}\right) = -kt \tag{5.126}$$

$$M_R = \exp\left(-kt\right) \tag{5.127}$$

The K values can be found by simple or multivariable regression with the measured drying curves. More than one hundred empirical or semi-theoretical equations (Iglesias and Chirife 1982; Wolf et al. 1985; Jian and Jayas 2018b) have been published in literature. Erbay and Icier (2010) and Ertekin and Firat (2017) provided reviews. Since there is no universal semi-theoretical or empirical model that can generalize the drying behaviour of all products, researchers usually try several semi-theoretical and empirical models to evaluate their goodness of fit and then choose the best one to represent the drying kinetics. The widely used empirical and semi-theoretical equations are: two-term, Henderson and Pabis, Lewis, Page, and modified Page (Kucuk et al. 2014) (Table 5.2). Most of the equations are mathematically

Table 5.2 Semi-Theoretical and Empirical Models of Thin Layer Drying, Wetting, and Soaking

Model	*Equation*	*Most Application*	*Source*[a]
Henderson and Pabis	$M_R = ae^{-kt}$	Drying, wetting, and soaking	Henderson and Pabis (1961)
Kaleemullah and Kailappan	$M_R = e^{-k_1 T} + ct^{(k_2 T + d)}$	Drying	Kaleemullah and Kailappan (2006)
Modified Henderson and Pabis	$M_R = ae^{-k_1 t} + ce^{-k_2 t} + de^{-k_3 t}$	Drying, wetting, and soaking	Karathanos (1999)
Lewis (Newton)	$M_R = e^{-kt}$	Drying, wetting, and soaking	Lewis (1921)
Logarithmic	$M_R = ae^{-kt} + c$	Drying, wetting, and soaking	Chandra and Singh (1994)
Page	$M_R = e^{-kt^i}$	Drying, wetting, and soaking	Page (1949)
Modified page	$M_R = e^{(-kt)^i}$	Drying, wetting, and soaking	Overhults et al. (1973)
Thompson Model	$t = a\ln(M_R) + c(\ln(M_R))^2$	Drying	Thompson et al. (1968)
Two-term exponential	$M_R = ae^{-kt} + (1-a)e^{-kct}$	Drying, wetting, and soaking	Sharaf-Eldeen et al. (1980)
Two-term model (binominal)	$M_R = ae^{-k_1 t} + ce^{-k_2 t}$	Drying, wetting, and soaking	Henderson (1974)
Jian et al.	$M_R = \frac{a}{T + 273.15} e^{-kTt} + c$	Soaking	Jian et al. (2017a)
Midilli and Kucuk	$M_R = ae^{-kt^i} + ct$	Drying, wetting, and soaking	Midilli et al. (2002)
Peleg	$M_t - M_0 = \frac{t}{k_1 + k_2 t}$	Soaking	Peleg (1988)
Sigmoidal	$M_t = \frac{M_e}{1 + e^{-k(t-\tau)}}$	Soaking	(Miano and Augusto 2015)
Modified Two-Term Exponential Models	$M_R = ae^{-k_1 t} + (1-a)e^{-k_2 t}$	Drying, wetting, and soaking	Verma et al. (1985)
Wang Singh	$M_r = 1 + at + ct^2$	Drying and wetting	Wang and Singh (1978)
Weibull	$M_R = e^{-(t/a)^c}$	Soaking	Cunha et al. (1998)

[a] Original source of the published model.

the same equation. For example, Henderson and Pabis, Modified Henderson and Pabis, Lewis (Newton), Logarithmic, Page, Modified page, Two-term model (binominal), Modified Binomial, and Midilli and Kucuk equations can be written in the following format:

$$M_R = ae^{-k_1 t} + ce^{-k_2 t} + de^{-k_3 t} + Bt + \mathrm{C} \tag{5.128}$$

5.2.3.2.3 Disadvantages and Advantages of Empirical Models The disadvantages are:

1. Model is obtained by mass balancing only, without considering the effect of the solid temperature change. The empirical models developed at low temperatures usually cannot be used at high temperatures;
2. Models are only valid within the process condition applied without revealing any physical insights of drying;
3. Need to determine critical water content, or drying stages to allow more accurate curve fitting;
4. Parameters in the models, which are influenced by ambient temperatures and moistures, have no physical meaning and are not fully understood and predictable, although they may describe the drying curve well for an experimental condition; and
5. This modelling concept generally cannot deal with the change in process between drying and wetting, and hysteresis due to the above mentioned disadvantages.

Even though the empirical and semi-theoretical models have the above mentioned disadvantages, these models are practical for use because these are simplified enough to describe the drying curve by requiring less data, so that these can be used to: (1) simulate the drying process; (2) conduct model comparisons; (3) design new drying systems; (4) select optimum drying conditions; and (5) characterize the drying system and time.

5.2.4 Models of Soaking (Wetting)

Soaking (wetting) of grain kernels is generally assumed to occur by water diffusion caused by the moisture gradient between the surface and core of the kernel. Therefore, water sorption is a special case of wetting and mainly influenced by the water diffusivity of the material. Fick's law of diffusion can be used to model the kinetic process of water sorption inside grain kernels (Jian et al. 2017a,b). The solution of this water diffusion depends on whether the diffusion coefficient is assumed as a constant or a variable and the boundary conditions considered. Equation 5.114 can be used to model the soaking process if the quasi isothermal conditions for a regular geometric material are assumed. Therefore, some of the developed semi-theoretical and empirical models for drying can also be used for soaking (Table 5.2).

Shafaei and Masoumi (2014) compared 14 published empirical and semi-theoretical models and found all 14 models were appropriate and Binomial model had a higher prediction accuracy. Peleg model is the best model to describe the hydration kinetics of chickpea splits (Prasad et al. 2010). Jian et al. (2017a) compared seven published models and found all the published models had the similar prediction accuracy. Jian et al. (2017a) also found their developed model, which considered the effect of temperature, was the best equation to fit the water sorption data of kidney beans.

Numerical simulation techniques allow modelling of the complex non-linear phenomena occurring during water sorption (such as the non-linear moving boundary of a grain kernel). Most developed numerical models are finite element models (Bakalis et al. 2009) and a few are finite difference models (Bello et al. 2010). Jian et al. (2017a) developed a finite element model and assumed (1) the kernels had different swelling ratios in different directions and at different soaking times; (2) effective diffusivity was the same in any direction, but different at different time; and (3) water movement inside a kernel was mainly controlled by diffusion, and bulk movement of water (capillary flow) inside a kernel was negligible. Jian et al. (2017a) found the trend of the effective water diffusivity calculated by the semi-theoretical model was inconsistent with the water sorption of the beans, while there was no significant difference between the measured and predicted moisture contents by the finite element model.

5.3 Models of Deep Bed Drying and Aeration

5.3.1 Relationship of Mathematical Modelling between Thin Layer and Deep Bed Drying

Deep bed drying is to convectively dry bulk materials much thicker than that of a thin layer. Natural air drying or aeration of grain in storage bins or hay in bales are examples of deep bed drying. Deep bed drying can be assumed as the drying of multiple thin layers in one dimension because the existing air condition of a layer will be the entering air conditions of the above layer along the airflow direction. Gradients of temperature and moisture content exist in the deep bed drying, while by the definition there are no gradients of temperature and moisture content within a thin layer. Therefore, any of the developed mathematical models describing thin layer drying could be theoretically used to simulate the deep bed drying by assuming that each layer is a thin layer after the deep bed is artificially separated into multiple layers. However, this approach is not successful because:

1. Deep bed drying is a low temperature and low airflow drying process and much longer drying time required than the thin layer drying. Thin layer drying models might not require all of the balance equations of mass, heat, heat transfer, and drying rate. Deep bed drying model which assumes either one of the balance equations is negligible could result in a low prediction accuracy because small error in a short time period could result an unreasonable prediction of the drying in a long time deep bed drying.
2. The air will equilibrate with grain at a higher temperature in much shorter time than that at a lower temperature. Low temperature drying is mostly conducted by using ambient air which is influenced by weather. Low temperature drying model needs to consider rewetting, hence hysteresis of the grain, because the weather might not always be at drying condition and grain might experience several cycles of water sorption and desorption. However, it is usually not considered in the deep bed drying for simplification the developed models.
3. Low temperature drying model needs to consider whether the grain equilibrates with the air in the time interval because this assumption influences the energy and mass balance and rate of drying, while it is usually not needed to consider these equilibriums for the high temperature drying because energy and mass balance in the high temperature models could be ignored due to the reason explained in 1.
4. Thin layer drying equations usually assume no gradient of temperature and moisture content in the drying bed, while these gradients must be considered in the model development of deep bed drying. Therefore, the thin layer models cannot properly describe the transfer phenomena in a convective deep bed dryer because of changing conditions experienced by thin layers of a deep bed.

Due to the above reasons, the thin layer drying equations (as the only one equation) without modification usually cannot describe the deep bed drying process except at some extreme conditions such as high drying temperatures and airflow rates. Yang and Milota (2013) used a one parameter Newton model to estimate the drying condition of deep bed of wood dried at 50–200°C and 0.3–0.9 m/s airflow rate. They found drying times at depths up to 23 cm were within ±4% of the experimental results. Morey and Li (1984) used the Page model to simulate the corn drying in a deep bed with drying temperature of 93°C, and found prediction of temperature profiles had a low accuracy, particularly near the top of the bed.

5.3.2 Types of Deep Bed Drying Models

Deep bed drying models are arbitrarily categorized as non-equilibrium, equilibrium, and logarithmic models. More than 20 deep bed drying models were published in literature and several research groups reviewed these developed models (Sharp 1982; Parry 1985; Cenkowski et al. 1993; Parde et al. 2003). Therefore, this book does not intend to present an exhaustive list of any of these models, but focuses on the introduction of fundamental theories used to develop these models. The reader can refer to these publications for more detailed information.

5.3.2.1 Non-Equilibrium Models

5.3.2.1.1 Mathematical Models Non-equilibrium model assumes no heat and mass equilibrium between the drying air and grain throughout the deep bed. Therefore, four PDEs are required to describe the moisture (mass) balance, heat balance (energy balance of air), heat transfer (energy balance of grain), and drying rate in the direction (one dimension) of air movement. Sharp (1982) summarised these four PDEs in the elemental bed (x, $x+\partial x$) over a time interval (t, $t+\partial t$) as follows:

Moisture (mass) balance:

Change in moisture in air across element = moisture leaving the grain – change in moisture in air within element.

$$G_a\,\partial t\left[Y_{\text{abs}}(x+\partial x,t)-Y_{\text{abs}}(x,t)\right]$$
$$=-\frac{\partial M}{\partial t}\partial t\rho\,\partial x-\varepsilon\rho_{\text{gas}}\,\partial x\left[Y_{\text{abs}}(x,t+\partial t)-Y_{\text{abs}}(x,t)\right] \quad (5.129)$$

where Y_{abs} is the absolute humidity of air (kg/kg dry air), and x is the transverse distance by the concentration gradient or layer thickness (m).

If $\partial t\to 0$, $\partial x\to 0$, then

$$G_a\frac{\partial Y_{\text{abs}}}{\partial x}=-\rho\frac{\partial M}{\partial t}-\varepsilon\rho_{\text{gas}}\frac{\partial Y_{\text{abs}}}{\partial t} \quad (5.130)$$

Heat balance (energy balance of air):

Change in enthalpy of air across element = Energy to increase temperature of moisture lost by grain to air temperature + Energy to heat grain by convective heat transfer + Energy to change enthalpy of air within element

After rearranging,

$$G_a \partial t\left(C_a + C_v Y_{\text{abs}}\right)\left[T_a(x+\partial x,t) - T_a(x,t)\right]$$
$$= \frac{\partial M}{\partial t}\partial t \rho \partial x C_v\left(T_a - T_g\right) - h_a\left(T_a - T_g\right)\partial t \partial x \quad (5.131)$$
$$- \rho_{\text{gas}}\varepsilon \partial x\left(C_a + C_v Y_{\text{abs}}\right)\left[T_a(x,t+\partial t) - T_a(x,t)\right]$$

where C_v is the specific heat of water vapour ($J \cdot kg^{-1} \cdot K^{-1}$), and T_g is the grain temperature (K).

If $\partial t \to 0$, $\partial x \to 0$, then

$$G_a\left(C_a + C_v Y_{\text{abs}}\right)\frac{\partial T_a}{\partial x} = \rho_{\text{gas}} C_v\left(T_a - T_g\right)\frac{\partial M}{\partial t} - h_a\left(T_a - T_g\right) - \rho_{\text{gas}}\varepsilon\left(C_a + C_v Y_{\text{abs}}\right)\frac{\partial T_a}{\partial t} \quad (5.132)$$

Energy balance of grain (heat transfer equation):

Change in enthalpy of grain across element = Energy to heat grain by convective heat transfer – Energy lost by grain for evaporation of moisture

$$\rho \partial x\left(C_g + C_{ws} M\right)\left[T_g(x,t+\partial t) - T_g(x,t)\right] = h_a\left(T_a - T_g\right)\partial t \partial x + \frac{\partial M}{\partial t}\partial t \rho \partial x h_d \quad (5.133)$$

where C_{ws} is the specific heat of water ($J \cdot kg^{-1} \cdot K^{-1}$), and h_d is the heat of desorption of water (J/kg).

If $\partial t \to 0$, $\partial x \to 0$, then

$$\rho\left(C_g + C_{ws} M\right)\frac{\partial T_g}{\partial t} = h_a\left(T_a - T_g\right) + h_d \frac{\partial M}{\partial t}\rho \quad (5.134)$$

Drying rate equation:

Most researchers (Sharp 1982; Cenkowski et al. 1993) used the drying rate equation of thin layer drying of the materials such as:

$$\frac{\partial M}{\partial t} = -K\left(M_i - M_e\right) \quad (5.135)$$

where K is the drying rate constant (s^{-1}).

5.3.2.1.2 Modification of the Non-Equilibrium Model The non-equilibrium models could be modified by using one of the following methods:

1. Lots of non-equilibrium models are developed by modifying the four PDEs because one PDE may generate many algebraic operational grain drying models depending on the method of solution and assumptions. The assumptions consider the following factors: shrinkage of grain, temperature gradients within the individual

kernels of grain, thermal properties of the grain, conduction heat and mass transfer between kernels, condensation within the deep bed, airflow and grain type, changes in air temperature and specific humidity over lime, heat transfer to and from bin wall, heat capacity of moist air and grain, flow direction of the grain and air, and equilibrium isotherm. Most of the early published models (Spencer 1969; Brooker et al. 1974; Parry 1983) assumed some of these factors were negligible. Some researchers assumed that $\frac{\partial T_g}{\partial T}$ and $\frac{\partial Y_{\text{abs}}}{\partial T}$ were small and negligible (Bakker-Arkema et al. 1967; Spencer 1969). Some research groups (Spencer 1969; Ingram 1976; Sharp 1982) neglected accumulation terms in the energy and mass balance equations. Sun et al. (1995) considered the rewetting and condensation effects. In their models, the energy accumulation terms in the air energy balance equation were conserved. Wang (1993) found significant differences between the predictions of the models which include those terms and those do not. He also remarked the neglecting of $\frac{\partial T_g}{\partial T}$ and $\frac{\partial Y_{\text{abs}}}{\partial T}$ terms was not fully justified. Cenkowski and Sokhansanj (1988) modelled the continuous crossflow drying of the grain using equations of heat and moisture balances for grain and air.

2. Different researchers used different drying rate equations because different materials have different drying rate even at the same drying conditions. Some used diffusion equation to describe the drying rate and rewetting was allowed (Hamdy and Barre 1970). Liu et al. (2015) considered the water potential difference between the outside and inside of grain kernels in the drying rate equation, so vacuum deep bed drying could be modelled. Ingram (1976) introduced an intra-grain diffusion equation to represent the drying rate.
3. Different researchers used different auxiliary equations and correlations to characterise various quantities appearing in the model equations (Parry 1983; Cenkowski and Sokhansanj 1988; Liu et al. 2015). These quantities include drying rate constant, heat of desorption of water, grain bed heat transfer coefficient, equilibrium moisture content, and condensation calculation.
4. Istadi and Sitompul (2002) extended the non-equilibrium models by adding momentum equations, which were modelled using Navier–Stoke momentum equations. The mass equations and energy equations were also expanded by considering the balance in air and grain, separately. The model also considered the moisture transfer inside grain kernels and between a grain kernel and convected air. The interphase transfer was modified by using empirical equations. Therefore, Istadi and Sitompul (2002) model

is the most comprehensive model developed so far, however, there is no other researcher using this model because of its complexity and extensive computation.

5.3.2.1.3 Solution of the Non-Equilibrium Model The four PDEs must be solved simultaneously. Therefore, numerical technique such as finite difference method, finite element method, or finite volume method should be used. Zare et al. (2012) used dimensionless analysis of the Buckingham theory to simplify the four PDEs to one general equation and they claimed the model was capable of predicting the paddy moisture content with good accuracy. Messai et al. (2014) developed a 2D and 1D non-equilibrium models and found 2D model was more realistic.

The four PDEs must be solved layer by layer, and the out leaving air will be the entering air of the above layer along the airflow direction. Although non-equilibrium models are likely to be more accurate than other model types, and are able to cope with a wider range of grain and air conditions, the solution of the equations is computationally intensive. Therefore, it is not widely used for software development, while widely used by researchers to characterize the fundamental drying principles.

5.3.2.2 Equilibrium Models

5.3.2.2.1 Mathematical Model Equilibrium models assume near equilibrium conditions between the drying air and the grain in each layer during each time interval. Therefore, the equilibrium models are a simplification of the non-equilibrium models even though the equilibrium models were not directly developed from the non-equilibrium models. Under the equilibrium condition, the drying rate equation is not needed, and the energy balance of both air and grain could be calculated in one equation. Therefore, the equilibrium models can be simplified into two equations: mass and energy balance equations.

Energy balance in each layer:

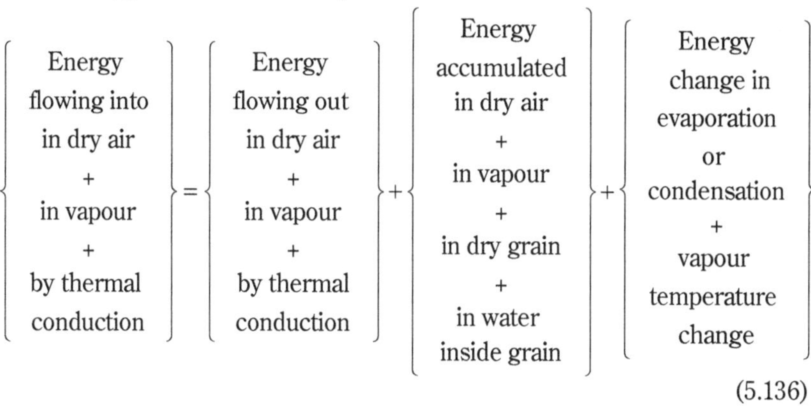

$$\left\{\begin{array}{c}\text{Energy}\\\text{flowing into}\\\text{in dry air}\\+\\\text{in vapour}\\+\\\text{by thermal}\\\text{conduction}\end{array}\right\}=\left\{\begin{array}{c}\text{Energy}\\\text{flowing out}\\\text{in dry air}\\+\\\text{in vapour}\\+\\\text{by thermal}\\\text{conduction}\end{array}\right\}+\left\{\begin{array}{c}\text{Energy}\\\text{accumulated}\\\text{in dry air}\\+\\\text{in vapour}\\+\\\text{in dry grain}\\+\\\text{in water}\\\text{inside grain}\end{array}\right\}+\left\{\begin{array}{c}\text{Energy}\\\text{change in}\\\text{evaporation}\\\text{or}\\\text{condensation}\\+\\\text{vapour}\\\text{temperature}\\\text{change}\end{array}\right\} \quad (5.136)$$

Jian et al. (2019) further simplified this energy conservation equation by assuming that energy transfer by thermal conduction in the airflow direction was negligible and there was only heat transfer in the normal direction to the airflow at the boundary of a grain column (bin walls). After reorganization, the energy balance equation is:

$$
\begin{aligned}
&C_aT_{a,en}+Y_{en}C_vT_{a,en}+\frac{\varepsilon xC_a}{\Delta tu_a}T_{a,en}+\frac{\varepsilon xC_v}{\Delta tu_a}(Y_{en}-Y_{out})T_{a,en}+\alpha C_gT_{g,i}\\
&+\alpha(M_f-M_i)C_{ws}T_{g,i}+(Y_{en}-Y_{out})H_{\text{vap}}+\left(Y_{out}-Y_{en}\right)C_vT_{a,en}\\
&+2\pi x(T_{amb}-T_{g,i})/\tau_{therm}=C_aT_{a,out}+Y_{out}C_vT_{a,out}+\frac{\varepsilon xC_a}{\Delta tu_a}T_{a,out}\\
&+\frac{\varepsilon xC_V}{\Delta tu_a}(Y_{en}-Y_{out})T_{a,out}+\alpha C_gT_{a,out}+\left(Y_{out}-Y_{en}\right)C_vT_{a,out}
\end{aligned}
\tag{5.136}
$$

where $T_{a,\,en}$ and $T_{a,out}$ are the entering and leaving out air temperature, respectively (°C); Y_{en} and Y_{out} are the absolute humidity of entering and leaving air (kg water/kg dry air), respectively; u_a is the apparent or superficial air velocity (m/s); M_f is the final grain moisture content (wet basis, decimal); $T_{g,i}$ is the initial grain temperature (°C); T_{amb} is the ambient temperature (°C); τ_{therm} is the total thermal resistance (m·°C·W^{-1}); and C_g is the specific heat of grain (J·kg^{-1}·K^{-1}).

$$
\alpha=\frac{\rho x}{\rho_{\text{gas}}u_a\Delta t}
\tag{5.137}
$$

Jian et al. (2019) suggested that the following equation can be used if the hysteresis status was reached:

$$
\begin{aligned}
&C_aT_{a,en}+Y_{en}C_vT_{a,en}+\frac{\varepsilon xC_a}{\Delta tu_a}T_{a,en}+\frac{\varepsilon xC_v}{\Delta tu_a}(Y_{en}-Y_{out})T_{a,en}+\alpha C_gT_{g,i}\\
&+(Y_{en}-Y_{out})H_{\text{vap}}+\left(Y_{out}-Y_{en}\right)C_vT_{a,en}+2\pi x(T_{amb}-T_{g,i})/\\
&\tau_{therm}=C_aT_{a,out}+Y_{out}C_vT_{a,out}+\frac{\varepsilon xC_a}{\Delta tu_a}T_{a,out}+\frac{\varepsilon xC_V}{\Delta tu_a}(Y_{en}-Y_{out})T_{a,out}\\
&+\alpha C_gT_{a,out}+\left(Y_{out}-Y_{en}\right)C_vT_{a,out}
\end{aligned}
\tag{5.138}
$$

Equations 5.136 and 5.137 could be used to calculate the temperature of the out leaving air ($T_{a,out}$) at desorption or sorption condition, and Eq. 5.138 was used to calculate the temperature of out leaving air at a hysteresis condition.

Water mass balance in each layer in each time interval:

$$\left\{\begin{array}{c}\text{Mass flowing into}\\-\\\text{Mass flowing out}\end{array}\right\}=\left\{\begin{array}{c}\text{Mass changed in the grain}\\+\\\text{Mass changed in the air}\end{array}\right\}$$

$$\Delta t\dot{A}\rho_{\text{gas}}u_aY_{en}-\Delta t\dot{A}\rho_{\text{gas}}u_aY_{out}=V_L\rho\left(M_f-M_i\right)/100+\rho_{\text{gas}}\varepsilon V_L\left(Y_{out}-Y_{en}\right) \tag{5.139}$$

where V_L is the layer volume (m^3).

This equation was simplified as:

$$Y_{en}-Y_{out}=\alpha\left(M_f-M_i\right)/100+\frac{\varepsilon x\left(Y_{out}-Y_i\right)}{u_a\Delta t} \tag{5.140}$$

5.3.2.2.2 Modification of the Equilibrium Model Most developed models made different assumptions based on the assumption of equilibrium or near equilibrium condition. Most models assume the final equilibrium temperature in each layer would fall between the initial grain and air temperatures after each time interval (Thompson 1972; Jian et al. 2019). Pierce and Thompson (1980) modified Thompson (1972) model by incorporating a complete set of thin-layer drying equations to cover a wide range of temperatures, so the drying with a high airflow rate could be simulated.

Jian et al. (2019) considered the hysteresis, sorption, and rewetting in their developed models. They calculated the equilibrium RH (ERH) of the grain by using the sorption and desorption isotherms, and the grain temperature and moisture content which was determined by Eqs. 5.136, 5.138, and 5.140. The calculated ERH was used to calculate vapour pressure of the water inside grain kernels at desorption ($P_{g,de}$) and sorption ($P_{g,so}$) conditions, separately. This calculated vapour pressure was compared with the vapour pressure ($P_{a,en}$) of the entering air. If $P_{a,en} > P_{g,so}$, the grain was at sorption condition. If $P_{a,en} < P_{g,de}$, the grain was at desorption condition. If $P_{a,en} \geq P_{g,de}$ and $P_{a,en} \leq P_{g,so}$, the grain was at hysteresis status. Incorporation of this hysteresis effect significantly increased the model prediction accuracy (Jian et al. 2019).

5.3.2.2.3 Solution of Equilibrium Models Mass and energy conservation equations must be solved simultaneously in each layer. The calculation is processed layer by layer, and the layer thickness influences the prediction accuracy because α value is influenced by the layer thickness. The computational time of the equilibrium model is much less than that of the non-equilibrium models because Eqs. 5.136, 5.138, and 5.140 can be directly calculated after each parameter in the equations is determined. The simulation time of a grain bin with less than 10 m of grain depth is less than 3 min by using a PC with 2.7 GHz CPU (unpublished data).

The most used models are the equilibrium or near equilibrium models because the assumption of equilibrium condition between grain and air simplifies the calculation and also has been shown to provide a reasonable prediction of moisture content in a deep bed for low temperature drying with natural air or aeration. For high airflow, dynamic weather, and shallow bed, the equilibrium assumption might have a low prediction accuracy. However, the equilibrium model is mostly used to develop commercial software due to its low computation intensity.

5.3.2.3 Logarithmic Models

Even though logarithmic models were developed before the development of the non-equilibrium models (Parry 1985), the Logarithmic model is another simplification of the non-equilibrium models. It assumes a unique relationship connecting $\frac{\partial M}{\partial t}$ and $\frac{\partial T_a}{\partial x}$. After this assumption, the model can be developed by combining the heat balance (Eq. 5.132) and heat transfer equations (Eq. 5.134) and eliminating $(T_a - T_g)$ and $\frac{\partial T_g}{\partial T_a}$.

$$G_a\left(C_a + C_v Y_{\text{abs}}\right)\frac{\partial T_a}{\partial x} = -\rho\left(C_g + C_{ws}M\right)\frac{\partial T_g}{\partial t} + \rho\left[h_d + C_v\left(T_a - T_g\right)\right]\frac{\partial M}{\partial t} \tag{5.141}$$

Logarithmic model also assumes grain density is constant and the following three sensible heats are negligible: grain temperature increase, temperature increase of the moisture removed from the grain, and temperature increase of the air due to the increase of the air relative humidity. These assumptions give:

$$G_a C_a \frac{\partial T_a}{\partial x} = \rho h_d \frac{\partial M}{\partial T_a} \tag{5.142}$$

Logarithmic model further assumes the travel rate of the drying zone (Q_t) may be related to airflow rate as:

$$C_a G_a \frac{\partial T_a}{\partial x} = -Q_t h_d \frac{\partial M}{\partial x} \tag{5.143}$$

$$Q_t = \left(\frac{T_{a0} - T_{wb}}{M_i - M_e}\right)\frac{C_a G_a}{h_d} \tag{5.144}$$

where Q_t is the travel rate of the drying zone (kg·m^{-2}·s^{-1}), T_{a0} is the initial air temperature (K or °C), and T_{wb} is the wet bulb air temperature (K).

Analytical solution of Eqs. 5.142, 5.143, and 5.144 together with a thin-layer drying rate (Eq. 5.135) yields the expression of the logarithmic model.

$$\frac{M-M_e}{M_i-M_e}=\frac{e^D}{e^D+e^{T_a}-1} \tag{5.145}$$

$$D=\frac{K\rho h_d\left(M_i-M_e\right)_x}{Q_t C_a\left(T_{a0}-T_{wb}\right)} \tag{5.146}$$

Even though the logarithmic models are modified by several researchers (Parry 1985), the low prediction accuracy of the logarithmic models still exists (Sharp 1982) due to its several unrealistic assumptions. It is occasionally used due to its simplicity.

5.3.3 Numerical Modelling and CFD Analysis

5.3.3.1 Numerical Modelling and CFD

Numerical methods use one of the following techniques to solve PDEs governing the mass and heat transfer during drying: finite element (Alagusundaram et al. 1990; Montross et al. 2002; Jian et al. 2013; Jian et al. 2015), finite difference (Thompson et al. 1968; Thompson 1972; Muir et al. 1980; Metzger and Muir 1983), finite volume (Khankari et al. 1995; Lawrence and Maier 2011; Rocha et al. 2013), and finite discrete (Tsory et al. 2013; Haddad et al. 2014; Rusinek and Kobyłka 2014) methods. In the simulation of drying, the PDEs can be any PDEs characterizing drying processes. These numerical techniques have been extensively used to simulate almost all of the topics discussed in this chapter. The above listed references provide the models developed for entire stored grain bins. There are extensive review papers related to different topics dealing with drying and aeration (Luikov 1975; Briun and Luyben 1980; Sharp 1982; Parry 1985; Jayas et al. 1991; Claussen et al. 2007; Kucuk et al. 2014; Esther et al. 2016; Vu and Tsotsas 2018). Many mathematical models have been developed to simulate thin layer drying in different dryer types and different drying methods (Cenkowski and Sokhansanj 1988; Ramachandran et al. 2017), deep bed in-bin drying (Cenkowski et al. 1993; Lawrence et al. 2012), and soaking (Jian et al. 2017b).

Due to the heavy calculation involved in numerical modelling, computer is required to simulate the changes in the drying medium (grain seed or bulk) and supplied fluid (air for drying or liquid for soaking), as well as the interactions of fluid with surfaces defined by boundary conditions and inside the drying medium (among or inside grain kernels). With the advance of high-speed supercomputers and comprehensive set of PDEs governing heat, mass and momentum transfers in porous media, most complex problems can be simulated.

To develop a numerical model, the tedious work is to transfer the PDEs into a system of linear equations and computer codes for both the transformation and solution. This requires the researcher to have significant knowledge of the drying theory, flow dynamics and heat transfer, mass transfer, momentum transfer, mathematical modelling, computer program coding, and method of solution, i.e., finite element, finite difference, or finite volume. This needs many years of training and tedious study. To focus on the development of drying theories, avoid the tedious computer program coding, and minimize the time of leaning numerical methods, many researchers now use off-the-shelf computer software to help with this modelling process. There is a trend now towards using software of Computational Fluid Dynamics (CFD) analysis to replace the classical method of numerical analyses of drying processes.

Computational Fluid Dynamics uses numerical methods to analyze and solve problems that involve fluid motion and heat and mass transfer over a multi-dimensional domain. CFD is rooted in classical fluid mechanics and designed to use advanced numerical methods for solving PDEs of mass, momentum and energy conservation in fluid flow and transfer of heat and mass. CFD software was original developed by Richardson (1922) to solve the Navier–Stokes equations numerically. Since then, this approach had been further advanced by different researchers (Thom 1933) especially during 1950s–1970s (Harlow 1956; Spalding 1960; Fromm and Harlow 1963; Harlow and Welch 1965; Gentry et al. 1966; Patankar and Spalding 1966; Hess and Smith 1967; Murman and Cole 1970; Patankar and Spalding 1972; Elgobashi and Spalding 1977). Along with the advancement of computer capacity, this method and software have been gradually enriched and used by researchers in different fields. At the beginning of the 1980s, the solution of the first 2D and later also 3D Euler equations became feasible. Initially, CFD was almost exclusively used in the fields of aerospace and mechanical industries to simulate the processes in combustion chambers of rocket engines; around rocket airframe and supersonic aircrafts. Subsequently its application was extended to chemical, civil and environmental engineering (Karpinska and Bridgeman 2016). Malin (2020) provided an extensive review on the history of the CFD development. Recent development of commercial software makes CFD modelling available for grain storage engineers. Their computational power for solving mathematical equations has been proven to be an effective tool.

CFD is now widely used in many fields including aerodynamics and aerospace analysis, biological engineering, engine and combustion analysis, fluid flows and heat transfer, industrial system design and analysis, natural sciences and environmental engineering, and weather simulation. Several review papers have been published since 2000 (Xia and Sun 2002;

Anandharamakrishnan 2003; Norton et al. 2013; Malekjani and Jafari 2018). Karpinska and Bridgeman (2016) provided an extensive review on models of numerical simulation of the turbulent flow with CFD codes. There are books specially discussing CFD (Ferziger and Peric 2002; Verboven et al. 2004; Norton and Sun 2007). Therefore, this section only introduces the basic principles of CFD and its application to grain drying.

5.3.3.2 Governing Equations of CFD

CFD provides the numerical solution of the Navier-Stokes equations. The Navier–Stokes equations mathematically express conservation of momentum, mass and energy for Newtonian and incompressible fluids. The continuity equation under conservation of mass is:

$$\frac{\partial \rho}{\partial t} + \nabla(\rho u_a) = 0 \tag{5.147a}$$

Equation 5.147a states that the mass flows entering a drying solid or system must balance exactly with the mass leaving the solid or system. When density of the compressed fluid does not change, Eq. 5.147a can be simplified as:

$$\nabla(u_a) = 0 \tag{5.147b}$$

The momentum equation (Cauchy momentum equation) in convective form is:

$$\rho\frac{\partial u_a}{\partial t} = -\nabla \mathrm{P} + \nabla \cdot \tau_\sigma + \rho g \tag{5.147c}$$

where τ_σ is the deviatoric stress tensor, which has order 2.

Equation 5.147c states that the sum of the external forces acting on a fluid particle is equal to its rate of change of linear momentum. The Euler equation for fluid motion in the x direction (the fluid is incompressible) is given by:

$$\rho\frac{\partial u_a}{\partial t} + \nabla \cdot (\rho u_x \bar{u}) = -\nabla P + \rho g \tag{5.147d}$$

where u_x is the air velocity in x coordinate (m/s), $\bar{u}$ is the velocity vector in all (x, y, z) directions (m/s).

When applying Newton's second law to fluid motion of viscous flow, this fluid motion can be modified by introducing viscous transport into the equation.

$$\rho\frac{\partial u_a}{\partial t} + \nabla \cdot (\rho u_x \bar{u}) = -\nabla P + \mu\nabla^2\bar{u} + \rho g \tag{5.147e}$$

The first, second, and third terms on the right side are the surface force, molecular dependent momentum exchange (diffusion), and the mass force (gravitational force), respectively. This modified equation is termed as the Navier–Stokes equation for Newtonian fluids. This Navier–Stokes equation forms the basis for CFD.

The conservation of energy as the first law of thermodynamics states that there is an equality between the energy changing rate of a fluid element and added heat or work done on it. This energy flow is governed by the driving force of temperature gradients. The energy balance equation in the x direction can be written as:

$$\frac{\partial}{\partial t}(\rho C_a T_k)+\frac{\partial}{\partial x}(\rho u_x C_a T_k)-\frac{\partial}{\partial x}\left(K_T\frac{\partial T_k}{\partial x}\right)=S_T \tag{5.147f}$$

where S_T is the thermal sink or source (W/m^3). The third term on the left side is the heat flux (diffusion) and S_T is the source term for heat addition or removal (transfer of mechanical energy to heat). Fourier equation can be used to determine heat exchange in an isotropic solid as follows (Norton et al. 2013):

$$\frac{\partial}{\partial t}(\rho C_a T_k)-\frac{\partial}{\partial x}\left(K_T\frac{\partial T_k}{\partial x}\right)=S_T \tag{5.147g}$$

Fick's law of moisture diffusion (Eq. 5.63) can also be solved in drying process with the combination of the mass and momentum conservations (Eqs. 5.147a–5.147g). Different equations can be used to model the boundary conditions such as convection, conduction, and radiation (Norton et al. 2013). All these equations can be used in the CFD to simultaneously simulate the temperature and moisture of the drying materials.

Equation 5.147 relates pressure, temperature and density of drying solid and fluid. When density variation in the drying medium is considered, one of the following three methods is usually used: (1) Boussinesq approximation (the density differentials in the flow are only considered in the momentum equation) (Ghani et al. 1991); (2) ideal gas approximation (deriving density difference from the ideal gas equation) (Ramachandran et al. 2018), and (3) high temperature differential (Ferziger and Peric 2002). After simplification with different assumptions, Eq. 5.147 is represented by different mathematical formations such as Eulerian–Eulerian or two-fluid/two-phase models (interpenetrating continuum) and Eulerian–Lagrangian or discrete particle models. Jamaleddine and Ray (2010) reported that almost all equations used in fluid flow were derived from the Reynolds transport theory. The viscosity, another important fluid property, is usually quantified via the mathematical correlation between the rate of deformation of the fluid and its shear stress (Norton and Sun 2007). Turbulence models can be solved by using Reynolds averaged Navier-Stokes (RANS) model, the Eddy Viscosity model, Large eddy simulation (LES), and hybrid RANS-LES (Jamaleddine and Ray 2010). Malekjani and Jafari (2018) reviewed these methods.

The differences between Navier–Stokes equations and Euler equations are that Navier–Stokes equations take viscosity into account while the Euler equations deal only with inviscid flow. Current commercial software such as FLUENT, a CFD software package from ANSYS, Inc., can solve both equations based on different assumptions. CFD codes developed by different companies can efficiently solve the PDEs governing all fluid flow, heat transfer (such as radiation), and many other physical phenomena. Even though CFD software has the capability of using finite element and finite difference methods, the most used method is finite volume method (Defraeye 2014) because of less calculation time and the same prediction accuracy as the finite element method.

5.3.3.3 Procedure of CFD Model Development

The procedure of CFD model development is the same as the numerical modelling process, which includes mesh discretizing, mathematical formations, applying the boundary conditions, computer program coding, conducting simulation, verifying and calibrating with published literature and experimental data, validating the developed model with experimental data, and applying the validated model. The only difference between using CFD and developing a numerical model is that the researcher will not need to code the mathematical formations into computer codes and debug the codes if CFD software is used. Therefore, the researcher only needs to know how to use the CFD software and do not need to code the finite element, finite difference, or finite volume models. The researcher still need the knowledge to select the appropriate commercial CFD package, choose and simplify suitable models (such as laminar or turbulence model, Eulerian–Eulerian or two-fluid/two-phase models, and Eulerian–Lagrangian or discrete particle models), input suitable parameters, simplify the geometry of problem domains, select empirical correlations, determine the degree of details expected from the model, and understand the physics to be solved in the fluid and solid domains. Selection of these parameters will not only influence the prediction accuracy, but also save time because CFD modelling usually needs extreme computational time (one simulation might need a good computer to run for days or weeks). There are several review publication on this topic (Ramachandran et al. 2018).

5.3.3.4 Advantages and Disadvantages of CFD

The advantages of using CFD are the same as using numerical modelling which are outlined below:

1. The heat and mass flow can be quantified throughout space and time. Simulations can be performed with the actual scale of the dryer, which may not be feasible with the experimental measurements.

2. There is no limitation for the operating conditions (e.g., limitation of measurements).
3. Lower cost than most experimental measurements and less limitation of model development. For example, to determine effects of only single physical factor on grain temperature inside stored grain bins, such as bin size, could require at least 10 bins ranging diameters from 4 to 30 m and grain height from 3 to 30 m. Such an experimental approach is not practical or economically feasible. Using CFD will only need one experienced person for less than one week.
4. Numerical models have the capability to give a deeper insight of the drying process with a 3D configuration which is not possible with most drying experiments. For example, the models without using numerical techniques are only capable to describe the effective/average drying characteristics of materials in most cases. With the use of numerical techniques, the resolution of the information on the temporal and spatial variation in transport phenomena and material properties during drying could be improved significantly. CFD have the versatility of application in any solid subjected to any operating condition of the dryer. A reliable CFD model could simulate the drying processes at any level of details, if variables are properly selected, mathematical equations are appropriately formulated, and boundary, initial conditions, empirical correlations are well defined (Ramachandran et al. 2018), and computer running time is allowed.

Even though advantages of using CFD software are obvious, it has the following disadvantages:

1. Computational cost and running time constrain detailed calculations for complex models and geometries using numerical simulation, especially CFD software. Compared with numerical models coded by individual researchers, CFD simulation might use more computation time because the researcher can easily modify his/her code by simplifying the assumptions in simulation without too much compromising its accuracy, but not possible to do this easily when CFD software is used. When CFD software is used, user must write different codes and compile with the CFD software to modify the CFD software. Lots of data are copied and pasted between these codes. More computation time is required.
2. Complex flow simulations are challenging and error-prone, and it takes a lot of engineering expertise to obtain realistic solutions. It is easy to make mistakes when the user does not know the code of the CFD.

3. It is difficult sometimes to explain the main mechanisms and kinetic of drying when complex models (such as using the PDEs provided by CFD) are applied. Compared with CFD models, the simplified models can directly explain the main mechanisms and kinetic of drying at the same accuracy level as the CFD simulation. The physics behind drying mechanisms is rich and complex and not entirely captured by CFD methods due to their reliance on experimental observations and correlated equations. For example, changes of drying from constant to falling drying rate require that equations governing the drying kinetics outside the porous material should be changed to equations governing the drying kinetics inside the porous material. Even though using conjugate approach can partially solve this problem, but this approach has limitations such as low computational efficiency, not available for all situations, explicit implementation, additional modelling effort or computational cost, and not always high accuracy (Defraeye 2014).
4. The accuracy to predict the drying of irregularly shaped materials and turbulent flow is still questioned by researchers when CFD is used (Defraeye 2014; Ramachandran et al. 2018). CFD equations should be modified and customised by approximation, thus lacks robustness. It is a challenge to develop a new model by modifying the CFD code. There are two reasons for this challenge. The first reason is that most researchers using CFD might have not enough time to learn the coding of computer language (such as C or C++) and fully understand the numerical method. For such researchers, modifying CFD code will be difficult obviously. The second reason is that CFD software is a complex package which has more than thousand pages of computing code. Lots of modification should be based on the fully understanding of the CFD code. Compared with coding the entire model by researcher himself, modifying part of the CFD software without fully understanding CFD code increases difficulty. Even though lots of CFD software increases their flexibility allowing users use different equations, different packages, and adding codes, some modification still have certain difficulty. For example, modelling shrinkage (especially non-free shrinkage) and internal pore development during drying by using CFD will result in much more calculation time and have a low accuracy if the user want to reduce the total computer running time (Defraeye 2014) because it is extremely time consuming when fine mesh is used for the shrinking solid. Jian and Jayas (2018a) developed a finite element model and considered the shrinkage factor during thin layer drying of red kidney beans. The simulation required less than 30 min. The same approach by using CFD is difficult (unpublished data).

Researchers usually add codes to expand CFD application, update some parameters, or customise its application (Defraeye 2014; Ranjbaran et al. 2014). Thorpe (2008) and Ranjbaran et al. (2014) updated the calculated grain moisture content after each time step. This updating might use lots of computing time because the newly developed program needs to read the calculation result from the CFD software, and then input it back into the CFD software after updating. This process is usually needed for every time step. Therefore, even though the CFD software can provide complex models, these models should be simplified in order to fit the drying conditions and characterize the drying mechanisms, kinetics and dynamics, and reduce running time. These simplified models should be validated because no model becomes usable before experimental validations, also, suitable models should be selected, verified, and optimized.

5.3.3.5 Application of CFD in Grain Drying

A wide variation of commercial CFD software is available today and different CFD codes provided by different companies have different features and capabilities (Xia and Sun 2002; Jamaleddine and Ray 2010; Norton et al. 2013; Ramachandran et al. 2018). Advantages and disadvantages of these CFD codes are available in review papers (Xia and Sun 2002; Jamaleddine and Ray 2010; Ramachandran et al. 2018). These software have been used to simulate grain drying and storage (Thorpe 2008; Ranjbaran et al. 2014).

Even though food is one of the main materials studied by CFD modelling (Defraeye 2014), there are a few studies of grain drying by using CFD software. Thorpe (2008) demonstrated CFD can be used to conduct aeration simulation after the PDE and the software were customised. Liu et al. (2016) used CFD to simulate aeration for cooling paddy rice in a Warehouse. Ranjbaran et al. (2014) simulated the deep bed paddy drying by using CFD software. Panigrahi et al. (2020) simulated the velocity and airflow resistance in a peaked stored grain bin with an aeration ducting system. All these simulations gave reasonable accuracy (at the similar level as the published numerical models). However, using CFD to simulate grain drying and aeration did not receive a wide attention because: (1) extra codes must be developed and compiled with CFD software (Thorpe 2008; Ranjbaran et al. 2014); (2) Eq. 5.147 must be modified to cope with the special requirement of grain drying and aeration; (3) these modification significantly increase the computation time; (4) days (even months) are required to complete a simulation of deep bed grain drying or aeration and 3D models are usually needed, this requires extreme computer running time which discourages researchers; and (5) there are difficulties to modify CFD codes (refer Section 3.3.4).

Challenges of using CFD for simulating grain aeration and drying include: sensible to latent heat (or vice versa) conversion due to grain moisture changes in the process of aeration and drying (Ranjbaran et al. 2014; Liu et al. 2016), differences of airflow resistance in horizontal and vertical directions (Jayas et al. 1987; Liu et al. 2016), some physical property (such as specific heat) must be input as constant (Thorpe 2008); and CFD does not specifically account for the moisture vapour in the interstitial air that is transported around grain kernels (Thorpe 2008).

Liu et al. (2016) simulated aeration by using CFD and found the simulated temperatures agreed with the measured values, with relative differences less than 6.3%. In their CFD model, the PDE included moisture transfer equations, heat transfer equations, momentum transfer equations, equations for boundary conditions, properties of ambient air and paddy rice. To decrease the computer running time, they assumed the constant physical properties of the paddy rice. They claimed that without this assumption, some governing equations would be implicit and too difficult to solve. They tested different time steps of simulation and found 6 s was suitable. Defraeye (2014) also concluded that the propagation of the variability (or uncertainty) on the material properties into the numerical simulation results was also a key aspect to be addressed because it could have a large impact on the outcome of drying calculation. Jian et al. (2019) developed an aeration model based on energy and mass conservation without considering the momentum. The prediction accuracy of Jian et al. (2019) model is slightly higher than Liu et al. (2016) model. The simulation time step in Jian et al. (2019) model is 2 h. Therefore, the computer run time of Jian et al. (2019) model is very short (less than 30 s when aeration time is less than 30 h). Only using 2 s step time will increase the running time by at least 3600-fold for 30 h simulation (unpublished data). Jian et al. (2019) also considered the different physical properties of canola and hysteresis of the canola due to moisture and temperature change during aeration, and it is difficult to complete this target by using any of current CFD software. Therefore, more studies should be conducted to develop simplified model and customise the CFD model (Panigrahi et al. 2020).

5.4 Remarks

The vast majority of the studies have been devoted to the drying of different materials, such as determination of effective water diffusivity, and drying rate of materials under different drying conditions. Despite these extensive studies, the heat and mass transfer on the surface of the material is not fully understood. For example, the mass and heat transfer during drying will be in opposite directions. It is not fully understood how these opposite transfers influence or promote each other. Most developed models only assumed the equilibrium condition at the surface of the material between the material and supplied air. However, Jian and Jayas (2018a) found the Lewis number was 27 during the thin layer drying of red kidney beans. It is not known whether there are temperature and RH discontinuity at the surface of materials. The velocity distribution of the vapour and air at the surface of the material is not known. Therefore, a more fundamental model understanding of these phenomena is required. Understanding these phenomena will not only help to develop robust models, but also provide theoretical basis for the development of new drying models.

It is usually assumed that practical aspects of drying and wetting theories are more important than the cognition of theories because hypothesis must be tested by validation, and predicted values of the hypothesis (presented in modelling format) must be close to the measured values. Unfortunately, validation of the developed model is costly and might be a challenging work sometimes because it is difficult to measure some parameters. For example, water diffusivity is widely used in the mathematical modelling of grain drying and wetting. However, water diffusivity cannot be directly measured. Under this condition, mathematical models are developed with a few indirectly measured parameters. The models are usually validated by comparing the measured temperatures and moistures of materials. After this validation, it is still not known whether the parameters are correctly estimated or not. This might be one of the reasons that many theories (presented in modelling format) are developed. It will be critical that the parameters are correctly estimated, and this correct estimation will depend on the correct understanding of the drying theories and making assumptions. Therefore, the function of the mathematical modelling should be both cognitive and practical. To advance the grain drying, researchers should focus more on the cognition of theories sometimes.

Even though technical limitations of the current experimental technique and devices are one of the reasons explaining the above-mentioned issues, the few and scattered studies might be another reason. Extensive simplification of the heat and mass transfer process might also be one of the reasons. The models developed based on the simplification of the drying process and neglected the heat and mass transfer completely, might not only lead to wrong predictions but also might lead to misinterpretation of experimental results in some cases. Although all of the studies described in this chapter provide valuable insight into the problem of the evaporation of water into supplied air and have led to achieving significant advances in the field, further theoretical and experimental studies are needed to understand these findings. These findings can help in the dryer design to minimize the energy and capital cost while maintaining high product quality.

Nomenclature

α: parameter (or effective frequency of collision);

α_i (i = 1, 2, ...): roots of the Bessel function of the first kind and zero order;

a_{mL}: surface concentration of Langmuir-type primary adsorption sites (mol/g);

β: parameter (rate of dehydration constant, or Boltzmann's constant $1.38064852 \times 10^{-23}$ J/K);

β_n: factor dependent on the geometry of a body and the Biot number;

Υ: constant or ratio of the saturated vapour pressure of the interstitial water to the saturated vapour pressure of pure liquid water at the same temperature;

$f(r)$: pore volume density function;

λ: constant;

θ: fractional occupancy of adsorption sites (fraction of surface already covered by a layer of bound moisture);

φ: fraction of surface covered by two or more layers of molecules;

P_o: Boltzmann's constant ($1.38064852 \times 10^{-23}$ J/K);

ρ: material density (kg/m^3);

ρ_g: apparent (bulk) density of grain kernel (kg/m^3);

ρ_{gas}: gas density (kg/m^3);

ρ_L: density of the moisture liquid (kg/m^3);

ρ_s: mass concentration of solid (kg/m^3);

ρ_v: density of vapour (kg/m^3);

$\rho_{v,\infty}$: density of vapour at boundary (kg/m^3);

ρ_w: water density (kg/m^3);

ρ_0: initial density of the material (kg/m^3);

μ: dynamic viscosity (Pa·s);

μ_w: liquid dynamic viscosity (Pa·s);

τ: tortuosity of capillary paths;

τ_σ: deviatoric stress tensor, which has order 2;

τ_e: tortuosity factor allowing for extra path length;

ε: porosity of material (fraction of total volume occupied by air and water vapour);

ε_g: porosity occupied by vapour;

ϕ: fractionality relative to the saturation vapour concentration at the interface;

ψ_w: liquid phase potential (Pa);

Ψ_w: gravity potential (Pa/m);

δ_m: overall isothermal mass diffusivity ($\delta_m = \delta_{wm} + \delta_{vm}$) (m^2/s);

δ_T: overall thermal mass diffusivity ($\delta_T = \delta_{wT} + \delta_{vT}$) (m^2/s);

δ_{vm}: isothermal diffusivity of vapour (m^2/s);

δ_{vT}: thermal diffusivity of vapour (m^2/s);
δ_{wm}: isothermal diffusivity of water (m^2/s);
δ_{wT}: thermal diffusivity of water (m^2/s);
ω: frequency of precession (rad/s);
ω_v: vapour mass fraction (kg/kg);
χ: depth scalar (m);
η: factor dependent on the geometry of a body and the Biot number;
τ_{therm}: total thermal resistance ($m \cdot {}^{\circ}C \cdot W^{-1}$);
σ: surface tension (J/m^2);
A: constant related to surface area;
$\dot{A}$: surface area (m^2);
A_1: geometric constant;
A_2: geometric constant;
a: constant;
a_{prim}: value of adsorption on primary sites;
a_w: water activity (decimal);
a_{wj}: water activity of the binary mixture with the j component (decimal);
a_{wmax}: maximum value of water activity to which the material has been subjected (decimal);
a_1, a_2: constant;
B: constant;
b: Langmuir constant (Pa) or constant;
C, c: constants;
C_a: specific heat capacity of the air ($J \cdot kg^{-1} \cdot K^{-1}$);
C_g: specific heat of grain ($J \cdot kg^{-1} \cdot K^{-1}$);
C_p: specific heat of the material ($J \cdot kg^{-1} \cdot K^{-1}$);
C_v: specific heat of water vapour ($J \cdot kg^{-1} \cdot K^{-1}$);
C_W: water concentration (%);
C_{ws}: specific heat of water ($J \cdot kg^{-1} \cdot K^{-1}$);
C_0: constant (or pre-exponential factor);
D, d: constants;
D': effective vapour space diffusion coefficient (m^2/s);
D_{AB}: diffusivity of the water and vapour (m^2/s);
D_{atm}: molecular diffusivity of water vapour in air (m^2/s);
D_b: diffusivity of bound liquid (m^2/s);
D_{eff}: effective diffusivity (m^2/s);
D_f: diffusion coefficient (m^2/s);
D_L: capillary conductivity (m^2/s);
D_{sorb}: bound water conductivity (m^2/s);
D_T: thermal diffusivity (m^2/s);
D_v: diffusivity of vapour (m^2/s);

D_{vsorb}: vapour transfer coefficient in sorb zone ($kg \cdot m^{-2} \cdot s^{-1} \cdot Pa^{-1}$);
D_{vw}: vapour transfer coefficient in wet zone (m^2/s);
D_x, D_y, D_z: diffusion coefficient in x, y, and z directions, respectively (m^2/s);
E_v: correction factor in apparent activation energy for drying due to the increasing difficulty of removing water at low water content levels;
F_0: Fourier number;
f: parameter;
G_a: air flow rate ($kg \cdot s^{-1} \cdot m^{-2}$);
G_c: geometric constant;
g: gravitational acceleration constant (9.806 m/s^2);
H, H': parameters;
ΔH: enthalpy change (J/mol);
H_a: the enthalpy of air (J/mol);
H_b: enthalpy of the bound water (J/mol);
H_d: molar heat of desorption (J/kg);
H_e: enthalpy of liquid water (J/mol);
H_k: heat of condensation of water vapour (J/mol);
H_l: heat of condensation of pure water (J/mol);
$H_{s,mono}$: total heat of adsorption of mono layer (J/mol);
H_{va}: enthalpy of water vapour (J/mol);
H_{vap}: latent heat of vaporization (J/mol, J/Kg);
H_w: differential heat of sorption (J/kg);
h: dimensionless constant;
h_a: grain bed heat transfer coefficient ($J \cdot m^{-3} \cdot K^{-1}$);
h_d: heat of desorption of water (J/kg);
h_h: convective heat transfer coefficient ($W \cdot m^{-2} \cdot K^{-1}$);
h_m: mass transfer coefficient ($kg \cdot m^{-2} \cdot s^{-1}$);
H_s: enthalpy of the solid (J/mol);
h_v: convective heat transfer coefficient of vapour ($W \cdot m^{-2} \cdot K^{-1}$);
i: integer number;
J: flux ($kg \cdot m \cdot s^{-1}$) or density ($kg \cdot s^{-1} \cdot m^{-2}$) of moisture flow rate;
J_e: flux of heat (W/m^2);
J_l: liquid flux ($kg \cdot s^{-1} \cdot m^{-2}$);
J_v: vapour flux ($kg \cdot s^{-1} \cdot m^{-2}$);
J_w: flux of water ($kg \cdot s^{-1} \cdot m^{-2}$);
j: integer number;
K, k, k': constant;
K_a: absolute permeability (m^2);
K_{absw}: absolute (intrinsic) liquid permeability (m^2);
K_b: adsorption equilibrium constants;
K_c: capillary (liquid) conductivity (m^2/s);
K_{eff}: effective thermal conductivity ($W \cdot m^{-1} \cdot K^{-1}$);

K_{eq}: equilibrium constant;
K_g: relative permeability tensor of gas (decimal);
K_H: solubility coefficient of the Henry's law;
K_h: permeability (m^2);
K_L: Langmuir adsorption equilibrium constant (Pa);
K_r: relative permeability of porous material (decimal);
K_{rw}: relative liquid permeabilities (decimal);
K_s: single phase permeability of porous material (m^2);
K_{sorb}: thermal conductivity in the dry zone ($W{\cdot}m^{-1}{\cdot}K^{-1}$);
K_T: thermal conductivity ($W{\cdot}m^{-1}{\cdot}K^{-1}$);
K_w: permeability of liquid (m^2);
K_0, K_1,
K_2, K_3, K_4: constant;
L: number of layers;
L_h: half length of material (m);
L_{hx}, L_{hy},
L_{hz}: half length of material in x, y, and z spatial coordinates, respectively (m);
L_v: latent heat of water vaporization (J/kg);
M: moisture content of solids (kg/kg of the solid);
M_a: molecular weight of water;
M_b: maximum water adsorption at secondary sites;
M_{bound}: moisture content of bound water (kg/kg);
M_{cr}: critical moisture content (dry basis, %);
M_{de}: equilibrium moisture content in desorption process (dry basis, %);
M_e: equilibrium moisture content at the drying condition (dry basis, %);
M_{ev}: vapour content (kg/kg);
M_f: final grain moisture content (wet basis, decimal);
M_i: initial moisture content (dry basis, %);
M_L: Langmuir capacity constant;
M_m: monolayer moisture content (kg/kg);
M_{max}: maximum mass of the adsorbent (kg);
m_{mp}: maximum sorption on primary sites (%);
M_{ms}: maximum sorptive moisture content (kg/kg);
$M_{0.5}$: moisture content at $a_w = 0.5$ (kg/kg);
M_1, M_2,: molecular weight of solvent 1, and 2, respectively;
M_R: dimensionless moisture ratio;
M_s: an average value of the molecular weight of a polar unit;
M_{sob}: equilibrium moisture content in sorption process (dry basis, %);
$M_{sorb,}$: adsorbed water content (kg/kg);
M_{suf}: moisture at the surface of the material (kg/kg);
M_t: moisture content (dry basis, %) at time t;
M_v: water vapour concentration (dry basis, %);

M_w: moisture content in the wet zone (kg/kg);
m: mass of the adsorbent or mass of the material (kg);
m_{max}: maximum mass of adsorbent (kg);
m_{ev}: evaporation rate in the wet zone (kg/s);
m_v: transferred water vapour (kg/s);
N: mean number of water molecules per cluster;
N_A: Avogadro's number (6.023×10^{23});
N_{Fi}: Fick number ($D_{eff} t / X_{\Theta}^2$);
N_o: number of anhydrous groups;
N_1: number of monohydrate groups;
n: number of water molecules in the solid solution;
$\hat{n}$: outward-pointing normal vector at the boundary surface;
n_1, n_2: constant;
P: total (vapour) pressure (Pa);
PDE: partial differential equations;
P_g: pressure of gas (Pa/m);
P_l: pressure of liquid (Pa/m);
P_m: moisture transfer potential (Pa/m);
P_s: saturation vapour pressure (Pa);
P_v: vapour pressure (Pa);
P_{va}: vapour pressure in the air (Pa);
$P_{v,\infty}$: vapour pressure at the boundary (Pa);
P_w: partial pressure of water (Pa);
P_w^0: vapour pressure of water (Pa);
Q: the sorption energy (J/mol);
Q_s: net isosteric heat of sorption (J/mol);
Q_t: travel rate of the drying zone ($kg \cdot m^{-2} \cdot s^{-1}$);
Q_w: liquid water transfer in bulk flow ($kg \cdot m^{-2} \cdot s^{-1}$);
q: sorption energy (W);
q_L: heat of condensation (J/kg or J/mol);
q_{sorb}: heat production (heat source) in the dry zone (W/m^3);
q_w: heat production (heat source) in the wet zone (W/m^3);
R: gas constant ($8.31451 \times 10^{-7}\ J \cdot K^{-1} \cdot mol^{-1}$);
r: constant that involves net isosteric heat with moisture content adjustable parameters;
$\dot{r}$: radius (m);
r_m, r_{max}: minimum and maximum radius of capillaries, respectively (m);
S_T: thermal sink or source (W/m^3);
T: temperature (K);
T_a: supplied air temperature (K);
$T_{a,en}$, $T_{a,out}$: entering and leaving out air temperature, respectively (°C);
T_{amb}: ambient temperature (°C);
T_{a0}: initial air temperature (K);
T_r: temperature at referred condition (K);

T_{wb}:	wet bulb air temperature (K);
T_c:	temperature (°C);
T_e:	temperature at equilibrium (°C);
T_g:	grain temperature (K);
$T_{g,i}$:	initial grain temperature (°C);
T_k:	temperature (K);
T_R:	dimensionless temperature ratio;
T_s:	surface temperature (K);
T_{sorb}:	temperature of the dry zone (K);
T_w:	temperature of the wet material (°C);
T_0:	initial temperature of the material (°C);
T_1:	temperature at condition 1 (K);
T_∞:	temperature of bulk drying air (K);
t:	time (s);
$\bar{u}$:	velocity vector in all (x, y, z) directions (m/s);
u_a:	apparent or superficial air velocity (m/s);
u_x:	air velocity in x coordinate (m/s);
V_{ad}:	adsorbed volume at saturation (m^3);
V_g:	air velocity (m/s);
V_L:	layer volume (m^3);
V_v:	volumetric air content of medium (m^3/m^3);
V_W:	liquid phase velocity (m/s);
w:	ratio of water molecules adsorbed on the primary sites converted into the secondary adsorption sites;
X_θ:	semi-thickness of the drying material (m);
X_s:	mole fraction of solute;
X_w:	mole fraction of water;
x:	distance (m) or spatial coordinates;
x:	transverse distance by the concentration gradient or layer thickness (m);
y:	spatial coordinates (m);
Y_{abs}:	absolute humidity of air (kg/kg dry air);
Y_i:	absolute humidity of air at the initial condition (kg water/kg dry air);
Y_{en}:	absolute humidity of entering air (kg water/kg dry air);
Y_{out}:	absolute humidity of leaving air (kg water/kg dry air);
Y_s:	relative humidity at the surface of the material under the surface temperature (%);
Y_{sat}:	saturation humidity at the droplet surface under the surface temperature (%);
Y_∞:	surrounding gas relative humidity (%);
Z:	grain depth (m);
z:	spatial coordinates (m).

References

Abramowitz, M., and I. A. Stegun. 1972. Handbook of Mathematical Functions, Dover Publications, Inc., New York.

Abu-Ghannam, N., and B. McKenna. 1997. The application of Peleg's equation to model water absorption during the soaking of red kidney beans (*Phaseolus vulgaris* L.). Journal of Food Engineering 32: 391–401.

Alagusundaram, K., D. S. Jayas, N. D. G. White, and W. E. Muir. 1990. Three-dimensional, finite element, heat tranfer model of temperature distribution in grain storage bins. Transactions of the ASAE 33: 577–584.

Al-Muhtaseb, A. H., W. A. M. McMinn, and T. R. A. Magee. 2002. Moisture sorption isotherm characteristics of food products: a review. Food and Bioproducts Processing 80: 118–128.

Al-Muhtaseb, A., W. A. M. McMinn, and T. R. A. Magee. 2004. Water sorption isotherms of starch powders: part 1: mathematical description of experimental data. Journal of Food Engineering 61: 297–307.

Anandharamakrishnan, C. 2003. Computational fluid dynamics (CFD) applications for the food industry. Indian Food Industry 22: 62–68.

Arslan, N., and H. Togrul. 2006. The fitting of various models to water sorption isotherms of tea stored in a chamber under controlled temperature and humidity. Journal of Stored Products Research 42: 112–135.

ASABE. 2016. ASAE D245.6 OCT2007. Moisture Relationships of Plant-based Agricultural Products, ASABE Standard. American Society of Agricultural and Biological Engineers, St. Joseph, MI.

Ayawei, N., A. N. Ebelegi, and D. Wankasi. 2017. Modelling and interpretation of adsorption isotherms. Journal of Chemistry 2017: 1–11.

Bakalis, S., A. Kyritsi, V. T. Karathanos, and S. Yanniotis. 2009. Modeling of rice hydration using finite elements. Journal of Food Engineering 94: 321–325.

Bakker-Arkema, F. W., W. G. Bickert, and R. J. Patterson. 1967. Simultaneous heat and mass transfer during the cooling of a deep bed of biological products under varying inlet conditions. Journal of Agricultural Engineering Research 12: 297–307.

Barre, H. J., G. R. Baughman, and M. R. Hamdy. 1971. Application of the logarithmic model to cross-flow deep bed grain drying. Transactions of the ASAE 14: 1061–1064.

Basu, S., U. S. Shivhare, and A. S. Mujumdar. 2006. Models for sorption isotherms for foods: a review. Drying Technology 24: 917–930.

Basunia, M. A., and T. Abe. 2001. Moisture desorption isotherms of medium-grain rough rice. Journal of Stored Products Research 37: 205–219.

Becker, H. A., and H. R. Sallans. 1956. A study of the desorption isotherms of wheat at 25°C and 50°C. Cereal Chemistry 33: 79–85.

Bello, M. P., M. P. Tolaba, R. J. Aguerre, and C. Suarez. 2010. Modeling water uptake in a cereal grain during soaking. Joumal of Food Engineering 97: 95–100.

Berger, D., and D. C. T. Pei. 1973. Drying of hygroscopic capillary porous solids-a theoretical approach. International Journal of Heat and Mass Transfer 16: 293–302.

Bingol, G., B. Prakash, and Z. Pan. 2012. Dynamic vapor sorption isotherms of medium grain rice varieties. LWT - Food Science and Technology 48: 156–163.

Bizot, H. 1983. Using the 'GAB' model to construct sorption isotherms, pp. 43–54. In R. Jowitt, F. Escher and G. Vos (eds.), Physical Properties of Food. Applied Science Publishers, London.

Blahovec, J., and S. Yanniotis. 2008. GAB generalized equation for sorption phenomena. Food Bioprocess Technology 1: 82–90.

Boquet, R., J. Chirife, and H. A. Iglesias. 1978. Equations for fitting water sorption isotherms of foods. Part II: evaluation of various two-parameter models. Journal of Food Technology 13: 319–327.

Boquet, R., J. Chirife, and H. A. Iglesias. 1979. Equations for fitting water sorption isotherms of foods. Part III: evaluation of various three-parameter models. Journal of Food Technology 14: 527–532.

Boquet, R., J. Chirife, and H. A. Iglesias. 1980. On the equivalence of isotherm equations. Journal of Food Technology 15: 345–349.

Bradley, R. S. 1936. Polymolecular adsorbed films. Part II. The general theory of the condensation of vapours on finely divided solids. Journal of Chemical Society 1936: 1788–1804.

Bramhall, G. 1979. Sorption diffusion in wood. Wood Science 12: 3–13.

Briun, S., and K. A. M. Luyben. 1980. Drying of food materials: a review of recent developments, pp. 155–215. In A. S. Mujumdar (ed.), Advances in Drying, vol. 1. Hemisphere Publishing Corp., Washington, DC.

Brooker, D. B., F. W. Bakker-Arkema, and C. W. Hall. 1974. Drying Cereal Grains, AVI Publishing Company, Westport, Conn.

Brunauer, S. 1945. The Adsorption of Gases and Vapors, Princeton University Press, Princeton, New Jersey.

Brunauer, S., P. Emmett, and E. Teller. 1938. Adsorption of gases in mutimolecular layers. Journal of American Chemistry Society 60: 309–320.

Caurie, M. 1970. A new model equation for predicting safe storage moisture levels for optimum stability of dehydrated foods. Journal of Food Technology 5: 301–307.

Ceaglske, N. H., and O. A. Hougen. 1937. Drying of granular solids. Industrial and Engineering Chemistry 29: 805–813.

Cenkowski, S., D. S. Jayas, and S. Pabis. 1993. Deep-bed grain drying – a review of particular theories. Drying Technology 11: 1553–1582.

Cenkowski, S., and S. Sokhansanj. 1988. Mathematical model of radial continuous crossflow agricultural dryers. Journal of Food Processing Engineering 10: 165–182.

Chandra, P. K., and R. P. Singh. 1994. Applied Numerical Methods for Food and Agricultural Engineers, CRC Press, Boca Raton, FL.

Chen, C. 1988. A Study of Equilibrium Relative Humidity for Yellow-Dent Corn Kernels. Ph.D., University of Minnesota St. Paul.

Chen, C. S., and J. T. Clayton. 1971. The effect of temperature on sorption isotherms of biological materials. Transactions of the ASAE 14: 927–929.
Chen, C. S., and W. H. Johnson. 1969. Kinetics of moisture movement in hygroscopic materials I. Theoretical considerations of drying phenomena. Transaction of the ASAE 12: 109–113.
Chen, C. C., and R. V. Morey. 1989. Comparison of four EMC/ERH equations. Transactions of the ASAE 32: 983–990.
Chen, P., and D. C. T. Pei. 1989. A mathematical model of drying processes. International Journal of Heat and Mass Transfer 32: 297–310.
Chen, P., and P. S. Schmidt. 1990. An integral model for drying of hygroscopic and non-hygroscopic materials with dielectric heating. Drying Technology 8: 907–930.
Chen, X. D., and G. Z. Xie. 1997. Fingerprints of the drying behaviour of particulate or thin layer food materials established using a reaction engineering model. Transactions of the Institution of Chemical Engineers 75 (Part C): 213–222.
Chinachoti, P. 1990. Isotherm equations for starch, sucrose and salt for calculation of high system water activities. Journal of Food Science 55: 265–266.
Chirife, J., and H. A. Iglesias. 1978. Equations for fitting water sorption isotherms of foods: part 1 – a review. Journal of Food Technology 13: 159–174.
Chirife, J., S. L. Resnik, and C. Ferro-Fontan. 1985. Application of Ross equation for prediction of water activity in intermediate moisture food system containing a nonsolute solids. Journal of Food Technology 20: 773–779.
Chirife, J., E. O. Timmermann, H. A. Iglesias, and R. Boquet. 1991. Some features of the parameter k of the GAB equation as applied to sorption isotherms of selected food materials. Journal of Food Engineering 15: 75–82.
Chung, D. S., and H. B. Pfost. 1967a. Adsorption and desorption of water vapour by cereal grains and their products. Part II. Hypothesis for explaining the hysteresis effect. Transactions of the ASAE 10: 552–555.
Chung, D. S., and H. B. Pfost. 1967b. Adsorption and desorption of water vapour by cereal grains and their products. Part I. Heat and free energy changes of adsorption and desorption. Transactions of the ASAE 10: 549–551, 555.
Ciurzyńska , A., and A. Lenart. 2009. The influence of temperature on rehydration and sorption properties of freeze-dried strawberries. Journal of Food Science and Technology 1: 15–23.
Claussen, I. C., T. S. Ustad, I. Strommen, and P. M. Walde. 2007. Atmospheric freeze drying—a review. Drying Technology 25: 957–967.
Crank, J. 1970. The Mathematics of Diffusion, 2nd ed. Clarendon Press, Oxford, UK.
Cunha, L. M., F. A. R. Oliveira, and J. C. Oliveira. 1998. Optimal experimental design for estimating the kinetic parameters of processes described by the Weibull probability distribution function. Journal of Food Engineering 37: 175–191.

Dalgic, A. C., H. Pekmez, and K. B. Belibagh. 2012. Effect of drying methods on the moisture sorption isotherms and thermodynamic properties of mint leaves. Journal of Food Science and Technology 49: 439–449.

D'Arcy, R. L., and I. C. Watt. 1970. Analysis of sorption isotherms of non-homogeneous sorbents. Transacrions of the Faraday Society 66: 1236–1245.

Day, D. L., and G. L. Nelson. 1965. Desorption isotherm for wheat. Transaction of the ASAE 8: 293–297.

de Boer, J. H., and C. Zwikker. 1929. Adsorption als folge von polarisation. Zeitschrift für Physikalische Chemie B3: 407–420.

De Vries, D. A. 1958. Simultaneous transfer of heat and moisture in porous media. Transactions of the American Geophysical Union 39: 9098–9916.

de Vries, U., P. Sluimer, and A. H. Blocksma. 1989. A quantitative model for heat transport in dough and crumb during baking. In N. G. Asp (ed.), Cereal Science and Technology in Sweden. STU Lund University, Lund, Ystad, Sweden.

Defraeye, T. 2014. Advanced computational modeling for drying processes—a review. Applied Energy 131: 323–344.

Dietl, C., O. P. George, and N. K. Bansal. 1995. Modeling of diffusion in capillary porous materials during the drying process. Drying Technology 13: 267–293.

Dincer, T. D., and A. Esin. 1996. Sorption isotherms for macaroni. Journal of Food Engineering 27: 211–228.

Dyer, D. F., D. K. Carpenter, and J. E. Sunderland. 1966. Equilibrium vapor pressure of frozen bovine muscle. Journal of Food Science 31: 196–206.

Ece, M. C., and A. Cihan. 1993. A liquid diffusion model for drying rough rice. Transactions of the ASAE 36: 837–840.

Efremov, G., and T. Kudra. 2004. Calculation of the effective diffusion coefficients by applying a quasi-stationary equation for drying kinetics. Drying Technology 22: 2273–2279.

Efremov, G., and T. Kudra. 2005. Model-based estimate for time-dependent apparent diffusivity. Drying Technology 23: 2513–2522.

Elgobashi, S., and D. B. Spalding. 1977. Equilibrium Chemical Reaction of Supersonic Hydrogen-Air Jets (The ALMA Computer Program), CR-2725. NASA.

Erbay, Z., and F. Icier. 2010. A review of thin layer drying of foods: theory, modeling, and experimental results. Critical Reviews in Food Science and Nutrition 50: 441–464.

Ertekin, C., and M. Z. Firat. 2017. A comprehensive review of thin-layer drying models used in agricultural products. Critical Reviews in Food Science and Nutrition 57: 701–717.

Esther, M. S. M., P. E. P. Banuu, and S. Subhashini. 2016. Thin layer and deep bed drying basic theories and modelling: a review. CIGR Journal 18: 314–325.

Farroni, A. E., and M. P. Buera. 2014. Cornflake production process: state diagram and water mobility characteristics. Food Bioprocess Technology 7: 2902–2911.

Ferziger, J. F., and M. Peric. 2002. Computational Methods for Fluid Dynamics, 3rd ed. Springers publications, New York.

Fontan, F. C., J. Chirife, E. Sancho, and H. A. Iglesias. 1982. Analysis of a model for water sorption phenomena in foods. Journal of Food Science 47: 1590–1594.

Foo, K. Y., and B. H. Hameed. 2010. Insights into the modeling of adsorption isotherm systems. Chemical Engineering Journal 156: 2–10.

Fortes, M., and M. R. Okos. 1980. Drying theories: their bases and limitations as applied to foods and grains, pp. 119–154. In A. S. Mujumdar (ed.), Advances in Drying, vol. 1. Hemisphere Publishing Corp., Washington, DC.

Freundlich, H. 1926. Colloid and Capillary Chemistry, Methuen, London.

Fromm, J. E., and F. H. Harlow. 1963. Numerical solution of the problem of vortex street development. Physics of Fluids 6: 975.

Fugassi, P., and G. Ostapchenko. 1959. Sorption of polar vapore on swelling gels. Fuel 38: 271–276.

Furmaniak, S., A. P. Terzyk, P. A. Gauden, and G. Rychlicki. 2007. Applicability of the generalised D'Arcy and Watt model to description of water sorption on pineapple and other foodstuffs. Journal of Food Engineering 48: 103–107.

Gentry, R. A., R. E. Martin, and J. B. Daly. 1966. A Eulerian differencing method for unsteady compressible flow problems. Journal of Computational Physics 1: 87–118.

Ghani, J. A., R. Supnick, and P. Rooney. 1991. The experience of flow in computer-mediated and in face-to-face groups. In Proceedings of the Twelfth International Conference on Information Systems, New York, NY.

Haddad, H., M. Guessasma, and J. Fortin. 2014. Heat transfer by conduction using DEM-FEM coupling method. Computational Materials Science 81: 339–347.

Halsey, G. 1948. Physical adsorption on non-uniform surfaces. Journal of Chemical Physics 16: 931–937.

Hamdy, M. Y., and M. J. Barre. 1970. Analysis and hybrid simulation of deep-bed drying of grain. Transactions of the ASAE 13: 752–757.

Hansen, J. R. 1976. Hydration of soybean protein. Journal of Agricultural Food Chemistry 24: 1136–1141.

Harlow, F. H. 1956. A Machine Calculation Method for Hydrodynamic Problems, Los Alamos Scientific Laboratory Report LAMS-1956.

Harlow, F. H., and J. E. Welch. 1965. Numerical calculation of time-dependent viscous incompressible flow of fluid with a free surface. Physics of Fluids 8: 2182–2189.

Harlwood, A. J., and S. Horrobin. 1946. Absorption of water by polymers: analysis in terms of a simple model. Transactions of Farady Society 42B: 84–92.

Henderson, S. M. 1952. A basic concept of equilibrium moisture. Agricultural Engineering 33: 29–32.
Henderson, S. M. 1974. Progress in developing the thin layer drying equation. Transactions of the ASAE 17: 1167–1172.
Henderson, S. M., and S. Pabis. 1961. Grain drying theory. I. Temperature effect on drying coefficients. Journal of Agricultural Engineering Research 6: 169–174.
Henry, P. S. H. 1948. Diffusion of moisture and heat through textiles. Discussions of the Faraday Society 3: 243–257.
Herman, E., D. L. Cruz, and M. A. Garcia. 1999. Prediction of pineapple isotherms using the Ross equation. Drying Technology 17: 1161–1172.
Hess, J. L., and A. M. O. Smith. 1967. Calculation of potential flow about arbitrary bodies. Progress in Aerospace Sciences 8: 1–138.
Hill, J. E., and J. E. Sunderland. 1967. Equilibrium vapour pressure and latent heat of sublimation for frozen meats. Food Technology 21: 112–114.
Hoover, S. R., and E. F. Mellon. 1950. Application of polarization theory to sorption of water vapor by high polymers. Journal of America Chemistry Society 72: 2562.
Hougen, O. A., H. J. Cauley, and W. R. Marshall. 1939. Limitations of Diffusion Equations in Drying, American Institute of Chemical Engineers, Houston, TX, USA.
Hussain, A., C. S. Chen, J. T. Clayton, and L. F. Whitney. 1972. Mathematical simulation of mass and heat transfer in high moisture foods. Transaction of the ASAE 15: 732–736.
Iglesias, H. A., and J. Chirife. 1976a. Isosteric heats of water vapor sorption on dehydrated foods. Part I: analysis of the differential heat curves. Lebensmittel-Wissenschaft und Technologie 9: 116–122.
Iglesias, H. A., and J. Chirife. 1976b. A model for describing the water sorption behaviour of foods. Journal of Food Science 41: 984–992.
Iglesias, H. A., and J. Chirife. 1976c. Prediction of effect of temperature on water sorption of food materials. Journal of Food Technology 11: 109–116.
Iglesias, H. A., and J. Chirife. 1982. Handbook of Food Isotherms: Water Sorption Parameters for Food and Food Components, Academic Press, New York.
Iglesias, H. A., and J. Chirife. 1995. An alternative to the Guggenheim, Anderson and De Boer for the mathematical description of moisture sorption isotherms of foods. Food Research International 28: 317–321.
Iglesias, H. A., J. Chirife, and J. L. Lombardi. 1975. An equation for correlating equilibrium moisture content in foods. Food Technology 10: 289–294.
Ingram, G. W. 1976. Deep bed drier simulation with intra-particle moisture diffusion. Journal of Agricultural and Engineering Research 21: 263–272.

Istadi, I., and J. P. Sitompul. 2002. A comprehensive mathematical and numerical modeling of deep-bed grain drying. Drying Technology 20: 1123–1142.

Jamaleddine, T., and M. B. Ray. 2010. Application of computational fluid dynamics for simulation of drying processes: a review. Drying Technology 28: 120–154.

Jayas, D. S., S. Cenkowski, and S. Pabis. 1991. Review of thin-layer drying and wetting equations. Drying Technology 9: 551–588.

Jayas, D. S., and G. Mazza. 1993. Comparison of five, three-parameter equations for the description of adsorption data of oats. Transactions of the ASAE 36: 119–125.

Jayas, D. S., S. Sokhansanj, E. B. Moysey, and E. M. Barber. 1987. The effect of airflow direction on the resistance of canola (rapeseed) to airflow. Canadian Agricultural Engineering 29: 189–192.

Jian, F., V. Chelladurai, D. S. Jayas, and N. D. G. White. 2015. Three-dimensional transient heat, mass, and momentum transfer model to predict conditions of canola stored inside silo bags under Canadian prairie conditions: part II. Model of canola bulk temperature and moisture content. Transactions of the ASABE 58: 1135–1144.

Jian, F., and D. S. Jayas. 2018a. Characterization of isotherms and thin-layer drying of red kidney beans, part II: three dimensional finite element models to estimate transient mass and heat transfer coefficients and water diffusivity. Drying Technology 36: 1707–1718.

Jian, F., and D. S. Jayas. 2018b. Characterization of isotherms and thin-layer drying of red kidney beans, part I: choosing appropriate empirical and semi theoretical models. Drying Technology 36: 1696–1706.

Jian, F., D. S. Jayas, P. G. Fields, and N. D. G. White. 2017a. Water sorption and cooking time of red kidney beans: part II – mathematical models of water sorption. International Journal of Food Science & Technology 52: 2412–2421.

Jian, F., D. S. Jayas, P. G. Fields, and N. D. G. White. 2017b. Water sorption and cooking time of red kidney beans (*Phaseolus vulgaris* L.): Part I – effect of freezing and drying conditions on water sorption and cooking time. International Journal of Food Science & Technology 52: 2031–2039.

Jian, F., D. S. Jayas, N. D. G. White, and K. Alagusundara. 2013. A three–dimensional, asymmetric, and transient model to predict grain temperatures in grain storage bins. Transactions of the ASAE 48: 263–271.

Jian, F., J. Liu, and D. S. Jayas. 2019. A new mathematical model to simulate sorption, desorption and hysteresis of stored canola during aeration. Drying Technology 38: 2190–2201.

Johnny, S., S. M. A. Razavi, and D. Khodaei. 2015. Hydration kinetics and physical properties of split chickpea as affected by soaking temperature and time. Journal of Food Science and Technology 52: 8377–8382.

Kahveci, K., and A. Cihan. 2008. Drying of Food Materials: Transport Phenomena, Nova Science Publishers Inc., New York.

Kaleemullah, S., and R. Kailappan. 2006. Modelling of thin-layer drying kinetics of red chillies. Journal of Food Engineering 76: 531–537.

Karathanos, V. T. 1999. Determination of water content of dried fruits by drying kinetics. Journal of Food Engineering 39: 337–344.

Karpinska, A. M., and J. Bridgeman. 2016. CFD-aided modelling of activated sludge systems – a critical review. Water Research 88: 861–879.

Kaymak-Ertekin, F., and A. Gedik. 2004. Sorption isotherms and isosteric heat of sorption for grapes, apricots, apples and potatoes. LWT - Food Science and Technology 37: 429–438.

Keey, R. B. 1980. Theoretical foundations of drying technology, pp. 1–22. In A. S. Mujumdar (ed.), Advances in Drying, vol. 1. Hemisphere Publishing Corporation, Washington, New York, London.

Keey, R. B. 1991. Kinetics of drying, pp. 215–254, Drying of Loose and Particulate Materials. Hemisphere Publishing Co., New York, Washington, Philadelphia, London.

Ketelaars, A. A. J., L. Pel, W. J. Coumans, and P. J. A. M. Kerkhof. 1995. Drying kinetics: a comparison of diffusion coefficients from moisture concentration profiles and drying curves. Chemical Engineering Science 50: 1187–1191.

Khankari, K. K., S. V. Patankar, and R. V. Morey. 1995. A mathematical model for natural convection moisture migration in stored grain. Transactions of the ASAE 38: 1777–1787.

King, C. J. 1968. Rates of moisture sorption and desorption in porous dried foodstuffs. Food Technology 22: 165–171.

Krishna, R., and J. A. Wesselingh. 1997. The Maxwell-Stefan approach to mass transfer. Chemical Engineering Science 52: 861–911.

Kucuk, H., A. Midilli, A. Kilic, and I. Dincer. 2014. A review on thin-layer drying-curve equations. Drying Technology 32: 753–773.

Kumar, K. R., and N. Balasubrahmanyam. 1986. Moisture sorption and the applicability of the Brunauer-Emmett-Teller equation for some dry food products. Journal of Stored Products Research 22: 205–209.

La Mer, V. K. 1967. The calculation of thermodynamic quantities from hysteresis data. Journal of Colloid Interface Science 23: 297–301.

Labuza, T. P., and B. Altunakar. 2007. Water activity prediction and moisture sorption isotherms, pp. 109–154. In G. V. Barbosa-Cánovas (ed.), Water Activity in Foods: Fundamentals and Applications. Blackwell Publishing and IFT Press, Ames, IA.

Labuza, T. P., A. Kaanane, and J. Y. Chen. 1985. Effect of temperature on the moisture sorption isotherms and water activity shift of two dehydrated foods. Journal of Food Science 50: 385–391.

Labuza, T. P., S. Mizrahi, and M. Karel. 1972. Mathematical models for optimization of flexible film packaging of foods for storage. Transactions of the ASAE 15: 150–155.

Langmuir, I. 1916. The constitution and fundamental properties of solids and liquids. Part I. solids. Journal of the American Chemical Society 38: 2221–2295.

Langmuir, I. 1918. The adsorption of gases on plane surfaces of glass, mica and platinum. Journal of the American Chemical Society 40: 1361–1402.

Langrish, T. A. G., R. E. Bahu, and D. Reay. 1991. Drying kinetics of particles from thin layer drying experiments. Transactions of the Institution of Chemical Engineers 69: 417–424.

Lawrence, J., and D. E. Maier. 2011. Development and validation of a model to predict air temperatures and humidities in the headspace of partially filled stored grain silos. Transactions of the ASABE 54: 1809–1817.

Lawrence, J., D. E. Maier, J. Hardin, and C. L. Jones. 2012. Development and validation of a headspace model for a stored grain silo filled to its eave. Journal of Stored Products Research 49: 176–183.

Leonardi, E., and C. Angeli. 2010. On the Maxwell-Stefan approach to diffusion: a general resolution in the transient regime for one-dimensional systems. Journal of Physical Chemistry B 114: 151–164.

Lewicki, P. P. 2000. Raoult's law based food water sorption isotherm. Journal of Food Engineering 43: 31–40.

Lewis, W. K. 1921. The rate of drying of solid materials. Journal of Industrial & Engineering Chemistry 13: 427–432.

Li, Z., and N. Kobayashi. 2005. Determination of moisture diffusivity by thermo-gravimetic analysis under non-isothermal condition. Drying Technology 23: 1331–1342.

Liu, H. L., Z. Y. Jin, X. M. Xu, Z. J. Xie, B. H. Pan, and J. G. Li. 2011. Introducing temperature effect into peleg, generalized D'Arcy-Watt and Blahovec-Yanniotis model to simulate moisture sorption isotherms of a typical broiler feed. Applied Engineering in Agriculture 27: 115–125.

Liu, Q., G. Yang, Q. Zhang, and C. Ding. 2016. CFD simulations of aeration for cooling paddy rice in a warehouse-type storage facility. Transactions of the ASABE 59: 1873–1882.

Liu, X. J., Y. F. Shi, M. A. Kalbassi, R. Underwood, and Y. S. Liu. 2013. Water vapor adsorption isotherm expressions based on capillary condensation. Separation and Purification Technology 95–100: 95–100.

Liu, Z., Z. Wu, X. Wang, J. Song, and W. Wu. 2015. Numerical simulation and experimental study of deep bed corn drying based on water potential. Mathematical Problems in Engineering 2015: 1–13.

Lomauro, C. J., A. S. Bakshi, and T. P. Labuza. 1985a. Evaluation of food moisture sorption isotherm equations. Part I. fruit vegetable and meat products. Lebensmittel-Wissenschaft und Technologie 18: 111–117.

Lomauro, C. J., A. S. Bakshi, and T. P. Labuza. 1985b. Moisture transfer properties of dry and semimoist foods. Journal of Food Science 50: 397–400.

Luikov, A. V. 1966a. Application of irreversible thermodynamics methods to investigation of heat and mass transfer. International Journal of Heat and Mass Transfer 9: 139–152.
Luikov, A. V. 1966b. Heat and Mass Transfer in Capillary Porous Bodies, Pergamon, New York.
Luikov, A. V. 1975. Systems of differential equations of heat and mass transfer in capillary-porous bodies (review). International Journal of Heat and Mass Transfer 18: 1–14.
Malekjani, N., and S. M. Jafari. 2018. Simulation of food drying processes by computational fluid dynamics (CFD); recent advances and approaches. Trends in Food Science and Technology 78: 206–223.
Malin, M. R. 2020. Brian Spalding: some contributions to computational fluid dynamics during the period 1993 to 2004, pp. 3–39. In A. SRunchal (ed.), 50 Years of CFD in Engineering Sciences. Springer, Singapore.
McGavack, J. J., and W. A. Patrick. 1920. The adsorption of sulfer dioxide by the gel of silicic acid. Journald of America Chemistry Society 42: 946–978.
Meinders, M. B. J., and T. van Vliet. 2009. Modeling water sorption dynamics of cellular solid food systems using free volume theory. Food Hydrocolloids 23: 2234–2242.
Messai, S., M. E. Ganaoui, J. Sghaier, L. Chrusciel, and S. Gabsi. 2014. Comparison of 1D and 2D models predicting a packed bed drying. International Journal for Simulation and Multidisciplinary Design Optimization, EDP sciences/NPU (China) 5: 1–10.
Metzger, J. F., and W. E. Muir. 1983. Aeration of stored wheat in the Canadian Prairies. Canadian Agricultural Engineering 25(1): 127–137.
Miano, A. C., and P. E. D. Augusto. 2015. From the sigmoidal to the downward concave shape behavior during the hydration of grains: effect of the initial moisture content on Adzuki beans (*Vigna angularis*). Food and Bioproducts Processing 96: 43–51.
Midilli, A., H. Kucuk, and Z. Yapar. 2002. A new model for single-layer drying. Drying Technology 20: 1503–1513.
Mizralii, S., T. P. Labuza, and M. Karel. 1970. Computer-aided predictions of extent of browning in dehydrated cabbage. Journal of Food Science 35: 799–803.
Montross, M. D., D. E. Maier, and K. Haghighi. 2002. Validation of a finite-element stored grain ecosystem model. Transactions of the ASAE 45: 1465–1474.
Morey, R. V., and H. Li. 1984. Thin-layer equation effects on deep-bed drying prediction. Transaction of the ASAE 27: 1924–1928.
Mousa, W., F. M. Ghazali, S. Jinap, H. M. Ghazali, and S. Radu. 2014. Sorption isotherms and isosteric heats of sorption of Malaysian paddy. Journal of Food Science and Technology 51: 2656–2663.
Moyne, C., and P. Perre. 1991. Processes related to drying: part i, theoretical model. Drying Technology 9: 1135–1152.

Muir, W. E., B. M. Fraser, R. N. Sinha, and J. Shejbal. 1980. Simulation model of two-dimensional heat transfer in controlled-atmosphere grain bins, pp. 385–398, Controlled Atmosphere Storage of Grains. Elsevier Scientific Plub. Co., Amsterdam.

Murman, E., and J. Cole. 1970. Calculation of Plane Steady Transonic Flow, AIAA 8th Aerospace Sciences Meeting, New York.

Nasrallah, S. B., and P. Perre. 1988. Detailed study of a model of heat and mass transfer during convective drying of porous media. International Journal of Heat and Mass Transfer 31: 957–967.

Norrish, R. S. 1966. An equation for the activity coefficients and equilibrium relative humilities of water in confectionary syrups. Journal of Food Technology 1: 25–39.

Norton, T., and D. W. Sun. 2007. An overview of CFD applications in the food industry, pp. 1–43. In D. W. Sun (ed.), Computational Fluid Dynamics in Food Processing. CRC Press, Boca Raton.

Norton, T., B. Tiwari, and D. W. Sun. 2013. Computational fluid dynamics in the design and analysis of thermal processes: a review of recent advances. Critical reviews in food science and nutrition 53: 251–275.

Oswin, C. R. 1946. The kinetics of package life. III. Isotherm. Journal of Society Chemical Industry 65: 419–421.

Overhults, D. G., G. M. White, H. E. Hamilton, and I. J. Ross. 1973. Drying soybeans with heated air. Transactions of the ASAE 16: 112–113.

Page, G. E. 1949. Factors Influencing the Maximum Rates of Air Drying Shelled Corn in Thin-Layers. Department of Mechanical Engineering, Purdue University, West Lafayette, IN.

Palou, E., A. López-Malo, and A. Argaiz. 1997. Effect of temperature on the moisture sorption isotherms of some cookies and corn snacks. Journal of Food Engineering 31: 85–93.

Pan, Z. 2003. Adsorption characteristics of functional soy protein products. Journal of food process engineering 25: 499–514.

Panigrahi, S. S., C. B. Singh, and J. Fielke. 2020. CFD modelling of physical velocity and anisotropic resistance components in a peaked stored grain with aeration ducting systems. Computers and Electronics in Agriculture 179: 1–10.

Parde, S. R., D. S. Jayas, and N. D. G. White. 2003. Grain drying: a review. Sciences Des Aliments - SCI ALIMENT 23: 589–622.

Park, G. S. 1986. Transport principles—solution, diffusion and permeation in polymer membranes, pp. 57–107. In P. M. Bungay, H. K. Lonsdale and M. N. de Pinho (eds.), Synthetic Membranes: Science, Engineering and Applications. Springer, Reidel, Holland.

Parry, J. L. 1983. Mathematical Modelling and Computer Simulation of Heat and Mass Transfer in Agricultural Grain Drying. PhD, Cranfield Institute of Technology.

Parry, J. L. 1985. Mathematical modelling and computer simulation of heat and mass transfer in agricultural grain drying: a review. Journal of Agriculture Engineering Research 32: 1–29.

Patankar, S. V., and D. B. Spalding. 1966. A calculation procedure for heat transfer by forced convection through two-dimensional uniform-property turbulent boundary layers on smooth impermeable walls, vol. 2, Processing of the 3rd International Heat Transfer Conference, Chicago.

Patankar, S. V., and D. B. Spalding. 1972. A calculation procedure for heat, mass and momentum transfer in three-dimensional parabolic flows. International Journal of Heat and Mass Transfer 15: 1787–1806.

Peirce, F. T. 1929. A two-phase theory of the absorption of water vapour by cellulose. Journal of Textile Institute 20: T133–150.

Peleg, M. 1988. An empirical model for the description of moisture sorption curves. Journal of Food Science 53: 1216–1219.

Peleg, M. 1993. Assessment of a semi-empirical four parameter general model for sigmoid moisture sorption isotherms. Journal of Food Process Engineering 16: 21–37.

Peng, G. L., X. G. Chen, W. F. Wu, and X. J. Jiang. 2007. Modeling of water sorption isotherm for corn starch. Journal of Food Engineering 80: 562–567.

Perre, P. 1997. Image analysis, homogenization, numerical simulation and experiment as complementary tools to enlighten the relationship between wood anatomy and drying behavior. Drying Technology 15: 2211–2238.

Perre, P., and C. Moyne. 1991. Processes related to drying: part II use of the same model to solve transfers both in saturated and unsaturated porous media. Drying Technology 9: 1153–1179.

Perre, P., and I. W. Turner. 2002. A heterogeneous wood drying computational model that accounts for material property variations across growth rings. Chemical Engineering Journal 86: 117–131.

Pfost, H. B., S. G. Mourer, D. S. Chung, and G. A. Milliken. 1976. Summarizing and reporting equilibrium moisture data for grains. In Annual Conference of ASAE. ASAE, ASAE, St. Joseph, MI.

Philip, J. R., and D. A. De Vries. 1957. Moisture movement in porous materials under temperature gradient. Transactions, American Geophysical Union 38: 222–232.

Phillips, T. J. 1989. An Investigation of the Combined Heat and Mass Transfer Processes in the Drying of Hygroscopic Porous Media with Two Disparate Length Scales. PhD, University of Tennessee Knoxville.

Pierce, R. D., and T. L. Thompson. 1980. Management of solar and low-temperature grain drying systems. Part II. Layer drying and solution to the overdrying problem. Transactions of the ASAE 23: 1024–1027, 1032.

Poyet, S. 2009. Experimental investigation of the effect of temperature on the first desorption isotherm of concrete. Cement and Concrete Research 39: 1052–1059.

Poyet, S., and S. Charles. 2009. Temperature dependence of the sorption isotherms of cement-based materials: heat of sorption and Clausius-Clapeyron formula. Cement and Concrete Research 39: 1060–1067.

Prasad, K., P. R. Vairagar, and M. B. Bera. 2010. Temperature dependent hydration kinetics of Cicer arietinum splits. Food Research International 43: 483–488.

Ramachandran, R. P., M. Akbarzadeh, J. Paliwal, and S. Cenkowski. 2017. Three-dimensional CFD modelling of superheated steam drying of a single distillers' spent grain pellet. Journal of Food Engineering 212: 121–135.

Ramachandran, R. P., M. Akbarzadeh, J. Paliwal, and S. Cenkowski. 2018. Computational fluid dynamics in drying process modelling—a technical review. Food and Bioprocess Technology 11: 271–292.

Ranjbaran, M., B. Emadi, and D. Zare. 2014. CFD simulation of deep-bed paddy drying process and performance. Drying Technology 32: 919–934.

Raoult, F. M. 1886. Loi générale des tensions de vapeur des dissolvants (General law of vapor pressures of solvents). Comptes Rendus 104: 1430–1433.

Richardson, L. F. 1922. Weather Prediction by Numerical Process, Cambridge University Press, New York.

Rocha, K. S. O., J. H. Martins, M. A. Martins, J. A. O. Saraz, and A. F. L. Filho. 2013. Three-dimensional modeling and simulation of heat and mass transfer processes in porous media: an application for maize stored in a flat bin. Drying Technology 31: 1099–1106.

Roman, A. D., E. Herman-y-Lara, M. A. Salgado-Cervantes, and M. A. García-Alvarado. 2004. Food sorption isotherms prediction using the Ross equation. Drying Technology 22: 1829–1843.

Ross, K. D. 1975. Estimation of water activity in intermediate moisture foods. Food Technology 3: 26–54.

Rusinek, R., and R. Kobyłka. 2014. Experimental study and discrete element method modeling of temperature distributions in rapeseed stored in a model bin. Journal of Stored Products Research 59: 254–259.

Sakaki, T. 1953. On the Harkins – Jura's adsorption isotherm ad its constants. Journal of America Chemistry Society 26: 213–218.

Schrader, G. W., and J. B. Litchfield. 1992. Moisture profiles in a model food gel during drying: measurement using magnetic resonance imaging and evaluation of the Hckian model. Drying Technology 10: 295–332.

Shafaei, S. M., and A. A. Masoumi. 2014. Studying and modeling of hydration kinetics in chickpea seeds (*Cicer arietinum* L.). Agricultural Communications 2: 15–21.

Sharaf-Eldeen, Y. I., J. L. Blaisdell, and M. Y. Hamdy. 1980. A model for ear corn drying. Transactions of the ASAE 23: 1261–1271.

Sharma, H. N., S. J. Harley, Y. Sun, and E. A. Glascoe. 2017. Dynamic triple-mode sorption and outgassing in materials. Scientific Reports 7: 2942 (2941–2912).

Sharp, J. R. 1982. A review of low temperature drying simulation models. Journal of Agricultural and Engineering Research 27: 169–190.

Sherwood, T. K. 1929. The drying of solids – II. Industrial and Engineering Chemistry 21: 976–980.

Singh, R. S., and T. P. Ojha. 1974. Equilibrium moisture content of groundnut and chillies. Journal of the Science of Food and Agriculture 25: 451–459.

Smith, S. E. 1947. The sorption of water vapour by high polymers. Journal of America Chemistry Society 69: 646–651.

Soleimanifard, S. S., and N. Hamdami. 2018. Modelling of the sorption isotherms and determination of the isosteric heat of split pistachios, pistachio kernels and shells. Czech Journal of Food Science 36: 1–8.

Sopade, P. A., E. S. Ajisegiri, and M. H. Badau. 1992. The use of Peleg's equation to model water absorption in some cereal grains during soaking. Journal of Food Engineering 15: 269–283.

Sopade, P. A., and J. A. Obekpa. 1990. Modelling water absorption in soybean, cowpea and peanuts at three temperatures using Peleg's equation. Journal of Food Science 55: 1084–1087.

Spalding, D. B. 1960. A standard formulation of the steady convective mass transfer problem. International Journal of Heat and Mass Transfer 1: 192–207.

Spencer, H. B. 1969. A mathematical simulation of grain drying. Journal of Agricultural Engineering Research 14: 226–235.

Srikiatden, J., and J. S. Roberts. 2005. Measuring moisture diffusivity of potato and carrot (core and cortex) during convective hot air and isothermal drying. Journal of Food Engineering 74: 143–152.

Staudt, P. B., C. P. Kechinski, I. C. Tessaro, L. D. F. Marczak, R. d. P. Soares, and N. S. M. Cardozo. 2013. A new method for predicting sorption isotherms at different temperatures using the BET model. Food Engineering 114: 139–145.

Storey, R. M., and G. Stainsby. 1970. The equilibrium water vapor pressure of frozen cod. Journal of Food Technology 5: 157–163.

Strohman, R. D., and R. R. Yoerger. 1967. A new equilibrium moisture-content equation. Transactions of the ASAE 10: 675–677.

Suarez, C., P. Viollaz, and J. Chirife. 1980. Diffusional analysis of air drying of grain sorghum. Journal of Food Technology 15: 523–531.

Sun, D. 1998. Selection of EMC/ERH isotherm equations for shelled corn based on fitting to available data. Drying Technology 16: 779–797.

Sun, D., and C. Byrne. 1998. Selection of EMC/ERH isotherm equations for rapeseed. Journal of Agricultural Engineering Research 69: 307–315.

Sun, D. W. 1999. Comparison and selection of EMC ERH isotherm equations for rice. Journal of Stored Products Research 35: 249–264.

Sun, D.-W., and J. L. Woods. 1994. The selection of sorption isotherm equations for wheat based on the fitting of available data. Journal of Stored Products Research 30: 27–43.

Sun, Y., C. C. Pantelides, and Z. S. Xhalabi. 1995. Mathematical modelling and simulation of near-ambient grain drying. Computers and Electronics in Agriculture 13: 243–271.

Thom, A. 1933. The Flow Past Circular Cylinders at Low Speeds, Royal Society, London.

Thompson, T. L. 1972. Temporary storage of high-moisture shelled corn using continuous aeration. Transactions of the ASAE 15: 333–337.

Thompson, T. L., R. M. Peart, and G. H. Foster. 1968. Mathematical simulation of corn drying—a new model. Transactions of the ASAE 11: 582–586.

Thorpe, G. R. 2008. The application of computational fluid dynamics codes to simulate heat and moisture transfer in stored grains. Journal of Stored Products Research 44: 21–31.

Thorvaldsson, K., and H. Janestad. 1999. A model for simultaneous heat, water and vapour diffusion. Journal of Food Engineering 40: 167–172.

Timmermann, E. O., and J. Chirife. 1991. The physical state of water sorbed at high activities in starch in terms of the GAB sorption equation. Journal of Food Engineering 13: 171–179.

Timmermann, E. O., J. Chirife, and H. A. Iglesias. 2001. Water sorption isotherms of foods and foodstuffs: BET or GAB parameters? Journal of Food Engineering 48: 19–31.

Truscott, S. L., and I. W. Turner. 2005. A heterogeneous three dimensional computational model for wood drying. Applied Mathematical Modelling 29: 381–410.

Tsory, T., N. Ben-Jacob, T. Brosh, and A. Levy. 2013. Thermal DEM-CFD modeling and simulation of heat transfer through packed bed. Powder Technology 244: 52–60.

Turhan, M., S. Sayar, and S. Gunasekaran. 2002. Application Peleg model to study water absorption in chickpea during soaking. Journal of Food Engineering 53: 153–159.

Van Meel, D. A. 1958. Adiabatic convection batch drying with recirculation of air. Chemical Engineering Science 9: 36–44.

Van Milligen, B. P., P. D. Bons, B. A. Carreras, and R. Sanchez. 2005. On the applicability of Fick's law to diffusion in inhomogeneous systems. European Journal of Physics 26: 913–925.

Verboven, P., J. de Baerdemaeker, and B. M. Nicola. 2004. Using computational fluid dynamics to optimize thermal processes, pp. 82–102. In P. Richardson (ed.), Improving the Thermal Processing of Foods. Woodhead Publishing, Cambridge, U.K.

Verma, L. R., R. A. Bucklin, J. B. Ednan, and F. T. Wratten. 1985. Effects of drying air parameters on rice drying models. Transactions of the ASAE 28: 296–301.

Viollaz, P. E., and C. O. Rovedo. 1999. Equilibrium sorption isotherms and thermodynamic properties of starch and gluten. Journal of Food Engineering 40: 287–292.

Vu, H. T., and E. Tsotsas. 2018. Mass and heat transport models for analysis of the drying process in porous mediaL a review and numerical implementation. International Journal of Chemical Engineering: 1–13.

Waananen, K. M., and M. R. Okos. 1996. Effect of porosity on moisture diffusion during drying of pasta. Journal of Food Engineering 28: 121–137.

Wang, J. J. 1993. Mathematical modelling of the drying process in fixed-bed drying. Numerical Heat Transfer, Part B 24: 229–241.

Wang, C. Y., and R. P. Singh. 1978. A single layer drying equation for rough rice, Annual Meeting of ASAE. ASAE, St. Joseph, MI.

Whitaker, S. 1977. Simultaneous heat, mass, and momentum transfer in porous media: a theory of drying. Advances in Heat Transfer 13: 119–203.

Whitaker, S., and W. T.-H. Chou. 1983. Drying granular porous media-theory and experiment. Drying Technology 1: 3–33.

Whitaker, T. B., and J. H. Young. 1972. Application of the vapor-diffusion equation for concentric spheres in describing moisture movement in peanut pods. Transaction of the ASAE 15: 167–171,174.

White, H. J., and H. Eyring. 1947. The adsorption of water by swelling high polyrneric materials. Textile Research Journal 17: 523–553.

Wolf, W., W. E. L. Spiess, and G. Jung. 1985. Sorption Isotherms and Water Activity of Food Materials, Elsevier, Amsterdam.

Xia, B., and D. W. Sun. 2002. Applications of computational fluid dynamics (CFD) in the food industry: a review. Computers and Electronics in Agriculture 34: 5–24.

Yang, H., and M. R. Milota. 2013. Modeling the process of drying biomass in a fixed bed. Forest Products Society 63: 148–154.

Young, J. H. 1976. Evaluation of models to describe sorption and desorption equilibrium moisture content isotherms of Virginia-type peanuts. Transaction of the ASAE 19: 146–150.

Young, J. H., and G. L. Nelson. 1967. Theory of hysteresis between sorption and desorption isotherms in biological materials. Transactions of ASAE 10: 260–263.

Zare, D., D. S. Jayas, and C. B. Singh. 2012. A generalized dimensionless model for deep bed drying of paddy. Drying Technology 30: 44–51.

Zhao, X., W. Li, H. Zhang, X. Li, and W. Fang. 2019. Reaction–diffusion approach to modeling water diffusion in glutinous rice flour particles during dynamic vapor adsorption. Journal of Food Science and Technology 56: 4605–4615.

Zhao, X., H. Zhang, R. Duan, and Z. Feng. 2017. The states of water in glutinous rice flour characterized by interpreting desorption isotherm. Journal of Food Science and Technology 54: 1491–1501.

Chapter 6

Moisture Migration and Safe Moisture Content for Storing Grain in Bins

Introduction

After harvesting, grains are dried to their recommended safe storage moisture contents for a region and this grain is generally considered safe for a storage period of up to one year under regional temperature conditions. However, the grain moisture content might change during storage due to seasonal weather variations, water entering into the grain, exchange between headspace air and the air in the grain bulk, moisture migration, and multiplication of insects and fungi. These changes can cause the grain at the safe moisture contents to become unsafe and localized hotspots might develop in the stored bin. The qualitative and quantitative nature of moisture movement or migration can vary with quality of stored grain, storage structure, size of grain bulk, and weather conditions. This chapter explains these moisture movements and migration. The development of safe storage moisture contents of different crops will also be presented.

DOI: 10.1201/9781003186199-6

6.1 Hygroscopicity of Grains

6.1.1 Hygroscopicity of Different Crop Kernels

Hygroscopicity of biomaterials is the tendency of the dry materials to absorb moisture from the surrounding atmosphere due to the affinity of its chemical components, such as hemicellulose, starch, sugar, cellulose, and protein, to water. Both the morphological structure and composition of materials play a key role in water sorption (Cozzolino et al. 2014). Stored grain kernels usually have a strong hygroscopicity although different crop kernels have different water sorption behaviours (Fig. 6.1) due to their different material components and structures. Moisture content of grain kernels will equilibrate with its environment, if maintained constant; however, this process is usually a slow process and equilibrium moisture content may reach in days or months depending on seed type. But due to continuous changes in weather conditions, equilibrium with the changing surroundings is rarely established in a grain bin. The equilibrium moisture contents of different crop seeds are different under the same storage conditions due to different chemical components and morphological structures (Fig. 6.1). This is the reason that the equilibrium moisture contents of each crop at given environmental conditions should be determined (Fig. 6.1, Appendix A).

Different crop kernels need different times to reach the equilibrium moisture content at a given environmental condition. For example, the rate of moisture absorption by canola is much faster than wheat (Coleman and Fellows 1955; Pichler 1957). Grosh and Max (1959) used dye technique to study the water penetration into wheat kernel and concluded that a tempering period of at least 8 h was required to ensure even distribution of moisture within the kernel. Sokhansanj et al. (1983) recommended at least 46, 48, 4, and 122 h of tempering period for wheat, barley, canola, and corn to reach approximate equilibrium ($M_R = 1$), respectively, at 2°C. At 21°C, the recommended tempering times were reduced to 32, 48, 1, and 96 h for wheat, barley, canola, and corn, respectively. However, Flood and White (1984) found the moisture dissipation inside popcorn was very slow at lower than 10°C, and the popcorn did not reach equilibrium after 67 days. Jian et al. (2017) found most of water (more than 80%) was absorbed in the first 6 h and took more than 12–13 h to reach to equilibrium moisture content when kidney bean was soaked in water. These differences might be caused by different methods of equilibrium evaluation and different processes of water sorption. During grain storage studies, tempering time is at least one week if the material is used for drying study.

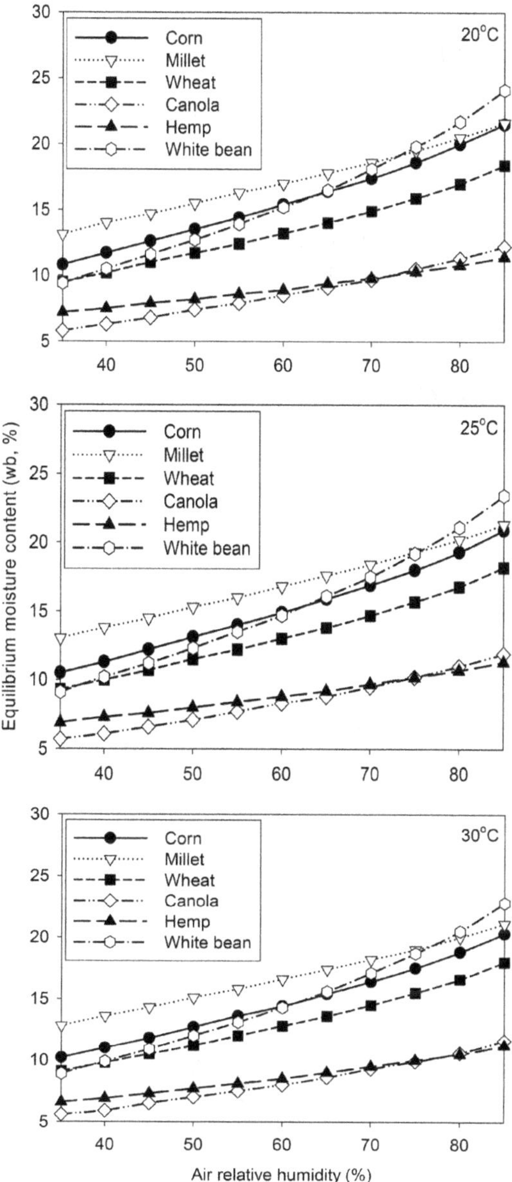

Figure 6.1 Equilibrium moisture content and relative humidity curves of various grain seeds at different temperatures under desorption condition. (Data sources: corn, millet, wheat and canola [ASABE 2016], hemp [Jian et al. 2018], and white bean [Hutchinson and Otten 1984]).

6.1.2 Hygroscopicity of Different Classes and Cultivars of the Same Crop Type

Researchers reported inconsistent conclusions on the equilibrium moisture content of a given crop based on different classes (e.g., soft wheat or hard wheat, two-row or six-row barley), cultivars, harvest years, and maturities under the same storage condition. Also, researchers misuse term "varieties" to mean different things. Sometimes varieties are used for types of crops (e.g., wheat, barley, canola are different types of crops), sometimes for different classes within a crop type (e.g., Canada Western Red Spring, Canada Western Amber Durum are classes of wheat in Western Canada) or sometimes for different cultivars within the same class (e.g., AC Barrie and Roblin are cultivars of Canada Western Red Spring wheat). In consistent use of terms makes it difficult to compare results among different studies. In this book, variety represents a distinct phenotype (how it looks/grows) of a species that differs from other varieties across the species' geographic range. Therefore, it is the taxonomic rank found in nature. The term cultivar is short for cultivated variety and is selectively bred for traits by humans. Pfost et al. (1976) concluded that sorption isotherms of two maize cultivars were significantly different. Bielewicz et al. (1993) reported that the same cultivar of canola harvested at different locations in the same year had up to 1 percentage point difference (db) in the equilibrium moisture content at the same temperature and relative humidity, but different years slightly influenced the hysteresis. The isotherms of corn are significantly different for different cultivars and harvest years (Chen and Morey 1989). However, Sokhansanj et al. (1986) concluded for different cultivars of canola published in different studies had similar isotherms. Castilloa et al. (2003) found five commercial black bean cultivars harvested in different provinces of Argentina could use the same isotherm. Chen (2000) also concluded cultivar and growing location did affect the equilibrium relative humidity (ERH) property of corn and soybean, and there is no difference among different published data for the isotherm of rough rice, brown rice, corn cobs, and red beans. Jayas and Mazza (1993) reported there was no difference in the isotherm between two classes of oats or between two cultivars of common oats. Pixton and Henderson (1981) found that the difference in the equilibrium moisture content (EMC)/ERH relationships of five cultivars of Canadian wheat and two cultivars of rapeseed was small and significant only at low moisture contents.

This inconsistency might be caused by the differences of determining methods of moisture content, evaluation times (whether the seeds reach true equilibrium), and pre-treatment of the samples. Osborn et al. (1989) concluded that obvious differences existed between two studies due to different test cultivars or the ERH determining methods. However, Mazza et al. (1990) found that the cultivar and method for determining of equilibrium moisture content had no significant effect on the sorption properties of flax seed. After comparing different published data, Chen (2000) concluded that the ERH determining method did not influence ERH properties, the drying temperature was significant for rough rice, brown rice, corn kernels, and corn cobs while not significant for soybeans and red beans. Effect of hysteresis on these properties was statistically significant for rough rice, brown rice, corn kernels, and corn cobs.

The difference range among classes of the same crop can be larger than 4 percentage points of the moisture content. Different cultivars of yellow-dent corn can have larger than 4 percentage points difference in EMC among different studies (Chen and Morey 1989). Bielewicz et al. (1993) reported the maximum difference among different cultivars of canola was less than 0.9 percentage points. Jian et al. (2019a) found the minimum and maximum average absolute differences between the measured and predicted (by using the equation and parameters recommended by (ASABE 2016)) moisture contents of canola cultivars were 0.5 and 1.6 percentage points, and different field treatments such as desiccant application did not influence the EMC. They speculated that the high oil content of new developed canola cultivars might result in these differences. Therefore, the isotherm of a similar cultivar of a crop should be used when the calculated EMC is used to estimate safe storage time.

6.1.3 Hygroscopicity of Different Parts of a Kernel

Different parts of a grain kernel have different water sorption rates. Corn germ absorbs water much faster than the endosperm (Kumar 1973). There is a high rate of water absorption at the beginning of water sorption by the embryo and the shell, followed by the scutellum and then by the endosperm (Reynolds and MacWilliam 1966). During water sorption, there is significant resistance to the transfer of moisture from the protective shielding tissues (pericarp, testa, and aleurone) to the nucleus (endosperm and embryo) of barley seeds. Water diffusivities in the whole barley seeds vary between 5.28 and 7.61×10^{-12} m^2/s, and for endosperm are 5–7 fold higher than whole seeds, showing that the external tissues have high resistance to moisture transfer (Mayolle et al. 2012).

6.1.4 Factors Influencing Hygroscopicity

In addition to different crops and their classes and cultivars, postharvest treatment also influences water sorption. Cracked or damaged grain kernels usually have a high hygroscopicity. Chung et al. (1972) found that heat damaged corn kernels had lower EMC than sound kernels due to microorganism development. However, Throne (1991) found cracked and whole corn had the same EMC at the same temperature and RH. The water adsorbing ability of yellow dent corn drops as the drying temperature increases (Chen and Morey 1989). Intermittency and ultrasound treated barley achieves higher hydration without changing the quality of the malt (Alvarez et al. 2020). Bielewicz et al. (1993) found compression due to grain weight inside bins and irradiation with dose of 10 kGy did not influence isotherm and hysteresis of the canola. Bielewicz et al. (1993) found recycling of drying and rewetting widens the hysteresis loop from 2 to 2.5 times in the range of 20–70% RH. However, Hubbard et al. (1957) found hysteresis in wheat decreased with repeated drying and rewetting cycles.

Water sorption process can be sped up by increasing temperature and humidity. At 30°C and higher than 70% RH, canola can reach equilibrium with the air in about one week (Sun et al. 2014). Canola isotherm increases at a constant rate up to about 80% RH, and capillary condensation results in a rapid increase of moisture content at >85% RH (Pichler 1957). After 6 h of steeping, a moisture content increase of 20% is observed when the conditioning water temperature is raised from 1°C to 27°C (Fan et al. 1963).

6.2 Initial Grain Moisture Content and Modes of Water Entering into Bins

6.2.1 Initial Moisture Content

6.2.1.1 Moisture Content at Harvest

The grain moisture content will be different at different places on the same field due to different maturities of the plants caused by different application of fertilizer, weed growth, and soil moisture. Seeds at different places of a plant also have different maturities. In a dense head of canola seeds, moisture content of seeds on the sunny side is lower than that at the shady side. During harvest, the weather can vary day to day and night to night and the kernels can be continually dried by sun and rewetted by rain or dew. These phenomena result in different initial moisture content at different locations in the bin after the harvested grain is loaded. Prasad et al. (1978) measured the dockage distribution, microflora, moisture content and temperature of freshly harvested wheat and rapeseed on farms in Manitoba. They found the moisture content of whole wheat kernels was higher than that of the associated dockage while the reverse was found for rapeseed. Navarro et al. (1969) measured moisture content of wheat sampled from trucks during loading of a bin, and the moisture content of the wheat samples varied from 8.5 to 12.8%.

6.2.1.2 Moisture Content after Filling

When bins are filled, grain segregates according to density and kernel size. Moisture content is the most important factor influencing the segregation and there is a high chance that the wet or damp kernels are located at the similar locations. A central column under the filling spout can form that has a different moisture content than the remainder of the bulk. Loschiavo (1985) found the initial wheat moisture content inside six uninfested granaries in 1982 was 13.2–15.6% at the centre, while it was 13.0–15.1% at the edge of the bins. In 1983, the initial wheat moisture content was 9.8–12.7% at the centre, while it was 9.6–12.4% at the edge of the bins. Therefore, uniformity of moisture content in bins is a concept, not a fact.

6.2.1.3 Moisture Content after Drying

The moisture content of kernels dried by the same drier is different due to different initial grain moisture contents, variation of whether temperature and relative humidity, and unstable energy supply. Grain dried by in-bin drying usually has a large grain moisture variation after the drying is completed because the weather condition changes during the drying period (natural air drying can take few weeks to few months). After the completion of the near ambient drying at an airflow rate of 12.2 $L{\cdot}s^{-1}{\cdot}m^{-3}$, the average moisture content of the wheat was reduced to 11.3%, while the moisture content of grain at the top of the bulk was 14.8% (Sinha et al. 1985). If the in-bin drying is to produce an average bulk moisture content instead of moving the drying front completely through the grain bulk, the top grain can be at its initial moisture content, while the grain at the bottom can be over dried. Phillips et al. (1988) loaded 2283.9 tonnes of rough rice into a bin for natural air drying study at an airflow rate of 0.025 $m^3{\cdot}tonne^{-1}{\cdot}s^{-1}$ (about 12.5 $L{\cdot}m^{-3}{\cdot}s^{-1}$), and the moisture content ranged from 21 to 26% with a mean of 22.4%. The rice was dried over a period of 3 months at 10–12% moisture content. The rice at the bottom was over dried.

6.2.1.4 Moisture Content after Aeration

Aeration does not result in even moisture distribution because the main purpose of aeration is to even the grain temperature. At an aeration airflow rate of 0.8 $L{\cdot}s^{-1}{\cdot}m^{-3}$, the maximum moisture content remained about 3.5 percentage points above the average and the minimum moisture content remained about 4.0 percentage points below the average moisture content (Sinha et al. 1985). Ghaly (1984) aerated two steel silos each filled with 44 tonnes of 9.4% MC wheat at 33°C at the rate of 2 $L{\cdot}tonne^{-1}{\cdot}s^{-1}$ (about 1.4 $L{\cdot}m^{-3}{\cdot}s^{-1}$) (88 h/week at the beginning and 15 h/week after 193 h of aeration) for 18 months. The moisture front moved up during the first 3 months aeration and the moisture content of the grain at the duct increased from 9.7 to 12.7%. The moisture content increased to 15.1% at the duct while that of the top half of the bulk remained below 10% after 425 h of fan operation. Moisture content near the duct dropped during spring and summer. In the meantime, the wetting front moved up through the grain. At the end of the trial, the moisture distribution in the bulk ranged between 14.4% at the duct and about 9.5% at the surface.

6.2.2 Entrance of Moisture into Grain Bulks

6.2.2.1 Snow Water

Granaries are usually built to prevent entering of rain water. However, snow might fly into the bins because snow can fly in any direction and there are gaps or openings in the walls or roofs of most metal bins or thin wall granaries because many bins are constructed to allow natural ventilation through these gaps. Snow also enters bins through vents and fan openings. In Canada, one of the most common source of increased moisture content of stored grain is snow. The authors visited more than 100 farm bins in Canada from 2005 to 2010 and found about 5% of the bins had snow piled up at the top of the grain bulks between November and April. Even during winter, the snow melts slowly because the headspace temperature is higher than the ambient temperature. Jian et al. (2009b) reported that the headspace temperature in a galvanized steel silo (3.7 m diameter) was on average 2.9 ± 0.2°C higher than that of the ambient air with a maximum of 18°C during a 15-month period. In a day, the average difference between the highest and lowest temperatures inside the headspace over the 15-month period was 4.3 ± 0.4°C with a maximum of 36.0°C. During winter, the melted snow is gradually absorbed by the grain at the bottom of the snow and 30 cm of the top grain bulk, and some water could trickle down through the bulk to the floor because kernels absorb liquid water slowly, so the grain from the top to floor under the snow could have expressively increased moisture content. Depending on the amount of snow, these wetted grains could develop a hotspot in most cases, and some grain kernels could sprout. Muir et al. (1978) reported snow entered the galvanized-steel bin (filled with 14% MC wheat) through the closed lid gaps. Grain moisture contents in March were 24% in the samples taken from the peak and 16% in samples taken 30 cm below the peak of the grain cone. In June, the moisture content at 30 cm down had further increased to 18% and became mouldy.

6.2.2.2 Grain Wetting Due to Humid Air

The main purpose of aeration and chilling is to cool grain to a uniform temperature. If cool air forced through the grain has high humidity, it can increase grain moisture content. Navarro et al. (1973b) reported an increase in grain moisture, reaching 13.0–13.4% at the location about 20 cm from the aeration duct after 47 h of cooling, and the initial grain moisture content was 11.9–12.3%. At the end of cooling (94 h) the maximum moisture content close to the duct reached 16%. In a similar experiment to cool soybean (Navarro et al. 1973a), there was a marked drop in moisture content in the depth range of 3.5–6.9 m after the first cooling (from May 5, 1971 to May 7, 1971 for 45.4 h), while in the upper depth range (0–2.3 m) and the lower depth range (8.1–16.1 m) moisture contents dropped slightly. The average moisture content of the bulk excluding the region around the aeration duct, dropped from an initial 13.9% to a final 12.8%. Humid air can blow into granaries through fan inlets and roof vents if these openings are not properly sealed and moisture content can change due to the same reason as the aeration and chilling.

6.3 Grain Moisture Migration

Except the water entrance into the grain bulk (explained in the previous section), grain moisture content inside bins is not directly influenced by ambient weather relative humidity and temperature. Inside grain bulks without ventilation, there are generally three modes of moisture transfer: Diffusion of moisture through grain kernels and intergranular air under temperature gradients or moisture gradients or both, and translocation of water vapour through free convection currents caused by temperature gradients established over time in large bulks of grain.

6.3.1 Diffusion under Moisture Gradients at a Uniform Temperature

6.3.1.1 Mechanism

Under a moisture gradient with a uniform temperature, water will move from high moisture locations to low moisture locations either between kernels through contact points, or through intergranular air or a combination of both. The driving force of this moisture movement is the difference in vapour pressure, and the mechanism of this water movement is the water diffusion under vapour pressure gradients. For example, wheat at 20°C and moisture contents of 10, 14, and 18% has equilibrium RH of 35, 62, and 84% (Table 6.1), respectively. The free water at this temperature has the saturation vapour pressure of 2.33 kPa. The vapour pressure of the grain with 10, 14, and 18% moisture contents is 0.82 (= 0.35×2.33), 1.45, and 1.96 kPa, respectively. The driving force for diffusion from the 14% wheat to the

Table 6.1 Vapour Pressure and Equilibrium Relative Humidity of Wheat at Different Temperatures and Moisture Contents (Data Were Generated by Using the Parameters Reported by ASABE (2016) for Hard Red Winter Wheat and Modified Henderson Equation)

T (°C)[a]	*MC (%)*[b]	*RH (%)*[c]	*VP (kPa)*[d]
10	10	30.3	0.37
	14	55.4	0.68
	18	78.1	0.95
	22	92.4	1.13
	26	98.3	1.20
20	10	35.0	0.82
	14	61.9	1.45
	18	83.7	1.96
	22	95.4	2.23
	26	99.3	2.32
30	10	39.4	1.67
	14	67.4	2.86
	18	87.9	3.73
	22	97.2	4.12
	26	99.7	4.23

[a] Temperature.
[b] Moisture content.
[c] Equilibrium relative humidity.
[d] Vapour pressure.

10% wheat (1.45 – 0.82 = 0.63 kPa) is 1.2 times that from the 18% wheat to 14% wheat (0.51 kPa). Therefore, the wetter grain will equilibrate slower than drier grain at the same moisture content difference. This water diffusion process can be increased by raising grain temperature. The saturation vapour pressure at 30°C is 4.24 kPa. This 10°C increase results in the vapour pressure difference between 18 and 14% wheat of 0.85, which is 1.7 times greater than that at 20°C.

6.3.1.2 Process

Moisture migration inside grain bulks due to moisture gradients is a slow process because water diffusion through kernels or intergranular air without air movement is a very slow process. The diffusion coefficient of moisture through wheat bulks at 5 and 22°C is 2.5×10^{-10} and 8.0×10^{-10} m²/s, respectively (Pixton and Griffiths 1971; Pixton and

Warburton 1977). The transfer of water vapour through air at 25°C is approximately 32,000 times faster than through a wheat bulk because the diffusion coefficient of water vapour in air is 2.6×10^{-5} m^2/s at 25°C. Dry corn (8% moisture content) intimately mixed with wet corn (25% moisture content) will reach 95% of their final equilibrium in 9 days at 4.5°C, 5 days at 21°C, and 2 days at 38°C (White et al. 1972). When barley at 7% moisture content is mixed with 23% barley at a ratio 1:3 at 20°C, about 83% of the total moisture change occurs in 1 day, and 95% change in 3–5 days (Henderson 1987). After 7 days, the original wet and dry individual kernels have standard deviations up to 0.6% to each of their means. After 28 days, a difference of 1 percentage point moisture content between original wet and dry individual kernels still exists. When small parcels (120 g) of wet wheat with 22.1% moisture content are embedded in 550 g lots of dry wheat with 13.7% moisture content at 5 and 22.5°C, although the rate of change at 22.5°C is more rapid, complete equality between the moisture content of the "wet" and "dry" grain is not established even after 140 days due to the hysteresis effect. Equilibrium of moisture content still had not been attained at either 5 or 22.5°C up to 409 days (Pixton and Griffiths 1971). A few mathematical models of moisture diffusion have been developed (Jayas 1995), but are not widely used because water migration through grain bulks is a slow process and negligible compared to moisture migration caused by convection currents.

6.3.1.3 Application

To achieve the desired grain moisture content, mixing wet grain with dry grain is a common process in grain elevators. If grain is not well mixed or there are layers or pockets of wet grain kernels, the water has to move a larger distance. Even though the overall RH after grain mixing is less than 65%, the RH inside single kernels or the pocket of wet grain might be higher than 75% if the grain is not well mixed. This could result in spoilage of individual kernels or pockets of grain kernels.

6.3.2 Diffusion under Temperature Gradients at a Uniform Moisture Content

6.3.2.1 Mechanism

A grain bulk at a uniform moisture content will have different vapour pressures if different parts of the grain bulk are at different temperatures. The difference in vapour pressure is caused by two factors acting simultaneously. The saturated vapour pressure of free water increases with the increase of temperature. At 80% RH, the vapour pressure increases from 1.36 kPa at

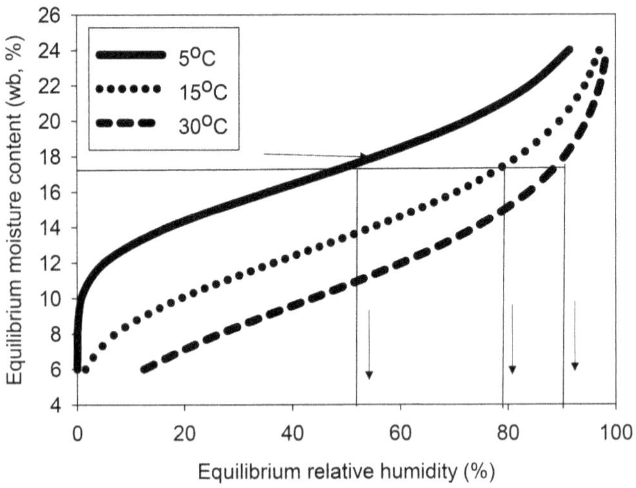

Figure 6.2 Equilibrium moisture content and equilibrium RH of soft winter wheat (CV: Hobbit) under desorption condition. Graph is generated by using the modified Chung-Pfost equation and the constants provided in the ASABE standard (2016).

15°C to 3.40 kPa at 30°C. The second factor is the increase of equilibrium RH of grain with the increase of temperatures. Wheat at 18% moisture content has about 80% RH at 15°C and about 88% RH at 30°C (Fig. 6.2), and the vapour pressure is 1.36 and 3.82 kPa, respectively. The vapour pressure difference of wheat with 18% moisture content between at 15 and 30°C is 2.46 kPa, which includes 2.04 kPa due to the effect of temperature on saturated pressure and 0.42 kPa due to the effect of temperature on the equilibrium RH (Fig. 6.3). Therefore, moisture diffusion due to temperature gradients is mainly a function of the relationship between temperature and vapour pressure rather than between temperature and RH.

6.3.2.2 Process

Stewart (1975) showed that the gradient of vapour pressure could be developed under gradients of moisture and temperature, and the driving forces created by the temperature gradients could be 13 times larger than those created by the moisture gradients. Anderson et al. (1943) measured moisture migration in wheat in a 1.8 m long, 0.6 m wide, and 0.5 m high container. The horizontal container was insulated around its length and two ends of the container were kept at 0°C and 35°C, respectively. The initial grain moisture content of the 450 kg wheat was 14.6%. At the end of the experiment, which lasted 316 days, the maximum moisture content recorded at

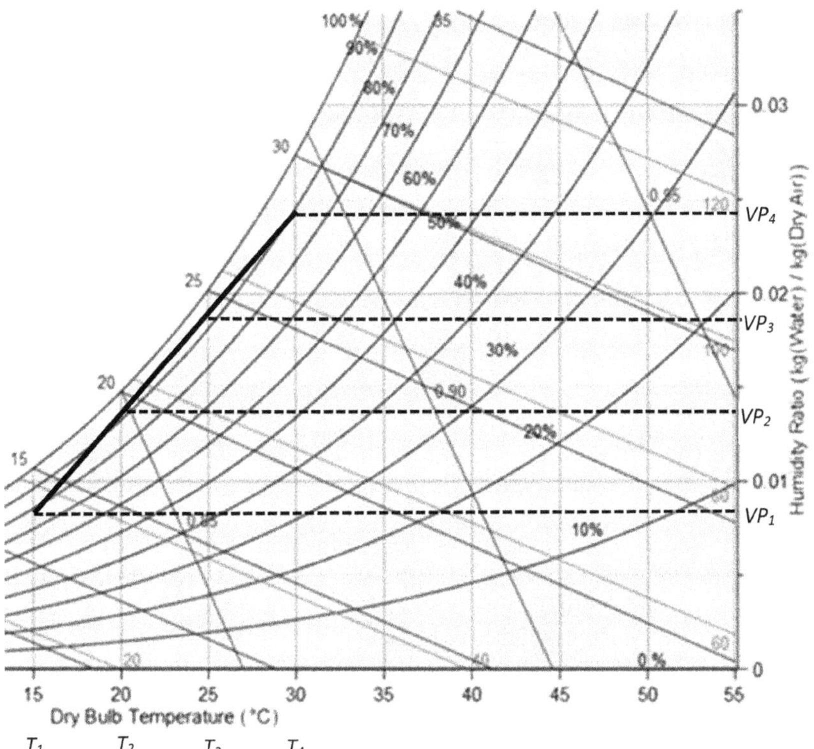

Figure 6.3 Changes in humidity ratio $\left(H = \dfrac{0.6219P_e}{P_{atm} - P_v}\right)$ when the temperature of wheat with 18% moisture content (wb) increases from 15 to 30°C, where H is the humidity ratio (kg water/kg dry air), P_{atm} is the atmosphere pressure (Pa), and P_v is the vapour pressure (Pa).

the cold end (0°C) was 29.6%, and the minimum recorded at the warm end (35°C) was 10.9%. To prevent the influence of the free convection, Griffiths (1964) measured the moisture migration due to temperature gradient in a vertical column by applying the warmer temperature (35°C) at the top of column, while the bottom was kept at 25°C, and the wall of the column was insulated. The initial moisture content of the wheat was 11.9%, and the moisture content at the end of 7 months of the experiment was 14.8% at the colder end and 10.0% at the warmer end of the column.

For equal differences in temperature between T_1 and T_2 and between T_3 and T_4, the difference in vapour pressure between the first pair is smaller than that between the second pair because the driving force increases with the increase of temperature (Fig. 6.3). This will result in a larger rate of water diffusion when the temperature is at the second pair than that the first pair. Therefore, moisture migration occurs faster at higher temperature than that at lower temperatures. This is one of the reasons that moisture migration increases more rapidly in regions where warm grain is located. Harvested grain is usually loaded into storage bins directly from the field or from a heated air drier. If the grain is not aerated and there are temperature gradients, water migration might occur inside grain bin. After the diffusion starts, a net diffusion flux of water vapour is the sum of the two opposite fluxes due to temperature and moisture gradients. Therefore, the moisture migration will be stopped after these two forces reach equilibrium.

6.3.2.3 Application

Moisture content variations of the grain throughout the bin have little effect on the vapour pressure differences and the rate of moisture diffusion because the vapour pressure change due to temperature change is larger than that due to moisture change. Inside bins, moisture will diffuse from warm grain to cool grain even if the moisture content of the warm grain is slightly less than the moisture content of the cold grain. Moisture migration due to moisture gradients only has effect when wet grain is close to dry grain. Therefore, moisture migration due to moisture gradients inside grain bin is usually negligible. Compared with the moisture migration due to convection currents, moisture migration due to temperature gradients is also negligible. Khankari et al. (1995) proved that in the absence of natural convection, the effect of moisture migration is limited mainly near the outer walls of the grain bin.

6.3.3 Moisture Migration Due to Free Convection Currents

6.3.3.1 Mechanism

In northern hemisphere, the weather temperature gradually decreases in the fall which cools down the stored grain from outside to inside of bins. In the spring, the weather temperature gradually warms up the stored grain bulk. These temperature fluctuations results in temperature gradients inside storage bins (Jian et al. 2009b), which produce the free convective currents (Figs. 6.4 and 2.10). Different free convection currents and moisture migrations might occur in different storage structures. This section describes the patterns of moisture migration in a metal hopper silo (bin), which is the most common storage structure in the world.

In the fall or winter, the air temperature and relative humidity do not change much when the air flows down near the walls because a uniform

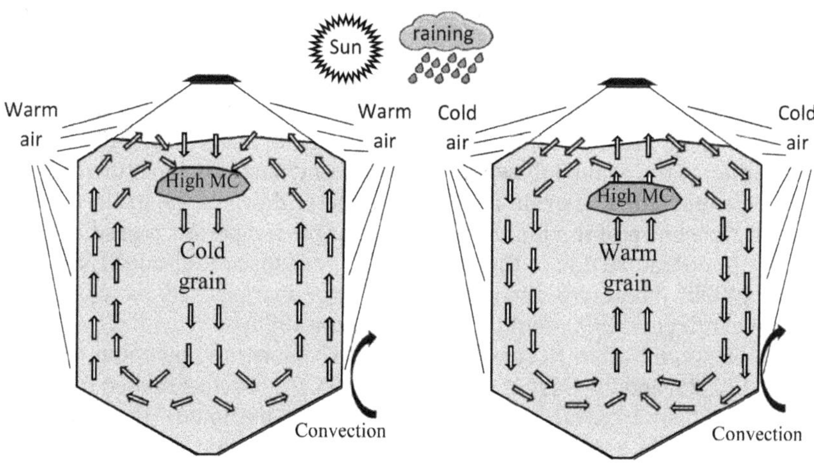

Figure 6.4 Probable free convection paths of air and moisture migration due to the convection currents during the fall/winter (right) and spring/summer (left). In the graph, MC = moisture content. (Figures are modified from Jayas (1995) and Navarro et al. (2001)).

cooling might occur along the bin walls. All the air near the wall is approximately at a low temperature, e.g., T_1 in Fig. 6.3. The air will gradually be warmed up as it flows to the centre of the bin, e.g., from T_1 to T_2, T_3, and T_4. Because the convection currents have a slow velocity, the air will be or close to be in equilibrium with the grain at each location. The water holding capacity increases with the increase of the air temperature. Therefore, the air must pick up water from the warmer grain as it moves from cold places to warmer places (e.g., T_1 to, T_2, T_3, and then T_4), and the grain with a warmer temperature loses water. After the air reach to the centre of the bin, the air is pushed or buoyed up through the central column of grain (Fig. 6.4), and cools from T_4, T_3, T_2, and to T_1 (Figs. 6.3 and 2.10). The water holding capacity of the air decreases as it is cooled and water in the air is reabsorbed by the grain at the cooler locations. For equal temperature increment from T_1 to T_2 and from T_3 to T_4, the change in moisture holding capacity of the air is much higher between the pair of two higher temperatures than the two lower temperatures (Fig. 6.3). Most grain at the top of the bin centre and about one meter down to the grain bulk surface has a lower temperature than that at the centre, but has a higher temperature than that at the surface of the grain bulk. The greatest change in moisture content of the grain does not necessarily occur where the air undergoes it greatest changes. The amount of change depends on the relative amount of air passing through the grain. Where the isothermal lines are close together and the temperature gradient is high, it is the location where the air usually passes through. The air current moves through the path with least resistance. The change in moisture content of the grain at this location is high. In the galvanized-steel bin holding wheat with 13.6 ± 0.1% moisture content, the temperature gradients are much greater at the top centre of the grain mass (Fig. 2.10). This is the region of the grain bulk where the greatest changes in moisture content can be expected. The grain at about one meter down of the grain surface and at the centre of the bin usually has a larger temperature gradient and cooler temperature due to the effect of the headspace (Schmidt 1955; Jian et al. 2009b). Therefore, the grain at this place increases its moisture content (Gough et al. 1990; Smith and Sokhansanj 1990).

The equilibrium RH of wheat with 13.6% moisture content at –5, 0, 5, 10, 15, 20, 25, and 27°C is 58, 60, 61, 62, 64, 65, 67, 68% (calculated from the ASABE (2016) and Hard Red Spring wheat was assumed). In November, when air at 10°C moves to the grain at 25°C (the bottom centre of the bin, Fig. 2.10), the relative humidity of the air changes from 62 to 67%, and the moisture content inside the air changes from 0.0043 to 0.0135 kg water per kg of dry air. This increased water in the air current comes from the grain. Therefore, grain at this location (the bottom centre of the bin) loses moisture and will be gradually dried. After the current moves up farther and the moving air will equilibrate with the grain at 27°C and 68% RH (Fig. 2.10). The water holding capacity of the air at this higher temperature location increases and air will pick up more water from the grain. The grain at the top centre location will also gradually be dried. After the current moves up farther into the grain at 20°C and 65% RH, the water holding capacity of the air decreases from 0.016 to 0.0095 kg water per kg of dry air. The air current will lose water to grain, and the grain at the top centre (usually a few centimetres to 1 m down below the surface of the grain bulk) will increase its moisture. Compared with the amount of grain which loses moisture content, the mass of grain gaining moisture content is usually small due to the special pattern of the temperature gradients. Therefore, a small pocket of grain at this location significantly increases its moisture content.

In spring and summer, the direction of the convection currents reverses due to the reversal of the temperature gradient directions (Figs. 6.4 and 2.10). In the galvanized-steel bin holding canola with 8.5–9.4% moisture content in late April, warm air in the headspace enters the warm grain in the top surface. The convection currents move from the grain at 5°C and 61% RH to the grain at 0°C at 60% RH, and then to the grain at –5°C and 58% RH (centre of the bin). The water holding capacity gradually decreases as the air moves down, so the grain gains water and the air current is gradually dried. The same grain (or close to the same location) that absorbs moistures during fall and winter gains the water due to the special pattern of the temperature gradients. In a bin with different configurations as presented in Fig. 2.10, temperature gradients and distributions may be different, which might result in moisture migrations to different locations such as the bottom of grain bulks (Jayas 1995) or about 2 m down the grain surface (Noyes and Navarro 2001).

6.3.3.2 Experimental Proof

In 16 farm bins in North Dakota from April to August, Hellevang (1987) reported that the moisture content of the grain in the top surfaces decreased about 2.6%, while the grain at 0.6–1.8 m depth below the surface increased moisture content by up to 0.4%, and the grain at the bottom centre increased by 0.2%. The grain up to 1.2 m from the walls did not change moisture content. This result is consistent with the theoretical explanation outlined above. Muir et al. (1980) conducted measurements in five 25-tonne and one 50-tonne bins holding 14% moisture content of wheat. In all six bins, the moisture content in the peak of the cone decreased 1% while 0.3 m below the peak the moisture content increased by about 1.5%. Jian et al. (2009b) concluded that even a small silo could produce notable temperature gradients to induce moisture migration. They also reported that grain at the surface of the grain mass underwent larger changes than those inside the grain mass, and grain at one location lost moisture in one-time period and gained moisture in another time. The similar temperature distribution and moisture migration patterns inside different small storage structures have been reported (Montross et al. 2002). There is no report on the distribution of convection currents and moisture migration inside a large bin (larger than 6 m diameter) due to the cost of a large-scale experiment.

Several tests and mathematical models proved the temperature distribution and the moisture migration patterns (Schmidt 1955; Muir et al. 1973; Smith and Sokhansanj 1990; Jian et al. 2005; Jian et al. 2015), while some reports did not find these patterns (Reed 1992). The moisture and temperature distribution for many rectangular – and bunker-type storages has been simulated (Nguyen 1986, 1987). Jian et al. (2015) simulated the temperature and moisture distribution in silo bags, and found the measured canola temperature and moisture content due to moisture migration were similar as that predicted by the mathematical models. Similar observations were made in hermetically sealed bunker silos with 12,000–15,000 tonne capacity (Navarro et al. 1984, 1993). Reed (1992) found moisture accumulated nearly uniformly at the surface of a naturally cooling grain mass but not at the centre surface in three round metal bins containing about 270 tonne of wheat per bin located on farms in Kansas. This discrepancy might be caused by other factors, which will be discussed in the next section. Even though there are discrepancies in the published research, the moisture migration due to convective currents does occur and is widely accepted.

6.3.3.3 Application

The moisture migration can increase moisture content of grain at one location up to 1.5 percentage point due to the continuous air movement during the entire storage period. Most of the mathematical models that attempted to simulate these moisture migrations predicted moisture accumulation of up to 2 percentage points (Jian et al. 2009b). Moisture accumulation occurs not where the grain is cold but where it is slightly cooler. This location is also warm. Fungi, insects and mites can rapidly develop at this location. The respiration of these organisms and grain kernels produce sufficient respiration heat which results in a self-sustaining hotspot.

The above explanation of the moisture migration is based on the same moisture content in the entire bin. Different grain moisture contents in a bin and especially high moisture content can increase this moisture migration because the absolute water content in the air current will be higher when grain moisture content is high. When the RH in moving air is 100%, condensation occurs when warm air with 100% RH moves to cooler places. Under most conditions the effect of high moisture content and condensation is much smaller than the temperature gradient effect.

6.3.4 Condensation of Water on Grain Surface

Water condensation inside stored grain bulk under non aeration conditions is not a common phenomenon because: (1) grain has low temperature change rate than that of air due to the lower thermal diffusivity of the grain than the air; (2) grain will be or close to be in equilibrium with the intergranular air due to the low rate of temperature change inside grain bulk; and (3) grain will absorb or give up water due to this equilibrium process, so no water is condensed out from the air. However, under some special conditions, condensation can occur.

6.3.4.1 Condensation During Turning and Loading

Condensation of water can occur when the temperature of humid air inside bins or around the grain kernels is decreased by the cold grain or cold ambient environments. Condensation from warm and humid ambient air can occur on the cold surface of grain when grain is handled on conveyors, trucks, railcars, barges, and ships. This surface phenomenon is not influenced by the grain initial moisture content. During grain handling, less thickness of the handled grain bulk will have larger surface area of the grain bulk exposed to air, which may result in a larger condensation. Exposure time also influences the condensation. Corn with 25 mm thickness, at 5°C in 1 and 5 min of exposure to 25°C and 85% RH air changed moisture content from 13.6% to 13.7% and 13.9%, respectively; while when the corn is at 15°C, the moisture content changed to 13.7% and 13.8%, respectively (Aldis and Foster 1980). For storage applications, initial moisture content will become important. For example, adding this much moisture to 10% wheat may not be an issue for safe storage, but adding this much moisture to 15% wheat might result in grain spoilage during storage. However, in large mass of grain, a small in increase in moisture on the surface of the grain due to condensation will have negligible or no effect.

6.3.4.2 Condensation inside Bins

Stored grain on the inside surface of galvanized bin walls can become spoiled, which can be caused by condensation (Leitgeb et al. 1990). Galvanized roof was heated to a maximum of 45°C when outside ambient temperature was –15°C in winter at Winnipeg, Canada (Leitgeb et al. 1990). Moisture migrates to the headspace during day time and then condenses on the underside of the roof as it cools rapidly at night. This water may then run down the underside of the roof and drip on to the grain near the walls. This increased moisture can spoil small amount of grain near the walls. This phenomenon occurs mostly in metal bins at limited geographical locations where large temperature fluctuations can occur in the head space due to solar radiation. Jian et al. (2009b) did not find this condensation inside the galvanized bin in 1.5 year observation under Canadian weather conditions.

6.3.4.3 Water Sorption During Air Exchange

Warmer and humid outside air from the head space or outside of a bin entering cold grain will bring moisture to the grain for the same reason as moisture migration caused by the convection currents. Jian et al. (2009b) found grain moisture at the surface of the grain mass underwent larger changes than those inside the grain mass. When warm air inside the headspace moved into the grain, the slight cold grain gained the moisture. The average change of grain moisture at the grain bulk surfaces was 0.8 ± 0.7 percentage point per month with a maximum of 3.0 percentage points. The surface grain at the centre of the bin lost 5.0 percentage points moisture content from March to August 2004, while gained moisture during winter. They concluded that the grain moisture at the surface of the grain mass was mainly influenced by the headspace temperature and relative humidity due to the air exchange between the headspace and the grain at the top of the grain bulk. Between late August when the granaries were filled and early November, moisture content was significantly greater at the surface than at a depth of 20–130 cm (Loschiavo 1985). Therefore, it is recommended to seal the fan inlet when the fan is not used to stop this air movement because the fan ducts can aid the convective currents.

6.3.4.4 Freezing Due to Condensation

High moisture grain kernels usually do not freeze together if there is no dripping water inside stored grain bulk. Ice crystals inside stored grain bulk are also not common. However, water condensation inside stored grain bulk under aeration conditions can occur if grain temperature is lower than air because the water absorption by the grain at lower than freezing temperature cannot absorb the water condensed from the air. When grain temperature is lower than 0°C, this condensed water could be frozen on the surface of the grain kernels. This freezing phenomenon can occur if warmer air in the free convection currents meets with grain at <0°C. Normally this will occur only in a shallow layer of grain and only for a period of time until that grain has been warmed by the air. Such freezing grain could block or restrict the air ways of the ventilation. Therefore, it is not recommended to cool grain to lower than 0°C or aeration should not be conducted to the low temperature grain. It is also not recommended to aerate or dry grain when ambient temperature is lower than 0°C due to the risk of condensation on the under-roof of the headspace.

6.3.5 Factors Affecting Moisture Migration

6.3.5.1 Temperature and Moisture Distribution

Factors affecting the pattern of convection currents and moisture migration are the temperature and moisture distribution and the factors influencing the speed of the free convection currents. For example, the grain temperature at the beginning of the storage may not be the same at different locations due to different harvest conditions at different harvest times. Temperature difference increases due to high initial temperature, large grain bulk that changes grain temperature at its centre slowly, big temperature difference between summer and winter, and bin wall materials that have a high net rate of solar radiation absorption. This uneven temperature distribution will influence the temperature distribution pattern, which might result in different moisture migration patterns. High initial grain moisture content will result in a high moisture content in the air which also increases the amount of water carried by the air currents. The longer the temperature and moisture differences exist in the bin, the longer the convection currents can carry and translocate the moisture. Also, a small moisture accumulation at the place with a high initial moisture content could result in grain spoilage. For example, wheat with 12.5% moisture content can be safely stored at 20°C for at least six months, while 14.0% moisture content wheat (by adding 1.5 percentage point due to moisture migration) cannot be safely stored at 20°C for six months (Canadian Grain Commission 2019).

6.3.5.2 Resistance to Airflow

The more porous is the grain bulk the lower the resistance to airflow and the greater is the possibility of moisture migration. Khankari et al. (1995) proved that moisture migration in wheat was largely governed by diffusion while the effect of natural convection was more prominent in corn because permeability of the grain bed was a critical factor in determining natural convection moisture migration. Smith and Sokhansanj (1990) demonstrated that conduction normally dominates the process of heat transfer. However, convection was important if the resistance to airflow was low enough (if the Rayleigh number $R_a > \left(\frac{L}{R}+\frac{R}{L}\right)^2 k/(2\alpha)$) and if the radius of the storage bin is approximately equal to the height of the bin. Where; L is the height of the grain in a cylindrical bin (m), R is the radius of the cylindrical bin (m), k is the thermal conductivity of the bulk grain (mixture of air and grain) ($W{\cdot}m^{-1}{\cdot}K^{-1}$), and α is the thermal diffusivity of the bulk grain (m^2/s). The method to prevent moisture migration is to narrow down the temperature gradients. The following methods can decrease the temperature difference inside grain bulks: aeration, grain turning, coring grain, and control of insect and mold.

6.3.5.3 Infection by Fungi and Infestation by Insects

Respiration of fungi and insects in stored grain produces water and heat that can change moisture content and temperature in stored grain bins (Jian and Jayas 2012, 2014). Under grain storage conditions, most fungi can multiply when grain temperature is higher than 10°C and water activity is higher than 0.7, even though their multiplication is slow when grain temperature is low (e.g., lower than 20°C). Higher temperature and moisture content encourage a higher multiplication of the infecting fungi which results in the damage of stored grains. Insects can multiply at recommended safe storage moisture contents if the grain temperature is higher than 18°C (White 1995). Even though it is not certain whether insects can directly result in hotspot development (Sinha and Wallace 1966; Smith 1983), insect multiplication can result in mould multiplication, and mould multiplication can result in the increase of temperature and moisture content of stored grain (Sinha and Wallace 1965). These increased temperature and moisture content will boost higher multiplication of insects and mould (a chain action). The consequence of this chain action is the development of hotspots in stored grain bulk. The temperature in a developing hotspot in the centre of a wheat bin can increase to a maximum temperature of about 65°C, while the grain temperature only 40 cm away from the 65°C grain still be at the temperature of 10°C (Sinha and Wallace 1965).

At the same time, moisture migrates to the boundary of the hotspot. The size of the hotspot can be a few centimetres to a few metres. The development of hotspots will result in uneven distribution of grain temperature and moisture content. These increased temperature and moisture content and their uneven distributions result in moisture migration due to the combination of moisture migration mechanisms discussed in this chapter. Therefore, multiplication of insects and fungi will not only directly damage storage grain but will also result in moisture and temperature redistribution, which might further increase the spoilage of the stored grain.

6.4 Safe Storage Moisture Content

6.4.1 Moisture Content Required by Pests Infesting Grain

Insects, mites, rodents, and birds can infest stored grain. Insect pests can infest grain even at a low moisture content (water activity) because stored products insects have adapted to dry food environments through evolution. Stored product insects have a Malpighian tubule at the end of the midgut, so metabolism water can be recovered by the Malpighian tubule before the digestion waste is excreted. Rice weevil adults can survive more than 7 weeks on 9% MC wheat at 29.4°C (Cotton et al. 1960). Confused flour beetle adults can survive more than 14 weeks on 8% MC wheat at 26.7°C. The water activity of the stored grain is lower than 0.3 under these surviving conditions. Therefore, reducing grain moisture contents to the recommended level of safe storage moisture contents cannot effectively control insect pests.

Mites prefer moist environments and require a high moisture content for multiplication. The favourable relative humidity of mite is 75–85%. At 60% RH and lower, a population of the flour mite *Acarus* will eventually extinguish in weeks. Some mites can survive at 62.5% RH as long as the temperature is not much below 10°C or above 20°C (Solomon 1962). At the optimum relative humidity, most mites can multiply at 4–30°C (Solomon 1962) and can survive from 2 to 40°C (Sinha et al. 1991). The equilibrium moisture content of hard red spring wheat at 80% RH and 4–30°C is from 17.5% to 16.6% (Appendix A). These moisture contents are higher than the recommended safe storage moisture contents. Therefore, reducing grain moisture contents to the recommended safe storage moistures can effectively control mites.

Rodents and birds can eat grain with any moisture contents. Effective methods of controlling rodents and birds are proper design of storage facilities and sound management of stored grain.

6.4.2 Moisture Content Required by Milling

Even though different crop types have different desired moisture contents for grinding, these desired moisture contents for grinding are close to the recommended safe storage moisture contents. The desired milling moisture content of wheat is about 15%. If moisture content is lower than 10%, extra stages of tempering must be conducted. The optimum moisture level of wheat flour milling is in the range of 15–16% and moisture levels outside this range will have a negative effect on flour yield. Most commercial wheat flour mills aim for moisture levels near 15% (MOG 2011). Desired moisture content before grinding determines: (1) total tempering time before grinding (more tempering time is needed for grain with low moisture contents); (2) total amount of products (less flour is produced from grain with low moisture contents); (3) powder properties (desired particle size cannot be reached if grain is too dry); (4) efficiency of the grinding process and methods used for grinding; (5) grinding time and energy consumption during grinding (high energy consumption for grain with low moisture contents); and (6) handling efficiency after grinding (Jung et al. 2018). Therefore, reducing grain moisture contents to the level, which insects cannot survive on, is not recommended after considering grinding requirements and feasibility and economical profits of grain drying and storage. However, selling Canadian grain to tropical climates will require drying of grains to lower moisture contents than what is typical in Canada due to low ambient temperatures, for safe storage in those regions.

6.4.3 Moisture Content Required by Microorganism

The main purpose of controlling grain moisture content is to control mould multiplication because growth and multiplication of microorganisms can be eliminated or reduced if the storage grain is at or lower than the recommenced safe storage moisture contents (Appendix A). If grain moisture content is higher than the recommended safe storage moisture content, microorganism might infect the stored grain. Infection by fungi can result in germination reduction, grain discoloration, and increase of grain moisture content and temperature. Infection by fungi and infestation by insects may occur at the same time. Insect and fungi species in a series of succession communities often exist in spoiled grain (Jian and Jayas 2012). The produced heat and water by the respiration of the infecting fungi further increase the multiplication of insects and fungi. Seeds at near 100% RH sprout and produce more water and heat. These interactions result in chain action and the result of these chain actions is the development of hotspot (Jian and Jayas 2014). Grain may smell musty and be caked and could cause total decay in the developed hotspot.

The growth and multiplication of microorganisms are mostly associated with temperatures and moisture contents of stored grain. It is not known exactly how many species of microorganisms attack stored grain because a few dozen common fungal species or more than 150 species of filamentous fungi and yeasts have been identified in stored grain (Nishimwe et al. 2020). Bacteria require near 100% RH, hence, are not involved in storage losses other than in very wet grain or in the final stage of heating of spoiled grain. The most common cause of spoiled grain is the growth of mould (or fungi) on the grain (Christensen and Kaufmann 1969). These fungi are further classified as pre-harvest or field fungi and post-harvest or storage fungi (Christensen 1957). In some situations, a definite distinction between these two groups is impossible because some species may grow in the field and during storage (Nishimwe et al. 2020). This classification is based mainly on the moisture requirements for the fungi growth. Field fungi attack on developing, naturally matured, or fresh harvested seeds at more than 90% RH. Field fungi will die under good grain storage conditions such as when water activity is lower than 0.65. The most common field fungi species are *Fusarium, Alternaria, Cladosporium*, and *Helminthosporium* (Bothast 1978). Certain *Penicillium* species also infect matured seeds at field, but it predominates more in storage, so is more commonly considered as storage fungi. *Aspergillus flavus* and *A. niger* are also found in both stored and freshly harvest grain. The infection by these species starts at the field and carries over through storage (Jedidi et al. 2018).

Compared to the field conditions, stored grain conditions are more defined and stable, and this results in the infection by similar microflora for all stored grains in the world. The main storage fungi in cereal grains, oilseeds, and legumes are *Penicillium*, *Aspergillus*, and *Alternaria* species (Bothast 1978). Even though different species have different optimum growth temperatures and water activities (Fig. 6.5), storage fungi grow well at 0.7–0.9 water activity (Figs. 6.5 and 6.6), and the wheat moisture content is higher than 13.7% when the temperature is 25°C and water activity is higher than 0.7 (Fig. 6.5). Sun et al. (2014) reported the predominant species infecting stored canola were *Penicillium* spp., *Aspergillus* glaucus group, and *Aspergillus candidus* Link. Even though these fungi can infect the stored canola at 10–40°C and 8–14% moisture content, these fungi prefer higher temperature (30–40°C) and moisture content (12–14%). *Alternaria alternata* (Fr.) Keissler and *Cladosporium* spp. were occasionally found in the spoiled canola. The following fungi are the predominant species infecting stored wheat: *A. glaucus* group, *A. flavus* Link, *A. candidus* Link, and *Penicillium* spp. (Karunakaran et al. 2001). *Aspergillus glaucus, Penicillium spp.*, and *A. Alternata* are dominant species infecting pinto bean (Rani et al. 2013). *Alternaria alternata* (Fr.) Keissler is most found in pinto bean with 18–20% MC at ≤20°C, while *Aspergillus ochraceous* and *Aspergillus candidus* are found in pinto bean at high temperatures (≥25°C) and moisture

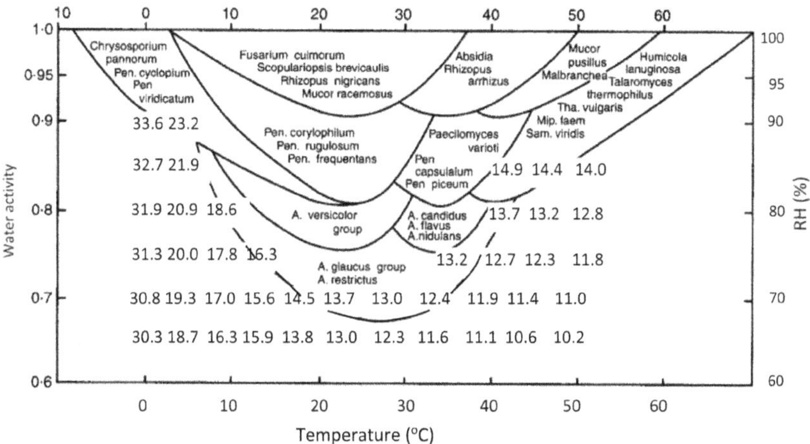

Figure 6.5 Temperature, moisture content, and water activity (relative humidity) for growth of main storage microorganisms. (Figure is adapted from Lacey et al. (1980)). Numbers inside the picture show the equilibrium moisture content (wb, %) of soft winter wheat (Hobbit) under re-wetting condition at the temperature and water activity. Overlap between EMC and microorganisms indicates the microorganism will infect the grain under the condition.

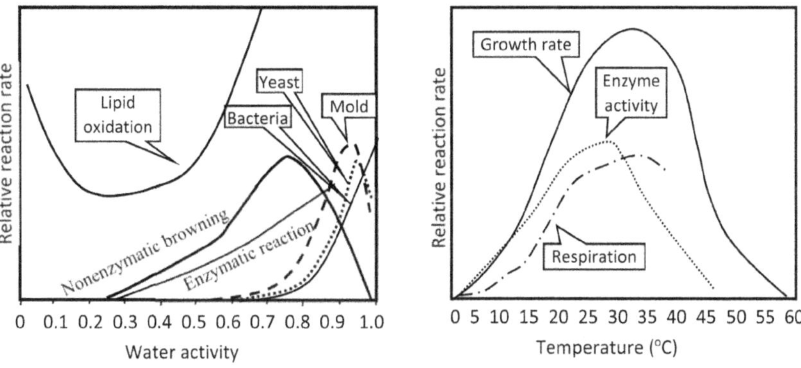

Figure 6.6 Food stability map as a function of water activity and temperature. In the graph, growth rate, enzyme activity, and respiration are the relative reaction of mold and seeds under optimum moisture content at the specified temperature. The water activity map was modified from Labuza et al. (1972).

contents (≥18% MC) after more than 8 weeks of storage. *Alternaria alternata* (Fr.) Keisler is the main species infecting hemp seeds at high temperature (≥25°C) and relative humidity (≥70% RH), while *Aspergillus flavus* Link, *Aspergillus candidus*, *A. ochraceus*, and *A. wentii* can infect hemp seeds at any tested temperatures when the RH is higher than 75% (Jian et al. 2019b). Even though *Penicillium* spp. infects hemp seeds at any temperatures at high RH, it prefers low temperatures.

The main factors influencing fungi multiplication in stored grain is the water activity (equilibrium relative humidity) and temperature of the stored product (Figs. 6.5 and 6.6). Most species of fungi and yeast require at least 0.65 water activity, while bacteria require 0.90 water activity (Woloshuk and Martinez 2012; Maciel et al. 2020). *Aspergillus spp.* can grow in grain at 0.65 water activity. The equilibrium moisture content of wheat and canola at 20°C and 65% RH is approximately 13.8–14.2% and 8.8–9.1%, respectively (Appendix A). These moisture contents are the recommended upper limits of safe storage moisture contents at low temperatures (20°C).

Even though the temperature ranges for continued but slow growth of storage fungi are much wider, the optimum growth temperatures of most microorganisms are between 20 and 40°C. The range of growth temperature of *Aspergillus* spp., *Penicillium* spp., yeasts, and bacteria are −8 to 58, −4 to 48, −2 to 47, and −8 to 80°C, respectively. These temperatures are the temperature of most stored grain in the world. Therefore, controlling grain temperature may be more expensive than controlling grain moisture to control microorganism infection. Properly managed chilled aeration in large grain masses can help manage fungal and insect infestations.

6.4.4 Moisture Content Required by Grain for Respiration and Germination

Compared with the respiration of mold, heat and water produced by the respiration of grain at less than 0.9 water activity is negligible (Seitz et al. 1982; Willcock et al. 2002; Jian et al. 2014a; Jian et al. 2014b). Cereal grains, oilseeds, and legumes sprout at close to1.0 water activity, and this water activity occurs in stored grain bins mostly after heavy mold infection. Therefore, keeping grain at the recommended safe storage moisture contents can effectively control grain respiration, hence grain quality and germination can be maintained. In some countries such as China, grain is stored at lower than the recommended safe storage moisture contents for long-term storage. The state-run grain depots in China store grains for more than five years without quality downgrade.

6.4.5 Moisture Content Required by Chemical Reactions

During storage especially at higher temperatures and moisture contents, chemical reactions, such as lipid oxidization and protein denaturation, occur. Spoiled cereal grains, oilseeds, and legumes have higher fatty acid value (FAV) than the sound grain seeds due to lipid oxidation (Sathya et al. 2008; Sun et al. 2014; Jian et al. 2009b). The hydrolysis of phospholipids leads to the release of glycerol and fatty acids, and this reaction hastens with increasing seed moisture content and temperature. Lipid oxidation can occur regardless whether the grain is infected by mold or not and whether it is processed dry food products (Barden 2014) or stored grain seeds, but fungi infection could hasten lipid oxidation. The loss of native structure and the death of cells can dramatically accelerate the deteriorative reactions of lipid oxidation. A decrease of water activity below 1.0 can initially increase in reaction rates of oxidation because substrates and reactants become more concentrated (Choe and Oh 2013), but a further decrease in a_w will significantly decrease reaction rates because most reactions become diffusion limited (Fig. 6.6). If water activity is low (e.g., less than 0.4), oxidization increases. Even though this phenomenon could be explained by monolayer theory and glass transition theory (Barden 2014), different researchers have different conclusions (Nelson and Labuza 1992; Johnson and Decker 2015) and this is out of scope of this chapter. Jian et al. (2009b) found free FAV increased more rapidly at higher temperatures than at lower temperatures because the oxidation of lipids by lipolytic enzymes will be accelerated by increasing temperatures (Fig. 6.6). Brown rice stored at low temperatures has lower FAV, peroxide value, and carbonyl value than those stored at high temperatures (Liu et al. 2016). The highest value of FAV in spoiled durum wheat is about 150 mg KOH/100 g dry seed which is found at 40°C and 17% moisture content (Nithya et al. 2011). The FAV in stored hemp seeds increases almost linearly to maximum and then keep constant except at 20°C and 95% RH (Jian et al. 2009b). The FAV in stored canola increases linearly to maximum and keep constant except at 30 and 40°C and 14% and 12% MC (Sun et al. 2014). The stored pinto bean also increases its FAV linearly and does not decrease the FAV after the FAV reaches its peak value (Rani et al. 2013). The main factors influencing the rate and extent of oxidative deterioration in food products include the lipid substrate, presence of oxygen, pro-oxidants, inhibitors, moisture content, and temperature.

Denaturation of protein occurs at high temperatures. When corn seeds are stored for 30 days at 40°C and 18% MC, enthalpy of protein denaturation decreases by about 80% for proteins with polar predominance (albumin and globulin), and within 50 days for proteins with hydrophobic predominance (prolamin and glutelin) (Angel et al. 2003). Nonenzymatic browning has been observed in different powdered protein hydrolysates during storage at medium to high a_w (Rao et al. 2016). Endogenous and exogenous enzymatic activities, fungi contamination, water activity, and temperature all contribute to protein deterioration. Below ambient temperature, proteins can be affected adversely with notable changes in cellular systems, solubility, and enzymatic activity. These alterations can affect storage time, salt concentration, and pH dehydration effects. At higher than ambient temperature, heat induced protein denaturation occurs. These chemical reactions usually occur over time as a function of increased storage temperature and relative humidity (moisture content).

In stored grain bins, hotspot can develop because the heat, produced by the respiration of microorganism and seeds, is not dissipated rapidly due to low thermal conductivity and diffusivity of the stored grain (Jian et al. 2014a) and low free convective currents inside the stored grain bulk. In a hotspot, the produced heat and water can induce a self-accelerating heating process, which can increase 15°C from 5 to 20°C in 2 weeks and then increase a further 45°C to a maximum about 65°C in only one more week (Sinha 1961; Sinha and Wallace 1965). Inside the hotspot, mold stop growing at about 65°C, and bacteria can continue to grow up to 80°C. After the biological respiration ceases under the high temperature, chemical reactions continue and increase temperature further to 380–400°C, the ignition temperature of most seeds. The development of hotspot is a complex procedure and mostly influenced by the grain temperature and moisture content. This complexity also explains why some developed hotspots die in the middle of their development (Sinha 1961; Sinha and Wallace 1965; Jian and Jayas 2014). Grain inside hotspots experiences a rapid deterioration and reduction of germinability and quality. In Canada and the USA, a few individual farmers and managers of grain elevators experience heavy losses from heating grain.

An increase in temperature typically increases the rate of chemical reaction. This is true for both processed products and seeds regardless the seed is dead or live. This reaction follows temperature coefficient Q_{10} which is developed from the Van't Hoff-Arrhenius temperature equation. The temperature coefficient Q_{10} states that a rise in 10°C results in around doubling of the reaction rate.

$$Q_{10} = \left(\frac{R_2}{R_1}\right)^{\frac{10}{T_1 - T_2}} \tag{6.1}$$

where Q_{10} is the number of times that the process rate increases with a 10°C rise in temperature, R_1 and R_2 are the chemical reaction rates at temperature T_1 and T_2, respectively, and $T_2 > T_1$. If grain seeds are live, the enzyme activity follows the temperature coefficient Q_{10}. If grain seed is dead, the chemical reaction follows temperature coefficient Q_{10}. Therefore, dead grain seed will decay quickly if temperature is higher than 40°C. It is usually difficult to directly quantify the relationship between moisture content (water activity) and chemical reaction rate because it is a complex process and involves multiple substrates under different conditions. However, grain temperature and moisture content are the main factors influencing this complex process. Therefore, reducing grain moisture content to lower than the recommended safe moisture content and stored grain at low temperatures (e.g., lower than 25°C) are the most recommended methods to safely store grain.

6.4.6 Safe Storage Time and Moisture Content

Safe storage time is the exposure period of a product at a particular moisture content (relative humidity) and temperature above which crop deterioration may occur. To develop a guideline of safe storage moisture content of a crop, all factors influencing grain deterioration and germination should be considered. These factors include microorganism multiplication, grain respiration and sprouting, grinding moisture content, and chemical reaction. Animals (insects, mites, rodent, and birds) infestation is not considered in this process. Therefore, the safe storage moisture content is not always at 65% RH because chemical reaction under different temperatures should be considered. For any stored crop seeds, the ERH of the recommended safe storage moisture content at 20–25°C is about 65% (Appendix A). The ERH of the recommended safe storage moisture content is lower than 65% when temperature is higher than 25°C, while is higher at lower than 20°C (Appendix A).

Considering the complex process of grain spoilage, therefore, the safe storage guideline is mostly developed by storing grain seeds at different storage moisture contents and temperatures. The measured grain temperatures are in the range of 10–40°C because: (1) respiration and chemical reaction is negligible when temperature is lower than 10°C; (2) grain is usually stored at lower than 40°C because storing grain at higher than 40°C has killing effect on seeds and chemical reactions are faster. The tested grain moisture contents are selected at different relative humidities such as 30, 50, 75, and 90%. The following factors are usually determined to evaluate the grain quality: germination, FAV, visible and invisible mold, and discoloration (Cheng et al. 1994; Karunakaran et al. 2001; Sathya et al. 2008). The time of a decrease of 20 percentage points from the initial germination is usually assigned as the safe storage time (Karunakaran et al. 2001; Jian et al. 2009b; Sun et al. 2014). This recommendation is usually used for about half year storage because: (1) a decrease of 20 percentage points from the initial germination occurs in about half year under lab conditions; (2) grain will experience higher temperature after half year storage if the harvest time is in autumn; and (3) moisture migration will result in high moisture content after half year storage. The recommended safe storage moisture

Table 6.2 Upper Limit of Safe Moisture Contents for Storing Grain up to One Year under Canadian Prairie Conditions

	Safe Storage MC (Wet Basis, %)	
Crop	*For Half Year*	*For One Year*
Barley	14.8	13
Canola/rapeseed	10	8
Corn/maize	15.5	13
Domestic buckwheat	16.0	16
Domestic mustard seed	9.5	9.5
Fababeans	16.0	16
Flaxseed and solin	10.0	10
Lentils	14.0	13.5
Oats	14.0	13
Peas	16.0	16
Rye	14.0	13
Safflower	9.5	9.5
Soybean	14.0	13
Sunflower	9.5	8.5
Triticale	14.0	
Wheat	14.5	13

Source: Canadian Grain Commission (1993) and Jayas and Ghosh (2006).

contents of most common cereal grains, oilseeds, and legumes are presented in Appendix A. The safe storage moisture content for up to one year storage is usually 1 percentage point less than that for half year storage (Table 6.2).

The main factors influencing the quality of the stored grain and oilseeds are moisture content and temperature. A decrease in either increases the safe storage time of the stored grain. Compared with decreasing grain temperature and keeping grain at a low temperature, reducing grain moisture content and keeping grain at a low moisture content during the entire storage period is usually less expensive, and more convenient and efficient because: (1) the stored grain usually has a large bulk volume and keeping all the stored grain in a large volume at low temperatures requires appropriately sized chilling unit, a huge storage structure with good insulation. This is costly. (2) The temperature of grain in storage structures is influenced by many factors and one of the main factors is the weather temperature and solar radiation on the structure. To keep grain in the huge structure at a low temperature, removing the energy from the grain received from the ambient weather and solar radiation should be conducted continuously during the entire storage period. This can be done by conducting aeration or chilled aeration, but these could result in rewetting or over-drying of some of the grain. This also adds cost. (3) After each loading, unloading, and grain turning, removing heat from grain should be conducted again. (4) Gain should be dried even though it is stored at low temperatures if the low temperature is not lower than 5–10°C. Therefore, the harvested grain is usually dried to the recommended safe storage moisture contents and stored in a structure. During storage, the grain moisture content will not be directly influenced by ambient weather if no rain or snow water directly moves into the structure. Therefore, one time drying the stored grain before the start of a long-term storage will be much cheaper than multiple cooling of the stored grain. If the grain moisture content is lower than the recommended safe storage moisture content, it is usually not required to control the grain temperature unless aeration is conducted to minimise temperature gradients which result in moisture migration.

6.5 Remark on Ecosystem Consideration

This chapter discussed the uneven distribution of moisture in stored grain bins due to following sources or mechanisms: uneven distribution of grain moisture contents at the very beginning of the storage, water entering into storage structures, and moisture migration due to temperature and moisture gradients and convection currents. This uneven distribution of grain moisture will result in further moisture increase at some locations. These increased moisture content might have a negligible influence on grain chemical and physical properties. However, this high moisture content will boost the multiplication of fungi and insects because the low thermal and water diffusivity of the stored grain bulk will provide niches for the development and multiplication of storage insects and fungi. Insects and fungi have developed the strategies to adapt to this grain storage ecosystem. By adapting to storage temperatures and moisture contents both physically and genetically some species of insects and fungi can circumvent some of the restriction imposed by grain temperature and moisture content. For example, insects move inside stored grain bins to find locations with suitable temperatures and moisture contents for their development and multiplication (Jian et al. 2009a). Most fungi can go to dormant state if water activity is lower than 0.65 in stored grain bulk, while can quickly multiply when water activity is higher than 0.70. The respiration of mould in wet canola (14%) can consume all oxygen in less than 3 days under airtight condition (Jian et al. 2014a), and there was no significant difference in concentration of CO_2 produced by microflora within different crops at 35°C, while there was significant difference at 15 and 25°C (Jian et al. 2009b). The multiplication of insects and fungi will further increase the gradient of grain moisture content. Therefore, controlling grain moisture content and their migration should consider the water activity requirement of insects and fungi for their multiplication and the chain reaction of hotspot development.

Many mechanisms and sources can result in higher moisture contents at some places than that at other locations inside a stored grain bin. Many factors influence grain temperatures as well as affect grain moisture distribution. Grain temperature and moisture content also influence each other. Therefore, initial low moisture content in bins does not mean a safe storage moisture content during later storage. The interaction of these biotic and abiotic factors creates a complex grain storage ecosystem (Jayas 1995) and also increases the difficulty to predict and control the moisture and temperature distributions inside stored grain bins. Regardless how complex is the ecosystem, control of temperature and moisture content is the basic requirement because all living agents require certain temperatures and water activities (grain moisture content) for their growth and multiplication. Keeping grain at a constant temperature during an entire storage period is more difficult than keeping grain in a certain range of moisture content. Therefore, controlling grain moisture content should be always practiced.

Nomenclature

α: thermal diffusivity of the bulk grain (mixture of air and grain) (m^2/s);

EMC: equilibrium moisture content (wet basis, %);

ERH: equilibrium relative humidity (%);

FAV: fatty acid value (mg KOH/100 g dry seed);

H: humidity ratio (kg water/kg dry air);

K: thermal conductivity of the bulk grain (mixture of air and grain) ($W \cdot m^{-1} \cdot K^{-1}$);

L: height of the grain in a cylindrical bin (m);

MC: moisture content (wet basis, %);

M_R: dimensionless moisture ratio;

P_{atm}: atmosphere pressure (Pa);

P_v: vapour pressure (Pa);

Q_{10}: number of times that the process rate increases with a 10°C rise in temperature;

R: radius of the cylindrical bin of grain (m);

Ra: Rayleigh number;

RH: relative humidity (%);

R_1, R_2: chemical reaction rate at temperature T_1 and T_2, respectively, $T_2 > T_1$;

T: temperature (°C).

References

Aldis, D. F., and G. H. Foster. 1980. Moisture change in grain from exposure to ambient air. Transactions of the ASAE 23: 753–760.

Alvarez, D. C., L. M. M. Jorge, and R. M. M. Jorge. 2020. The impact of periodic operation on barley hydration. Journal of Food Process Engineering 43: 1–11.

Anderson, J. A., J. D. Babbitt, and W. O. S. Meredith. 1943. The effect of temperature differential on the moisture content of stored wheat. Canadian Journal of Research 21: 297–306.

Angel, S. S., E. M. Martinez, and M. A. V. Lopez. 2003. Study of denaturation of corn proteins during storage using differential scanning calorimetry. Food Chemistry 83: 531–540.

ASABE. 2016. ASAE D245.6 OCT2007. Moisture Relationships of Plant-Based Agricultural Products, ASABE Standard. American Society of Agricultural and Biological Engineers, St. Joseph, MI.

Barden, L. 2014. Understanding Lipid Oxidation in Low-Moisture Food. PhD, University of Massachusetts Amherst.

Bielewicz, J., S. Cenkowski, and W. E. Muir. 1993. Pretreatment effects on equilibrium moisture contents of canola. Journal of Stored Products Research 29: 37–44.

Bothast, R. J. 1978. Fungal Deterioration and Related Phenomena in Cereals, Legumes and Oilseeds, pp. 462. In H. O. Hultin and M. Milner (eds.), Post-Harvest Biology and Biotechnology Food Nutrition Press Inc., Westport, CT.

Canadian Grain Commission. 1993. Grain Grading Handbook for Western Canada. Canadian Grain Commission, Winnipeg, MB.

Canadian Grain Commission. 2019. Official Grain Grading Guide. Canada Grain Commission, Winnipeg, MB.

Castilloa, M. D., E. J. Martínez, H. H. L. González, A. M. Pacin, and S. L. Resnik. 2003. Study of mathematical models applied to sorption isotherms of Argentinean black bean varieties. Journal of Food Engineering 60: 343–348.

Chen, C. 2000. Factors which affect equilibrium relative humidity of agricultural products. Transactions of the ASAE 43: 673–683.

Chen, C. C., and V. Morey. 1989. Equilibrium relativity humidity (ERH) relationships for yellow-dent corn. Transactions of the ASAE 32: 999–1006.

Cheng, F., B. Mei, J. Yu, H. Dou, and S. Tang 1994. Technical study of controlled atmosphere with carbon dioxide in brick silo for safe storage of wheat, pp. 68–70. In E. Highley, E. J. Wright, H. J. Banks and B. R. Champ (eds.), Proceedings of the 6th International Working Conference on Stored-Product Protection. CAB International, Canberra, Australia.

Choe, E., and S. Oh. 2013. Effects of water activity on the lipid oxidation and antioxidants of dried laver (*Porphyra*) during storage in the dark. Journal of Food Science 78: 1144–1151.

Christensen, C. M. 1957. Deterioration of stored grain by fungi. Botanical Review 23(2): 108–134.

Christensen, C. M., and H. H. Kaufmann. 1969. Grain Storage, the Role of Fungi in Quality Loss, pp. 153. University of Minnisota Press. Minneapolis, Minnesota.

Chung, D. S., S. W. Park, W. J. Hoover, and C. A. Watson. 1972. Sorption kinetics of water vapor by yellow dent corn. Part I. Analysis of kinetic data for damage corn. Cereal Chemistry 49: 598–604.

Coleman, D. A., and H. C. Fellows. 1955. Hygroscopic moisture of cereal grains and flaxseed exposed to different relative humidities. Cereal Chemistry 2: 275–297.

Cotton, R. T., H. H. Walkden, G. D. White, and D. A. Wilbur. 1960. Causes of Outbreaks of Stored-Grain Insects, Agriculture and Apply Science. Kansas State University. Manhattan, Kansas.

Cozzolino, D., S. Roumeliotis, and J. K. Eglinton. 2014. The role of total lipids and fatty acids profile on the water uptake of barley grain during steeping. Food Chemistry 151: 231–235.

Fan, L. T., P. S. Chu, and J. A. Shellenberger. 1963. Diffusion of water in kernels of corn and sorghum. Cereal Chemistry 40: 303–313.

Flood, C. A. J., and J. M. White. 1984. Desorption equilibrium moisture relationships for popcorn. Transactions of the ASAE 27: 561–571.

Ghaly, T. F. 1984. Aeration trial of farm-stored wheat for the control of insect infestation and quality loss. Journal of Stored Products Research 20: 125–131.

Gough, M. C., C. B. S. Uiso, and C. J. Stigter. 1990. Air convection current in metal silos storing maize grain. Tropical Science 30: 217–222.

Griffiths, H. J. 1964. Bulk Storage of Grain: A Summary of Factors Governing Control of Deterioration. CISRO Division of Mechanical Engineering, Melboume, Australia.

Grosh, G. M., and M. Max. 1959. Water penetration and internal cracking in tempered wheat grains. Cereal Chemistry 36: 260–273.

Hellevang, S. 1987. Field Study of Moisture Movement in Stored Grain During Spring and Summer, pp. 14. Annual Conference of the ASAE. ASAE, St. Joseph.

Henderson, S. 1987. Moisture transfer between mixed wet and dry barley grains. Journal of Agricultural and Engineering Research 37: 163–170.

Hubbard, J. E., F. R. Earle, and F. R. Senti. 1957. Moisture relation in wheat and corn. Cereal Chemistry 34: 422–433.

Hutchinson, D. H., and I. Otten. 1984. Equilibrium moisture content of white beans. Cereal Chemistry 61: 155–158.

Jayas, D. S. 1995. Mathematical modeling of heat, moisture, and gas transfer in stored-grain ecosystems, pp. 527–567. In S. Jayas, D. N, G. White, and W. E. Muir (eds.), Stored Grain Ecosystems. Marcel Dekker, New York.

Jayas, D. S., and P. K. Ghosh. 2006. Preserving quality during grain drying and techniques for measuring grain quality, pp. 969–982. In I. Lorini, B. Bacaltchuk, H. Beckel, D. Deckers, E. Sundfeld, J. P. dos Santos,

J. D. Biagi, J. C. Celaro, L. R.'A. Faroni, L. O. F. Bortolini, M. R. Sartori, M. C. Elias, C. Guedes, R. G. da Fonseca and V. M. Scussel (eds.), Proceedings of the 9th International Working Conference on Stored-Product Protection. Brazilian Post-harvest Association ABRAPOS, Campinas, Brazil.

Jayas, D. S., and G. Mazza. 1993. Comparison of five, three-parameter equations for the description of adsorption data of oats. Transactions of the ASAE 36: 119–125.

Jedidi, I., C. Soldevilla, A. Lahouar, and S. Marín. 2018. Mycoflora isolation and molecular characterization of *Aspergillus* and *Fusarium* species in Tunisian cereals. Saudi Journal of Biological Sciences 25: 868–874.

Jian, F., V. Chelladurai, D. S. Jayas, C. J. Demianyk, and N. D. G. White. 2014b. Interstitial concentrations of carbon dioxide and oxygen in stored canola, soybean, and wheat seeds under various conditions. Journal of Stored Products Research 57: 63–72.

Jian, F., V. Chelladurai, D. S. Jayas, and N. D. G. White. 2015. Three-dimensional transient heat, mass, and momentum transfer model to predict conditions of canola stored inside silo bags under Canadian prairie conditions: Part II. Model of canola bulk temperature and moisture content. Transactions of the ASABE 58: 1135–1144.

Jian, F., D. Divagar, J. Mhaiki, D. S. Jayas, P. G. Fields, and N. D. G. White. 2018. Static and dynamic methods to determine adsorption isotherms of hemp seed (*Cannabis sativa* l.) with different percentages of dockage. Food Science & Nutrition 6: 1629–1640.

Jian, F., and D. S. Jayas. 2012. The ecosystem approach to grain storage. Agricultural Research 1: 148–156.

Jian, F., and D. S. Jayas. 2014. Understanding the initiation and development of hotspots in storage-grain ecosystems. Journal of Applied Zoological Research 25: 1–10.

Jian, F., D. S. Jayas, P. G. Fields, and N. D. G. White. 2017. Water sorption and cooking time of red kidney beans (*Phaseolus vulgaris* L.): Part I – effect of freezing and drying conditions on water sorption and cooking time. International Journal of Food Science & Technology 52: 2031–2039.

Jian, F., D. S. Jayas, and N. D. G. White. 2009a. Optimal environmental search and scattered orientations during movement of adult rusty grain beetles, *Cryptolestes ferrugineus* (Stephens), in grain bulks – suggested movement and distribution patterns. Journal of Stored Products Research 45: 177–183.

Jian, F., D. S. Jayas, and N. D. G. White. 2009b. Temperature fluctuations and moisture migration in wheat stored for 15 months in a metal silo in Canada. Journal of Stored Products Research 45: 82–90.

Jian, F., D. S. Jayas, and N. D. G. White. 2014a. Heat production of stored canola seeds under airtight and non-airtight conditions. Transactions of the ASABE 57: 1151–1162.

Jian, F., D. S. Jayas, N. D. G. White, and K. Alagusundaram. 2005. A three-dimensional, asymmetric, and transient model to predict grain temperatures in grain storage bins. Transactions of the ASAE 48: 263–271.

Jian, F., M. A. A. Mamun, N. D. G. White, D. S. Jayas, P. G. Fields, and J. McCombe. 2019b. Safe storage times of FINOLA® hemp (*Cannabis sativa*) seeds with dockage. Journal of Stored Products Research 83: 34–43.

Jian, F., P. Tang, M. A. A. Mamun, and D. S. Jayas. 2019a. Effect of field treatment on microfloral respiration and storability of canola under different storage conditions. American Journal of Plant Sciences 10: 1989–2001.

Johnson, D. R., and E. A. Decker. 2015. The role of oxygen in lipid oxidation reactions: a review. Annual Review of Food Science and Technology 6: 171–190.

Jung, H., Y. J. Lee, and W. B. Yoon. 2018. Effect of moisture content on the grinding process and powder properties in food: a review. Processes 69: 1–16.

Karunakaran, C., W. E. Muir, D. S. Jayas, N. D. G. White, and D. Abramson. 2001. Safe storage time of high moisture wheat. Journal of Stored Products Research 37: 303–312.

Khankari, K. K., S. V. Patankar, and R. V. Morey. 1995. A mathematical model for natural convection moisture migration in stored grain. Transactions of the ASAE 38: 1777–1787.

Kumar, M. 1973. Moisture distribution between whole corn, endosperm and germ by various methods of conditioning. Food Technology 8: 407–417.

Labuza, T. P., L. McNally, D. Gallagher, J. Hawkes, and F. Hurtado. 1972. Stability of intermediate moisture foods. 1. Lipid oxidation. Journal of Food Science 37: 154–159.

Lacey, J., S. T. Hill, and M. A. Edwards. 1980. Micro-organisms in stored grains: their enumeration and significance. Tropical Stored Products Information 39: 19–32.

Leitgeb, J. M., M. G. Zhang, and Britton. 1990. Effects of change in ambient temperature on wall pressure in grain bins, pp. 13. Annual Conference of the ASAE. ASAE, St. Joseph, MI.

Liu, K., Y. Li, F. Chen, and F. Yong. 2016. Lipid oxidation of brown rice stored at different temperatures. International Journal of Food Science & Technology 52: 188–195.

Loschiavo, S. R. 1985. Post-harvest grain temperature, moisture, and insect infestation in steel granaries in Manitoba. The Canadian Entomologist 117: 7–14.

Maciel, G., D. Torre, L. M. Cardoso, M. G. Cendoya, and J. R. Wagner. 2020. Determination of safe storage moisture content of soybean expeller by means of sorption isotherms and product respiration. Journal of Stored Products Research 86: 1–7.

Mayolle, J. E., V. Lullien-Pellerin, F. Corbineau, P. Boivin, and V. Guillard. 2012. Water diffusion and enzyme activities during malting of barley grains: a relationship assessment. Journal of Food Engineering 109: 358–365.

Mazza, G., D. S. Jayas, and N. D. G. White. 1990. Moisture sorption isotherms of flax seed. Transactions of the ASAE 33: 1313–1318.

MOG, D. 2011. An Analysis of Factors Influencing Wheat Flour Yield. MSc, Kansas State University Manhatten, Kansas.

Montross, M. D., D. E. Maier, and K. Haghighi. 2002. Validation of a finite-element stored grain ecosystem model. Transactions of the ASAE 45: 1465–1474.

Muir, W. E., A. Kumar, B. M. Fraser, and M. G. Britton. 1978. Development of emergency structures for grain storage. Canadian Agricultural Engineering 20: 30–33.

Muir, W. E., R. N. Sinha, and H. A. H. Wallace. 1973. Abiotic and biotic characteristics of grain stored in temporary farm bins. Canadian Agricultural Engineering 15: 35–42.

Muir, W. E., R. N. Sinha, H. A. Wallace, and P. L. Sholberg. 1980. Emergency farm structures for storing grain – a multidisciplinary evaluation. Transactions of the ASAE 23: 208–213, 217.

Navarro, S., E. Donahaye, and M. Calderon. 1969. Observations on prolonged grain storage with forced aeration in Israel. Journal of Stored Products Research 5: 73–81.

Navarro, S., E. Donahaye, and M. Calderon. 1973a. Studies on aeration with refrigerated air–II. Chilling of soybeans undergoing spontaneous heating. Journal of Stored Products Research 9: 261–268.

Navarro, S., E. Donahaye, and M. Calderon. 1973b. Studies on aeration with refrigerated air–I. Chilling of wheat in a concrete elevator. Journal of Stored Products Research 9: 253–259.

Navarro, S., E. Donahaye, Y. Eashanchi, V. Pisarev, and O. Bulbul 1984. Airtight storage of wheat in a P.V.C. Covered bunker, pp. 601–614. In B. E. Ripp (ed.), Controlled Atmosphere an Fumigation in Grain Storages. Esevier, Amsterdam.

Navarro, S., R. Jayas, and Noyes. 2001. Stored grain ecosystem and heat, and moisture transfer in grain bulks. In S. Navarro and R. T. Noyes (eds.), The Mechanics and Physics of Modern Grain Aeration Management. CRC Press, Boca Raton, London.

Navarro, S., E. Varnava, and E. Donahaye 1993. Preservation of grain in hermetically sealed plastic liners with particular reference to storage of barley in Cyprus, pp. 223–234. In S. Navarro and E. Donahye (eds.), International Conference on Controlled Atmosphere and Fumigation in Grain Storages. Caspit Press, Winnipeg, Canada.

Nelson, K. A., and T. P. Labuza 1992. Relationship between water and lipid oxidation rates: water activity and glass transition theory, pp. 93–103. In A. J. St Angelo (ed.), Lipid Oxidation in Foods. American Chemical Society, Washington, DC.

Nguyen, T. V. 1986. Modelling temperatrue and moisture changes resulting from natural convection in grain stores, pp. 81–88. In B. R. Champ and E. Highley (eds.), Peserving Grain Quality by Aeration and In-Store Drying. Australia Centre Internation Agricultural Research, Canberra, Australia.

Nguyen, T. V. 1987. Natural convection effects in stored grains – a simulation study. Drying Technology 5: 541–560.

Nishimwe, K., J. A. L. Mandap, and G. Munkvold 2020. Advances in understanding fungal contamination in cereals, pp. 31–65. In D. E. Maier (ed.), Advances in Postharvest Management of Cereals and Grains. Burleigh Dodds Science Publishing Limited, Cambridge.

Nithya, U., V. Chelladurai, D. S. Jayas, and N. D. G. White. 2011. Safe storage guidelines for durum wheat. Journal of Stored Products Research 47: 328–333.

Noyes, R., and S. Navarro 2001. Operating aeratiuon systems, pp. 315–412. In S. Navarro and R. Noyes (eds.), The Mechanics and Physics of Modern Grain Aeration Management. CRC Press, Boca Raton, London, NY, Washington DC.

Osborn, G., G. White, H. Salaiman, and L. Walton. 1989. Predicting equilibrium moisture properties of soybeans. Transactions of the ASAE 32: 2109–2113.

Phillips, S., S. Widjaja, A. Wallbridge, and R. Cooke. 1988. Rice yellowing during post-harvest drying by aeration and during storage. Journal of Stored Products Research 24: 173–181.

Pfost, H. B., S. G. Mourer, D. S. Chung, and G. A. Milliken. 1976. Summarizing and reporting equilibrium moisture data for grains. Annual Conference of ASAE, December 6–8, St. Joseph, MI. Paper No. 76-3520.

Pichler, J. H. 1957. Sorption isotherms of grain and rape. Journal of Agricultural Engineering Research 2: 159–165.

Pixton, S. W., and H. J. Griffiths. 1971. Diffusion of moisture through grain. Journal of Stored Products Research 7: 133–152.

Pixton, S. W., and S. Henderson. 1981. The moisture content-equilibrium relative humidity relationships of five varieties of Canadian wheat and of candle rapeseed at different temperatures. Journal of Stored Products Research 17: 187–190.

Pixton, S. W., and S. Warburton. 1977. The moisture content/equilibrium relative humidity relationship of a dried yeast product. Journal of Stored Products Research 13: 35–37.

Prasad, D. C., W. E. Muir, and H. A. H. Wallace. 1978. Characteristics of freshly harvested wheat and rapeseed. Transactions of the ASAE 21: 782–784.

Rani, P. R., V. Chelladurai, D. S. Jayas, N. D. G. White, and C. V. Kavitha-Abirami. 2013. Storage studies on pinto beans under different moisture contents and temperature regimes. Journal of Stored Products Research 52: 78–85.

Rao, Q., A. K. Kamdar, and T. P. Labuza. 2016. Storage stability of food protein hydrolysates—a review. Critical Reviews in Food Science and Nutrition 56: 1169–1192.

Reed, C. 1992. Gain cooling, moisture transloation, and theoties of convection current, pp. 14–15. In D. S. Jayas, D. N. G. White, W. E. Muir and R. N. Sinha (eds.), Extended Abstracts: International Sympsium on Stored Grain Ecosystems. Deptartment of Agriculture Engineering, University of Manitoba, Winnipeg, MB.

Reynolds, T., and I. C. MacWilliam. 1966. Water uptake and enzymic activity during steeping of barley Journal of the Institute of Brewing 72.

Sathya, G., D. S. Jayas, and N. D. G. White. 2008. Safe storage guidelines for rye. Canadian Biosystems Engineering 50: 1–8.

Schmidt, J. L. 1955. Wheat Storage Research at Hutchinson, Kansas, and Jamestown, pp. 98. US Department of Agriculture, Washington DC.

Seitz, L. M., D. B. Sauer, and H. E. Mohr. 1982. Storage of high-moisture corn: fungal growth and dry matter loss. Cereal Chemistry 59: 100–105.

Sinha, R. N. 1961. Insects and mites associated with hot spots in farm stored grain. Canadian Entomologist 93: 609–621.

Sinha, R. N., W. E. Muir, and D. B. Sanderson. 1985. Quality assessment of stored wheat during drying with near-ambient temperature air. Canadian Journal of Plant Science 65: 849–866.

Sinha, R. N., W. E. Muir, D. B. Sanderson, and D. Tume. 1991. Ventilation of bin-stored moist wheat for quality preservation. Canadian Agricultural Engineering 33: 055–065.

Sinha, R. N., and H. A. H. Wallace. 1965. Ecology of a fungus-induced hot spot in stored grain. Canadian journal of plant science 45: 48–59.

Sinha, R. N., and H. A. H. Wallace. 1966. Ecology of insect-induced hot spots in stored grain in western Canada. Researches on Population Ecology VIII: 107–132.

Smith, E. A., and S. Sokhansanj. 1990. Moisture transport caused by natural convection in grain stores. Journal of Agricultural and Engineering Research 47: 23–34.

Smith, L. B. 1983. The relationship between wet grain, *Cryptolestes ferrugineus* (Coleoptera: Cucujidae) populations, and heating in wheat stored in granaries. Canadian Entomologist 115: 1383–1394.

Sokhansanj, S., W. P. Lampman, and J. D. MacAulay. 1983. Investigation of grain tempering on drying tests. Transacrions of the ASAE: 293–296.

Sokhansanj, S., W. Zhijie, D. S. Jayas, and T. Kameoko. 1986. Equilibrium relative humidity-moisture content of rapeseed (canola) from 5°C to 25°C. Transaction of ASAE 29: 837–839.

Solomon, M. E. 1962. Ecology of the flour mite, *Acarus siro L.* (Tyroglyphus farinae DeG.). Annals Applied Biology 50: 178–184.

Stewart, J. A. 1975. Moisture migration during storage of preserved, high moisture grains. Transactions of the ASAE 18: 387–393, 400.

Sun, K., F. Jian, D. S. Jayas, and N. D. G. White. 2014. Quality changes in high and low oil content canola during storage: part I – Safe storage time under constant temperatures. Journal of Stored Products Research 59: 320–327.

Throne, J. E. 1991. Equilibrium moisture contents of cracked and whole corn (maize). Journal of Stored Products Research 27: 129–130.

White, G. M., I. J. Ross, and J. D. Klaiber. 1972. Moisture equilibrium in mixing of shelled corn. Transaction of the ASAE 15: 508–509, 514.

White, N. D. G. 1995. Insects, mites, and insecticides in stored-grain ecosystems, pp. 527–567. In D. S. Jayas, G. White and W. E. Muir (eds.), Stored-Grain Ecosystems. Marcel Dekker, New York.

Willcock, J., D. Aldred, and N. Magan 2002. Effect of water activity and biocides on spoilage and dry matter losses of wheat straw, pp. 546–552. In P. F. Credland, D. Armitage, C. H. Bell, P. M. Cogan and E. Highley (eds.), Proceedings of the 8th International Working Conference on Stored-Product Protection. CAB International, York.

Woloshuk, C., and E. M. Martinez 2012. Molds and other microbes in stored products, pp. 63–68. In D. W. Hagstrum, T. W. Phillips and G. Cuperus (eds.), Stored Product Protection. Kansas State Research and Extension, Kansas.

Appendix A: Safe Storage Equilibrium Moisture Content (Wet Basis) of Stored Cereal Grains, Oilseeds, and Legumes (in Each Group, Crops Are Presented in the Alphabetical Order)

A1 Cereal Grains

Barley Seeds under Drying Condition

	Relative Humidity (%)[a]										
T (°C)	*35*	*40*	*45*	*50*	*55*	*60*	*65*	*70*	*75*	*80*	*85*
10	10.3	11.1	11.8	12.5	13.3	14.1	14.9	15.8	**16.8**	**18.0**	**19.4**
13	10.1	10.9	11.6	12.3	13.1	13.9	14.7	15.6	**16.7**	**17.8**	**19.2**
15	10.0	10.7	11.5	12.2	13.0	13.8	14.6	15.5	**16.5**	**17.7**	**19.1**
18	9.8	10.6	11.3	12.0	12.8	13.6	14.4	15.4	**16.4**	**17.6**	**19.0**
20	9.7	10.4	11.2	11.9	12.7	13.5	14.3	**15.3**	**16.3**	**17.5**	**18.9**
23	9.5	10.3	11.0	11.7	12.5	13.3	14.2	**15.1**	**16.1**	**17.3**	**18.8**
25	9.4	10.1	10.9	11.6	12.4	13.2	**14.1**	**15.0**	**16.0**	**17.2**	**18.7**
28	9.2	10.0	10.7	11.5	12.3	13.1	**13.9**	**14.9**	**15.9**	**17.1**	**18.5**
30	9.1	9.9	10.6	11.4	12.2	13.0	**13.8**	**14.8**	**15.8**	**17.0**	**18.4**
33	9.0	9.7	10.5	11.2	12.0	12.8	**13.7**	**14.6**	**15.7**	**16.9**	**18.3**
35	8.9	9.6	10.4	11.1	11.9	**12.7**	**13.6**	**14.5**	**15.6**	**16.8**	**18.2**
38	8.7	9.5	10.2	11.0	**11.8**	**12.6**	**13.4**	**14.4**	**15.4**	**16.6**	**18.1**
40	**8.6**	**9.4**	**10.1**	**10.9**	**11.7**	**12.5**	**13.4**	**14.3**	**15.3**	**16.6**	**18.0**
43	**8.5**	**9.2**	**10.0**	**10.7**	**11.5**	**12.3**	**13.2**	**14.2**	**15.2**	**16.4**	**17.9**
45	**8.4**	**9.1**	**9.9**	**10.6**	**11.4**	**12.3**	**13.1**	**14.1**	**15.1**	**16.4**	**17.8**
48	**8.2**	**9.0**	**9.7**	**10.5**	**11.3**	**12.1**	**13.0**	**14.0**	**15.0**	**16.2**	**17.7**
50	**8.1**	**8.9**	**9.6**	**10.4**	**11.2**	**12.0**	**12.9**	**13.9**	**14.9**	**16.2**	**17.6**
53	**8.0**	**8.7**	**9.5**	**10.3**	**11.1**	**11.9**	**12.8**	**13.7**	**14.8**	**16.0**	**17.5**
55	**7.9**	**8.7**	**9.4**	**10.2**	**11.0**	**11.8**	**12.7**	**13.7**	**14.7**	**16.0**	**17.4**

[a] Equilibrium moisture contents are calculated using the modified Chung-Pfost equation using constants recommended by ASABE (2016). Regular numbers are the moisture contents at which seeds can be safely stored for up to 6 months at given air conditions but at bolded moisture contents seeds will likely spoil within 6 months as recommended by Jacobsen and Fleurat-Lessard (2002).

Barley (*Hordium vulgare*) Seeds under Re-Wetting Condition

	Relative Humidity (%)[a]										
T (°C)	*45*	*50*	*55*	*600*	*65*	*70*	*75*	*80*	*85*	*90*	*95*
10	11.0	11.6	12.3	13.0	13.8	14.6	15.5	**16.5**	**17.8**	**19.5**	**22.2**
13	11.0	11.6	12.3	13.0	13.7	14.6	15.5	**16.5**	**17.8**	**19.5**	**22.1**
15	10.9	11.6	12.3	13.0	13.7	14.6	15.5	**16.5**	**17.8**	**19.5**	**22.1**
18	10.9	11.6	12.3	13.0	13.7	14.5	15.5	**16.5**	**17.8**	**19.5**	**22.1**
20	10.9	11.6	12.3	13.0	13.7	14.5	**15.4**	**16.5**	**17.8**	**19.5**	**22.1**
23	10.9	11.6	12.2	12.9	13.7	**14.5**	**15.4**	**16.5**	**17.8**	**19.5**	**22.1**
25	10.9	11.6	12.2	12.9	**13.7**	**14.5**	**15.4**	**16.5**	**17.8**	**19.4**	**22.1**
28	10.9	11.5	12.2	12.9	**13.7**	**14.5**	**15.4**	**16.5**	**17.7**	**19.4**	**22.1**
30	10.9	11.5	12.2	**12.9**	**13.7**	**14.5**	**15.4**	**16.5**	**17.7**	**19.4**	**22.1**
33	10.9	11.5	12.2	**12.9**	**13.6**	**14.5**	**15.4**	**16.4**	**17.7**	**19.4**	**22.1**
35	10.8	11.5	12.2	**12.9**	**13.6**	**14.5**	**15.4**	**16.4**	**17.7**	**19.4**	**22.1**
38	10.8	11.5	**12.2**	**12.9**	**13.6**	**14.4**	**15.4**	**16.4**	**17.7**	**19.4**	**22.0**
40	**10.8**	**11.5**	**12.2**	**12.9**	**13.6**	**14.4**	**15.4**	**16.4**	**17.7**	**19.4**	**22.0**

[a] Equilibrium moisture contents are calculated using the modified Chung-Pfost equation and constants from Basunia and Abe (2005). Regular numbers are the moisture contents at which seeds can be safely stored for up to 6 months at given air conditions but at bolded moisture contents seeds will likely spoil within 6 months as recommended by Jacobsen and Fleurat-Lessard (2002).

Buckwheat (*Miyazaki Ohtsubu*) Seeds under Drying Condition

	Relative Humidity (%)[a]										
T (°C)	*35*	*40*	*45*	*50*	*55*	*60*	*65*	*70*	*75*	*80*	*85*
10	12.1	12.9	13.7	14.4	15.2	16.0	16.8	17.6	18.5	**19.5**	**20.6**
13	11.9	12.7	13.4	14.2	14.9	15.7	16.5	17.3	18.1	**19.1**	**20.2**
15	11.8	12.5	13.3	14.0	14.7	15.5	16.2	17.0	17.9	**18.8**	**19.9**
18	11.6	12.3	13.0	13.7	14.5	15.2	15.9	16.7	17.5	**18.5**	**19.5**
20	11.4	12.2	12.9	13.6	14.3	15.0	**15.7**	**16.5**	**17.3**	**18.2**	**19.3**
23	11.2	12.0	12.6	13.3	14.0	14.7	**15.4**	**16.2**	**17.0**	**17.9**	**18.9**
25	11.1	11.8	12.5	**13.2**	**13.9**	**14.5**	**15.2**	**16.0**	**16.8**	**17.6**	**18.6**
28	10.9	11.6	12.3	**12.9**	**13.6**	**14.3**	**15.0**	**15.7**	**16.5**	**17.3**	**18.3**
30	10.8	**11.5**	**12.1**	**12.8**	**13.4**	**14.1**	**14.8**	**15.5**	**16.2**	**17.1**	**18.0**
33	**10.6**	**11.3**	**11.9**	**12.6**	**13.2**	**13.9**	**14.5**	**15.2**	**15.9**	**16.8**	**17.7**
35	**10.5**	**11.2**	**11.8**	**12.4**	**13.1**	**13.7**	**14.3**	**15.0**	**15.8**	**16.6**	**17.5**
38	**10.4**	**11.0**	**11.6**	**12.2**	**12.8**	**13.5**	**14.1**	**14.8**	**15.5**	**16.3**	**17.2**
40	**10.3**	**10.9**	**11.5**	**12.1**	**12.7**	**13.3**	**13.9**	**14.6**	**15.3**	**16.1**	**16.9**
43	**10.1**	**10.7**	**11.3**	**11.9**	**12.5**	**13.1**	**13.7**	**14.3**	**15.0**	**15.8**	**16.6**
45	**10.0**	**10.6**	**11.2**	**11.8**	**12.4**	**12.9**	**13.5**	**14.2**	**14.8**	**15.6**	**16.4**
48	**9.8**	**10.4**	**11.0**	**11.6**	**12.2**	**12.7**	**13.3**	**13.9**	**14.6**	**15.3**	**16.2**
50	**9.8**	**10.3**	**10.9**	**11.5**	**12.0**	**12.6**	**13.2**	**13.8**	**14.4**	**15.1**	**16.0**
53	**9.6**	**10.2**	**10.7**	**11.3**	**11.8**	**12.4**	**13.0**	**13.5**	**14.2**	**14.9**	**15.7**
55	**9.5**	**10.1**	**10.6**	**11.2**	**11.7**	**12.3**	**12.8**	**13.4**	**14.0**	**14.7**	**15.5**

[a] Equilibrium moisture contents are calculated using the Day and Nelson equation and constants from Tagawa et al. (1993). Regular numbers are the moisture contents at which seeds can be safely stored for up to 6 months at given air conditions but at bolded moisture contents seeds will likely spoil within 6 months (estimated based on authors' experience).

Buckwheat (Dialog) Seeds under Desorption or Sorption Condition (No Hysteresis)

	Relative Humidity (%)[a]										
T (°C)	*45*	*50*	*55*	*60*	*65*	*70*	*75*	*80*	*85*	*90*	*95*
10	12.6	13.1	13.7	14.4	15.0	15.8	16.7	17.7	**19.0**	**20.9**	**24.2**
13	12.4	12.9	13.5	14.2	14.8	15.6	16.4	17.5	**18.8**	**20.7**	**23.9**
15	12.3	12.8	13.4	14.0	14.7	15.4	16.3	17.3	**18.6**	**20.5**	**23.8**
18	12.1	12.6	13.2	13.8	14.5	15.2	16.1	17.1	**18.4**	**20.2**	**23.5**
20	12.0	12.5	13.1	13.7	**14.3**	**15.1**	**15.9**	**16.9**	**18.2**	**20.0**	**23.3**
23	11.8	12.3	12.9	13.5	**14.1**	**14.9**	**15.7**	**16.7**	**18.0**	**19.8**	**23.0**
25	11.7	12.2	12.8	**13.4**	**14.0**	**14.7**	**15.5**	**16.5**	**17.8**	**19.6**	**22.8**
28	11.5	12.0	12.6	**13.2**	**13.8**	**14.5**	**15.3**	**16.3**	**17.5**	**19.3**	**22.4**
30	11.4	**11.9**	**12.4**	**13.0**	**13.7**	**14.4**	**15.2**	**16.1**	**17.4**	**19.1**	**22.2**
33	**11.2**	**11.7**	**12.3**	**12.8**	**13.4**	**14.1**	**14.9**	**15.9**	**17.1**	**18.8**	**21.9**
35	**11.1**	**11.6**	**12.1**	**12.7**	**13.3**	**14.0**	**14.8**	**15.7**	**16.9**	**18.7**	**21.7**
38	**10.9**	**11.4**	**11.9**	**12.5**	**13.1**	**13.8**	**14.5**	**15.5**	**16.7**	**18.4**	**21.4**
40	**10.8**	**11.3**	**11.8**	**12.3**	**12.9**	**13.6**	**14.4**	**15.3**	**16.5**	**18.2**	**21.2**

[a] Equilibrium moisture contents are calculated using the modified Oswin equation and constants from Menkov et al. (2009). Regular numbers are the moisture contents at which seeds can be safely stored for up to 6 months at given air conditions but at bolded moisture contents seeds will likely spoil within 6 months (estimated based on authors' experience).

Corn (Hybrid 704) Seeds under Drying Condition

	Relative Humidity (%)[a]										
T (°C)	*35*	*40*	*45*	*50*	*55*	*60*	*65*	*70*	*75*	*80*	*85*
10	11.6	12.6	13.5	14.5	15.5	16.5	17.6	18.7	**20.0**	**21.4**	**23.1**
13	11.3	12.3	13.2	14.2	15.1	16.1	17.2	18.3	**19.5**	**21.0**	**22.6**
15	11.2	12.1	13.0	14.0	14.9	15.9	16.9	18.0	**19.3**	**20.7**	**22.3**
18	11.0	11.9	12.8	13.7	14.6	15.6	16.6	**17.7**	**18.9**	**20.2**	**21.8**
20	10.8	11.7	12.6	13.5	14.4	15.4	**16.4**	**17.4**	**18.6**	**20.0**	**21.5**
23	10.6	11.5	12.4	13.2	14.1	15.1	**16.1**	**17.1**	**18.3**	**19.6**	**21.1**
25	10.5	11.3	12.2	13.1	14.0	**14.9**	**15.9**	**16.9**	**18.0**	**19.3**	**20.9**
28	10.3	11.1	12.0	12.8	**13.7**	**14.6**	**15.6**	**16.6**	**17.7**	**19.0**	**20.5**
30	10.2	11.0	11.8	**12.7**	**13.6**	**14.4**	**15.4**	**16.4**	**17.5**	**18.8**	**20.3**
33	10.0	10.8	11.6	**12.5**	**13.3**	**14.2**	**15.1**	**16.1**	**17.2**	**18.5**	**19.9**
35	9.9	10.7	**11.5**	**12.3**	**13.2**	**14.0**	**15.0**	**16.0**	**17.0**	**18.3**	**19.7**
38	9.7	10.5	**11.3**	**12.1**	**13.0**	**13.8**	**14.7**	**15.7**	**16.8**	**18.0**	**19.4**
40	**9.6**	**10.4**	**11.2**	**12.0**	**12.8**	**13.7**	**14.6**	**15.5**	**16.6**	**17.8**	**19.2**
43	**9.5**	**10.3**	**11.0**	**11.8**	**12.6**	**13.5**	**14.4**	**15.3**	**16.3**	**17.5**	**18.9**
45	**9.4**	**10.2**	**10.9**	**11.7**	**12.5**	**13.3**	**14.2**	**15.2**	**16.2**	**17.3**	**18.7**
48	**9.3**	**10.0**	**10.8**	**11.6**	**12.3**	**13.2**	**14.0**	**14.9**	**16.0**	**17.1**	**18.5**
50	**9.2**	**9.9**	**10.7**	**11.4**	**12.2**	**13.0**	**13.9**	**14.8**	**15.8**	**16.9**	**18.3**
53	**9.0**	**9.8**	**10.5**	**11.3**	**12.1**	**12.9**	**13.7**	**14.6**	**15.6**	**16.7**	**18.0**
55	**9.0**	**9.7**	**10.4**	**11.2**	**12.0**	**12.7**	**13.6**	**14.5**	**15.5**	**16.6**	**17.9**

[a] Equilibrium moisture contents are calculated using the modified Henderson equation and constants from ASABE (2016). Regular numbers are the moisture contents at which seeds can be safely stored for up to 6 months at given air conditions but at bolded moisture contents seeds will likely spoil within 6 months as recommended by Sumner and Williams (2009).

Corn (Hybrid 704) Seeds under Re-Wetting Condition

	Relative Humidity (%)[a]										
T (°C)	*45*	*50*	*55*	*60*	*65*	*70*	*75*	*80*	*85*	*90*	*95*
10	12.5	13.4	14.4	15.4	16.4	17.5	18.8	**20.2**	**21.8**	**24.0**	**27.2**
13	12.2	13.1	14.1	15.0	16.0	17.1	18.3	**19.7**	**21.3**	**23.4**	**26.6**
15	12.0	12.9	13.9	14.8	15.8	16.9	**18.1**	**19.4**	**21.0**	**23.1**	**26.2**
18	11.8	12.7	13.6	14.5	15.5	**16.5**	**17.7**	**19.0**	**20.6**	**22.6**	**25.7**
20	11.6	12.5	13.4	14.3	15.3	**16.3**	**17.5**	**18.8**	**20.3**	**22.3**	**25.3**
23	11.4	12.2	13.1	14.0	15.0	**16.0**	**17.1**	**18.4**	**19.9**	**21.9**	**24.8**
25	11.3	12.1	12.9	13.8	**14.8**	**15.8**	**16.9**	**18.2**	**19.7**	**21.6**	**24.5**
28	11.1	11.9	12.7	**13.6**	**14.5**	**15.5**	**16.6**	**17.8**	**19.3**	**21.2**	**24.0**
30	10.9	11.7	**12.6**	**13.4**	**14.3**	**15.3**	**16.4**	**17.6**	**19.1**	**20.9**	**23.8**
33	10.7	11.5	**12.3**	**13.2**	**14.1**	**15.0**	**16.1**	**17.3**	**18.7**	**20.6**	**23.3**
35	10.6	**11.4**	**12.2**	**13.0**	**13.9**	**14.9**	**15.9**	**17.1**	**18.5**	**20.3**	**23.1**
38	10.4	**11.2**	**12.0**	**12.8**	**13.7**	**14.6**	**15.7**	**16.8**	**18.2**	**20.0**	**22.7**
40	**10.3**	**11.1**	**11.9**	**12.7**	**13.6**	**14.5**	**15.5**	**16.7**	**18.0**	**19.8**	**22.5**

[a] Equilibrium moisture contents are calculated using the modified Henderson equation and constants from ASABE (2016). Regular numbers are the moisture contents at which seeds can be safely stored for up to 6 months at given air conditions but at bolded moisture contents seeds will likely spoil within 6 months as recommended by Sumner and Williams (2009).

Millet (Bx-Borno) Seeds under Drying Condition

	Relative Humidity (%)[a]										
T (°C)	35	40	45	50	55	60	65	70	75	80	85
10	13.5	14.4	15.2	15.9	16.7	17.5	18.3	19.2	20.1	**21.0**	**22.2**
13	13.4	14.2	15.0	15.8	16.6	17.4	18.2	19.0	19.9	**20.9**	**22.0**
15	13.3	14.2	14.9	15.7	16.5	17.3	18.1	18.9	19.8	**20.7**	**21.9**
18	13.2	14.0	14.8	15.6	16.3	17.1	17.9	18.7	19.6	**20.6**	**21.7**
20	13.1	14.0	14.7	15.5	16.3	17.0	17.8	18.6	**19.5**	**20.5**	**21.6**
23	13.0	13.8	14.6	15.4	16.1	16.9	17.7	18.5	**19.4**	**20.3**	**21.4**
25	13.0	13.8	14.5	15.3	16.0	16.8	**17.6**	**18.4**	**19.3**	**20.2**	**21.3**
28	12.9	13.7	14.4	15.2	15.9	16.7	**17.4**	**18.3**	**19.1**	**20.1**	**21.2**
30	12.8	13.6	14.3	**15.1**	**15.8**	**16.6**	**17.4**	**18.2**	**19.0**	**20.0**	**21.1**
33	12.7	13.5	14.2	**15.0**	**15.7**	**16.5**	**17.2**	**18.0**	**18.9**	**19.8**	**20.9**
35	12.6	**13.4**	**14.2**	**14.9**	**15.6**	**16.4**	**17.2**	**17.9**	**18.8**	**19.7**	**20.8**
38	**12.5**	**13.3**	**14.1**	**14.8**	**15.5**	**16.3**	**17.0**	**17.8**	**18.7**	**19.6**	**20.7**
40	**12.5**	**13.2**	**14.0**	**14.7**	**15.5**	**16.2**	**16.9**	**17.7**	**18.6**	**19.5**	**20.6**
43	**12.4**	**13.2**	**13.9**	**14.6**	**15.3**	**16.1**	**16.8**	**17.6**	**18.5**	**19.4**	**20.4**
45	**12.3**	**13.1**	**13.8**	**14.6**	**15.3**	**16.0**	**16.8**	**17.5**	**18.4**	**19.3**	**20.3**
48	**12.2**	**13.0**	**13.7**	**14.5**	**15.2**	**15.9**	**16.6**	**17.4**	**18.3**	**19.2**	**20.2**
50	**12.2**	**12.9**	**13.7**	**14.4**	**15.1**	**15.8**	**16.6**	**17.3**	**18.2**	**19.1**	**20.1**
53	**12.1**	**12.9**	**13.6**	**14.3**	**15.0**	**15.7**	**16.5**	**17.2**	**18.1**	**19.0**	**20.0**
55	**12.0**	**12.8**	**13.5**	**14.2**	**14.9**	**15.7**	**16.4**	**17.2**	**18.0**	**18.9**	**19.9**

[a] Equilibrium moisture contents are calculated using the modified Henderson equation and constants from ASABE (2016). Regular numbers are the moisture contents at which seeds can be safely stored for up to 6 months at given air conditions but at bolded moisture contents seeds will likely spoil within 6 months (estimated based on authors' experience).

Millet (Bx-Borno) Seeds under Re-Wetting Condition

	Relative Humidity (%)[a]										
T (°C)	*45*	*50*	*55*	*60*	*65*	*70*	*75*	*80*	*85*	*90*	*95*
10	13.2	13.9	14.6	15.3	16.1	16.8	17.6	**18.5**	**19.5**	**20.8**	**22.6**
13	13.0	13.7	14.4	15.1	15.8	16.6	17.4	**18.2**	**19.3**	**20.5**	**22.3**
15	12.9	13.6	14.3	14.9	15.7	16.4	17.2	**18.1**	**19.1**	**20.3**	**22.1**
18	12.7	13.4	14.0	14.7	15.4	16.2	**16.9**	**17.8**	**18.8**	**20.0**	**21.8**
20	12.6	13.2	13.9	14.6	15.3	16.0	**16.8**	**17.6**	**18.6**	**19.8**	**21.6**
23	12.4	13.1	13.7	14.4	15.1	15.8	**16.6**	**17.4**	**18.4**	**19.6**	**21.3**
25	12.3	12.9	13.6	14.3	**14.9**	**15.7**	**16.4**	**17.3**	**18.2**	**19.4**	**21.1**
28	12.1	12.8	13.4	14.1	**14.7**	**15.5**	**16.2**	**17.0**	**18.0**	**19.2**	**20.9**
30	12.0	12.7	13.3	13.9	**14.6**	**15.3**	**16.1**	**16.9**	**17.9**	**19.0**	**20.7**
33	11.9	12.5	**13.1**	**13.8**	**14.4**	**15.1**	**15.9**	**16.7**	**17.6**	**18.8**	**20.5**
35	11.8	12.4	**13.0**	**13.7**	**14.3**	**15.0**	**15.8**	**16.6**	**17.5**	**18.7**	**20.3**
38	**11.6**	**12.2**	**12.9**	**13.5**	**14.2**	**14.8**	**15.6**	**16.4**	**17.3**	**18.5**	**20.1**
40	**11.5**	**12.2**	**12.8**	**13.4**	**14.1**	**14.7**	**15.5**	**16.3**	**17.2**	**18.3**	**20.0**

[a] Equilibrium moisture contents are calculated using the modified Henderson equation and constants from ASABE (2016). Regular numbers are the moisture contents at which seeds can be safely stored for up to 6 months at given air conditions but at bolded moisture contents seeds will likely spoil within 6 months (estimated based on authors' experience).

Oats (Hulled, Dumont) Seeds under Re-Wetting Condition

	Relative Humidity (%)[a]										
T (°C)	*45*	*50*	*55*	*60*	*65*	*70*	*75*	*80*	*85*	*90*	*95*
10	10.9	11.6	12.2	13.0	13.8	14.6	15.6	16.8	**18.3**	**20.3**	**23.6**
13	10.6	11.3	12.0	12.7	13.5	14.4	15.4	16.5	**18.0**	**20.0**	**23.3**
15	10.5	11.1	11.8	12.5	13.3	14.2	15.2	**16.4**	**17.8**	**19.8**	**23.2**
18	10.2	10.9	11.6	12.3	13.1	14.0	**15.0**	**16.1**	**17.6**	**19.6**	**22.9**
20	10.1	10.7	11.4	12.1	12.9	**13.8**	**14.8**	**16.0**	**17.4**	**19.4**	**22.8**
23	9.9	10.5	11.2	11.9	12.7	**13.6**	**14.6**	**15.8**	**17.2**	**19.2**	**22.5**
25	9.7	10.4	11.1	11.8	**12.6**	**13.4**	**14.4**	**15.6**	**17.1**	**19.1**	**22.4**
28	9.5	10.2	10.9	**11.6**	**12.4**	**13.2**	**14.2**	**15.4**	**16.9**	**18.9**	**22.2**
30	9.4	10.0	**10.7**	**11.5**	**12.2**	**13.1**	**14.1**	**15.3**	**16.7**	**18.7**	**22.1**
33	9.2	9.8	**10.5**	**11.3**	**12.0**	**12.9**	**13.9**	**15.1**	**16.6**	**18.6**	**21.9**
35	**9.1**	**9.7**	**10.4**	**11.1**	**11.9**	**12.8**	**13.8**	**15.0**	**16.4**	**18.4**	**21.8**
38	**8.9**	**9.5**	**10.2**	**11.0**	**11.7**	**12.6**	**13.6**	**14.8**	**16.3**	**18.3**	**21.6**
40	**8.8**	**9.4**	**10.1**	**10.8**	**11.6**	**12.5**	**13.5**	**14.7**	**16.1**	**18.1**	**21.5**

[a] Equilibrium moisture contents are calculated using the modified Chung-Pfost equation and constants from Jayas and Mazza (1993). Regular numbers are the moisture contents at which seeds can be safely stored for up to 6 months at given air conditions but at bolded moisture contents seeds will likely spoil within 6 months as recommended by Prairie Oat Growers Association.

Oats (Hull-Less, AC Belmont) seeds under Drying Condition

	Relative Humidity (%)[a]										
T (°C)	*35*	*40*	*45*	*50*	*55*	*60*	*65*	*70*	*75*	*80*	*85*
10	11.5	12.3	13.2	14.1	15.0	16.0	17.1	**18.3**	**19.6**	**21.3**	**23.3**
13	11.2	12.1	13.0	13.9	14.8	15.8	16.9	**18.1**	**19.4**	**21.0**	**23.0**
15	11.1	12.0	12.8	13.7	14.7	15.7	**16.7**	**17.9**	**19.3**	**20.9**	**22.9**
18	10.9	11.8	12.6	13.5	14.5	15.4	**16.5**	**17.7**	**19.1**	**20.7**	**22.7**
20	10.8	11.6	12.5	13.4	14.3	**15.3**	**16.4**	**17.6**	**18.9**	**20.5**	**22.6**
23	10.6	11.4	12.3	13.2	14.1	**15.1**	**16.2**	**17.4**	**18.7**	**20.4**	**22.4**
25	10.4	11.3	12.2	**13.1**	**14.0**	**15.0**	**16.1**	**17.3**	**18.6**	**20.2**	**22.2**
28	10.2	11.1	**12.0**	**12.9**	**13.8**	**14.8**	**15.9**	**17.1**	**18.4**	**20.0**	**22.0**
30	10.1	**11.0**	**11.9**	**12.7**	**13.7**	**14.7**	**15.8**	**17.0**	**18.3**	**19.9**	**21.9**
33	**9.9**	**10.8**	**11.7**	**12.6**	**13.5**	**14.5**	**15.6**	**16.8**	**18.1**	**19.7**	**21.7**
35	**9.8**	**10.7**	**11.6**	**12.5**	**13.4**	**14.4**	**15.5**	**16.7**	**18.0**	**19.6**	**21.6**
38	**9.7**	**10.5**	**11.4**	**12.3**	**13.2**	**14.2**	**15.3**	**16.5**	**17.8**	**19.5**	**21.5**
40	**9.5**	**10.4**	**11.3**	**12.2**	**13.1**	**14.1**	**15.2**	**16.4**	**17.7**	**19.3**	**21.3**
43	**9.4**	**10.2**	**11.1**	**12.0**	**12.9**	**13.9**	**15.0**	**16.2**	**17.6**	**19.2**	**21.2**
45	**9.3**	**10.1**	**11.0**	**11.9**	**12.8**	**13.8**	**14.9**	**16.1**	**17.5**	**19.1**	**21.1**
48	**9.1**	**10.0**	**10.8**	**11.7**	**12.7**	**13.7**	**14.8**	**15.9**	**17.3**	**18.9**	**20.9**
50	**9.0**	**9.9**	**10.7**	**11.6**	**12.6**	**13.6**	**14.7**	**15.8**	**17.2**	**18.8**	**20.8**
53	**8.9**	**9.7**	**10.6**	**11.5**	**12.4**	**13.4**	**14.5**	**15.7**	**17.1**	**18.7**	**20.7**
55	**8.8**	**9.6**	**10.5**	**11.4**	**12.3**	**13.3**	**14.4**	**15.6**	**17.0**	**18.6**	**20.6**

[a] Equilibrium moisture contents are calculated using the modified Chung-Pfost equation and constants from Hulasare et al. (2001). Regular numbers are the moisture contents at which seeds can be safely stored for up to 6 months at given air conditions but at bolded moisture contents seeds will likely spoil within 6 months as recommended by Prairie Oat Growers Association.

Oats (Hull-Less, AC Belmont) seeds under Re-Wetting Condition

	Relative Humidity (%)[a]										
T (°C)	*45*	*50*	*55*	*60*	*65*	*70*	*75*	*80*	*85*	*90*	*95*
10	12.4	13.4	14.4	15.5	16.7	18.0	19.5	21.3	**23.5**	**26.5**	**31.5**
13	12.1	13.1	14.1	15.2	16.4	17.7	19.2	21.0	**23.2**	**26.2**	**31.2**
15	11.9	12.9	13.9	15.0	16.2	17.5	19.0	**20.8**	**23.0**	**26.0**	**31.0**
18	11.6	12.6	13.6	14.7	15.9	17.2	**18.7**	**20.5**	**22.7**	**25.7**	**30.7**
20	11.4	12.4	13.4	14.5	15.7	17.0	**18.5**	**20.3**	**22.5**	**25.5**	**30.5**
23	11.1	12.1	13.2	14.3	**15.4**	**16.8**	**18.3**	**20.0**	**22.2**	**25.2**	**30.2**
25	11.0	12.0	13.0	**14.1**	**15.3**	**16.6**	**18.1**	**19.8**	**22.1**	**25.1**	**30.1**
28	10.7	11.7	**12.7**	**13.8**	**15.0**	**16.3**	**17.8**	**19.6**	**21.8**	**24.8**	**29.8**
30	**10.6**	**11.5**	**12.6**	**13.7**	**14.9**	**16.2**	**17.7**	**19.4**	**21.6**	**24.7**	**29.7**
33	**10.3**	**11.3**	**12.3**	**13.4**	**14.6**	**15.9**	**17.4**	**19.2**	**21.4**	**24.4**	**29.4**
35	**10.2**	**11.2**	**12.2**	**13.3**	**14.5**	**15.8**	**17.3**	**19.0**	**21.3**	**24.3**	**29.3**
38	**9.9**	**10.9**	**12.0**	**13.1**	**14.2**	**15.6**	**17.1**	**18.8**	**21.0**	**24.0**	**29.0**
40	**9.8**	**10.8**	**11.8**	**12.9**	**14.1**	**15.4**	**16.9**	**18.7**	**20.9**	**23.9**	**28.9**

[a] Equilibrium moisture contents are calculated using the modified Chung-Pfost equation and constants from Hulasare et al. (2001). Regular numbers are the moisture contents at which seeds can be safely stored for up to 6 months at given air conditions but at bolded moisture contents seeds will likely spoil within 6 months as recommended by Prairie Oat Growers Association.

Popcorn (Purdue-410) under Drying Condition

	Relative Humidity (%)[a]										
T (°C)	*35*	*40*	*45*	*50*	*55*	*60*	*65*	*70*	*75*	*80*	*85*
10	9.4	10.2	10.9	11.7	12.5	13.4	14.3	15.4	**16.5**	**17.9**	**19.7**
13	9.1	9.9	10.6	11.4	12.2	13.1	14.0	15.0	**16.2**	**17.6**	**19.4**
15	8.9	9.7	10.4	11.2	12.0	12.9	13.8	**14.8**	**16.0**	**17.4**	**19.2**
18	8.6	9.4	10.1	10.9	11.7	12.6	13.5	**14.6**	**15.7**	**17.1**	**18.9**
20	8.4	9.2	10.0	10.7	11.5	12.4	**13.3**	**14.4**	**15.6**	**17.0**	**18.7**
23	8.2	8.9	9.7	10.5	11.3	12.1	**13.1**	**14.1**	**15.3**	**16.7**	**18.4**
25	8.0	8.8	9.5	10.3	11.1	**12.0**	**12.9**	**14.0**	**15.1**	**16.5**	**18.3**
28	7.8	8.5	9.3	10.1	10.9	**11.7**	**12.7**	**13.7**	**14.9**	**16.3**	**18.0**
30	7.6	8.4	9.1	9.9	**10.7**	**11.6**	**12.5**	**13.6**	**14.7**	**16.1**	**17.9**
33	7.4	8.1	8.9	9.7	**10.5**	**11.4**	**12.3**	**13.3**	**14.5**	**15.9**	**17.6**
35	7.2	8.0	8.8	**9.5**	**10.3**	**11.2**	**12.1**	**13.2**	**14.4**	**15.8**	**17.5**
38	**7.0**	**7.8**	**8.5**	**9.3**	**10.1**	**11.0**	**11.9**	**13.0**	**14.1**	**15.5**	**17.3**
40	**6.9**	**7.6**	**8.4**	**9.2**	**10.0**	**10.9**	**11.8**	**12.8**	**14.0**	**15.4**	**17.2**
43	**6.7**	**7.4**	**8.2**	**9.0**	**9.8**	**10.7**	**11.6**	**12.6**	**13.8**	**15.2**	**17.0**
45	**6.6**	**7.3**	**8.1**	**8.9**	**9.7**	**10.5**	**11.5**	**12.5**	**13.7**	**15.1**	**16.8**
48	**6.4**	**7.1**	**7.9**	**8.7**	**9.5**	**10.3**	**11.3**	**12.3**	**13.5**	**14.9**	**16.6**
50	**6.3**	**7.0**	**7.8**	**8.5**	**9.4**	**10.2**	**11.2**	**12.2**	**13.4**	**14.8**	**16.5**
53	**6.1**	**6.8**	**7.6**	**8.4**	**9.2**	**10.0**	**11.0**	**12.0**	**13.2**	**14.6**	**16.3**
55	**6.0**	**6.7**	**7.5**	**8.3**	**9.1**	**9.9**	**10.9**	**11.9**	**13.1**	**14.5**	**16.2**

[a] Equilibrium moisture contents are calculated using the Chung equation and constants from Flood and White (1984). Regular numbers are the moisture contents at which seeds can be safely stored for up to 6 months at given air conditions but at bolded moisture contents seeds will likely spoil within 6 months (estimated based on authors' experience).

Rough Rice Seeds under Drying Condition

	Relative Humidity (%)[a]										
T (°C)	*35*	*40*	*45*	*50*	*55*	*60*	*65*	*70*	*75*	*80*	*85*
10	8.4	9.1	9.8	10.5	11.1	11.9	12.6	13.4	14.2	**15.2**	**16.3**
13	8.3	8.9	9.6	10.3	11.0	11.7	12.4	13.2	14.0	**15.0**	**16.0**
15	8.2	8.8	9.5	10.2	10.9	11.6	12.3	13.0	13.9	**14.8**	**15.9**
18	8.0	8.7	9.4	10.0	10.7	11.4	12.1	12.8	**13.7**	**14.6**	**15.7**
20	8.0	8.6	9.3	9.9	10.6	11.3	12.0	12.7	**13.5**	**14.5**	**15.5**
23	7.8	8.5	9.1	9.8	10.4	11.1	11.8	12.5	**13.4**	**14.3**	**15.3**
25	7.8	8.4	9.1	9.7	10.3	11.0	**11.7**	**12.4**	**13.2**	**14.1**	**15.2**
28	7.7	8.3	8.9	9.6	10.2	10.8	**11.5**	**12.3**	**13.1**	**13.9**	**15.0**
30	7.6	8.2	8.8	9.5	**10.1**	**10.8**	**11.4**	**12.2**	**12.9**	**13.8**	**14.8**
33	7.5	8.1	8.7	9.3	**10.0**	**10.6**	**11.3**	**12.0**	**12.8**	**13.6**	**14.7**
35	7.4	8.0	8.7	9.3	**9.9**	**10.5**	**11.2**	**11.9**	**12.7**	**13.5**	**14.5**
38	7.3	7.9	**8.5**	**9.1**	**9.8**	**10.4**	**11.1**	**11.8**	**12.5**	**13.4**	**14.4**
40	**7.3**	**7.9**	**8.5**	**9.1**	**9.7**	**10.3**	**11.0**	**11.7**	**12.4**	**13.3**	**14.3**
43	**7.2**	**7.8**	**8.4**	**9.0**	**9.6**	**10.2**	**10.8**	**11.5**	**12.3**	**13.1**	**14.1**
45	**7.1**	**7.7**	**8.3**	**8.9**	**9.5**	**10.1**	**10.7**	**11.4**	**12.2**	**13.0**	**14.0**
48	**7.0**	**7.6**	**8.2**	**8.8**	**9.4**	**10.0**	**10.6**	**11.3**	**12.0**	**12.9**	**13.8**
50	**7.0**	**7.6**	**8.1**	**8.7**	**9.3**	**9.9**	**10.5**	**11.2**	**12.0**	**12.8**	**13.7**
53	**6.9**	**7.5**	**8.0**	**8.6**	**9.2**	**9.8**	**10.4**	**11.1**	**11.8**	**12.6**	**13.6**
55	**6.8**	**7.4**	**8.0**	**8.6**	**9.1**	**9.7**	**10.4**	**11.0**	**11.7**	**12.5**	**13.5**

[a] Equilibrium moisture contents are calculated using the modified Henderson equation and constants from Chen (2000). Regular numbers are the moisture contents at which seeds can be safely stored for up to 6 months at given air conditions but at bolded moisture contents seeds will likely spoil within 6 months (estimated based on authors' experience).

Rough Rice Seeds under Re-Wetting Condition

	Relative Humidity (%)[a]										
T (°C)	*45*	*50*	*55*	*60*	*65*	*70*	*75*	*80*	*85*	*90*	*95*
10	11.0	11.8	12.7	13.5	14.4	15.4	16.4	**17.6**	**19.0**	**20.6**	**23.1**
13	10.8	11.6	12.4	13.3	14.2	15.1	16.1	**17.3**	**18.6**	**20.3**	**22.7**
15	10.6	11.4	12.3	13.1	14.0	14.9	15.9	**17.1**	**18.4**	**20.0**	**22.5**
18	10.4	11.2	12.0	12.9	13.7	14.6	**15.6**	**16.8**	**18.1**	**19.7**	**22.1**
20	10.3	11.1	11.9	12.7	13.6	**14.5**	**15.5**	**16.6**	**17.9**	**19.5**	**21.9**
23	10.1	10.9	11.7	12.5	**13.3**	**14.2**	**15.2**	**16.3**	**17.6**	**19.2**	**21.5**
25	10.0	10.8	11.6	**12.3**	**13.2**	**14.1**	**15.0**	**16.1**	**17.4**	**19.0**	**21.3**
28	9.8	10.6	11.4	**12.1**	**13.0**	**13.8**	**14.8**	**15.9**	**17.1**	**18.7**	**21.0**
30	**9.7**	**10.5**	**11.2**	**12.0**	**12.8**	**13.7**	**14.6**	**15.7**	**16.9**	**18.5**	**20.8**
33	**9.6**	**10.3**	**11.1**	**11.8**	**12.6**	**13.5**	**14.4**	**15.5**	**16.7**	**18.2**	**20.5**
35	**9.5**	**10.2**	**10.9**	**11.7**	**12.5**	**13.4**	**14.3**	**15.3**	**16.5**	**18.1**	**20.3**
38	**9.3**	**10.1**	**10.8**	**11.5**	**12.3**	**13.2**	**14.1**	**15.1**	**16.3**	**17.8**	**20.0**
40	**9.2**	**10.0**	**10.7**	**11.4**	**12.2**	**13.0**	**13.9**	**15.0**	**16.2**	**17.7**	**19.9**

[a] Equilibrium moisture contents are calculated using the modified Henderson equation and constants from Chen (2000). Regular numbers are the moisture contents at which seeds can be safely stored for up to 6 months at given air conditions but at bolded moisture contents seeds will likely spoil within 6 months (estimated based on authors' experience).

Rye Seeds under Mixed Condition (Sorption and Desorption at Low and High Moisture Contents)

	Relative Humidity (%)[a]										
T (°C)	*35*	*40*	*45*	*50*	*55*	*60*	*65*	*70*	*75*	*80*	*85*
10	11.6	12.4	13.2	14.0	14.8	15.6	16.4	17.3	18.2	**19.3**	**20.5**
13	11.1	11.8	12.6	13.4	14.1	14.9	15.7	16.5	17.4	**18.4**	**19.6**
15	10.7	11.5	12.2	13.0	13.7	14.5	15.3	16.1	**17.0**	**17.9**	**19.0**
18	10.3	11.0	11.8	12.5	13.2	13.9	14.7	**15.5**	**16.3**	**17.3**	**18.4**
20	10.0	10.8	11.5	12.2	12.9	13.6	**14.3**	**15.1**	**15.9**	**16.9**	**17.9**
23	9.7	10.4	11.1	11.7	12.4	13.1	**13.8**	**14.6**	**15.4**	**16.3**	**17.4**
25	9.5	10.2	10.8	11.5	12.2	**12.8**	**13.6**	**14.3**	**15.1**	**16.0**	**17.0**
28	9.2	9.9	10.5	**11.2**	**11.8**	**12.5**	**13.2**	**13.9**	**14.7**	**15.5**	**16.5**
30	**9.0**	**9.7**	**10.3**	**10.9**	**11.6**	**12.2**	**12.9**	**13.6**	**14.4**	**15.3**	**16.2**
33	**8.8**	**9.4**	**10.0**	**10.6**	**11.3**	**11.9**	**12.6**	**13.3**	**14.0**	**14.9**	**15.8**
35	**8.6**	**9.2**	**9.9**	**10.5**	**11.1**	**11.7**	**12.4**	**13.1**	**13.8**	**14.6**	**15.6**
38	**8.4**	**9.0**	**9.6**	**10.2**	**10.8**	**11.4**	**12.1**	**12.7**	**13.5**	**14.3**	**15.2**
40	**8.3**	**8.9**	**9.5**	**10.1**	**10.7**	**11.3**	**11.9**	**12.6**	**13.3**	**14.1**	**15.0**
43	**8.1**	**8.7**	**9.3**	**9.8**	**10.4**	**11.0**	**11.6**	**12.3**	**13.0**	**13.8**	**14.7**
45	**8.0**	**8.5**	**9.1**	**9.7**	**10.3**	**10.9**	**11.5**	**12.1**	**12.8**	**13.6**	**14.5**
48	**7.8**	**8.4**	**8.9**	**9.5**	**10.1**	**10.6**	**11.2**	**11.9**	**12.6**	**13.3**	**14.2**
50	**7.7**	**8.3**	**8.8**	**9.4**	**9.9**	**10.5**	**11.1**	**11.7**	**12.4**	**13.2**	**14.0**
53	**7.5**	**8.1**	**8.6**	**9.2**	**9.7**	**10.3**	**10.9**	**11.5**	**12.2**	**12.9**	**13.8**
55	**7.5**	**8.0**	**8.5**	**9.1**	**9.6**	**10.2**	**10.8**	**11.4**	**12.0**	**12.8**	**13.6**

[a] Equilibrium moisture contents are calculated using the modified Henderson equation and constants recommended by ASABE (2016). Regular numbers are the moisture contents at which seeds can be safely stored for up to 6 months at given air conditions but at bolded moisture contents seeds will likely spoil within 6 months as recommended by Sathya et al. (2008).

Wheat (Hard Red Spring, Sinton) Seeds under Drying Condition

	Relative Humidity (%)[a]										
T (°C)	*35*	*40*	*45*	*50*	*55*	*60*	*65*	*70*	*75*	*80*	*85*
10	10.0	10.7	11.4	12.1	12.9	13.6	14.4	15.3	16.3	17.4	**18.8**
13	9.9	10.6	11.3	12.0	12.7	13.5	14.3	15.2	16.2	17.3	**18.7**
15	9.8	10.5	11.2	11.9	12.6	13.4	14.2	15.1	16.1	**17.2**	**18.6**
18	9.6	10.3	11.0	11.8	12.5	13.3	14.1	15.0	16.0	**17.1**	**18.5**
20	9.5	10.2	11.0	11.7	12.4	13.2	**14.0**	**14.9**	**15.9**	**17.0**	**18.4**
23	9.4	10.1	10.8	11.5	12.3	13.1	**13.9**	**14.8**	**15.8**	**16.9**	**18.3**
25	9.3	10.0	10.7	11.5	**12.2**	**13.0**	**13.8**	**14.7**	**15.7**	**16.8**	**18.2**
28	9.2	9.9	10.6	11.3	**12.1**	**12.8**	**13.7**	**14.6**	**15.6**	**16.7**	**18.1**
30	9.1	9.8	10.5	**11.2**	**12.0**	**12.8**	**13.6**	**14.5**	**15.5**	**16.6**	**18.0**
33	9.0	9.7	10.4	**11.1**	**11.9**	**12.6**	**13.5**	**14.4**	**15.4**	**16.5**	**17.9**
35	8.9	9.6	**10.3**	**11.0**	**11.8**	**12.6**	**13.4**	**14.3**	**15.3**	**16.5**	**17.9**
38	8.7	**9.5**	**10.2**	**10.9**	**11.7**	**12.5**	**13.3**	**14.2**	**15.2**	**16.4**	**17.8**
40	**8.7**	**9.4**	**10.1**	**10.8**	**11.6**	**12.4**	**13.2**	**14.1**	**15.1**	**16.3**	**17.7**
43	**8.5**	**9.3**	**10.0**	**10.7**	**11.5**	**12.3**	**13.1**	**14.0**	**15.0**	**16.2**	**17.6**
45	**8.5**	**9.2**	**9.9**	**10.7**	**11.4**	**12.2**	**13.0**	**13.9**	**15.0**	**16.1**	**17.5**
48	**8.4**	**9.1**	**9.8**	**10.5**	**11.3**	**12.1**	**12.9**	**13.8**	**14.9**	**16.0**	**17.4**
50	**8.3**	**9.0**	**9.7**	**10.5**	**11.2**	**12.0**	**12.9**	**13.8**	**14.8**	**16.0**	**17.4**
53	**8.2**	**8.9**	**9.6**	**10.4**	**11.1**	**11.9**	**12.8**	**13.7**	**14.7**	**15.9**	**17.3**
55	**8.1**	**8.8**	**9.6**	**10.3**	**11.1**	**11.8**	**12.7**	**13.6**	**14.6**	**15.8**	**17.2**

[a] Equilibrium moisture contents are calculated using the modified Chung-Pfost equation and constants recommended by ASABE (2016). Regular numbers are the moisture contents at which seeds can be safely stored for up to 6 months at given air conditions but at bolded moisture contents seeds will likely spoil within 6 months as recommended by Canadian Grain Commission (2020).

Wheat (Hard Red Spring, Waldron) Seeds under Drying Condition

	Relative Humidity (%)[a]										
T (°C)	*35*	*40*	*45*	*50*	*55*	*60*	*65*	*70*	*75*	*80*	*85*
10	11.2	11.8	12.4	13.1	13.8	14.5	15.2	16.0	17.0	**18.0**	**19.3**
13	10.8	11.5	12.1	12.8	13.5	14.2	14.9	15.8	16.7	**17.7**	**19.0**
15	10.6	11.3	12.0	12.6	13.3	14.0	14.8	15.6	**16.5**	**17.6**	**18.9**
18	10.4	11.0	11.7	12.4	13.0	13.8	14.5	**15.3**	**16.3**	**17.3**	**18.6**
20	10.2	10.9	11.5	12.2	12.9	13.6	**14.4**	**15.2**	**16.1**	**17.2**	**18.5**
23	9.9	10.6	11.3	11.9	12.6	13.3	**14.1**	**15.0**	**15.9**	**17.0**	**18.3**
25	9.8	10.4	11.1	11.8	**12.5**	**13.2**	**14.0**	**14.8**	**15.7**	**16.8**	**18.1**
28	9.5	10.2	10.9	11.5	**12.2**	**13.0**	**13.7**	**14.6**	**15.5**	**16.6**	**17.9**
30	9.4	10.0	10.7	**11.4**	**12.1**	**12.8**	**13.6**	**14.5**	**15.4**	**16.5**	**17.8**
33	9.1	9.8	10.5	**11.2**	**11.9**	**12.6**	**13.4**	**14.3**	**15.2**	**16.3**	**17.6**
35	9.0	**9.7**	**10.4**	**11.0**	**11.8**	**12.5**	**13.3**	**14.1**	**15.1**	**16.2**	**17.5**
38	**8.8**	**9.5**	**10.2**	**10.8**	**11.6**	**12.3**	**13.1**	**13.9**	**14.9**	**16.0**	**17.3**
40	**8.7**	**9.3**	**10.0**	**10.7**	**11.4**	**12.2**	**13.0**	**13.8**	**14.8**	**15.9**	**17.2**
43	**8.5**	**9.1**	**9.8**	**10.5**	**11.2**	**12.0**	**12.8**	**13.6**	**14.6**	**15.7**	**17.1**
45	**8.3**	**9.0**	**9.7**	**10.4**	**11.1**	**11.9**	**12.7**	**13.5**	**14.5**	**15.6**	**17.0**
48	**8.1**	**8.8**	**9.5**	**10.2**	**10.9**	**11.7**	**12.5**	**13.4**	**14.3**	**15.5**	**16.8**
50	**8.0**	**8.7**	**9.4**	**10.1**	**10.8**	**11.6**	**12.4**	**13.3**	**14.2**	**15.3**	**16.7**
53	**7.8**	**8.5**	**9.2**	**9.9**	**10.7**	**11.4**	**12.2**	**13.1**	**14.1**	**15.2**	**16.6**
55	**7.7**	**8.4**	**9.1**	**9.8**	**10.6**	**11.3**	**12.1**	**13.0**	**14.0**	**15.1**	**16.5**

[a] Equilibrium moisture contents are calculated using the modified Chung-Pfost equation and constants recommended by ASABE (2016). Regular numbers are the moisture contents at which seeds can be safely stored for up to 6 months at given air conditions but at bolded moisture contents seeds will likely spoil within 6 months as recommended by Canadian Grain Commission (2020).

Wheat (Soft Winter, Hobbit) Seeds under Drying Condition

	Relative Humidity (%)[a]										
T (°C)	*35*	*40*	*45*	*50*	*55*	*60*	*65*	*70*	*75*	*80*	*85*
10	13.4	13.9	14.4	15.0	15.5	16.1	16.7	17.4	18.1	19.0	**20.0**
13	12.4	12.9	13.5	14.0	14.5	15.1	15.8	16.4	17.2	18.1	**19.1**
15	11.8	12.4	12.9	13.5	14.0	14.6	15.2	15.9	16.7	**17.6**	**18.7**
18	11.1	11.7	12.2	12.8	13.3	13.9	14.6	15.3	16.0	**16.9**	**18.1**
20	10.7	11.2	11.8	12.3	12.9	13.5	**14.2**	**14.9**	**15.7**	**16.6**	**17.7**
23	10.1	10.7	11.2	11.8	12.4	13.0	**13.6**	**14.4**	**15.2**	**16.1**	**17.2**
25	9.8	10.3	10.9	11.5	12.0	12.7	**13.3**	**14.0**	**14.8**	**15.8**	**16.9**
28	9.3	9.9	10.4	11.0	11.6	12.2	**12.9**	**13.6**	**14.4**	**15.4**	**16.5**
30	9.0	9.6	**10.1**	**10.7**	**11.3**	**12.0**	**12.6**	**13.4**	**14.2**	**15.1**	**16.3**
33	8.6	9.2	**9.7**	**10.3**	**10.9**	**11.6**	**12.2**	**13.0**	**13.8**	**14.8**	**15.9**
35	**8.3**	**8.9**	**9.5**	**10.1**	**10.7**	**11.3**	**12.0**	**12.8**	**13.6**	**14.5**	**15.7**
38	**8.0**	**8.6**	**9.2**	**9.8**	**10.4**	**11.0**	**11.7**	**12.4**	**13.3**	**14.2**	**15.4**
40	**7.8**	**8.4**	**8.9**	**9.5**	**10.2**	**10.8**	**11.5**	**12.2**	**13.1**	**14.0**	**15.2**
43	**7.5**	**8.0**	**8.6**	**9.2**	**9.9**	**10.5**	**11.2**	**12.0**	**12.8**	**13.8**	**15.0**
45	**7.3**	**7.8**	**8.4**	**9.0**	**9.7**	**10.3**	**11.0**	**11.8**	**12.6**	**13.6**	**14.8**
48	**7.0**	**7.6**	**8.2**	**8.8**	**9.4**	**10.1**	**10.8**	**11.5**	**12.4**	**13.4**	**14.6**
50	**6.8**	**7.4**	**8.0**	**8.6**	**9.2**	**9.9**	**10.6**	**11.4**	**12.2**	**13.2**	**14.4**
53	**6.5**	**7.1**	**7.7**	**8.3**	**9.0**	**9.6**	**10.3**	**11.1**	**12.0**	**13.0**	**14.2**
55	**6.4**	**7.0**	**7.6**	**8.2**	**8.8**	**9.5**	**10.2**	**11.0**	**11.8**	**12.8**	**14.0**

[a] Equilibrium moisture contents are calculated using the modified Chung-Pfost equation and constants recommended by ASABE (2016). Regular numbers are the moisture contents at which seeds can be safely stored for up to 6 months at given air conditions but at bolded moisture contents seeds will likely spoil within 6 months as recommended by Canadian Grain Commission (2019).

Wheat (Soft Winter, Hobbit) Seeds under Re-Wetting Condition

	Relative Humidity (%)[a]										
T (°C)	*45*	*50*	*55*	*60*	*65*	*70*	*75*	*80*	*85*	*90*	*95*
10	14.1	14.6	15.2	15.7	16.3	17.0	17.8	18.6	**19.7**	**21.1**	**23.3**
13	13.1	13.6	14.2	14.8	15.4	16.1	16.8	17.7	**18.8**	**20.2**	**22.5**
15	12.5	13.1	13.7	14.2	14.9	15.6	16.3	**17.2**	**18.3**	**19.8**	**22.1**
18	11.8	12.4	13.0	13.6	14.2	14.9	15.7	**16.6**	**17.7**	**19.2**	**21.5**
20	11.4	12.0	12.6	13.2	**13.8**	**14.5**	**15.3**	**16.2**	**17.3**	**18.8**	**21.2**
23	10.9	11.4	12.0	12.6	**13.3**	**14.0**	**14.8**	**15.7**	**16.9**	**18.4**	**20.7**
25	10.5	11.1	11.7	**12.3**	**13.0**	**13.7**	**14.5**	**15.4**	**16.6**	**18.1**	**20.5**
28	10.1	10.6	11.2	**11.9**	**12.5**	**13.3**	**14.1**	**15.0**	**16.2**	**17.7**	**20.1**
30	**9.8**	**10.4**	**11.0**	**11.6**	**12.3**	**13.0**	**13.8**	**14.8**	**15.9**	**17.4**	**19.9**
33	**9.4**	**10.0**	**10.6**	**11.2**	**11.9**	**12.6**	**13.4**	**14.4**	**15.6**	**17.1**	**19.6**
35	**9.1**	**9.7**	**10.3**	**11.0**	**11.6**	**12.4**	**13.2**	**14.2**	**15.4**	**16.9**	**19.4**
38	**8.8**	**9.4**	**10.0**	**10.6**	**11.3**	**12.1**	**12.9**	**13.9**	**15.1**	**16.6**	**19.1**
40	**8.6**	**9.2**	**9.8**	**10.4**	**11.1**	**11.9**	**12.7**	**13.7**	**14.9**	**16.4**	**18.9**

[a] Equilibrium moisture contents are calculated using the modified Chung-Pfost equation and constants recommended by ASABE (2016). Regular numbers are the moisture contents at which seeds can be safely stored for up to 6 months at given air conditions but at bolded moisture contents seeds will likely spoil within 6 months as recommended by Canadian Grain Commission (2019).

A2 Oilseeds

Canola (Westar) Seeds under Drying Condition

	Relative Humidity (%)[a]										
T (°C)	*35*	*40*	*45*	*50*	*55*	*60*	*65*	*70*	*75*	*80*	*85*
10	6.3	6.7	7.3	7.8	8.4	9.1	9.7	10.4	**11.1**	**12.0**	**13.0**
13	6.1	6.6	7.1	7.7	8.3	8.9	9.5	10.2	**11.0**	**11.7**	**12.7**
15	6.0	6.5	7.1	7.6	8.2	8.8	9.4	10.1	**10.8**	**11.6**	**12.6**
18	5.9	6.4	6.9	7.5	8.0	8.6	9.3	**9.9**	**10.6**	**11.4**	**12.4**
20	5.8	6.3	6.8	7.4	7.9	8.5	9.1	**9.7**	**10.5**	**11.3**	**12.2**
23	5.7	6.2	6.7	7.2	7.8	8.3	8.9	**9.6**	**10.3**	**11.0**	**12.0**
25	5.7	6.1	6.6	7.1	7.7	8.3	8.8	**9.5**	**10.2**	**11.0**	**11.9**
28	5.6	6.0	6.5	7.1	7.6	8.1	**8.7**	**9.3**	**10.0**	**10.8**	**11.7**
30	5.6	5.9	6.5	7.0	7.5	**8.0**	**8.6**	**9.3**	**9.9**	**10.6**	**11.6**
33	5.5	5.8	6.4	6.9	7.4	**7.9**	**8.5**	**9.1**	**9.7**	**10.5**	**11.3**
35	5.4	5.8	6.3	6.8	**7.3**	**7.8**	**8.4**	**9.0**	**9.7**	**10.4**	**11.3**
38	5.3	5.7	6.2	6.7	**7.1**	**7.7**	**8.3**	**8.8**	**9.5**	**10.2**	**11.1**
40	**5.2**	**5.7**	**6.1**	**6.6**	**7.1**	**7.7**	**8.2**	**8.8**	**9.4**	**10.2**	**11.0**
43	**5.2**	**5.6**	**6.0**	**6.5**	**7.0**	**7.5**	**8.1**	**8.7**	**9.3**	**10.0**	**10.9**
45	**5.1**	**5.5**	**6.0**	**6.5**	**7.0**	**7.5**	**8.0**	**8.6**	**9.2**	**9.9**	**10.7**
48	**5.0**	**5.5**	**5.9**	**6.4**	**6.9**	**7.3**	**7.9**	**8.4**	**9.1**	**9.7**	**10.6**
50	**5.0**	**5.4**	**5.8**	**6.3**	**6.8**	**7.3**	**7.8**	**8.3**	**9.0**	**9.7**	**10.5**
53	**4.9**	**5.3**	**5.7**	**6.2**	**6.7**	**7.2**	**7.7**	**8.3**	**8.8**	**9.6**	**10.4**
55	**4.9**	**5.3**	**5.7**	**6.2**	**6.6**	**7.1**	**7.7**	**8.2**	**8.8**	**9.5**	**10.3**

[a] Equilibrium moisture contents are calculated using the modified Henderson equation and constants recommended by ASABE (2016). Regular numbers are the moisture contents at which seeds can be safely stored for up to 6 months at given air conditions but at bolded moisture contents seeds will likely spoil within 6 months as recommended by Sun et al. (2014).

Canola (Westar) Seeds under Re-Wetting Condition

	Relative Humidity (%)[a]										
T (°C)	*45*	*50*	*55*	*60*	*65*	*70*	*75*	*80*	*85*	*90*	*95*
10	6.9	7.5	8.0	8.6	9.2	9.8	**10.6**	**11.3**	**12.3**	**13.5**	**15.3**
13	6.8	7.3	7.9	8.5	9.1	9.7	**10.4**	**11.2**	**12.1**	**13.3**	**15.1**
15	6.7	7.3	7.8	8.4	9.0	9.7	**10.3**	**11.1**	**12.0**	**13.2**	**15.0**
18	6.6	7.1	7.7	8.3	8.8	**9.5**	**10.2**	**11.0**	**11.9**	**13.0**	**14.8**
20	6.6	7.1	7.7	8.2	8.8	**9.4**	**10.1**	**10.9**	**11.8**	**13.0**	**14.7**
23	6.5	7.1	7.6	8.1	8.7	**9.3**	**10.0**	**10.7**	**11.7**	**12.8**	**14.5**
25	6.5	7.0	7.5	8.0	8.6	**9.3**	**9.9**	**10.6**	**11.5**	**12.7**	**14.4**
28	6.4	6.9	7.4	7.9	**8.5**	**9.1**	**9.7**	**10.6**	**11.4**	**12.5**	**14.2**
30	6.3	6.8	7.3	7.8	**8.4**	**9.0**	**9.7**	**10.5**	**11.3**	**12.4**	**14.1**
33	6.3	6.7	7.2	**7.7**	**8.3**	**8.9**	**9.6**	**10.3**	**11.2**	**12.3**	**13.9**
35	**6.2**	**6.7**	**7.1**	**7.7**	**8.3**	**8.8**	**9.5**	**10.2**	**11.1**	**12.2**	**13.9**
38	**6.1**	**6.6**	**7.1**	**7.7**	**8.2**	**8.8**	**9.4**	**10.2**	**11.0**	**12.0**	**13.7**
40	**6.1**	**6.5**	**7.1**	**7.6**	**8.1**	**8.7**	**9.3**	**10.1**	**10.9**	**12.0**	**13.6**

[a] Equilibrium moisture contents are calculated using the modified Henderson equation and constants recommended by ASABE (2016). Regular numbers are the moisture contents at which seeds can be safely stored for up to 6 months at given air conditions but at bolded moisture contents seeds will likely spoil within 6 months as recommended by Sun et al. (2014).

Flax (Five Cultivars) Seeds under Mixed Condition (Sorption and Desorption at Low and High Moisture Contents)

	Relative Humidity (%)[a]										
T (°C)	*35*	*40*	*45*	*50*	*55*	*60*	*65*	*70*	*75*	*80*	*85*
10	6.3	6.7	7.1	7.6	8.2	8.8	9.5	10.3	11.4	**12.8**	**14.7**
13	6.1	6.5	7.0	7.5	8.0	8.6	9.3	10.1	11.2	**12.5**	**14.4**
15	6.0	6.4	6.9	7.3	7.9	8.5	9.2	10.0	**11.0**	**12.3**	**14.2**
18	5.9	6.3	6.7	7.2	7.7	8.3	9.0	9.8	**10.8**	**12.1**	**13.9**
20	5.8	6.2	6.6	7.1	7.6	8.2	8.8	9.6	**10.6**	**11.9**	**13.7**
23	5.7	6.1	6.5	6.9	7.4	8.0	8.6	**9.4**	**10.4**	**11.6**	**13.4**
25	5.6	6.0	6.4	6.8	7.3	7.9	**8.5**	**9.3**	**10.2**	**11.5**	**13.2**
28	5.5	5.8	6.2	6.7	7.1	7.7	**8.3**	**9.1**	**10.0**	**11.2**	**12.9**
30	5.4	5.7	6.1	**6.6**	**7.0**	**7.6**	**8.2**	**8.9**	**9.9**	**11.1**	**12.7**
33	5.3	5.6	6.0	**6.4**	**6.9**	**7.4**	**8.0**	**8.7**	**9.7**	**10.8**	**12.5**
35	5.2	5.5	**5.9**	**6.3**	**6.8**	**7.3**	**7.9**	**8.6**	**9.5**	**10.7**	**12.3**
38	**5.1**	**5.4**	**5.8**	**6.2**	**6.6**	**7.1**	**7.7**	**8.4**	**9.3**	**10.4**	**12.0**
40	**5.0**	**5.3**	**5.7**	**6.1**	**6.5**	**7.0**	**7.6**	**8.3**	**9.2**	**10.3**	**11.9**
43	**4.9**	**5.2**	**5.6**	**5.9**	**6.4**	**6.9**	**7.4**	**8.1**	**9.0**	**10.1**	**11.6**
45	**4.8**	**5.1**	**5.5**	**5.9**	**6.3**	**6.8**	**7.3**	**8.0**	**8.8**	**9.9**	**11.5**
48	**4.7**	**5.0**	**5.3**	**5.7**	**6.1**	**6.6**	**7.2**	**7.8**	**8.6**	**9.7**	**11.2**
50	**4.6**	**4.9**	**5.3**	**5.6**	**6.0**	**6.5**	**7.1**	**7.7**	**8.5**	**9.6**	**11.1**
53	**4.5**	**4.8**	**5.1**	**5.5**	**5.9**	**6.4**	**6.9**	**7.5**	**8.3**	**9.4**	**10.8**
55	**4.4**	**4.7**	**5.1**	**5.4**	**5.8**	**6.3**	**6.8**	**7.4**	**8.2**	**9.2**	**10.7**

[a] Equilibrium moisture contents are calculated using the modified Halsey equation and constants from Mazza et al. (1990). Regular numbers are the moisture contents at which seeds can be safely stored for up to 6 months at given air conditions but at bolded moisture contents seeds will likely spoil within 6 months as recommended by Canadian Grain Commission (2020).

Hemp (Finola®) Seeds under Drying Condition

	Relative Humidity (%)[a]										
T (°C)	*35*	*40*	*45*	*50*	*55*	*60*	*65*	*70*	*75*	*80*	*85*
10	7.5	7.8	8.1	8.4	8.7	9.1	9.5	9.9	**10.4**	**10.9**	**11.5**
13	7.5	7.7	8.0	8.4	8.7	9.1	9.5	**9.9**	**10.4**	**10.9**	**11.5**
15	7.4	7.7	8.0	8.3	8.7	9.1	9.5	**9.9**	**10.4**	**10.9**	**11.5**
18	6.9	7.2	7.9	8.3	8.6	8.9	**9.4**	**9.8**	**10.3**	**10.8**	**11.4**
20	7.2	7.5	7.9	8.2	8.6	**8.9**	**9.4**	**9.8**	**10.3**	**10.8**	**11.4**
23	7.1	7.4	7.7	8.1	**8.4**	**8.8**	**9.3**	**9.7**	**10.2**	**10.8**	**11.4**
25	6.9	7.3	7.6	8.0	**8.4**	**8.8**	**9.2**	**9.7**	**10.2**	**10.7**	**11.3**
28	6.7	7.1	7.4	**7.8**	**8.2**	**8.6**	**9.1**	**9.5**	**10.1**	**10.6**	**11.2**
30	6.6	6.9	**7.3**	**7.7**	**8.1**	**8.5**	**9.0**	**9.5**	**10.0**	**10.5**	**11.2**
33	6.3	6.7	**7.1**	**7.5**	**7.9**	**8.4**	**8.8**	**9.3**	**9.8**	**10.4**	**11.0**
35	6.2	**6.6**	**7.0**	**7.4**	**7.8**	**8.3**	**8.7**	**9.2**	**9.8**	**10.3**	**11.0**
38	6.0	**6.4**	**6.8**	**7.2**	**7.7**	**8.1**	**8.6**	**9.1**	**9.6**	**10.2**	**10.9**
40	**5.8**	**6.3**	**6.7**	**7.1**	**7.6**	**8.0**	**8.5**	9.0	**9.6**	**10.1**	**10.8**
43	**5.6**	**6.1**	**6.5**	**7.0**	**7.4**	**7.9**	**8.4**	**8.9**	**9.4**	**10.0**	**10.7**
45	**5.5**	**6.0**	**6.4**	**6.9**	**7.3**	**7.8**	**8.3**	**8.8**	**9.4**	**10.0**	**10.6**
48	**5.3**	**5.8**	**6.2**	**6.7**	**7.2**	**7.6**	**8.1**	**8.7**	**9.2**	**9.8**	**10.5**
50	**5.2**	**5.7**	**6.1**	**6.6**	**7.1**	**7.6**	**8.1**	**8.6**	**9.2**	**9.8**	**10.4**
53	**5.1**	**5.5**	**6.0**	**6.5**	**6.9**	**7.4**	**7.9**	**8.5**	**9.0**	**9.7**	**10.3**
55	**5.0**	**5.4**	**5.9**	**6.4**	**6.9**	**7.3**	**7.9**	**8.4**	**9.0**	**9.6**	**10.3**

[a] Equilibrium moisture contents are calculated using the GAB (Guggenheim, Anderson, de Boer) equation and constants from Jian et al. (2018). Regular numbers are the moisture contents at which seeds can be safely stored for up to 6 months at given air conditions but at bolded moisture contents seeds will likely spoil within 6 months as recommended by Jian et al. (2019).

Hemp (Finola®) Seeds under Re-Wetting Condition

	Relative Humidity (%)[a]										
T (°C)	*45*	*50*	*55*	*60*	*65*	*70*	*75*	*80*	*85*	*90*	*95*
10	7.0	7.3	7.6	8.0	8.3	8.7	9.2	9.6	**10.2**	**10.8**	**11.5**
13	7.0	7.3	7.6	8.0	8.3	8.7	9.2	9.6	**10.2**	**10.8**	**11.5**
15	7.0	7.3	7.6	8.0	8.3	8.7	9.2	**9.6**	**10.2**	**10.8**	**11.5**
18	7.0	7.3	7.6	8.0	8.3	8.7	**9.2**	**9.6**	**10.2**	**10.8**	**11.5**
20	7.0	7.3	7.6	7.9	8.3	8.7	**9.1**	**9.6**	**10.2**	**10.8**	**11.5**
23	7.0	7.3	7.6	7.9	**8.3**	**8.7**	**9.1**	**9.6**	**10.2**	**10.8**	**11.4**
25	6.9	7.2	7.5	7.9	**8.3**	**8.7**	**9.1**	**9.6**	**10.1**	**10.7**	**11.4**
28	6.8	7.1	7.5	**7.8**	**8.2**	**8.6**	**9.0**	**9.5**	**10.1**	**10.7**	**11.4**
30	6.8	7.1	**7.4**	**7.8**	**8.1**	**8.5**	**9.0**	**9.5**	**10.0**	**10.7**	**11.3**
33	6.6	**6.9**	**7.3**	**7.6**	**8.0**	**8.5**	**8.9**	**9.4**	**10.0**	**10.6**	**11.3**
35	**6.5**	**6.8**	**7.2**	**7.6**	**8.0**	**8.4**	**8.8**	**9.3**	**9.9**	**10.5**	**11.2**
38	**6.3**	**6.7**	**7.0**	**7.4**	**7.8**	**8.2**	**8.7**	**9.2**	**9.8**	**10.4**	**11.1**
40	**6.2**	**6.6**	**6.9**	**7.3**	**7.7**	**8.2**	**8.6**	**9.1**	**9.7**	**10.3**	**11.0**

[a] Equilibrium moisture contents are calculated using the GAB (Guggenheim, Anderson, de Boer) equation and constants from Jian et al. (2018). Regular numbers are the moisture contents at which seeds can be safely stored for up to 6 months at given air conditions but at bolded moisture contents seeds will likely spoil within 6 months as recommended by Jian et al. (2019).

Rapeseed (Tower) Seeds under Drying Condition

	Relative Humidity (%)[a]										
T (°C)	*35*	*40*	*45*	*50*	*55*	*60*	*65*	*70*	*75*	*80*	*85*
10	5.3	5.7	6.1	6.5	7.0	7.6	8.3	9.1	10.1	**11.4**	**13.2**
13	5.2	5.6	6.0	6.5	7.0	7.5	8.2	9.0	10.0	**11.3**	**13.1**
15	5.2	5.6	6.0	6.4	6.9	7.5	8.1	8.9	9.9	**11.2**	**13.0**
18	5.1	5.5	5.9	6.3	6.8	7.4	8.0	8.8	9.8	**11.0**	**12.8**
20	5.1	5.5	5.8	6.3	6.8	7.3	**8.0**	**8.7**	**9.7**	**11.0**	**12.7**
23	5.0	5.4	5.8	6.2	6.7	7.2	**7.9**	**8.6**	**9.6**	**10.8**	**12.6**
25	5.0	5.3	5.7	6.2	**6.6**	**7.2**	**7.8**	**8.6**	**9.5**	**10.7**	**12.5**
28	4.9	5.3	5.7	6.1	**6.5**	**7.1**	**7.7**	**8.5**	**9.4**	**10.6**	**12.3**
30	4.9	5.2	**5.6**	**6.0**	**6.5**	**7.0**	**7.6**	**8.4**	**9.3**	**10.5**	**12.2**
33	4.8	**5.2**	**5.5**	**6.0**	**6.4**	**6.9**	**7.5**	**8.3**	**9.2**	**10.4**	**12.1**
35	**4.8**	**5.1**	**5.5**	**5.9**	**6.4**	**6.9**	**7.5**	**8.2**	**9.1**	**10.3**	**12.0**
38	**4.7**	**5.1**	**5.4**	**5.8**	**6.3**	**6.8**	**7.4**	**8.1**	**9.0**	**10.2**	**11.9**
40	**4.7**	**5.0**	**5.4**	**5.8**	**6.2**	**6.7**	**7.3**	**8.1**	**9.0**	**10.1**	**11.8**
43	**4.6**	**5.0**	**5.3**	**5.7**	**6.2**	**6.7**	**7.2**	**8.0**	**8.8**	**10.0**	**11.6**
45	**4.6**	**4.9**	**5.3**	**5.7**	**6.1**	**6.6**	**7.2**	**7.9**	**8.8**	**9.9**	**11.6**
48	**4.5**	**4.9**	**5.2**	**5.6**	**6.0**	**6.5**	**7.1**	**7.8**	**8.7**	**9.8**	**11.4**
50	**4.5**	**4.8**	**5.2**	**5.5**	**6.0**	**6.5**	**7.0**	**7.7**	**8.6**	**9.7**	**11.3**
53	**4.4**	**4.8**	**5.1**	**5.5**	**5.9**	**6.4**	**7.0**	**7.6**	**8.5**	**9.6**	**11.2**
55	**4.4**	**4.7**	**5.1**	**5.4**	**5.9**	**6.3**	**6.9**	**7.6**	**8.4**	**9.5**	**11.1**

[a] Equilibrium moisture contents are calculated using the modified Halsey equation and constants from Sun and Byrne (1998). Regular numbers are the moisture contents at which seeds can be safely stored for up to 6 months at given air conditions but at bolded moisture contents seeds will likely spoil within 6 months as recommended by Sun et al. (2014).

Rapeseed (Tower) Seeds under Re-Wetting Condition

	Relative Humidity (%)[a]										
T (°C)	*45*	*50*	*55*	*60*	*65*	*70*	*75*	*80*	*85*	*90*	*95*
10	5.6	6.1	6.6	7.2	7.9	8.7	9.8	**11.2**	**13.1**	**16.3**	**22.9**
13	5.5	6.0	6.5	7.1	7.8	8.6	9.7	**11.0**	**13.0**	**16.2**	**22.7**
15	5.5	6.0	6.5	7.0	7.7	8.5	9.6	**11.0**	**12.9**	**16.0**	**22.6**
18	5.4	5.9	6.4	7.0	7.6	8.5	9.5	**10.8**	**12.8**	**15.9**	**22.4**
20	5.4	5.8	6.3	6.9	**7.6**	**8.4**	**9.4**	**10.8**	**12.7**	**15.8**	**22.2**
23	5.3	5.8	6.3	6.8	**7.5**	**8.3**	**9.3**	**10.6**	**12.6**	**15.6**	**22.0**
25	5.3	5.7	6.2	**6.8**	**7.4**	**8.2**	**9.2**	**10.6**	**12.5**	**15.5**	**21.9**
28	5.2	5.7	**6.1**	**6.7**	**7.4**	**8.1**	**9.1**	**10.5**	**12.3**	**15.4**	**21.7**
30	5.2	**5.6**	**6.1**	**6.7**	**7.3**	**8.1**	**9.1**	**10.4**	**12.2**	**15.2**	**21.5**
33	5.1	**5.6**	**6.0**	**6.6**	**7.2**	**8.0**	**9.0**	**10.3**	**12.1**	**15.1**	**21.3**
35	**5.1**	**5.5**	**6.0**	**6.5**	**7.2**	**7.9**	**8.9**	**10.2**	**12.0**	**15.0**	**21.2**
38	**5.0**	**5.5**	**5.9**	**6.5**	**7.1**	**7.8**	**8.8**	**10.1**	**11.9**	**14.8**	**21.0**
40	**5.0**	**5.4**	**5.9**	**6.4**	**7.0**	**7.8**	**8.7**	**10.0**	**11.8**	**14.7**	**20.9**

[a] Equilibrium moisture contents are calculated using the modified Halsey equation and constants from Sun and Byrne (1998). Regular numbers are the moisture contents at which seeds can be safely stored for up to 6 months at given air conditions but at bolded moisture contents seeds will likely spoil within 6 months as recommended by Sun et al. (2014).

Soybean (Nigerian) Seeds under Drying Condition

	Relative Humidity (%)[a]										
T (°C)	*35*	*40*	*45*	*50*	*55*	*60*	*65*	*70*	*75*	*80*	*85*
10	7.6	8.1	8.7	9.2	9.8	10.4	11.1	11.8	12.6	13.5	**14.7**
13	7.3	7.8	8.4	8.9	9.5	10.2	10.8	11.5	12.3	13.3	**14.4**
15	7.1	7.6	8.2	8.8	9.4	10.0	10.6	11.4	12.2	**13.1**	**14.2**
18	6.8	7.4	7.9	8.5	9.1	9.7	10.4	11.1	11.9	**12.9**	**14.0**
20	6.6	7.2	7.8	8.3	8.9	9.6	10.2	11.0	**11.8**	**12.7**	**13.9**
23	6.4	6.9	7.5	8.1	8.7	9.3	10.0	10.7	**11.6**	**12.5**	**13.7**
25	6.2	6.8	7.4	7.9	8.6	9.2	**9.9**	**10.6**	**11.4**	**12.4**	**13.5**
28	6.0	6.6	7.1	7.7	8.3	9.0	**9.6**	**10.4**	**11.2**	**12.2**	**13.3**
30	5.8	6.4	7.0	7.6	8.2	8.8	**9.5**	**10.3**	**11.1**	**12.0**	**13.2**
33	5.6	6.2	6.8	7.4	8.0	**8.6**	**9.3**	**10.1**	**10.9**	**11.9**	**13.0**
35	5.5	6.1	6.7	7.3	7.9	**8.5**	**9.2**	**9.9**	**10.8**	**11.7**	**12.9**
38	5.3	5.9	6.5	7.1	**7.7**	**8.3**	**9.0**	**9.8**	**10.6**	**11.6**	**12.7**
40	**5.2**	**5.8**	**6.4**	**7.0**	**7.6**	**8.2**	**8.9**	**9.6**	**10.5**	**11.5**	**12.6**
43	**5.0**	**5.6**	**6.2**	**6.8**	**7.4**	**8.0**	**8.7**	**9.5**	**10.3**	**11.3**	**12.5**
45	**4.9**	**5.5**	**6.1**	**6.7**	**7.3**	**7.9**	**8.6**	**9.4**	**10.2**	**11.2**	**12.4**
48	**4.7**	**5.3**	**5.9**	**6.5**	**7.1**	**7.8**	**8.5**	**9.2**	**10.1**	**11.0**	**12.2**
50	**4.6**	**5.2**	**5.8**	**6.4**	**7.0**	**7.7**	**8.4**	**9.1**	**10.0**	**11.0**	**12.1**
53	**4.4**	**5.0**	**5.6**	**6.2**	**6.9**	**7.5**	**8.2**	**9.0**	**9.8**	**10.8**	**12.0**
55	**4.3**	**4.9**	**5.5**	**6.1**	**6.8**	**7.4**	**8.1**	**8.9**	**9.7**	**10.7**	**11.9**

[a] Equilibrium moisture contents are calculated using the modified Chung-Pfost equation and constants recommended by ASABE (2016). Regular numbers are the moisture contents at which seeds can be safely stored for up to 6 months at given air conditions but at bolded moisture contents seeds will likely spoil within 6 months as recommended by Maciel et al. (2020).

Soybean (Nigerian) Seeds under Re-Wetting Condition

	Relative Humidity (%)[a]										
T (°C)	*45*	*50*	*55*	*60*	*65*	*70*	*75*	*80*	*85*	*90*	*95*
10	8.0	8.5	9.1	9.6	10.2	10.8	11.5	12.4	**13.4**	**14.7**	**16.9**
13	7.8	8.3	8.8	9.3	9.9	10.6	11.3	12.1	**13.1**	**14.5**	**16.7**
15	7.6	8.1	8.6	9.2	9.8	10.4	11.1	**12.0**	**13.0**	**14.4**	**16.5**
18	7.4	7.9	8.4	9.0	9.6	10.2	10.9	**11.8**	**12.8**	**14.2**	**16.3**
20	7.2	7.7	8.3	8.8	9.4	10.1	**10.8**	**11.6**	**12.7**	**14.0**	**16.2**
23	7.0	7.5	8.1	8.6	9.2	9.9	**10.6**	**11.4**	**12.5**	**13.8**	**16.0**
25	6.9	7.4	7.9	8.5	9.1	**9.7**	**10.5**	**11.3**	**12.4**	**13.7**	**15.9**
28	6.7	7.2	7.7	8.3	8.9	**9.6**	**10.3**	**11.1**	**12.2**	**13.6**	**15.8**
30	6.6	7.1	7.6	8.2	8.8	**9.4**	**10.2**	**11.0**	**12.1**	**13.5**	**15.7**
33	6.4	6.9	7.4	8.0	8.6	**9.3**	**10.0**	**10.9**	**11.9**	**13.3**	**15.5**
35	6.3	6.8	7.3	7.9	**8.5**	**9.2**	**9.9**	**10.8**	**11.8**	**13.2**	**15.4**
38	6.1	6.6	7.2	**7.7**	**8.3**	**9.0**	**9.7**	**10.6**	**11.7**	**13.1**	**15.3**
40	**6.0**	**6.5**	**7.1**	**7.6**	**8.2**	**8.9**	**9.6**	**10.5**	**11.6**	**13.0**	**15.2**

[a] Equilibrium moisture contents are calculated using the modified Chung-Pfost equation and constants recommended by ASABE (2016). Regular numbers are the moisture contents at which seeds can be safely stored for up to 6 months at given air conditions but at bolded moisture contents seeds will likely spoil within 6 months as recommended by Maciel et al. (2020).

A3 Legumes (Pulses)

Chickpea (Resurs 1) Seeds under Drying Condition

	Relative Humidity (%)[a]										
T (°C)	*35*	*40*	*45*	*50*	*55*	*60*	*65*	*70*	*75*	*80*	*85*
10	8.3	9.1	9.9	10.7	11.5	12.5	13.5	14.8	16.2	18.0	**20.4**
13	8.2	8.9	9.7	10.5	11.3	12.3	13.3	14.5	15.9	17.7	**20.1**
15	8.1	8.8	9.6	10.4	11.2	12.1	13.2	14.4	15.8	**17.5**	**19.8**
18	7.9	8.7	9.4	10.2	11.0	11.9	12.9	14.1	15.5	**17.2**	**19.5**
20	7.8	8.5	9.3	10.0	10.9	11.8	12.8	**13.9**	**15.3**	**17.0**	**19.3**
23	7.7	8.4	9.1	9.9	10.7	11.6	12.6	**13.7**	**15.1**	**16.7**	**19.0**
25	7.6	8.3	9.0	9.7	10.5	11.4	**12.4**	**13.5**	**14.9**	**16.5**	**18.8**
28	7.4	8.1	8.8	9.5	10.3	11.2	**12.2**	**13.3**	**14.6**	**16.2**	**18.4**
30	7.3	8.0	8.7	9.4	10.2	11.1	**12.0**	**13.1**	**14.4**	**16.0**	**18.2**
33	7.2	7.8	8.5	9.2	10.0	10.8	**11.8**	**12.9**	**14.1**	**15.7**	**17.9**
35	7.1	7.7	**8.4**	**9.1**	**9.9**	**10.7**	**11.6**	**12.7**	**14.0**	**15.5**	**17.7**
38	6.9	7.6	**8.2**	**8.9**	**9.7**	**10.5**	**11.4**	**12.4**	**13.7**	**15.2**	**17.3**
40	**6.8**	**7.5**	**8.1**	**8.8**	**9.5**	**10.3**	**11.2**	**12.3**	**13.5**	**15.0**	**17.1**
43	**6.7**	**7.3**	**7.9**	**8.6**	**9.3**	**10.1**	**11.0**	**12.0**	**13.2**	**14.7**	**16.8**
45	**6.6**	**7.2**	**7.8**	**8.5**	**9.2**	**9.9**	**10.8**	**11.8**	**13.0**	**14.5**	**16.5**
48	**6.4**	**7.0**	**7.6**	**8.3**	**9.0**	**9.7**	**10.6**	**11.6**	**12.7**	**14.2**	**16.2**
50	**6.3**	**6.9**	**7.5**	**8.1**	**8.8**	**9.6**	**10.4**	**11.4**	**12.6**	**14.0**	**16.0**
53	**6.2**	**6.7**	**7.3**	**7.9**	**8.6**	**9.3**	**10.2**	**11.1**	**12.3**	**13.7**	**15.6**
55	**6.1**	**6.6**	**7.2**	**7.8**	**8.5**	**9.2**	**10.0**	**10.9**	**12.1**	**13.5**	**15.4**

[a] Equilibrium moisture contents are calculated using the Modified Oswin equation and constants from Menkov (2000). Regular numbers are the moisture contents at which seeds can be safely stored for up to 6 months at given air conditions but at bolded moisture contents seeds will likely spoil within 6 months as recommended by Canadian Grain Commission (2019).

Chickpea (Resurs 1) Seeds under Re-Wetting Condition

	Relative Humidity (%)[a]										
T (°C)	*45*	*50*	*55*	*60*	*65*	*70*	*75*	*80*	*85*	*90*	*95*
10	8.5	9.3	10.2	11.2	12.3	13.5	15.1	17.0	**19.6**	**23.4**	**30.7**
13	8.4	9.2	10.1	11.0	12.1	13.4	14.9	16.8	**19.3**	**23.2**	**30.4**
15	8.3	9.1	10.0	10.9	12.0	13.3	14.8	**16.6**	**19.2**	**23.0**	**30.2**
18	8.2	9.0	9.8	10.8	11.9	13.1	14.6	**16.4**	**18.9**	**22.7**	**29.9**
20	8.1	8.9	9.8	10.7	11.7	**13.0**	**14.4**	**16.3**	**18.8**	**22.5**	**29.7**
23	8.0	8.8	9.6	10.5	11.6	**12.8**	**14.3**	**16.1**	**18.6**	**22.3**	**29.3**
25	7.9	8.7	9.5	10.4	**11.5**	**12.7**	**14.1**	**15.9**	**18.4**	**22.1**	**29.1**
28	7.8	8.6	9.4	10.3	**11.3**	**12.5**	**13.9**	**15.7**	**18.2**	**21.8**	**28.8**
30	7.8	8.5	9.3	10.2	**11.2**	**12.4**	**13.8**	**15.6**	**18.0**	**21.6**	**28.6**
33	7.6	8.4	9.2	10.1	**11.0**	**12.2**	**13.6**	**15.4**	**17.8**	**21.4**	**28.2**
35	7.6	8.3	**9.1**	**10.0**	**10.9**	**12.1**	**13.5**	**15.2**	**17.6**	**21.2**	**28.0**
38	**7.4**	**8.2**	**8.9**	**9.8**	**10.8**	**11.9**	**13.3**	**15.0**	**17.4**	**20.9**	**27.7**
40	**7.4**	**8.1**	**8.8**	**9.7**	**10.7**	**11.8**	**13.2**	**14.9**	**17.2**	**20.7**	**27.5**

[a] Equilibrium moisture contents are calculated using the Modified Oswin equation and constants from Menkov (2000). Regular numbers are the moisture contents at which seeds can be safely stored for up to 6 months at given air conditions but at bolded moisture contents seeds will likely spoil within 6 months as recommended by Canadian Grain Commission (2019).

Cowpea (Nigerian) Seeds under Drying Condition

	Relative Humidity (%)[a]										
T (°C)	*35*	*40*	*45*	*50*	*55*	*60*	*65*	*70*	*75*	*80*	*85*
10	10.9	11.8	12.6	13.5	14.4	15.3	16.2	17.2	18.2	**19.4**	**20.7**
13	10.6	11.4	12.3	13.1	14.0	14.8	15.7	16.7	17.7	**18.9**	**20.2**
15	10.4	11.2	12.1	12.9	13.7	14.6	15.5	**16.4**	**17.4**	**18.6**	**19.9**
18	10.1	10.9	11.8	12.6	13.4	14.2	15.1	**16.0**	**17.0**	**18.1**	**19.4**
20	9.9	10.8	11.6	12.4	13.2	14.0	14.9	**15.8**	**16.7**	**17.8**	**19.1**
23	9.7	10.5	11.3	12.1	12.9	13.7	14.5	**15.4**	**16.4**	**17.5**	**18.7**
25	9.6	10.3	11.1	11.9	12.7	13.5	**14.3**	**15.2**	**16.1**	**17.2**	**18.4**
28	9.3	10.1	10.9	11.6	12.4	13.2	**14.0**	**14.9**	**15.8**	**16.9**	**18.1**
30	9.2	10.0	10.7	11.5	12.2	13.0	**13.8**	**14.7**	**15.6**	**16.6**	**17.8**
33	9.0	9.8	10.5	11.2	12.0	**12.7**	**13.5**	**14.4**	**15.3**	**16.3**	**17.5**
35	8.9	**9.6**	**10.4**	**11.1**	**11.8**	**12.6**	**13.4**	**14.2**	**15.1**	**16.1**	**17.3**
38	**8.7**	**9.4**	**10.2**	**10.9**	**11.6**	**12.3**	**13.1**	**13.9**	**14.8**	**15.8**	**17.0**
40	**8.6**	**9.3**	**10.0**	**10.7**	**11.5**	**12.2**	**13.0**	**13.8**	**14.7**	**15.6**	**16.8**
43	**8.5**	**9.2**	**9.9**	**10.6**	**11.3**	**12.0**	**12.7**	**13.5**	**14.4**	**15.4**	**16.5**
45	**8.4**	**9.1**	**9.7**	**10.4**	**11.1**	**11.8**	**12.6**	**13.4**	**14.3**	**15.2**	**16.3**
48	**8.2**	**8.9**	**9.6**	**10.3**	**10.9**	**11.7**	**12.4**	**13.2**	**14.0**	**15.0**	**16.1**
50	**8.1**	**8.8**	**9.5**	**10.1**	**10.8**	**11.5**	**12.3**	**13.0**	**13.9**	**14.8**	**15.9**
53	**8.0**	**8.7**	**9.3**	**10.0**	**10.7**	**11.3**	**12.1**	**12.8**	**13.7**	**14.6**	**15.7**
55	**7.9**	**8.6**	**9.2**	**9.9**	**10.5**	**11.2**	**12.0**	**12.7**	**13.5**	**14.5**	**15.5**

[a] Equilibrium moisture contents are calculated using the modified Henderson equation and constants recommended by ASABE (2016). Regular numbers are the moisture contents at which seeds can be safely stored for up to 6 months at given air conditions but at bolded moisture contents seeds will likely spoil within 6 months as recommended by Canadian Grain Commission (2019).

Cowpea (Nigerian) Seeds under Re-Wetting Condition

	Relative Humidity (%)[a]										
T (°C)	*45*	*50*	*55*	*60*	*65*	*70*	*75*	*80*	*85*	*90*	*95*
10	12.2	13.0	13.9	14.8	15.7	16.7	17.8	**19.0**	**20.3**	**22.1**	**24.5**
13	11.8	12.7	13.5	14.4	15.3	16.3	17.3	**18.5**	**19.8**	**21.5**	**23.9**
15	11.6	12.4	13.3	14.1	15.0	**16.0**	**17.0**	**18.2**	**19.5**	**21.1**	**23.6**
18	11.3	12.1	12.9	13.8	14.7	**15.6**	**16.6**	**17.7**	**19.0**	**20.7**	**23.0**
20	11.1	11.9	12.7	13.6	14.4	**15.3**	**16.3**	**17.4**	**18.7**	**20.3**	**22.7**
23	10.9	11.6	12.4	13.2	14.1	**15.0**	**16.0**	**17.1**	**18.3**	**19.9**	**22.2**
25	10.7	11.5	12.2	13.0	**13.9**	**14.8**	**15.7**	**16.8**	**18.1**	**19.6**	**21.9**
28	10.4	11.2	12.0	12.8	**13.6**	**14.5**	**15.4**	**16.5**	**17.7**	**19.2**	**21.5**
30	10.3	11.0	11.8	12.6	**13.4**	**14.3**	**15.2**	**16.2**	**17.5**	**19.0**	**21.2**
33	10.1	10.8	11.6	**12.3**	**13.1**	**14.0**	**14.9**	**15.9**	**17.1**	**18.6**	**20.8**
35	**9.9**	**10.7**	**11.4**	**12.2**	**13.0**	**13.8**	**14.7**	**15.7**	**16.9**	**18.4**	**20.6**
38	**9.8**	**10.5**	**11.2**	**11.9**	**12.7**	**13.5**	**14.4**	**15.4**	**16.6**	**18.1**	**20.2**
40	**9.6**	**10.3**	**11.1**	**11.8**	**12.6**	**13.4**	**14.3**	**15.3**	**16.4**	**17.9**	**20.0**

[a] Equilibrium moisture contents are calculated using the modified Henderson equation and constants recommended by ASABE (2016). Regular numbers are the moisture contents at which seeds can be safely stored for up to 6 months at given air conditions but at bolded moisture contents seeds will likely spoil within 6 months as recommended by Canadian Grain Commission (2019).

Lentil (Larisa) Seeds under Drying Condition

	Relative Humidity (%)[a]										
T (°C)	*35*	*40*	*45*	*50*	*55*	*60*	*65*	*70*	*75*	*80*	*85*
10	9.6	10.2	10.9	11.6	12.4	13.2	14.0	15.0	16.2	17.6	**19.5**
13	9.4	10.1	10.7	11.4	12.2	13.0	13.8	14.8	16.0	17.4	**19.2**
15	9.3	10.0	10.6	11.3	12.0	12.8	13.7	14.7	**15.8**	**17.2**	**19.0**
18	9.2	9.8	10.5	11.1	11.9	12.6	13.5	14.5	**15.6**	**17.0**	**18.8**
20	9.0	9.7	10.3	11.0	11.7	12.5	13.3	**14.3**	**15.4**	**16.8**	**18.6**
23	8.9	9.5	10.2	10.8	11.5	12.3	13.1	**14.1**	**15.2**	**16.5**	**18.3**
25	8.8	9.4	10.1	10.7	11.4	12.2	**13.0**	**13.9**	**15.0**	**16.4**	**18.1**
28	8.6	9.3	9.9	10.5	11.2	12.0	**12.8**	**13.7**	**14.8**	**16.1**	**17.8**
30	8.5	9.1	9.8	10.4	11.1	**11.8**	**12.6**	**13.5**	**14.6**	**15.9**	**17.6**
33	8.4	9.0	9.6	10.2	10.9	**11.6**	**12.4**	**13.3**	**14.4**	**15.7**	**17.4**
35	8.3	8.9	**9.5**	**10.1**	**10.8**	**11.5**	**12.3**	**13.2**	**14.2**	**15.5**	**17.2**
38	8.1	8.7	**9.3**	**9.9**	**10.6**	**11.3**	**12.0**	**12.9**	**14.0**	**15.2**	**16.9**
40	**8.0**	**8.6**	**9.2**	**9.8**	**10.4**	**11.1**	**11.9**	**12.8**	**13.8**	**15.0**	**16.7**
43	**7.9**	**8.4**	**9.0**	**9.6**	**10.2**	**10.9**	**11.7**	**12.5**	**13.5**	**14.8**	**16.4**
45	**7.8**	**8.3**	**8.9**	**9.5**	**10.1**	**10.8**	**11.5**	**12.4**	**13.4**	**14.6**	**16.2**
48	**7.6**	**8.2**	**8.7**	**9.3**	**9.9**	**10.6**	**11.3**	**12.1**	**13.1**	**14.3**	**15.9**
50	**7.5**	**8.0**	**8.6**	**9.2**	**9.8**	**10.4**	**11.2**	**12.0**	**12.9**	**14.1**	**15.7**
53	**7.4**	**7.9**	**8.4**	**9.0**	**9.6**	**10.2**	**10.9**	**11.7**	**12.7**	**13.9**	**15.4**
55	**7.2**	**7.8**	**8.3**	**8.9**	**9.4**	**10.1**	**10.8**	**11.6**	**12.5**	**13.7**	**15.2**

[a] Equilibrium moisture contents are calculated using the modified Oswin equation and constants from Menkov (2000). Regular numbers are the moisture contents at which seeds can be safely stored for up to 6 months at given air conditions but at bolded moisture contents seeds will likely spoil within 6 months as recommended by Canadian Grain Commission (2020).

Lentil (Larisa) Seeds under Re-Wetting Condition

	Relative Humidity (%)[a]										
T (°C)	*45*	*50*	*55*	*60*	*65*	*70*	*75*	*80*	*85*	*90*	*95*
10	9.4	10.1	10.8	11.6	12.5	13.5	14.7	16.2	**18.2**	**21.0**	**26.4**
13	9.2	9.9	10.6	11.4	12.3	13.3	14.5	16.0	**17.9**	**20.8**	**26.0**
15	9.1	9.8	10.5	11.3	12.2	13.2	**14.4**	**15.8**	**17.8**	**20.6**	**25.8**
18	9.0	9.7	10.4	11.2	12.0	13.0	**14.2**	**15.6**	**17.5**	**20.3**	**25.5**
20	8.9	9.6	10.3	11.1	11.9	**12.9**	**14.0**	**15.5**	**17.4**	**20.1**	**25.3**
23	8.8	9.4	10.1	10.9	11.7	**12.7**	**13.8**	**15.3**	**17.1**	**19.9**	**25.0**
25	8.7	9.3	10.0	10.8	**11.6**	**12.6**	**13.7**	**15.1**	**17.0**	**19.7**	**24.8**
28	8.5	9.2	9.9	10.6	**11.4**	**12.4**	**13.5**	**14.9**	**16.7**	**19.4**	**24.4**
30	8.4	9.1	9.8	**10.5**	**11.3**	**12.3**	**13.4**	**14.7**	**16.5**	**19.2**	**24.2**
33	8.3	8.9	**9.6**	**10.3**	**11.1**	**12.1**	**13.2**	**14.5**	**16.3**	**18.9**	**23.9**
35	**8.2**	**8.8**	**9.5**	**10.2**	**11.0**	**11.9**	**13.0**	**14.4**	**16.1**	**18.8**	**23.7**
38	**8.1**	**8.7**	**9.3**	**10.0**	**10.8**	**11.7**	**12.8**	**14.1**	**15.9**	**18.5**	**23.3**
40	**8.0**	**8.6**	**9.2**	**9.9**	**10.7**	**11.6**	**12.7**	**14.0**	**15.7**	**18.3**	**23.1**

[a] Equilibrium moisture contents are calculated using the modified Oswin equation and constants from Menkov (2000). Regular numbers are the moisture contents at which seeds can be safely stored for up to 6 months at given air conditions but at bolded moisture contents seeds will likely spoil within 6 months as recommended by Canadian Grain Commission (2020).

Red Bean Seeds under Drying Condition

	Relative Humidity (%)[a]										
T (°C)	*35*	*40*	*45*	*50*	*55*	*60*	*65*	*70*	*75*	*80*	*85*
10	9.1	10.0	10.9	11.8	12.8	13.8	14.8	15.9	17.1	**18.5**	**20.1**
13	9.0	9.9	10.8	11.7	12.6	13.6	14.6	15.7	16.9	**18.3**	**19.9**
15	8.9	9.8	10.7	11.6	12.5	13.5	14.5	15.6	16.8	**18.2**	**19.7**
18	8.8	9.7	10.6	11.5	12.4	13.4	14.4	15.5	16.6	**18.0**	**19.5**
20	8.7	9.6	10.5	11.4	12.3	13.3	14.3	15.4	16.5	**17.9**	**19.4**
23	8.6	9.5	10.4	11.3	12.2	13.1	14.1	15.2	16.4	**17.7**	**19.2**
25	8.6	9.5	10.3	11.2	12.1	13.1	**14.0**	**15.1**	**16.3**	**17.6**	**19.1**
28	8.5	9.4	10.2	11.1	12.0	12.9	**13.9**	**15.0**	**16.1**	**17.4**	**18.9**
30	8.4	9.3	10.2	**11.0**	**11.9**	**12.8**	**13.8**	**14.9**	**16.0**	**17.3**	**18.8**
33	8.4	9.2	10.1	**10.9**	**11.8**	**12.7**	**13.7**	**14.7**	**15.9**	**17.1**	**18.6**
35	8.3	9.1	**10.0**	**10.8**	**11.7**	**12.6**	**13.6**	**14.6**	**15.8**	**17.0**	**18.5**
38	8.2	9.1	**9.9**	**10.7**	**11.6**	**12.5**	**13.5**	**14.5**	**15.6**	**16.9**	**18.4**
40	**8.2**	**9.0**	**9.8**	**10.7**	**11.5**	**12.4**	**13.4**	**14.4**	**15.5**	**16.8**	**18.3**
43	**8.1**	**8.9**	**9.7**	**10.6**	**11.4**	**12.3**	**13.3**	**14.3**	**15.4**	**16.6**	**18.1**
45	**8.0**	**8.9**	**9.7**	**10.5**	**11.4**	**12.3**	**13.2**	**14.2**	**15.3**	**16.5**	**18.0**
48	**8.0**	**8.8**	**9.6**	**10.4**	**11.3**	**12.1**	**13.1**	**14.1**	**15.2**	**16.4**	**17.8**
50	**7.9**	**8.7**	**9.5**	**10.3**	**11.2**	**12.1**	**13.0**	**14.0**	**15.1**	**16.3**	**17.7**
53	**7.8**	**8.6**	**9.4**	**10.3**	**11.1**	**12.0**	**12.9**	**13.9**	**15.0**	**16.2**	**17.6**
55	**7.8**	**8.6**	**9.4**	**10.2**	**11.0**	**11.9**	**12.8**	**13.8**	**14.9**	**16.1**	**17.5**

[a] Equilibrium moisture contents are calculated using the modified Henderson equation and constants from Chen (2000). Regular numbers are the moisture contents at which seeds can be safely stored for up to 6 months at given air conditions but at bolded moisture contents seeds will likely spoil within 6 months (estimated based on authors' experience).

Red Bean Seeds under Re-Wetting Condition

	Relative Humidity (%)[a]										
T (°C)	*45*	*50*	*55*	*60*	*65*	*70*	*75*	*80*	*85*	*90*	*95*
10	10.6	11.3	12.1	12.9	13.7	14.6	15.6	**16.6**	**17.9**	**19.4**	**21.7**
13	10.4	11.2	11.9	12.7	13.6	14.4	15.4	**16.4**	**17.7**	**19.2**	**21.5**
15	10.3	11.1	11.9	12.6	13.5	14.3	15.3	**16.3**	**17.5**	**19.1**	**21.3**
18	10.2	11.0	11.7	12.5	13.3	14.2	15.1	**16.1**	**17.3**	**18.9**	**21.1**
20	10.1	10.9	11.6	12.4	**13.2**	**14.1**	**15.0**	**16.0**	**17.2**	**18.7**	**21.0**
23	10.0	10.8	11.5	12.3	**13.1**	**13.9**	**14.8**	**15.8**	**17.0**	**18.5**	**20.7**
25	9.9	10.7	11.4	12.2	**13.0**	**13.8**	**14.7**	**15.7**	**16.9**	**18.4**	**20.6**
28	9.8	10.6	11.3	12.0	**12.8**	**13.7**	**14.6**	**15.6**	**16.7**	**18.2**	**20.4**
30	9.8	**10.5**	**11.2**	**11.9**	**12.7**	**13.6**	**14.5**	**15.5**	**16.6**	**18.1**	**20.3**
33	9.7	**10.4**	**11.1**	**11.8**	**12.6**	**13.4**	**14.3**	**15.3**	**16.5**	**17.9**	**20.1**
35	**9.6**	**10.3**	**11.0**	**11.7**	**12.5**	**13.3**	**14.2**	**15.2**	**16.4**	**17.8**	**20.0**
38	**9.5**	**10.2**	**10.9**	**11.6**	**12.4**	**13.2**	**14.1**	**15.1**	**16.2**	**17.7**	**19.8**
40	**9.4**	**10.1**	**10.8**	**11.5**	**12.3**	**13.1**	**14.0**	**15.0**	**16.1**	**17.5**	**19.7**

[a] Equilibrium moisture contents are calculated using the modified Henderson equation and constants from Chen (2000). Regular numbers are the moisture contents at which seeds can be safely stored for up to 6 months at given air conditions but at bolded moisture contents seeds will likely spoil within 6 months (estimated based on authors' experience).

Vetch Seeds under Re-Wetting Condition

	Relative Humidity (%)[a]										
T (°C)	*45*	*50*	*55*	*60*	*65*	*70*	*75*	*80*	*85*	*90*	*95*
10	9.2	9.8	10.5	11.3	12.2	13.3	14.7	**16.4**	**18.8**	**22.5**	**29.7**
13	9.0	9.7	10.3	11.1	12.0	13.1	14.5	**16.2**	**18.6**	**22.2**	**29.4**
15	8.9	9.6	10.3	11.0	11.9	13.0	14.3	**16.0**	**18.4**	**22.0**	**29.2**
18	8.8	9.4	10.1	10.9	11.8	12.8	14.1	**15.8**	**18.2**	**21.8**	**28.8**
20	8.7	9.3	10.0	10.8	11.7	12.7	**14.0**	**15.7**	**18.0**	**21.6**	**28.6**
23	8.6	9.2	9.9	10.6	11.5	12.5	**13.8**	**15.5**	**17.8**	**21.3**	**28.3**
25	8.5	9.1	9.8	10.5	**11.4**	**12.4**	**13.7**	**15.3**	**17.6**	**21.2**	**28.1**
28	8.4	9.0	9.6	10.4	**11.2**	**12.3**	**13.5**	**15.1**	**17.4**	**20.9**	**27.8**
30	8.3	8.9	9.6	10.3	**11.1**	**12.1**	**13.4**	**15.0**	**17.2**	**20.7**	**27.6**
33	8.2	8.8	9.4	**10.1**	**11.0**	**12.0**	**13.2**	**14.8**	**17.0**	**20.5**	**27.3**
35	**8.1**	**8.7**	**9.3**	**10.0**	**10.9**	**11.9**	**13.1**	**14.7**	**16.9**	**20.3**	**27.0**
38	**8.0**	**8.6**	**9.2**	**9.9**	**10.7**	**11.7**	**12.9**	**14.5**	**16.7**	**20.0**	**26.7**
40	**7.9**	**8.5**	**9.1**	**9.8**	**10.6**	**11.6**	**12.8**	**14.4**	**16.5**	**19.9**	**26.5**

[a] Equilibrium moisture contents are calculated using the Modified Halsey equation and constants recommended by ASABE (2016). Regular numbers are the moisture contents at which seeds can be safely stored for up to 6 months at given air conditions but at bolded moisture contents seeds will likely spoil within 6 months (estimated based on authors' experience).

Vetch Seeds under Drying Condition

	Relative Humidity (%)[a]										
T (°C)	*35*	*40*	*45*	*50*	*55*	*60*	*65*	*70*	*75*	*80*	*85*
10	9.2	9.8	10.5	11.2	12.0	13.0	14.0	15.3	16.9	**18.9**	**21.6**
13	9.0	9.6	10.3	11.0	11.8	12.7	13.7	15.0	16.5	**18.5**	**21.2**
15	8.8	9.5	10.1	10.8	11.6	12.5	13.5	14.8	16.3	**18.2**	**20.9**
18	8.7	9.2	9.9	10.6	11.4	12.2	13.3	14.5	16.0	**17.9**	**20.5**
20	8.5	9.1	9.7	10.4	11.2	12.1	13.1	14.3	**15.8**	**17.7**	**20.3**
23	8.3	8.9	9.5	10.2	11.0	11.8	12.8	14.0	**15.4**	**17.3**	**19.9**
25	8.2	8.8	9.4	10.1	10.8	11.7	**12.6**	**13.8**	**15.2**	**17.1**	**19.6**
28	8.0	8.6	9.2	9.8	10.6	11.4	**12.4**	**13.5**	**14.9**	**16.7**	**19.2**
30	7.9	8.5	9.1	9.7	**10.4**	**11.2**	**12.2**	**13.3**	**14.7**	**16.5**	**19.0**
33	7.7	8.3	8.9	**9.5**	**10.2**	**11.0**	**11.9**	**13.0**	**14.4**	**16.2**	**18.6**
35	7.6	8.2	**8.7**	**9.4**	**10.1**	**10.8**	**11.8**	**12.9**	**14.2**	**16.0**	**18.4**
38	7.5	8.0	**8.5**	**9.2**	**9.8**	**10.6**	**11.5**	**12.6**	**13.9**	**15.6**	**18.0**
40	**7.3**	**7.9**	**8.4**	**9.0**	**9.7**	**10.5**	**11.4**	**12.4**	**13.7**	**15.4**	**17.8**
43	**7.2**	**7.7**	**8.2**	**8.8**	**9.5**	**10.2**	**11.1**	**12.2**	**13.4**	**15.1**	**17.4**
45	**7.1**	**7.6**	**8.1**	**8.7**	**9.3**	**10.1**	**11.0**	**12.0**	**13.3**	**14.9**	**17.2**
48	**6.9**	**7.4**	**7.9**	**8.5**	**9.1**	**9.9**	**10.7**	**11.7**	**13.0**	**14.6**	**16.9**
50	**6.8**	**7.3**	**7.8**	**8.4**	**9.0**	**9.7**	**10.6**	**11.6**	**12.8**	**14.4**	**16.6**
53	**6.7**	**7.1**	**7.6**	**8.2**	**8.8**	**9.5**	**10.3**	**11.3**	**12.5**	**14.1**	**16.3**
55	**6.6**	**7.0**	**7.5**	**8.1**	**8.7**	**9.4**	**10.2**	**11.2**	**12.4**	**13.9**	**16.1**

[a] Equilibrium moisture contents are calculated using the Modified Halsey equation and constants recommended by ASABE (2016). Regular numbers are the moisture contents at which seeds can be safely stored for up to 6 months at given air conditions but at bolded moisture contents seeds will likely spoil within 6 months (estimated based on authors' experience).

White Bean (Seafarer) Seeds under Desorption Condition

	Relative Humidity (%)[a]										
T (°C)	*35*	*40*	*45*	*50*	*55*	*60*	*65*	*70*	*75*	*80*	*85*
10	10.0	11.1	12.3	13.5	14.8	16.1	17.6	19.2	21.0	**23.1**	**25.7**
13	9.8	10.9	12.1	13.3	14.5	15.8	17.3	18.8	20.6	**22.7**	**25.2**
15	9.7	10.8	11.9	13.1	14.3	15.6	17.1	18.6	**20.4**	**22.4**	**24.9**
18	9.5	10.6	11.7	12.9	14.1	15.4	16.7	18.3	**20.0**	**22.0**	**24.4**
20	9.4	10.5	11.6	12.7	13.9	15.2	16.5	**18.1**	**19.8**	**21.7**	**24.1**
23	9.2	10.3	11.4	12.5	13.7	14.9	**16.3**	**17.7**	**19.4**	**21.3**	**23.7**
25	9.1	10.2	11.2	12.3	13.5	14.7	**16.1**	**17.5**	**19.2**	**21.1**	**23.4**
28	9.0	10.0	11.0	12.1	13.3	14.5	**15.8**	**17.2**	**18.9**	**20.8**	**23.0**
30	8.9	9.9	10.9	12.0	13.1	**14.3**	**15.6**	**17.1**	**18.7**	**20.5**	**22.8**
33	8.7	9.7	10.7	11.8	12.9	**14.1**	**15.4**	**16.8**	**18.4**	**20.2**	**22.4**
35	**8.6**	**9.6**	**10.6**	**11.7**	**12.8**	**14.0**	**15.2**	**16.6**	**18.2**	**20.0**	**22.2**
38	**8.5**	**9.5**	**10.5**	**11.5**	**12.6**	**13.7**	**15.0**	**16.4**	**17.9**	**19.7**	**21.9**
40	**8.4**	**9.4**	**10.4**	**11.4**	**12.5**	**13.6**	**14.8**	**16.2**	**17.7**	**19.5**	**21.6**
43	**8.3**	**9.2**	**10.2**	**11.2**	**12.3**	**13.4**	**14.6**	**16.0**	**17.5**	**19.2**	**21.3**
45	**8.2**	**9.2**	**10.1**	**11.1**	**12.2**	**13.3**	**14.5**	**15.8**	**17.3**	**19.0**	**21.1**
48	**8.1**	**9.0**	**10.0**	**11.0**	**12.0**	**13.1**	**14.3**	**15.6**	**17.0**	**18.7**	**20.8**
50	**8.0**	**8.9**	**9.9**	**10.9**	**11.9**	**13.0**	**14.1**	**15.4**	**16.9**	**18.6**	**20.6**
53	**7.9**	**8.8**	**9.7**	**10.7**	**11.7**	**12.8**	**14.0**	**15.2**	**16.7**	**18.3**	**20.3**
55	**7.8**	**8.7**	**9.7**	**10.6**	**11.6**	**12.7**	**13.8**	**15.1**	**16.5**	**18.2**	**20.2**

[a] Equilibrium moisture contents are calculated using the modified Henderson equation and constants from Hutchinson and Otten (1984). Regular numbers are the moisture contents at which seeds can be safely stored for up to 6 months at given air conditions but at bolded moisture contents seeds will likely spoil within 6 months as recommended by Canadian Grain Commission (2020).

References

ASABE. 2016. ASAE D245.6 OCT2007. Moisture Relationships of Plant-Based Agricultural Products, ASABE Standard. American Society of Agricultural and Biological Engineers, St. Joseph, MI.

Basunia, M. A., and T. Abe. 2005. Adsorption isotherms of barley at low and high temperatures. Journal of Food Engineering 66: 129–136.

Canadian Grain Commission. 2019. Official Grain Grading Guide. Canada Grain Commission.

Canadian Grain Commission. 2020. Safe Storage Guidelines. Winnipeg.

Chen, C. 2000. Factors which affect equilibrium relative humidity of agricultural products. Transactions of the ASAE 43: 673–683.

Flood, C. A. J., and J. M. White. 1984. Desorption equilibrium moisture relationships for popcorn. Transactions of the ASAE 27: 561–571.

Hulasare, R., M. N. N. Habok, D. S. Jayas, and N. D. G. White. 2001. Near equilibrium moisture content values for hull–less oats. Applied Engineering in Agriculture 17: 325–328.

Hutchinson, D. H., and I. Otten. 1984. Equilibrium moisture content of white beans. Cereal Chemistry 61: 155–158.

Jacobsen, E. E., and F. Fleurat-Lessard 2002. Estimation of safe storage periods for malting barley using a model of heat production based on respiration experiments, pp. 456–463. In P. F. Credland, D. Armitage, C. H. Bell, P. M. Cogan and E. Highley (eds.), Proceedings of the 8th International Working Conference on Stored-Product Protection. CAB International, York.

Jayas, D. S., and G. Mazza. 1993. Comparison of five, three-parameter equations for the description of adsorption data of oats. Transactions of the ASAE 36: 119–125.

Jian, F., D. Divagar, J. Mhaiki, D. S. Jayas, P. G. Fields, and N. D. G. White. 2018. Static and dynamic methods to determine adsorption isotherms of hemp seed (*Cannabis sativa* L.) with different percentages of dockage. Food Science & Nutrition 6: 1629–1640.

Jian, F., M. A. A. Mamun, N. D. G. White, D. S. Jayas, P. G. Fields, and J. McCombe. 2019. Safe storage times of FINOLA® hemp (*Cannabis sativa*) seeds with dockage. Journal of Stored Products Research 83: 34–43.

Maciel, G., D. Torre, L. M. Cardoso, M. G. Cendoya, and J. R. Wagner. 2020. Determination of safe storage moisture content of soybean expeller by means of sorption isotherms and product respiration. Journal of Stored Products Research 86: 1–7.

Mazza, G., D. S. Jayas, and N. D. G. White. 1990. Moisture sorption isotherms of flax seed. Transactions of the ASAE 33: 1313–1318.

Menkov, N. D. 2000. Moisture sorption isotherms of lentil seeds at several temperatures. Journal of Food Engineering 44: 205–211.

Menkov, N. D., K. Dinkov, A. Durakova, and N. Toshkov. 2009. Sorption characteristics of buckwheat grain. Bulgarian Journal of Agricultural Science 15: 281–285.

Sathya, G., D. S. Jayas, and N. D. G. White. 2008. Safe storage guidelines for rye. Canadian Biosystems Engineering 50: 1–8.

Sumner, P. E., and E. J. Williams. 2009. Grain and Soybean Drying on Georgia Farms, pp. 12. Cooperative Extension, The University of Georgia, Athens, Georgia.

Sun, D., and C. Byrne. 1998. Selection of EMC/ERH isotherm equations for rapeseed. Journal of Agricultural Engineering Research 69: 307–315.

Sun, K., F. Jian, D. S. Jayas, and N. D. G. White. 2014. Quality changes in high and low oil content canola during storage: part I – safe storage time under constant temperatures. Journal of Stored Products Research 59: 320–327.

Tagawa, A., S. Murata, and H. Hayashi. 1993. Latent heat of vaporization in buckwheat using the data of equilibrium moisture content. Transactions of the ASAE 36: 113–118.

Appendix B: Psychrometric Chart

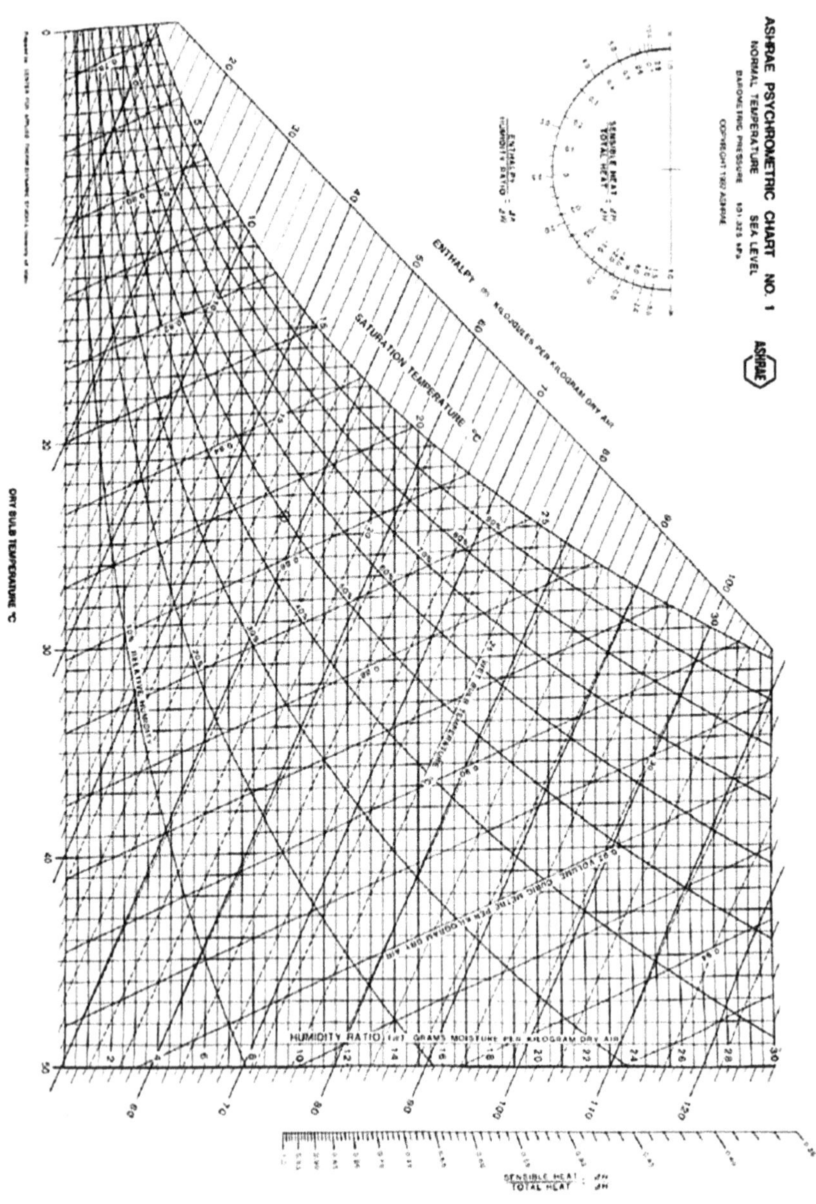
ASHRAE PSYCHROMETRIC CHART NO. 1
NORMAL TEMPERATURE
SEA LEVEL
ENTHALPY
SATURATION TEMPERATURE °C
HUMIDITY RATIO
DRY BULB TEMPERATURE °C

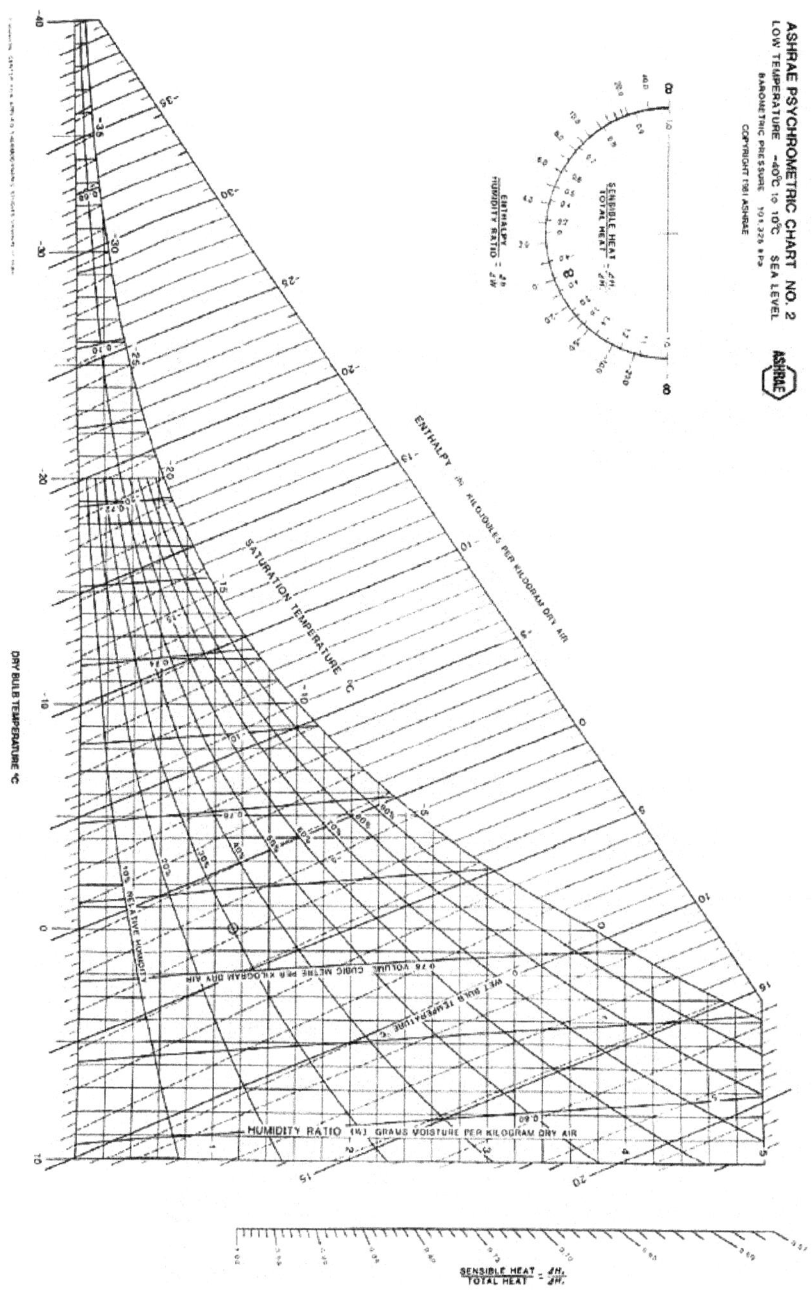
ASHRAE PSYCHROMETRIC CHART NO. 2
LOW TEMPERATURE -40°C to 10°C SEA LEVEL
BAROMETRIC PRESSURE 101.325 kPa
COPYRIGHT 1981 ASHRAE
ASHRAE
SENSIBLE HEAT / TOTAL HEAT
ENTHALPY / HUMIDITY RATIO
ENTHALPY (h) KILOJOULES PER KILOGRAM DRY AIR
SATURATION TEMPERATURE °C
RELATIVE HUMIDITY
VOLUME CUBIC METRE PER KILOGRAM DRY AIR
WET BULB TEMPERATURE °C
HUMIDITY RATIO (W) GRAMS MOISTURE PER KILOGRAM DRY AIR
DRY BULB TEMPERATURE °C

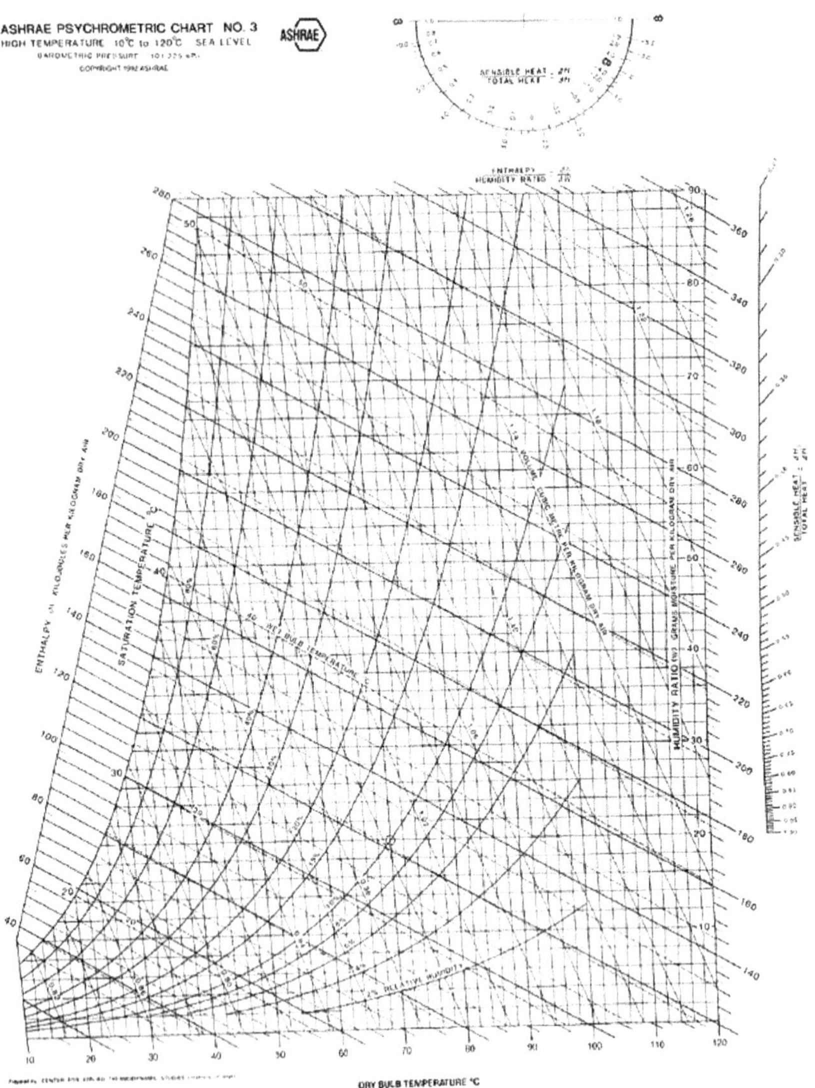
ASHRAE PSYCHROMETRIC CHART NO. 3
HIGH TEMPERATURE 10°C to 120°C SEA LEVEL
ASHRAE
SENSIBLE HEAT / TOTAL HEAT
ENTHALPY / HUMIDITY RATIO
ENTHALPY (h) KILOJOULES PER KILOGRAM DRY AIR
SATURATION TEMPERATURE °C
WET BULB TEMPERATURE
VOLUME CUBIC METER PER KILOGRAM DRY AIR
HUMIDITY RATIO (W) GRAMS MOISTURE PER KILOGRAM DRY AIR
RELATIVE HUMIDITY
DRY BULB TEMPERATURE °C

Index

Note: Locators in *italics* represent figures and **bold** indicate tables in the text.

A

Absorption, 180–181
Activation energy, 247
Adiabatic/constant wet-bulb process, 266
Adiabatic drying processes, 223
Adsorption, 180–181, *193*, 194
Aeration, 267–269, *268*, *269*, *270*, 272
Aerodynamic drag force, 40
Airflow
  Brinkman form, Darcy law, 55–56
  Darcy–Forchheimer law, 53–55, *54*
  Darcy law, 51–53, *52*
  Ergun equation, 55
  fluid acceleration, 56
  resistance in bulk grains
    equations used, 57–58, **59**
    grain conditions influencing, 60–68, **61**, *62*, **63**
    pressure drops, 59–60, *60*
  velocity through porous medium, 50
  wall friction, 56
Alternaria alternata, 406, 408
Ambient drying, *see* Natural air drying
American Association of Cereal Chemists (AACC) Approved Methods, 23
American Society of Agricultural and Biological Engineers (ASABE), 58, 60, 315
Angular frequency, 73
Archimedes' principle, 9
Arching, 49
Arrhenius relationship, 123–124
Aspergillus
  candidus, 406, 408
  flavus, 406, 408
  glaucus, 406
  niger, 406
  ochraceous, 406, 408
  wentii, 408

B

Barley seeds, 202, 387, 393, 424–425
Bed configuration, 250
Bessel function, 129
BET equation, *see* Brunauer-Emmett-Teller (BET) equation
Bin diameter, 144–146, *145*, *146*
Bins
  diameter, 144–146, *145*, *146*
  geographical location, 149–150, *150*
  grain depth, 144–146, *145*, *146*

headspace, 144
shape, *147*, 147–149, **148–149**
wall material, *147*, 147–149, **148–149**
Biot number, 238–239
Bradley equation, 292
Brown rice, 409
Brunauer–Emmett–Teller (BET) theory, 120, 288–292
Buckwheat, 426–427
Buckwheat hay, *231*
Bulk density, 2
measurement of, 2, 4–6
physical properties, 6–8
porosity of seeds, **3–4**

C

Calorimetric technique, 122–123
Canadian Grain Commission, 5, *5*
Canola oil, 106–107
Canola seeds, 7, 8, 64, 139, *146*, 267–269, *268*, *269*, *270*, 386, 388, 399, 443–444
Capacitor, *68*, 68–69
Capillary theory, 178–179, 223–224, *224*, 321–322
"Case-hardened" phenomena, 190
Cereal grains
barley seeds, 424–425
Buckwheat, 426–427
corn, 428–429
millet seeds, 430–431
oats (hull-less, AC Belmont), 433–434
oats (hulled, dumont) seeds, 432
popcorn, 435
rough rice seeds, 436–437
rye seeds, 438
wheat (hard red spring, Sinton) seeds, 439–440
wheat (soft winter, Hobbit) seeds, 441–442
Chemical changes, kernals, 260
Chickpea seeds, 452–453
Chilled mirror dew point, 174–175
Chung-Pfost equation, *394*
Clausius–Clapeyron equation, 121–122, 180, 188, 292–294, 313
Cohesion-tension theory, 194
Cohesive arching, 49
Colligative effect, 178
Combination drying, *272*, 273–274
Compression, 56
Computational fluid dynamics, 352–353
advantages of, 355–356
application in grain drying, 358–359
disadvantages of, 356–358
governing equations, 353–355
procedure of, 355
Conducting/chilled aeration, 413
Conduction, 218
Convection drying energy source, 233–234
Convective heat transfer, 127–128
measurement of coefficient, 128
crop grain, 136
determination methods, 134–136, **135**
dimensionless analysis, 131–133
dimensionless model, 133–134
Schumann method, 128–131, *130*
Convective mass transfer coefficient, 238
Corn, 8, 64, 74, 75, 77, 126, 167, *272*, 272–274, 386–387, 400, 409, 428–429
Coulomb's law, 71
Cowpea seeds, 454–455
Crack, 255, 257; *see also* Fissure
after drying, *258*, 258–259
during drying, 257–258
swelling, 259
Cultivars, 386–387

D

D'Arcy and Watt equation, 295–296
Darcy–Forchheimer law, 53–55, *54*
Darcy law, 51–53, *52*
Brinkman form, 55–56
DDI method, *see* Dynamic dew-point isotherm (DDI) method
De Boer and Zwikker equation, 309
Deep bed drying, 263, *264*, 267
mathematical modelling
computational fluid dynamics, 352–359
equilibrium models, 347–350
logarithmic models, 350–351
non-equilibrium models, 344–347
numerical models, 351–352
vs. thin layer, 342–344
Delayed cooling, 258
Denaturation, 409
Density-independent moisture calibration function, 75

Desorption, 181, *182*, *193*, 194
  cycles, 196–197
  isotherm, 183–185, *184*
Dielectric conductivity, 71–72
Dielectric constant, 75
Dielectric loss factor, 74
Dielectric properties, 68–70
  applications, 77–78
  factors influencing grain, 73–74
    bulk density, 75
    chemical composition, 76
    moisture content, 74–75
    temperature, 74
  measurement of, 76–77
  parameters of, 70–73
Differential scanning calorimeter (DSC), 103, *251*, 251–252
Direct methods, moisture content measurement, 165
Displacement current, 70
Drag force, 40–41
Dry basis percentage moisture content, 164
Dryeration, *272*, 272–274
Drying condition
  barley seeds, 424
  Buckwheat seeds, 426
  canola seeds, 443
  chickpea seeds, 452
  corn seeds, 428
  cowpea seeds, 454
  hemp seeds, 446
  lentil seeds, 456
  millet seeds, 430
  oats seeds, 433
  popcorn, 435
  rapeseed seeds, 448
  red bean seeds, 458
  rough rice seeds, 436
  soybean seeds, 450
  vetch seeds, 461
  wheat seeds, 439–441
Drying curve method, 245–246
Drying mechanism, 223
  airflow rate, 234–235
  capillary theory, 223–224, *224*
  combination of the mechanisms, 226–227
  drying materials, physical property of, 233
  effusion flow, 226
  initial grain condition, 234
  intermittent, 235
  liquid diffusion, 225
  liquid movement, gravitational effects, 224
  liquid or vapour or both migrations, 226
  supplied air, property of, 234
  tempering, 235
  vapour diffusion, 225
Drying methods
  application of, 274
  crack formation, 257–259, *258*
  types, 217
Drying rate
  thin layer drying, 228
Drying resistance, 235–236
  constant rate of drying phase period, 236
  evaluation parameters
    Biot number, 238–239
    convective mass transfer coefficient, 238
    Lewis number, 239–241
    water diffusivity, 241–250
  falling rate period, 236–237
  low efficiency of mass and heat transfer, 237
DSC enthalpy thermogram, 123
DVS method, *see* Dynamic vapour sorption (DVS) method
Dye technique, 384
Dynamic dew-point isotherm (DDI) method, 200–201
Dynamic vapour sorption (DVS) method, 199–200

E

Effective mass transfer coefficient, 338
Effusion flow, 226
Electric hygrometers, 175, 198
Electric susceptibility, 71
Electromagnetic energy-based drying, 256
Elongation, 16–17
Empirical model, 335–341
Entrance of moisture, grain bulks
  snow water, 390
  wetting due to humid air, 390–391
Enzymatic inactivation, 260
Equilibrium models, 347–350
Equilibrium moisture content (EMC), *195*, *197*, 253, 384, *385*, 386–387

INDEX

Equilibrium relative humidity (ERH), 176, 190, *195*, *197*, *266*, 266–267, 349, 386–387, **392**, 411
Ergun equation, 55
Evaporation-condensation model, 330–331
Evaporative cooling, 262–263
Expansion, kernels, 254, 256
Extrinsic factors, kernel volume changes, 254–255

F

Fan warming, 271–272
Fatty acid value (FAV), 409
Federal Grain Inspection Service (FGIS), 6
Fick's Law of Diffusion, 317–319, 354
Fissure, 257, 261; *see also* Crack
Fixed bed drying, *see* Deep bed drying
Flakiness, 16–17
Flax seeds, 64, 445
Flip-flop motion, 70
Fluid acceleration, 56
Fokker-Planck diffusivity law, 318
Food stability map, *407*
Free convection currents, 140–141, *141–143*
Freundlich adsorption isotherm, 296–297
Freundlich equation, 292
Frictional properties, grain bulks
  sliding friction
    coefficient of, 31–32
    measurement of, *32*, 32–33
  surface difference among crops and cultivars, 28–31
Fungi infection, 178, 403–404, 413

G

GAB (Guggenheim, Anderson, de Boer) equation, 124, 297–300
Gas pycnometer, 9
Geographical location, 149–150, *150*
Geometrical properties, grain kernels
  cross section, 17
  dimensions, 13
  measurement, 13–15, *14*
  roundness, 19–21, **20**
  shape factor, 16–17
  sphericity, *17*, 17–18
  surface area and volume, 15–16
Gibbs free energy, 176
Glass transition, 255–256
  DSC method, 252
  factors influencing, 252
  grain drying, 252–254, *253*, *254*
  measurement of, 251–252
  principle of, 250–251
Grain bulk, 267
  in silos
    lateral stress ratio, 48–49
    stress in bins, 49–50
    stress in silos without grain discharge, 46–48, *47*
    unloading flow pattern, *45*, 45–46
Grain depth, 144–146, *145*, *146*
Grain drying, 267–268, *268*
Grain moisture migration, 391
  condensation of water, grain surface, 400
    freezing, 402
    inside bins, 401
    during turning and loading, 400–401
  factors affecting
    fungi infection, 403–404
    insects infestation, 403–404
    resistance to airflow, 403
    temperature, 402
  free convection currents
    application, 400
    experimental proof, 399
    mechanism, 396–399, *397*
  moisture gradients at a uniform temperature
    application, 393
    mechanism, 391, **392**
    process, 392–393
  temperature gradients, uniform moisture content
    application, 396
    mechanism, 393–394, *394*
    process, 394–396, *395*, *396*
  water sorption, during air exchange, 401
Grain storage bins, 79
Granular materials, 1
Gravimetric method, 198
Gravimetric oven method, 166–167

H

Hagen–Poiseuille law, 323
Halsey equation, 309–310
Hardness, 21–23
  flour yield, 26
  measurement of, 23
  methods measuring the force, 25–26
  near-infrared spectroscopy method, 24–25
  particle size, 26
  particle size index method, 24
  single kernel characterization system, 25
  starch properties, 26
  surface area, 26
Hard to cook (HTC), 204–205
Harkins, 300
Harkins–Jura equation, 300–301
Hausner ratio (H_R), 8
Headspace, 144
Heat capacity, 102–103, *103*
Heat content, 102–103, *103*
Heat supply
  adding heater, 222, **222**
  air properties, ventilation, 220–222
  energy consumption, grain drying, 219–220
  energy efficiency of convection, 219
  methods, 218
Hemp (Finola®) seeds, 8, 222, 446–447
Henderson equation, 310–312
Henry's law, 306
N-hexane, 106–107
H-H equation, 290, 291
High temperature drying, 263
  effects of, 259–261
Hotspot, 410
Hukill and Ives model, 58
Humid air, 390–391
Hygrometric method, 198
Hygroscopicity
  crop kernels, 384, *385*
  definition, 384
  different classes and cultivars, same crop type, 386–387
  different parts of a kernel, 387
  factors influencing, 387–388
Hysteresis, food materials
  desorption cycles, 196–197
  mechanism of, 192–196, 206
  types of, 196

I

Ice sublimation, 256
Image processing, 259
In-bin drying, 263, 389
Indirect methods, moisture content measurement, 165–166
Industrial performance, kernals, 260–261
Initial grain moisture content, 261
  after aeration, 389–390
  after drying, 389
  after filling, 389
  at harvest, 388
Ink bottle, 193
Insects infestation, 403–404
In-storage stage drying, 274
Intermolecular attractive force, 171
Internal coefficient of friction, 31
Intrinsic factors, kernel volume changes, 254
Isotherm, 183–185, *184*
  factors influencing
    food composition effect, 191, **192**
    pressure effect, 190–191
    RH effect, *187–189*, 187–190
    temperature, *187–189*, 187–190
  hygroscopic property, 185–187, *186*
  shape, 185–187, *186*

K

Kezdi equation, 48–49

L

Laboratory determination, sorption and desorption isotherms, 197–198
  categories, 198
  DDI method, 200–201
  difference between isotherm methods, 201–202
  DVS method, 199–200
  SSS method, 198–199
  thermal cell method, 201
  thin layer dynamic method, 200
Langmuir and Freundlich model, 289–290
Langmuir isotherm equation, 301–302
Langmuir kinetics, 289

Latent heat of vaporization, 118–119
  cereal grain, 126–127
Lateral stress ratio, 48–49
Legumes
  chickpea seeds, 452–453
  cowpea seeds, 454–455
  lentil seeds, 456–457
  red bean seeds, 458–459
  vetch seeds, 460–461
  white bean (seafarer) seeds, 462
Lentil seeds, 456–457
Lewicki model, 303
Lewis model, 339
Lewis number, 239–241
Linear falling-rate period, 338
Line heat source method, 113–114
Lipid oxidation, 409
Liquid diffusion, 225
Logarithmic models, 330, 350–351
Low temperature drying model, 343
Luikov's model, 331–332

M

Maize, 386
Mass balance, 328–330
Mathematical modelling, 287
  assumptions, 327–328
  types of, 326–327
Mean flow velocity, 51
Medium temperature drying, 263
Membrane damage, 260
Microwave dielectric properties, 77
Millet seeds, 195, *195*, 430–431
Milling, 405
Mites, 404
Modified Chung–Pfost equation, 311
Moisture accumulation, 399, 400
Moisture content, biomaterial
  calculation, 164
  definition, 164
  in grain sample, 164–165
  gravimetric oven method, 166–167
  measurement types, 165
  status of water, 167–170, **168**
  vs. water activity, 176–178, **177**
  water distribution, 167
  water mobility, 170–171
Moisture effect, 247–248
Monolayer adsorption, 314
Monomeric method, 198

N

Natural air drying, 217, 220, 263, 270–272, *271*
Navier–Stoke momentum equations, 346, 353–354, 355
Near-Infrared (NIR) Spectroscopy method, 24–25
Net isosteric heat of sorption, 119
  cereal grain, 126–127
  determination of, 120–121
    Clausius–Clapeyron equation, 121–122
    common and speciality crops, **125**, 125–126
  factors influencing, 120
Non-adiabatic drying processes, 223
Nonenzymatic browning, 409
Non-equilibrium models, 344–347
Nuclear magnetic resonance (NMR) technique, 168, 169

O

Oats seeds, 432, 433–434
Oilseeds
  canola (Westar) seeds, 443–444
  flax seeds, 445
  hemp (Finola®) seeds, 446–447
  rapeseed, 448–449
  soybean (Nigerian) seeds, 450–451
Oswin equation, 311–312
Othmer equation, 124

P

Page model, 343–344
Park equation, 303
Particle Size Index (PSI) method, 24
Peleg equation, 303–305
Penicillium spp., 406
Physical changes, kernals, 260
Popcorn, 435
Pore formation, kernel, 254–255
  fissure, 257–259, *258*
  glass transition, 255–256
  puffing, 256
  stress-cracked kernels, determination, 259
  surface tension, 255
  swelling, 259

Porosity
- measurement, 11–12
- physical properties influencing, 12–13
- tank method, 9–10, *10*

Poyet equation, 294
Pressure drops, 59–60, *60*
- chaff concentration, 64
- direction, 60, **61**, *62*, **63**, 64
- drop height, 67–68
- filling method, 67–68
- fine concentration, 64
- grain conditions influencing, 60
- grain depth, 64–66
- moisture content, 66–67

Psychrometric chart, *230*, 264–266, *265*, *266*, 463–465
Psychrometric method, 174
Psychrometric processes, 267–268, *268*
Puffing, 256

R

Raoult law, 295, 297, 305
Rapeseed, 270–272, *271*, 448–449
Rayleigh number, 141
Reaction – diffusion approach, 288
Receding front theory, 322–326
Red bean seeds, 458–459
Red kidney bean, *203*, 204, *229*, 250
Relative complex permittivity, 72
Relative humidity (RH), 173–175, *174*, 176, 202, 393–394, *394*, 398, 400
Relaxation time, 73
Repose angle, 33, **34–35**, 35–37
- relationship between other physical properties, 37–39, *38*
- relationship with internal friction, 39–40, **40**

Residence time, 262
Resistance to airflow, 403
Re-wetting, 270–271
- barley seeds, 425
- canola seeds, 444
- chickpea seeds, 453
- corn seeds, 429
- cowpea seeds, 455
- hemp seeds, 447
- lentil seeds, 457
- millet seeds, 431
- oats seeds, 432, 434
- rapeseed seeds, 449
- red bean seeds, 459
- rough rice seeds, 437
- soybean seeds, 451
- vetch seeds, 460
- wheat seeds, 442

Reynolds number (Re), 131–132
Rice, 115–116, 126
Riedel equation, 122–123
Rodents, 404
Roff milling index, 26
Rolling resistance, 32
Rough rice seeds, 64, 436–437
Roundness, 19–21, **20**
Rubber elastomers, 250
Rye seeds, 438

S

Safe drying temperature, 261–263, **262**
Safe storage moisture content
- chemical reactions, 409–411
- microorganism, 405–408, *407*
- milling, 405
- by pests, 404
- respiration and germination, 408
- time, 411–413, **412**

Salt solution static (SSS) method, 198–199
Schumann method, 128–131, *130*, 134, 136
Seed vigour, 260
Semi-theoretical model, 335–341
Shedd equation, 58
Shrinkage, kernels, 26–28, *27*, 249–250, 254, 256
Sigmoidal isotherm, 184–185
Single component models, 313–314
Single Kernel Characterization System (SKCS), 25
Smith equation, 306
Snow water, 390
Soaking, 202, *203*, 204
- models of, 341–342

Sorption isotherm equations
- empirical models
  - Chen equation, 307, 309
  - Halsey equation, 309–310
  - Henderson equation, 310–312
  - modified Chung–Pfost equation, 311
  - Oswin equation, 311–312
  - sporadically used models, **308**

model selection
models with more or less parameters, 314–315
most used equations, 315
single component models, 313–314
temperature effect, 312–313
theoretical and semi-theoretical models, 288
BET equation, 288–292
Bradley equation, 292
Clausius–Clapeyron equation, 292–294
D'Arcy and Watt equation, 295–296
Freundlich adsorption equation, 296–297
GAB (Guggenheim, Anderson, de Boer) equation, 297–300
Harkins–Jura equation, 300–301
Henry's law, 306
Langmuir isotherm equation, 301–302
Lewicki model, 303
Park equation, 303
Peleg equation, 303–305
Raoult's law, 305
Smith equation, 306
Young and Nelson equation, 306–307
Soybean (Nigerian) seeds, 7, 112, 168, 170, 450–451
Specific heat, 103–104
freshly harvested canola, *108*
measurement methods, 104–107, *105*
prediction of, 107
relationship with other physical properties, *107*, 107–109, **108**
Sphericity, *17*, 17–18
SSS method, *see* Salt solution static (SSS) method
Steady-state method, 112
Stenvert method, 26
Stress-cracked kernels, determination, 259
Sublimation, 102
Surface area and volume, 15–16
Surface interaction, 179–180
Surface tension, 255
Swelling, 26–28, *27*, 259

T

Tangential Abrasion Dehulling Device (TADD), 25
Temperature distribution, grain moisture migration, 402
Temperatures, grain bulks, 137, 152
factors affecting in bins, 143–144
bin diameter, 144–146, *145*, *146*
geographical location, 149–150, *150*
grain depth, 144–146, *145*, *146*
headspace, 144
shape, bin, *147*, 147–149, **148–149**
wall material, *147*, 147–149, **148–149**
patterns
free convection currents, 140–141, *141–143*
initial temperature, 138–140, *139*
log, 137–138, *138*
temperature gradients, 140
Terminal velocity, 42
cereal grains/pulses (legumes)/ oilseeds/speciality crops, **44**
empirical equations, 42–43
factors influencing, 43–44
measurement, 42
Theoretical model
evaporation-condensation model, 330–331
logarithmic model, 330
Luikov's model, 331–332
mass and energy on surface, 328–330
Whitaker's model, 332–335
Thermal cell method, 201
Thermal conductivity, 109, 112
grain bulks, **110**, **111**
measurement method, 112–114
prediction of, 114
relationship with other physical properties, 114–116, **115**
Thermal diffusivity, 116–117, **117**, 150, 152
measurement method of, 117–118
relationship with other physical properties, 118
Thermal properties, grains
heat capacity, 102–103, *103*
heat content, 102–103, *103*
specific heat, 103–104
freshly harvested canola, *108*

measurement methods, 104–107, *105*
prediction of, 107
relationship with other physical properties, *107*, 107–109, **108**
thermal conductivity, 109, 112
grain bulks, **110**, **111**
measurement method, 112–114
prediction of, 114
relationship with other physical properties, 114–116, **115**
thermal diffusivity, 116–117, **117**
measurement method of, 117–118
relationship with other physical properties, 118
Thermodynamic freezing, 170
Thin layer drying, 217, *227*, 227–228
convection drying energy source, 233–234
drying rate, 228
material temperature, 231–232, *232*
phase, 228–229, *229*
water activity, 229–230, *230–231*
Thin layer dynamic method, 200
Toluene, 106–107
True density, 1–2
measurement, 9–11, *10*
physical factors influencing, 11
porosity of seeds, **3–4**
Type I, II, III, IV, V isotherms, 183–184, *184*
Typical grain drying, 224

U

United States Department of Agriculture (USDA), 6, *6*

V

Van't Hoff-Arrhenius temperature equation, 410
Vapour diffusion, 225
Vapour pressure, 171–173, **392**
Vetch seeds, 460–461
Visco Analyser (RVA), 26
Volume changes, kernel
extrinsic factors, 254–255
fissure, 257–259, *258*
glass transition, 255–256
intrinsic factors, 254
puffing, 256
stress-cracked kernels, determination, 259
surface tension, 255
swelling, 259

W

Wall friction, 56
Water activity, 175–176
map, *407*
measurement, 180
microorganism multiplication, 178
vs. moisture content, 176–178, **177**
surface properties, 178–180
thin layer drying, 229–230, *230–231*
Water diffusivity, 241–242
effective, **243**
factors influencing, 247–250
measurement of, 244–247
mechanism of, 242–244
Water distribution, 167
Water mobility, 170–171
Water movement, solids, 316
capillary theory, 321–322
effective diffusivity, 319–321
Fick's Law of Diffusion, 317–319
receding front theory, 322–326
Water sorption, 181, *182*, 204, 386–388, 401; *see also* Laboratory determination, sorption and desorption isotherms
isotherm, 183–185, *184*, 287
Wet and dry bulb method, 174
Wet basis percentage moisture content, 164
Wetting, 267–268, *268*
Wheat, 7, *50*, 77, *105*, 126, *138*, *139*, *142*, **148**, 167, 388, 389, **392**, 393, *394*, *395*, 398, 399, 439–440, 441–442
Whitaker's model, 332–335
White bean (seafarer) seeds, 462
The Winchester bushel weight, 6
Wisconsin type breakage susceptibility tester, 259

Y

Young and Nelson equation, 306–307

Z

Zones and fronts, 264–267, *265*, *266*

For Product Safety Concerns and Information please contact our EU
representative GPSR@taylorandfrancis.com
Taylor & Francis Verlag GmbH, Kaufingerstraße 24, 80331 München, Germany

www.ingramcontent.com/pod-product-compliance
Lightning Source LLC
Chambersburg PA
CBHW060423220326
41598CB00021BA/2268

* 9 7 8 1 0 3 2 0 2 9 9 9 3 *